The Physics and Chemistry of the Interstellar Medium

This work provides a comprehensive overview of our current theoretical and observational understanding of the interstellar medium of galaxies. With emphasis on the microscopic physical and chemical processes in space, and their influence on the macroscopic structure of the interstellar medium of galaxies, the book includes the latest developments in this area of molecular astrophysics. The various heating, cooling, and chemical processes relevant for the rarefied gas and submicron-sized dust grains that constitute the interstellar medium are discussed in detail. This provides a firm foundation for an in-depth understanding of the ionized, neutral atomic, and molecular phases of the interstellar medium. The physical and chemical properties of large polycyclic aromatic hydrocarbon molecules and their role in the interstellar medium are highlighted, and the physics and chemistry of warm and dense photodissociation regions are discussed. This is an invaluable reference source for advanced undergraduate and graduate students and research scientists. Related resources for this book can be found at www.cambridge.org/9780521826341.

ALEXANDER TIELENS is a professor of astrophysics at the Kapteyn Astronomical Institute in the Netherlands, and a senior scientist with the Dutch space agency, SRON. Prior to this, he has worked as an assistant researcher in the Astronomy Department of the University of California, Berkeley, and as a senior scientist at NASA Ames Research Center, California. He has published extensively on various aspects of the physics and chemistry of the interstellar medium.

THE PHYSICS AND CHEMISTRY OF THE INTERSTELLAR MEDIUM

A.G.G.M. TIELENS

Kapteyn Astronomical Institute

CAMBRIDGE UNIVERSITY PRESS
Cambridge, New York, Melbourne, Madrid, Cape Town, Singapore
São Paulo, Delhi, Dubai, Tokyo, Mexico City

Cambridge University Press
The Edinburgh Building, Cambridge CB2 8RU,UK

Published in the United States of America by Cambridge University Press, New York

www.cambridge.org
Information on this title: www.cambridge.org/9780521533720

First published 2005
Reprinted with corrections 2006
First paperback edition 2010

A catalogue record for this publication is available from the British Library

ISBN 978-0-521-82634-1 Hardback
ISBN 978-0-521-53372-0 Paperback

Contents

Preface

When, upon my return to Holland, I started to teach an advanced course on the interstellar medium in 1998, I quickly realized that there was no suitable textbook available. There is, of course, the incomparable monograph by Spitzer, *Physics of the Interstellar Medium* (1978, New York: Wiley and Sons). But that book is quite challenging and not very suitable for a student course. Moreover, by now, it is very dated. Over the intervening years, our insights into the basic physics of the interstellar medium have much improved thanks, for example, to the opening up of the infrared and submillimeter windows. In particular, molecules, which we now know to be deeply interwoven into the fabric of the Universe, play only a little role in Spitzer's book. When Eddington made his famous remark, "Atoms are physics but molecules are chemistry," he merely expressed, on the one hand, the dream of a physicist of a simple universe, which can be caught in a single equation, and, on the other hand, the dread of a reality where solutions are never clean and simple. The latter is of course obvious to a chemist and it is now abundantly clear that Eddington's fear has turned into reality, even for astronomy. Present-day graduate students will require an intimate knowledge of molecular astrophysics in order to be active in the field of the interstellar medium of our own or other galaxies whether it is in the here and now or all the way back in the early Universe. This will become even more the case with the launch of the submillimeter space mission, Herschel, in 2007, when the Atacama Large Millimeter Array is finished in 2011, and with the launch of the James Webb Space Telescope in the next decade. Together these missions will push the frontier of the molecular Universe all the way back to the initial pollution of the Universe with the first metals by the first generation of luminous objects, which forever spoiled the physicist's Garden of Eden.

This book covers both the physics and the chemistry of the interstellar medium. Chapters on heating, cooling, and chemical processes provide the students with the necessary toolbox for the astrophysics and astrochemistry of the interstellar medium. This background is rounded off with chapters on the physics and chemistry of interstellar dust and large molecules. Once the students have mastered these subjects, they are well prepared for an in-depth discussion of classical topics of the interstellar medium: HII regions, the phases of the interstellar medium, shocks, and the dynamical interaction of HII regions, supernova remnants, and stellar winds with the ISM. The chemistry of the interstellar medium is covered in chapters on diffuse clouds, photodissociation regions, and molecular clouds. All together, this forms a comprehensive course, covering most current aspects of the interstellar medium, which will prepare students well for the future.

Over the years, this book grew from the course that I taught in Groningen. Indeed, in many ways, writing this book carried me through those dark Dutch days. Fortunately, I have many good friends who understand that, when the Sun sets in October not to appear again until May, it is a good time to leave Holland and visit other institutes. I owe a deep debt of gratitude to the Miller Institute and the Astronomy Department of the University of California in Berkeley and my hosts Imke de Pater and Chris McKee, to the Space Sciences Division of NASA Ames Research Center and my hosts David Hollenbach and Lou Allamandola, to the Institute for Geophysics and Planetary Physics of the Lawrence Livermore National Laboratory and my hosts Wil van Breugel and John Bradley, to the Centre d'Etudes Spatiale des Rayonnements in Toulouse and my host Emmanuel Caux, and to the Laboratoire d'Astrophysique de l'Observatoire de Grenoble and my host Cecilia Ceccarelli for their hospitality and for providing an environment conducive to great science. Most of the chapters of this book were conceived and written during these extended visits. Of course, much of this book reflects a lifetime spent in discovering the molecular Universe. I want to thank Harm Habing without whose human touch I would have left astronomy before even finishing the first stage of my journey. Also, I owe much to Lou Allamandola and David Hollenbach, with whom I have spent so many wonderful hours on the trail of discovery: not only for sharing their deep insights and understanding of physical and chemical processes of relevance to studies of the interstellar medium but, particularly, for their friendship. I am also deeply indebted to the many graduate students, who carried me through the many stages of this course, solved the many LaTeX problems, and always succeeded in making the right figures, as well as for their careful proofreading of the manuscript. Most of all, their enthusiasm always managed to perk me up. In this regard, I specifically want to thank Adwin Boogert, Rense Boomsma, Jan Cami, Stephanie Cazaux, Sasha Hony, Jacquie Keane, Leticia Martín-Hernández, Chris Ormel, Els Peeters, and Henrik Spoon for their help. Finally, Marion understands like no other what it is to live away from what feels like home. Her encouragement to follow my dream and her support during these difficult years have made this possible. To my girls–Anneke, Saskia, and Elske–the only thing I can say is that, now, it is really done.

Constants

Physical constants

Symbol	Description	SI Value	SI Unit	cgs Value	cgs Unit
c	Speed of light	2.9979 (8)	m s^{-1}	2.9979 (10)	cm^{-1} s^{-1}
h	Planck's constant	6.6261(−34)	J s	6.6261(−27)	erg s
k	Boltzmann's constant	1.3807(−23)	J/K	1.3807(−16)	erg/K
σ_{SB}	Stefan–Boltzmann constant	5.6704 (−8)	W m^{-2} K^{-4}	5.6704 (−5)	erg s^{-1} cm^{-2} K^{-4}
G	Gravitational constant	6.674 (−11)	N m^{-2} kg^{-2}	6.674 (−8)	dyn cm^{-2} g^{-2}
N_A	Avogadro's constant	6.0221 (23)	mol^{-1}	6.0221 (23)	mol^{-1}
m_e	Electron rest mass	9.1094(−31)	kg	9.1094(−28)	g
m_p	Proton rest mass	1.6726(−27)	kg	1.6726(−24)	g
m_u	Atomic mass unit	1.6605(−27)	kg	1.6605(−24)	g
e	Electron charge	1.602 (−19)	C	4.803 (−10)	esu
α	Fine-structure constant	7.2974 (−3)		7.2974 (−3)	

Values $a \times 10^b$ are given as a (b).

Astronomical constants

Symbol	Description	SI Value	SI Unit	cgs Value	cgs Unit
AU	Astronomical unit	1.496 (11)	m	1.496 (13)	cm
ly	Light year	9.463 (15)	m	9.463 (17)	cm
pc	Parsec	3.086 (16)	m	3.086 (18)	cm
pc^2	Square parsec	9.5234 (32)	m^2	9.5234 (36)	cm^2
kpc^2	Square kiloparsec	9.5234 (38)	m^2	9.5234 (42)	cm^2
$L_\odot$	Solar luminosity	3.85 (26)	J s^{-1}	3.85 (33)	erg s^{-1}
$M_\odot$	Solar mass	1.989 (30)	kg	1.989 (33)	g
$R_\odot$	Solar radius	6.96 (8)	m	6.96 (10)	cm
$T_\odot$	Solar effective temperature	5.78 (3)	K	5.78 (3)	K
Jy	Jansky	1.00 (−26)	W m^{-2} H z^{-1}	1.00 (−23)	erg s^{-1} cm^{-2} Hz^{-1}

Values $a \times 10^b$ are given as a (b).

Conversion factors

Angles and lengths

Unit/symbol	Description	SI		cgs	
		Value	Unit	Value	Unit
deg	degree	1.745 3 (−2)	rad	1.745 3 (−2)	rad
arcmin	arcminute	2.908 88 (−4)	rad	2.908 88(−4)	rad
arcsec	arcsecond	4.848 1 (−6)	rad	4.848 1 (−6)	rad
sq deg	degree^2	3.046 (−4)	sr	3.046 (−4)	sr
Å	angstrom	1.0 (−10)	m	1.0 (−8)	cm
μm	micrometer	1.0 (−6)	m	1.0 (−4)	cm

Values $a \times 10^b$ are given as a (b).

SI and cgs units

Description	SI		cgs	
	Value	Unit	Value	Unit
Time	1	s	1	s
	1	year	3.16 (7)	s
Length	1	m	1 (2)	cm
Velocity	1	m s^{-1}	1 (2)	cm s^{-1}
Force	1	N	1 (5)	dyne
Pressure	1	Pa	1 (−1)	dyne cm^{-2}
Energy	1	J	1 (7)	erg
Charge	1	C	2.9979 (9)	esu
Magnetic flux density	1	T	1 (4)	gauss

Values $a \times 10^b$ are given as a (b).

Energy conversion factors

	erg	eV	K	cm^{-1}	Hz
erg	1.00	6.242 (11)	7.243 (15)	5.034 (15)	1.509 (26)
eV	1.602 (−12)	1.00	1.1604 (4)	8064.4	2.418 (14)
K	1.3806 (−16)	8.617 (−5)	1.00	0.695	2.084 (10)
cm^{-1}	1.9865 (−16)	1.240 (−4)	1.4389	1.00	2.9970(10)
Hz	6.626 (−27)	4.136 (−15)	4.798 (−11)	3.336 (−11)	1.00

Values $a \times 10^b$ are given as a (b). To convert from unit in column 1 to units above the rows, multiply by value; e.g., $1\,\text{eV} = 1.602 \times 10^{-12}$ erg.

A useful compendium of constants can be found in C. W. Allen, *Astrophysical Quantities*, (London: The Athlone Press). The website http://physics.nist.gov/cuu/, maintained by the National Institute of Standards and Technology, provides a wealth of information on constants.

1

The galactic ecosystem

The Milky Way is largely empty. Stars are separated by some 2 pc in the solar neighborhood ($\rho_\star = 6 \times 10^{-2}\,\mathrm{pc}^{-3}$). If we take our Solar System as a measure, with a heliosphere radius of $\simeq$235 AU, stars and their associated planetary systems fill about 3×10^{-10} of the available space. This book deals with what is in between these stars: the interstellar medium (ISM). The ISM is filled with a tenuous hydrogen and helium gas and a sprinkling of heavier atoms. These elements can be neutral, ionized, or in molecular form and in the gas phase or in the solid state. This gas and dust is visibly present in a variety of distinct objects: HII regions, reflection nebulae, dark clouds, and supernova remnants. In a more general sense, the gas is organized in phases – cold molecular clouds, cool HI clouds, warm intercloud gas, and hot coronal gas – of which those objects are highly visible manifestations. This gas and dust is heated by stellar photons, originating from many stars (the so-called average interstellar radiation field), cosmic rays (energetic [$\sim$GeV] protons), and X-rays (emitted by local, galactic, and extragalactic hot gas). This gas and dust cools through a variety of line and continuum processes and the spectrum will depend on the local physical conditions. Surveys in different wavelength regions therefore probe different components of the ISM. This first chapter presents an inventory of the ISM with an emphasis on prominent objects in the ISM and the global structure of the ISM.

The interstellar medium plays a central role in the evolution of the Galaxy. It is the repository of the ashes of previous generations of stars enriched by the nucleosynthetic products of the fiery cauldrons in the stellar interiors. These are injected either with a bang, in a supernova explosion, or with a whimper, in the much slower moving winds of low-mass stars on the asymptotic giant branch. In this way, the abundances of heavy elements in the ISM slowly increase. This is part of the cycle of life for the stars of the Galaxy, because the ISM itself is the birthplace of future generations of stars. It is this constant recycling and its

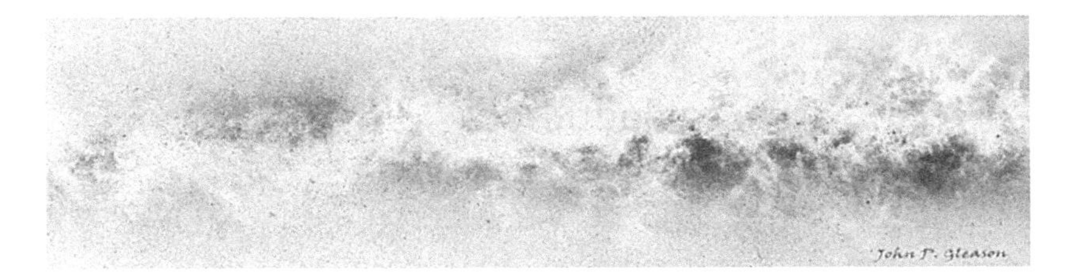

Figure 1.1 A panoramic image of a (southern) portion of the Milky Way's disk. The image has been inverted and dark corresponds to emission from ionized gas and reflection nebulae. The light band stretching irregularly across the whole image is due to absorption by dust clouds. Image courtesy of J. P. Gleason.

associated enrichment that drives the evolution of the Galaxy, both physically and in its emission characteristics.

1.1 Interstellar objects

1.1.1 HII regions

Ionized gas nebulae feature prominently in the Milky Way as bright visible nebulous objects. The Great Nebula in Orion (M42; Fig. 1.2) and the Lagoon Nebula (M8) are well-known examples. HII regions span a range in brightness, however, and fainter examples are the California Nebula and IC 434 (Fig. 1.3). The gas in these regions is ionized and has a temperature of about 10^4 K. Densities range from 10^3–10^4 cm^{-3} for compact ($\sim$0.5 pc) HII regions such as the Orion Nebula to $\sim$10 cm^{-3} for more diffuse and extended nebulae such as the North America Nebula ($\sim$10 pc). The optical spectra of these regions are dominated by H and He recombination lines and collisionally excited, optical (forbidden) line emission from trace ions such as [OII], [OIII], and [NII]. HII regions are also strong sources of thermal radio emission (free–free) from the ionized gas and of infrared emission due to warm dust. HII regions are formed by young massive stars with spectral type earlier than about B1 ($T_{\rm eff} > 25\,000$ K), which emit copious amounts of photons beyond the Lyman limit ($h\nu > 13.6$ eV) and ionize and heat their surrounding, nascent molecular clouds. They are, therefore, signposts of sites of massive star formation in the Galaxy.

1.1.2 Reflection nebulae

Reflection nebulae are bluish nebulae that reflect the light of a nearby bright star. NGC 2023 in the Orion constellation (see Fig. 1.3) and the striated nebulosity associated with the Pleiades are familiar cases. In this case, the observed light

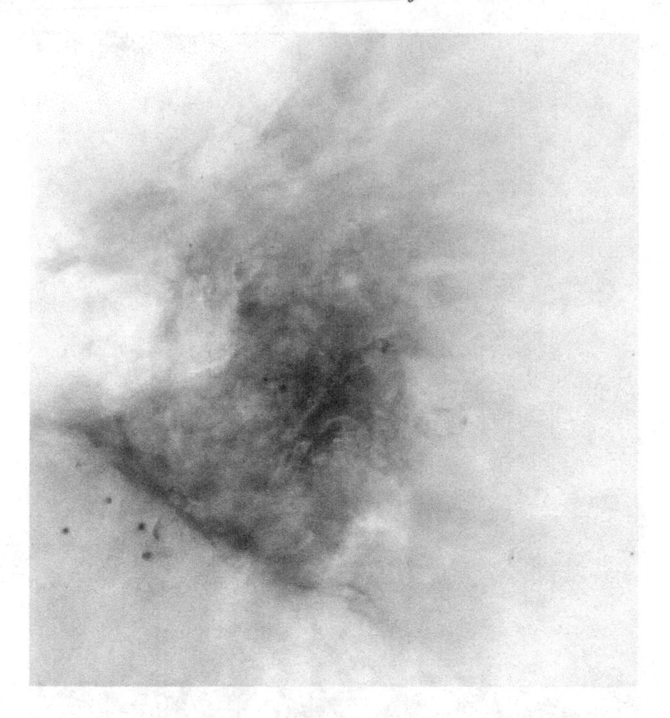

Figure 1.2 A black and white representation of the Orion Nebula as observed by the Hubble Space Telescope in [OIII], Hα, and [NII]. Light and dark have been inverted. The gas is ionized by the Trapezium cluster, in particular θ^1 C Ori, in the center of the image. The bright bar in the south-west is an ionization front eating its way into the surrounding neutral material in the photodissociation region known as the Orion Bar. The dark bay (light in this image), a cloud of foreground obscuring material, is also evident to the east. This image gives a clear view of the complex topography created by the interaction of a newly formed massive star with its surrounding natal cloud. Image courtesy of R. O'Dell.

is not due to hot gas but rather reflected starlight. There is no radio emission but there is infrared emission from warm dust, although this is less luminous than for HII regions. For the compact reflection nebulae, densities are typically a little smaller ($\simeq 10^3\,\mathrm{cm}^{-3}$) than for compact HII regions. Reflection nebulae are illuminated by stars with spectral types later than about B1. Regions around hotter stars also show (faint) reflected light emission but the spectrum is then dominated by the emission from the ionized gas. For the earlier stellar types, the surrounding nebulosity may be the material from which the star was formed (e.g., NGC 2023; NGC 7023). Often, however, the nebulosity is due to a chance encounter between the star and a cloud (e.g., the Pleiades). Reflection nebulae can also be associated with the ejecta of a late-type star (e.g., IC 2220; the Red Rectangle).

Figure 1.3 Part of the Orion molecular cloud containing the Horse Head Nebula. The diffuse glow behind the horse head is IC 434 ionized by the bright star, σ Ori. The horse head is a protrusion of the molecular cloud obvious in the lower part of this image. The nebula to the south-east of the horse head is the reflection nebula, NGC 2023. Image courtesy of the Canadian-France-Hawaii telescope, J.-C. Cuillandre, Coelem.

1.1.3 Dark nebulae

A striking aspect of the all-sky optical view of the Milky Way is the presence of many dark regions in which few stars are seen (cf. Fig. 1.1). The direction towards the center of the Galaxy is rampant with such dark clouds, which actually seem to divide the galactic plane in two. The Coalsack near the Southern Cross is a particularly nice example of a roundish dark cloud. Dark clouds are readily apparent when backlighted. The Horse Head Nebula (see Fig. 1.3) silhouetted against the reddish glow of the HII region, IC 434, and the dark bay in the Orion HII region (see Fig. 1.2) are two famous examples. Individual dark clouds come in a range of sizes from tens-of-parsecs large to the tiny ($\sim 10^{-2}$ pc) Bok globules associated with HII regions such as the Orion Nebula. Likewise, some dark clouds are completely black ($A_v > 10$ magnitudes) while others are hardly discernible. While dark clouds are outlined by the absence of stars, they do show faint optical reflected light. Also, they become bright at mid- and far-infrared wavelengths. Some really dense clouds are opaque even at mid-IR wavelengths and appear as infrared dark clouds (IRDC) in absorption against background galactic mid-IR emission.

1.1.4 Photodissociation regions

While HII regions and reflection nebulae dominate the Galaxy at visible wavelengths, in the infrared, photodissociation regions dominate the sky. Originally, the name photodissociation regions (PDRs; sometimes also called photodominated regions with fortunately the same abbreviation) was given to the atomic–molecular zones that separate ionized and molecular gas near bright luminous O and B stars (e.g., surrounding HII regions and reflection nebulae) and the Orion Bar (Fig. 1.2) and NGC 2023 (Fig. 1.3) are prime examples of classical PDRs. In these regions, penetrating far-ultraviolet (FUV) photons (with energies between 6 and 13.6 eV) dissociate and ionize molecular species. Most of the FUV photons are absorbed by the dust, but a small fraction heat the gas through the photoelectric effect to a few hundred degrees. Photodissociation regions are thus bright in IR dust continuum and atomic fine-structure cooling lines as well as molecular lines. In essence, of course, everywhere where FUV photons strike a cloud, a PDR will ensue. Indeed, the term PDRs has now expanded to include all regions of the ISM where FUV photons dominate the physical and chemical processes. As such, PDRs include the neutral atomic gas of the ISM as well as much of the gas in molecular clouds (except, e.g., for dense starless cores).

1.1.5 Supernova remnants

Supernova remnants (SNRs) are formed when the material ejected in the explosion that terminates the life of some stars shocks surrounding ISM material and an SNR's spectrum is that of a high velocity shock. About 100 supernova remnants are visible in our Galaxy; they are generally characterized by long, delicate filaments radiating in-line radiation (Fig. 1.4). Supernova remnants are prominent sources of radio emission due to relativistic electrons spiraling around a magnetic field (synchroton emission) and some 200 have been identified at radio wavelengths. Supernova remnants also stick out at X-ray wavelengths because of emission by hot ($\simeq 10^6$ K) gas. Not all SNRs are wispy. The Crab Nebula is an example of a compact SNR.

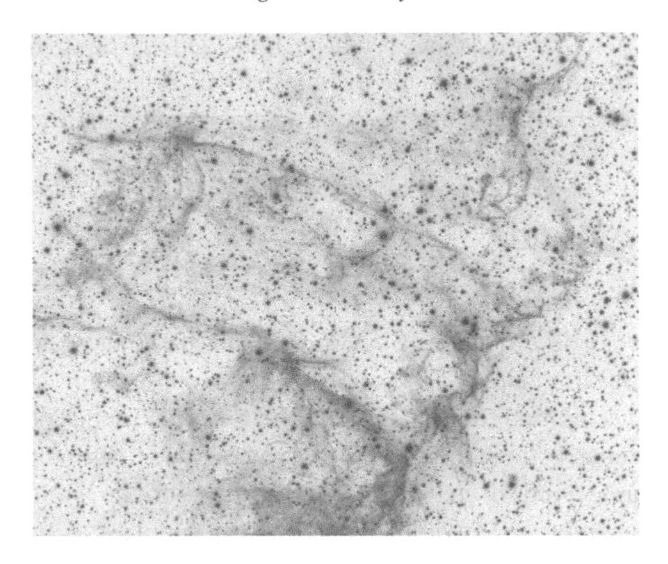

Figure 1.4 A small portion of the Cygnus Loop, the remnant of a supernova that exploded about 10 000 years ago. The image has been inverted to bring out the delicate structure of the nebulosity. The emission is due to a shock wave and is about 3 pc in size. Image courtesy of L. K. Tan, StarryScapes.

1.2 Components of the interstellar medium

The gas in the ISM is organized in a variety of phases. The physical properties of these phases are summarized in Table 1.1.

1.2.1 Neutral atomic gas

The 21 cm line of atomic hydrogen traces the neutral gas of the ISM. This neutral gas can also be observed in optical and UV absorption lines of various elements towards bright background stars. The neutral medium is organized in cold ($\simeq$ 100 K) diffuse HI clouds (cold neutral medium, CNM) and warm ($\approx$8000 K) intercloud gas (warm neutral medium, WNM). A standard HI cloud (often called a Spitzer-type cloud) has a typical density of 50 cm^{-3} and a size of 10 pc. The density of the WNM is much less ($\simeq$0.5 cm^{-3}). Between 4 and 8 kpc from the galactic center, 80% of the HI mass in the plane of the Galaxy is in diffuse clouds in a layer with a (Gaussian) scale height of about 100 pc. At higher latitudes, however, much of the HI mass is in the intercloud medium with a larger scale height of 220 pc but with an exponential tail extending well into the lower halo. These two neutral phases have, on average, similar surface densities. Because the Sun is located in the local bubble, the local, total WNM column density towards the North Galactic Pole is about 2.5 times that of the CNM. In the outer Galaxy, the HI scale height rapidly increases.

Table 1.1 *Characteristics of the phases of the interstellar medium*

Phase	n_0^a (cm^{-3})	T^b (K)	ϕ_v^c (%)	M^d (10^9 $M_\odot$)	$< n_0 >^e$ (cm^{-3})	H^f (pc)	Σ^g ($M_\odot$pc^{-2})
Hot intercloud	0.003	10^6	~50.0	—	0.0015	3000	0.3
Warm neutral medium	0.5	8000	30.0	2.8	0.1^h	220^h	1.5
					0.06^h	400^h	1.4
Warm ionized medium	0.1	8000	25.0	1.0	0.025^i	900^i	1.1
Cold neutral medium^j	50.0	80	1.0	2.2	0.4	94	2.3
Molecular clouds	>200.0	10	0.05	1.3	0.12	75	1.0
HII regions	1–10^5	10^4	—	0.05	0.015^k	70^k	0.05

[a] Typical gas density for each phase.
[b] Typical gas temperature for each phase.
[c] Volume filling factor (very uncertain and controversial!) of each phase.
[d] Total mass.
[e] Average mid-plane density.
[f] Gaussian scale height, $\sim \exp[-(z/H)^2/2]$, unless otherwise indicated.
[g] Surface density in the solar neighborhood.
[h] Best represented by a Gaussian and an exponential.
[i] WIM represented by an exponential.
[j] Diffuse clouds.
[k] HII regions represented by an exponential.

1.2.2 *Ionized gas*

Diffuse ionized gas in the ISM can be traced through dispersion of pulsar signals, through optical and UV ionic absorption lines against background sources, and through emission in the Hα recombination line (see Fig. 1.5). The first two can only be done in a limited number of selected sight-lines. The faintness and large extent of the galactic Hα hamper the last probe. While most of the Hα luminosity of the Milky Way is emitted by distinct HII regions, almost all of the mass of ionized gas (10^9 M$_\odot$) resides in a diffuse component. This warm ionized medium (WIM) has a low density ($\simeq$0.1 cm^{-3}), a temperature of $\approx$8000 K, a volume filling factor of $\simeq$0.25, and a scale height of $\simeq$1 kpc. The weakness of the [OI] λ6300 line (in a few selected directions) implies that the gas is nearly fully ionized. The source of ionization is not entirely clear. Energetically, ionizing photons from O stars are the most likely candidates but these photons have to "escape" from the associated HII regions and travel over large distances (hundreds of parsecs)

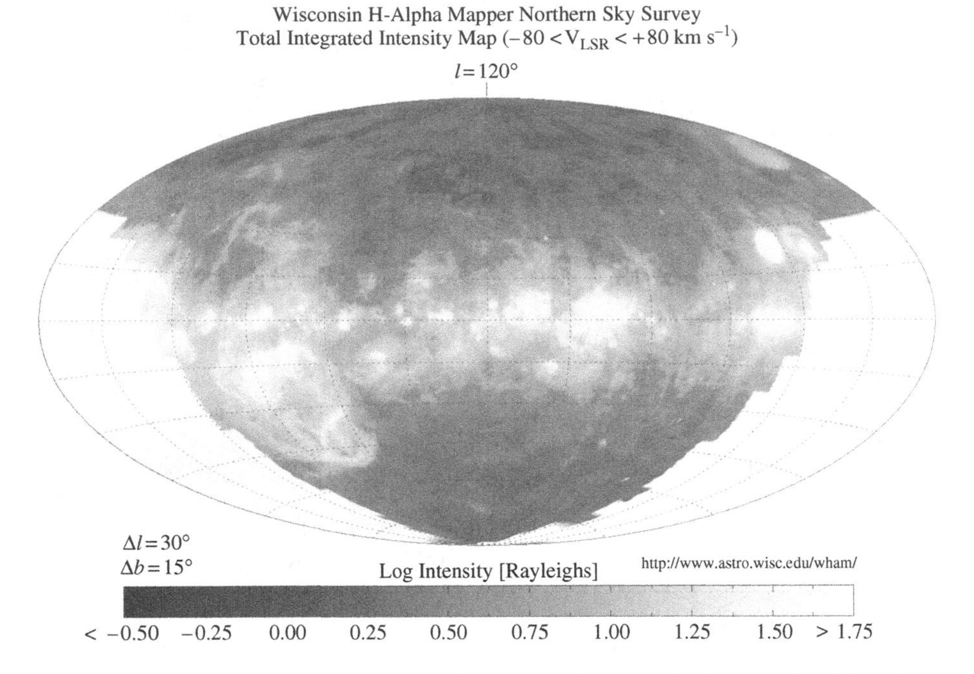

Figure 1.5 The integrated galactic (ℓ versus b) Hα emission obtained by the WHAM survey. This emission from ionized gas, about a million times fainter than the Orion Nebula, traces the warm ionized medium. Note the large filament (40°) sticking up out of the galactic plane. Image courtesy of R. Reynolds.

without being absorbed by omnipresent neutral hydrogen. Finally, the WIM shows a complex spatial structure including thin filaments sticking ~ 1 kpc out of the plane of the Galaxy, further compounding the ionization problem.

1.2.3 Molecular gas

The CO $J = 1-0$ transition at 2.6 mm is commonly used as a tracer of molecular gas in the Galaxy (Fig. 1.6). Surveys in this line have shown that much of the molecular gas in the Milky Way is localized in discrete giant molecular clouds with typical sizes of 40 pc, masses of 4×10^5 $M_\odot$, densities of $\simeq 200$ cm^{-3}, and temperatures of 10 K. However, it should be understood that molecular clouds show a large range in each of these properties. Molecular clouds are characterized by high turbulent pressures as indicated by the large linewidths of emission lines. Molecular clouds are self-gravitating rather than in pressure equilibrium with other phases in the ISM. While they are stable over time scales of $\simeq 3 \times 10^7$ years, presumably because of a balance of magnetic and turbulent pressure and gravity, molecular clouds are the sites of active star formation. Observations of molecular

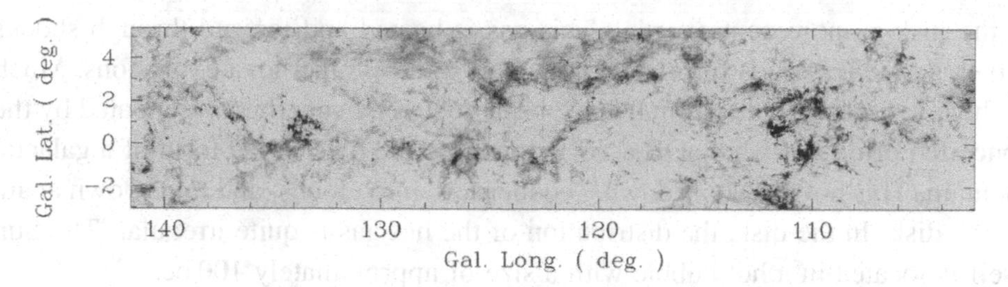

Figure 1.6 The emission of CO in the outer Galaxy obtained by the FCRAO survey. Much of the emission is local, stemming from molecular clouds between 0.5 and 1 kpc. In addition, giant molecular clouds associated with the NGC 7538-Sharpless 156-Sharpless 152 region ($\ell = 110°$) and the W3-W4-W5 region ($\ell = 135°$) are also present. The image has been inverted and dark corresponds to emission. Image courtesy of Chris Brunt (FCRAO).

clouds in the rotational transitions of a variety of species allow a detailed study of their physical and chemical properties. These studies show that molecular clouds have spatial structure on all scales. In particular, molecular clouds contain cores with sizes of $\simeq 1$ pc, densities in excess of 10^4 cm^{-3}, and masses in the range 10–10^3 $M_\odot$ in which star formation is localized. While CO is commonly used to trace interstellar molecular gas, H_2 is thought to be the dominant molecular species, with a H_2/CO ratio of 10^4–10^5. In addition, some 200 different molecular species have been detected – mainly through their rotational transitions in the submillimeter wavelength regions – in the shielded environments of molecular clouds. In general, these species are relatively simple, often unsaturated, radicals, or ions. Acetylenic carbon chains and their derivatives figure prominently on the list of detected molecules. However, this may merely reflect observational bias, since such species possess large dipole moments and relatively small partition functions, both of which make them readily detectable at microwave wavelengths.

1.2.4 Coronal gas

Hot ($\sim 10^5$–10^6 K) gas can be traced through UV absorption lines of highly ionized species (e.g., CIV, SVI, NV, OVI) seen against bright background sources. Such hot plasmas also emit continuum (bremsstrahlung, radiative recombination, two photon) and line (collisionally excited and recombination) radiation in the extreme ultraviolet and X-ray wavelength regions. These observations have revealed the existence of a pervasive hot (3–10×10^5 K) and tenuous ($\simeq 10^{-3}$ cm^{-3}) phase (the hot intercloud medium, HIM) of the ISM. The observations indicate a range of temperatures where the higher ionization stages probe hotter gas. The hot gas fills most of the volume of the halo (scale height $\simeq 3$ kpc) but the volume filling factor in the disk is more controversial. This gas is heated and ionized through shocks driven by stellar winds from early type stars and by supernova explosions. Much of the hot, high-latitude gas may have been vented by superbubbles created by the concerted efforts of whole OB associations into the halo in the form of a galactic fountain. This hot gas cools down, "condenses" into clouds, and rains down again on the disk. In the disk, the distribution of the hot gas is quite irregular. The Sun itself is located in a hot bubble with a size of approximately 100 pc.

1.2.5 Interstellar dust

The presence of dust in the interstellar medium manifests itself in various ways. Through their absorption and scattering, small dust grains give rise to a general reddening and extinction of the light from distant stars (Fig. 1.1). Moreover, polarization of starlight is caused by elongated large dust grains aligned in the galactic magnetic field (dichroic absorption). Furthermore, near bright stars, scattering of starlight by dust produces a reflection nebula. Finally, the interstellar medium is bright in the infrared because of continuum emission by cold dust grains. Analysis of the wavelength dependence of interstellar reddening implies a size distribution, $n(a) \sim a^{-3.5}$, which ranges from $\simeq$3000 Å all the way to the molecular domain ($\sim$5 Å). The number density of grains with sizes $\sim$1000 Å is $\simeq 10^{-13}$ per H atom. Most of the mass of interstellar dust is thus in the larger grains but the surface area is in the smallest grains. Abundance studies in the ISM show that many of the refractory elements (e.g., C, Si, Mg, Fe, Al, Ti, Ca) are locked up in dust; i.e., dust contains about 1% by mass of the gas.

Large ($\gtrsim$ 100 Å) interstellar dust grains are in radiative equilibrium with the interstellar radiation field at temperatures of $\simeq$15 K and the absorbed stellar photons are reradiated as infrared and submillimeter continuum emission. Near bright stars, the dust temperature is higher; typically some 75 K for a compact HII region. Emission by interstellar dust dominates these wavelength regions. Rotating interstellar dust grains also give rise to emission at radio wavelengths. Very small ($\lesssim$ 100 Å) dust grains undergo fluctuations in their temperature upon the absorption of a single UV photon and emit at mid-IR (25–60 μm) wavelengths.

1.2.6 Large interstellar molecules

Besides dust grains, the interstellar medium also contains a population of large molecules. These molecules are particularly "visible" at mid-IR wavelengths. The IR spectrum of most objects – HII regions, reflection nebulae, surfaces of dark clouds, diffuse interstellar clouds and cirrus clouds, galactic nuclei, the interstellar

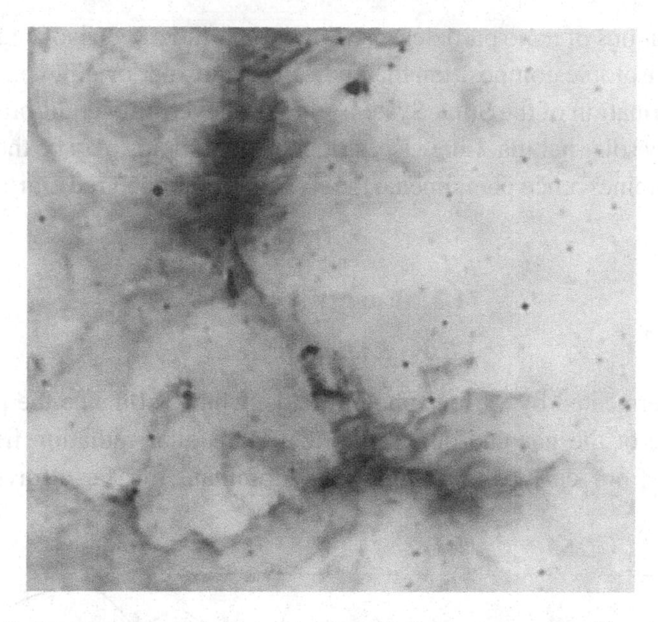

Figure 1.7 An infrared view of the Eagle Nebula (image inverted) obtained with ISOCAM on the Infrared Space Observatory. The dark emission that dominates the image is due to IR fluorescence of large polycyclic aromatic hydrocarbon molecules. Near the center of the image some of the faint emission is due to large, cool dust grains present in the ionized gas. The well-known pillars are visible just under the center of this image as the three-fingered hand. Image courtesy of the ISOGAL team, especially Andrea Moneti and Frederic Schuller.

medium of galaxies as a whole, and star burst galaxies – are dominated by broad infrared emission features (Fig. 1.7). These IR emission features are characteristics for polycyclic aromatic hydrocarbon (PAH) materials. These bands represent the vibrational relaxation process of FUV-pumped PAH species, containing some 50 C atoms. These species are very abundant, $\sim 10^{-7}$ relative to H, locking up about 10% of the elemental carbon.

These PAHs may just be one – very visible – representative of the molecular Universe. In fact, visible spectra of stars generally show prominent absorption features that are too broad to be atomic in origin. These so-called diffuse interstellar bands (DIBs) are generally attributed to electronic absorption by moderately large molecules (10–50 C atoms). Unsaturated carbon chains, containing 10–20 C atoms, are leading candidates for these DIBs. There are now in excess of 200 DIBs known, of which some 50 are moderately strong. Because, typically, the visible spectrum of a molecular species is dominated by at most one strong transition, the DIBs implicate the presence of a large number of different molecular species.

These large molecules seem to represent the extension of the interstellar grain size distribution into the molecular domain. Interstellar grains are known to contain several populations of nano particles: nano diamonds have been isolated from meteorites with an isotopic composition that indicates a presolar origin; i.e., these grains predate the formation of the Solar System and they never fully equilibrated with the gas in the early solar nebula. Likewise, silicon nanoparticles may be the carrier of a widespread luminescence phenomena, the so-called extended red emission (ERE).

1.3 Energy sources

1.3.1 Radiation fields

The ISM is permeated by various photon fields, which influence the physical and chemical state of the gas and dust (Fig. 1.8). The stellar radiation field contains contributions from early-type stars, which dominate the far-ultraviolet (FUV)

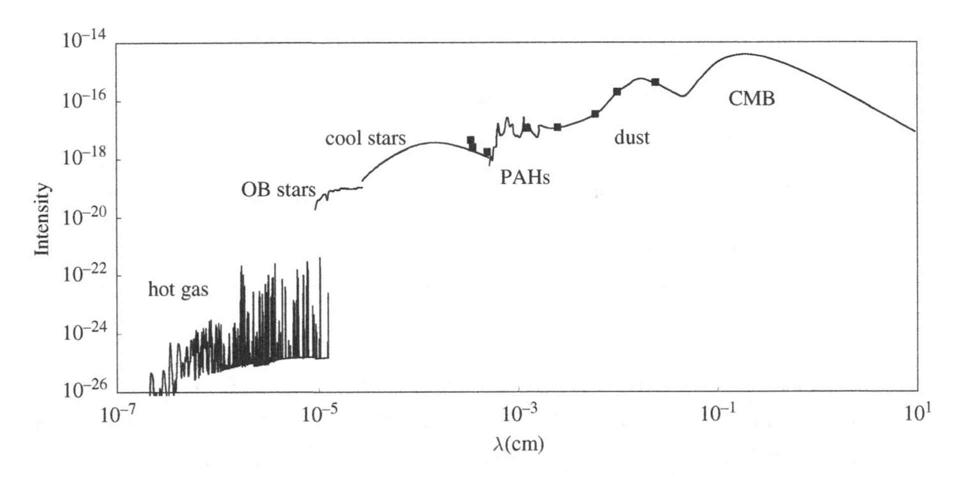

Figure 1.8 The mean intensity in units of erg cm^{-2} s^{-1} Hz^{-1} sr^{-1} of the interstellar radiation field in the solar neighborhood. Contributions by hot gas, OB stars, older stars, large molecules (PAHs), dust, and the cosmic microwave background are indicated. Figure adapted from J. Black, 1996, *First Symposium on the IR Cirrus and Diffuse Interstellar Clouds*, ed. R. M. Cutri and W. B. Latter (San Francisco: ASP), p. 355. The calculated X-ray/EUV emission spectrum and the FUV spectrum were kindly provided by J. Slavin. The dust emission is a fit to the COBE results for the galactic emission. The PAH spectrum was taken from ISO measurements of the mid-IR emission spectrum of the interstellar medium scaled to the measurements of the IR cirrus by IRAS (F. Boulanger, 2000, in *ISO Beyond Point Sources: Studies of Extended Infrared Emission*, ed. R. J. Laureijs, K. Leech, and M. F. Kessler, *E. S. A.-S. P.*, **455**, p. 3). The black squares at 12, 25, 60 and 100 μm are the IRAS measurements of the IR cirrus, the DIRBE/COM measurement at 240 μm, and those at 3.3, 3.5, and 4.95 μm are the balloon measurement by Proneas experiment (M. Giard, J. M. Lamarre, F. Pajot, and G. Serra, 1994, *A. & A.*, **286**, p. 203). Note that the latter have been superimposed on the stellar spectrum.

wavelengths, A-type stars, which control the visible region, and late-type stars, which are important at far-red to near-infrared wavelengths. The strength of the FUV average interstellar radiation is often expressed in terms of the Habing field, 1.2×10^{-4} erg cm^{-2} s^{-1} sr^{-1}, named after Harm Habing, a pioneer in this field. Current estimates put the average interstellar radiation field at $G_0 = 1.7$ Habing fields. Often, the radiation field in PDRs produced by a nearby star shining on a nearby cloud is expressed in terms of the equivalent one-dimensional average interstellar radiation field flux (e.g., 1.6×10^{-3} erg cm^{-2} s^{-1}).

These stellar photons are absorbed by dust grains and reradiated at longer wavelengths – in discrete emission bands in the mid-IR and in continuum emission in the far-IR and submillimeter regions (Fig. 1.8). The 2.7 K cosmological background takes over at millimeter wavelengths. At extreme ultraviolet wavelengths (EUV), even a small amount of neutral hydrogen absorbs all radiation and the intensity of the average radiation field due to stars drops precipitously at the Lyman edge (912 Å; Fig. 1.9). Stars do not contribute much at the shortest wavelengths (X-rays). Instead, emission by hot plasmas – the coronal gas in the halo and in SNRs – dominates the radiation field. This component shows numerous emission lines. There is also an extragalactic contribution at the hardest energies. These X-ray emission components are mediated by absorption by foreground

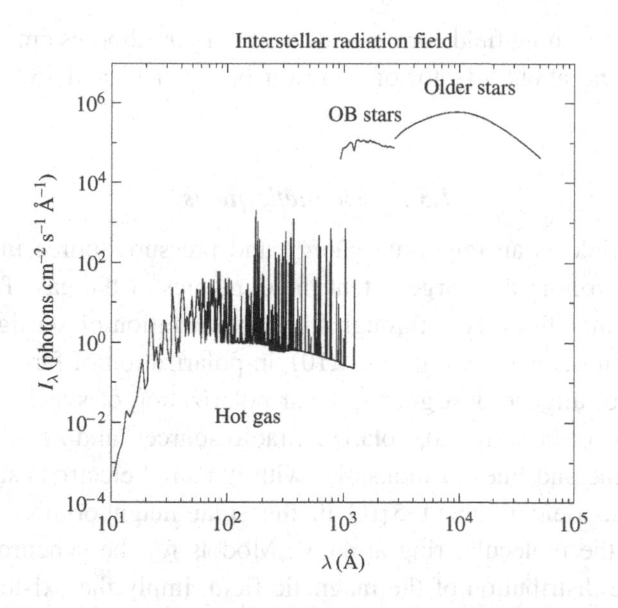

Figure 1.9 The average photon field in the solar neighborhood in units of photons cm^{-2} s^{-1} Å^{-1}. Contributions by hot gas, OB stars, and older stars are indicated. The calculated X-ray/EUV/FUV spectrum was kindly provided by J. Slavin. The visual spectrum was taken from J. S. Mathis, P. G. Mezger, and N. Panagia, 1983, *A. & A.*, **128**, p. 212.

Table 1.2 *Energy balance for diffuse clouds*

Source	P (10^{-12} dyne cm^{-2})	Energy density (eV cm^{-3})	Heating rate (erg s^{-1} H-atom^{-1})
Thermal	0.5	6.0	–5 (–26)[a]
UV	—	0.5	5 (–26)
Cosmic ray	1.0	2.0	3 (–27)
Magnetic fields	1.0	0.6	2 (–27)
Turbulence	0.8	1.5	1 (–27)
2.7 K background	—	0.25	—

[a] Energy loss rate.

gas and, because the ionization cross section decreases rapidly with increasing energy at EUV–X-ray wavelengths, this effect is more pronounced at the longer wavelengths.

Photodissociation rates and ionization rates are proportional to the photon intensity (Fig. 1.9). A simple polynomial fit for the mean photon intensity of the interstellar radiation field is

$$\mathcal{N}_{\rm ISRF} = 8.530 \times 10^{-5} \lambda^{-1} - 1.376 \times 10^{-1} \lambda^{-2} + 5.495 \times 10^{1} \lambda^{-3} \rm cm^{-2} s^{-1} Hz^{-1} sr^{-1}, \tag{1.1}$$

with λ in Å. The Habing field corresponds to about 10^8 photons cm^{-2} s^{-1} between 6 and 13.6 eV and about a factor of 10 fewer between 11 and 13.6 eV.

1.3.2 Magnetic fields

The magnetic field is an important energy and pressure source in the ISM (cf. Table 1.2), controlling to a large extent the dynamics of the gas. The interstellar magnetic field manifests itself through linear polarization of starlight by aligned dust grains (dichroic absorption; Fig. 1.10), in polarization of far-infrared continuum emission of aligned dust grains, linear polarization of synchroton emission, Faraday rotation of background, polarized, radio sources, and Zeeman splitting of the 21 cm HI line and lines of molecules with unpaired electrons such as OH.

The magnetic field is about 5 μG in the solar neighborhood, increasing to about 8 μG in the molecular ring at 4 kpc. Models for the synchroton emission, which trace the distribution of the magnetic field, imply the existence of a thin-disk component, associated with the gaseous disk, and a thick-disk component – the halo component – with a scale height of 1.5 kpc near the solar circle. The magnetic field consists of a uniform component and a non-uniform component. The uniform component of the magnetic field is roughly circular with a strength

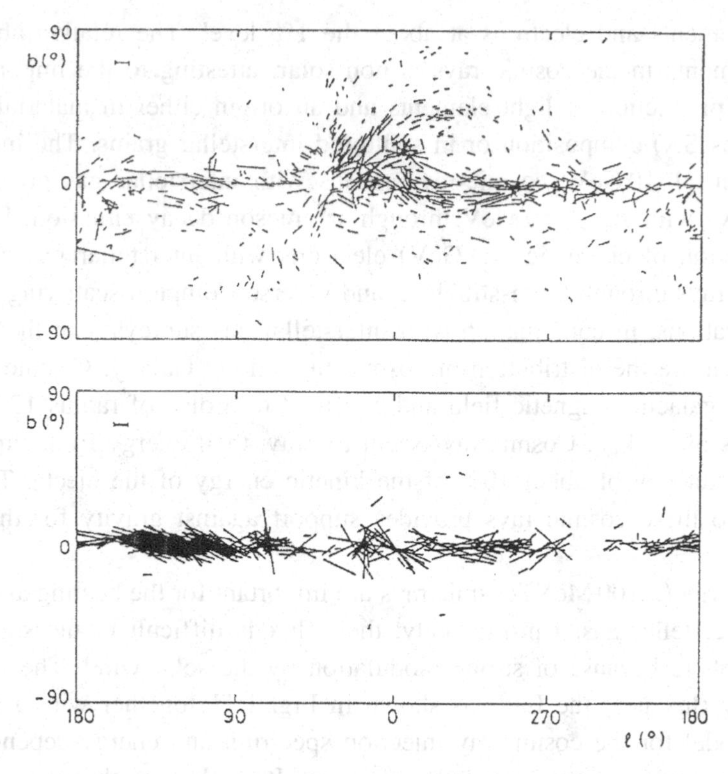

Figure 1.10 The direction of the magnetic field of the Galaxy as measured through optical polarization of starlight. Upper panel: stars within 400 pc, emphasizing the magnetic field associated with local clouds. Lower panel: stars between 2 and 4 kpc, showing some regions with magnetic field parallel to the galactic plane while others are more random. Figure reproduced with permission from D. S. Mathewson and V. L. Ford, 1970, *Mem. R. A. S.*, **74**, p. 139.

of about 1.5 μG in the solar neighborhood and has two field reversals within the solar circle and one outside. The field may show a spiral structure where the reversals occur in the interarm regions. There is also a considerable non-uniform magnetic field, partly associated with expanding interstellar shells (superbubbles) and their shocks. The strength of the magnetic field increases inside dense clouds, $B \sim n^{\alpha}$ with $\alpha \simeq 0.5$ and typically $B \simeq 30\,\mu$G at $n \simeq 10^4$ cm^{-3}. The direction of the field is correlated over the entire extent of the cloud from the diffuser outer parts to the denser cores.

1.3.3 Cosmic rays

High energy ($\gtrsim$100 MeV nucleon^{-1}) particles contribute considerably to the energy density of the ISM ($\simeq$2 eV cm^{-3}; Table 1.2). Cosmic rays consist mainly of relativistic protons with energies between 1 and 10 GeV, 10% helium, and heavier elements and electrons at about the 1% level. The relative abundance of the elements in the cosmic rays is non-solar, attesting to the importance of spallation–production of light elements and an origin either in material of stellar (perhaps SN) composition or in sputtered interstellar grains. The interaction of energetic (1–10 GeV) cosmic-ray protons with interstellar gas gives rise to gamma rays with $E_\gamma \gtrsim 50\,\text{MeV}$ through π^0 meson decay emission. Likewise, the interaction of energetic (<1 GeV) electrons with interstellar gas gives rise to gamma rays through bremsstrahlung and inverse Compton scattering. Gamma ray observations, in combination with interstellar gas surveys, can therefore be used to measure the distribution of cosmic rays in the Galaxy. Cosmic rays are tied to the galactic magnetic field and confined to a disk of radius 12 kpc with a thickness of ~2 kpc. Cosmic rays seem to draw their energy from supernovae with an efficiency of about 10% of the kinetic energy of the ejecta. The pressure due to these cosmic rays provides support against gravity for the gas in the ISM.

Low-energy (≃100 MeV) cosmic rays are important for the heating and ionization of interstellar gas. Unfortunately, their flux is difficult to measure within the heliosphere because of strong modulation by the solar wind. The measured cosmic-ray flux near the Earth is shown in Fig. 1.11 together with a fit based upon a model for the cosmic-ray injection spectrum and energy-dependence of the residence time in the interstellar medium. It is obvious that the correction is substantial at low energies. While the interstellar cosmic-ray flux is difficult to measure directly, the resulting ionization drives the build-up of simple molecules (e.g., OH), which can be studied (Section 8.7). These indirect measurements imply a primary cosmic-ray ionization rate in the ISM of $\zeta_{CR} \simeq 2 \times 10^{-16}$ (H atom)$^{-1}$ s^{-1}. Localized regions of much higher ionization rate may occur near associations of massive stars.

1.3.4 Kinetic energy of the ISM

Winds from early-type stars and supernovae explosions supply bulk kinetic energy to the ISM. Compared to the stellar radiative budget, the total mechanical energy output is only small, ~0.5 % (Table 1.2). However, the turbulent energy of the HI is about 6×10^{51} erg kpc^{-2} in the solar neighborhood and provides support for the HI gas against gravity. The HI also shows ordered vertical flow, at some 5 km s^{-1} in the Milky Way, as well as the infall of large gas complexes at higher latitudes.

The expanding shells blown by individual stars and the superbubbles blown by the concerted action of OB associations have an important influence on the morphology of the ISM. They sweep up and compress the surrounding ISM

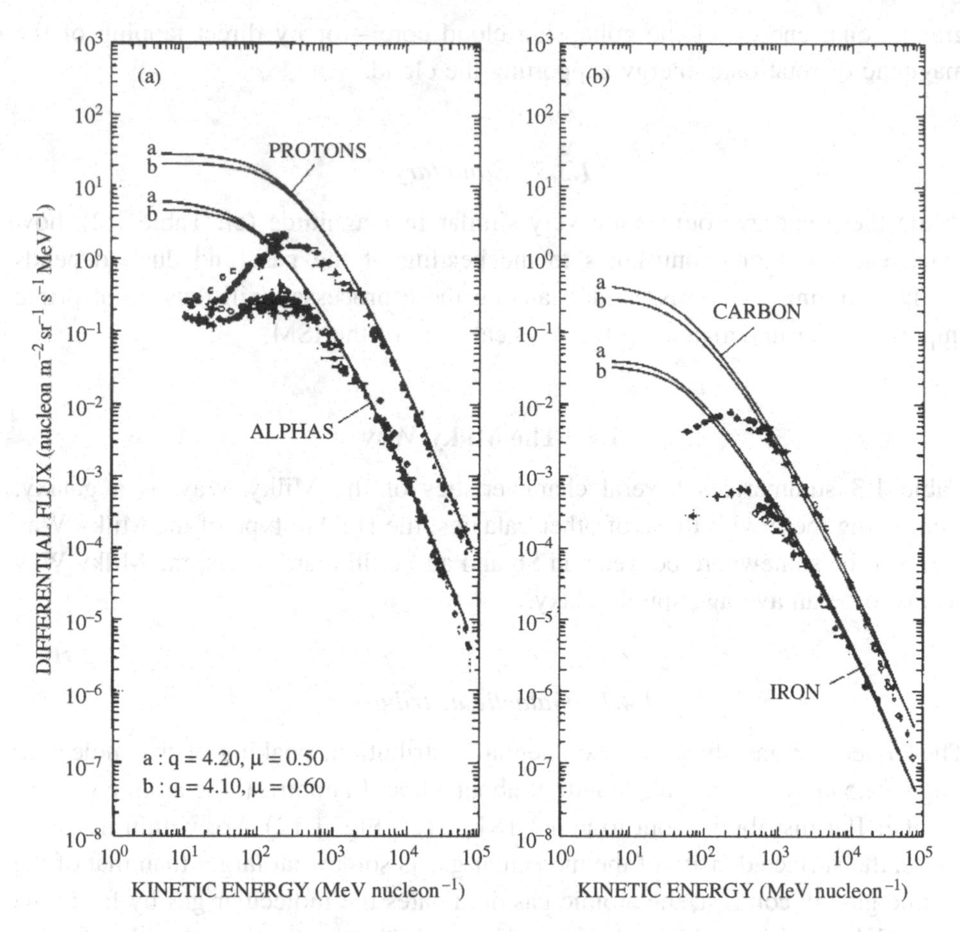

Figure 1.11 The cosmic-ray proton flux as a function of energy measured near the Earth and the inferred interstellar cosmic-ray flux after the effects of modulation by the solar wind have been taken into account. Figure reproduced with permission from W.-H. Ip and W. I. Axford, 1985, *A. & A.*, **149**, p. 71.

and set it into motion. These motions are often unstable to Rayleigh–Taylor and Kelvin–Helmholtz instabilities and, in general, turbulence is expected to take over. This kinetic energy decays through shock waves when clouds collide, emerging as line radiation, or through the excitation of plasma waves, which, eventually, also heat the gas. The average heating due to turbulence is, however, small (Table 1.2).

On a smaller scale, individual molecular clouds show linewidths in excess of the thermal width because of the presence of turbulent motions. This turbulence, which is probably of a magneto-hydrodynamic nature, supports these clouds against self-gravity. This turbulent energy is supplied by powerful outflows driven by newly formed stars into their surroundings – and hence derives from the gravitational energy of the collapsing cloud core – or by direct tapping of the magnetic or rotational energy supporting the cloud.

1.3.5 Summary

While these energy sources are very similar in magnitude (cf. Table 1.2), how much each of them contributes to the heating of the gas (and dust) depends on the coupling processes. Understanding these processes will thus be of prime importance for understanding the physical state of the ISM.

1.4 The Milky Way

Table 1.3 summarizes several characteristics of the Milky Way as a galaxy. Comparing these with those of other galaxies, the Hubble type of the Milky Way seems to be somewhere between an Sb and an Sc. In many ways, the Milky Way seems to be an average spiral galaxy.

1.4.1 Galactic distribution

The molecular gas shows an exponential distribution, peaking at the molecular ring ($\simeq$4.5 kpc) with a scale length of about 3 kpc. In contrast, the atomic gas has a rather flat distribution out to about 18 kpc (cf. Fig. 1.12). So, within the solar circle, the surface density of the molecular gas is somewhat larger than that of the atomic gas. In contrast, the atomic gas dominates the molecular gas by far in the outer Galaxy. Within 4 kpc, there is a hole in both the atomic and molecular gas distribution except for the nuclear ring in the inner Galaxy. The total HI mass is about 5 times that of H_2.

The most striking aspect of the gas and dust distribution of any galaxy is the thinness of the disk. The molecular gas has a thickness of $\simeq$75 pc compared with a radial scale length of 4 kpc. In the inner Milky Way, the HI scale height is two to three times larger, depending on which neutral gas component is considered. In the outer Galaxy, the neutral gas disk flares to a thickness of about 1 kpc. Nevertheless, this is still small compared to the radial scale length. The gas disk also shows a warp in the outer Galaxy.

1.4.2 Spiral structure

It is difficult to discern the spiral structure of our Galaxy from the atomic gas distribution because of the presence of non-circular velocity components. Molecular clouds are better tracers; not least because molecular clouds are more

Table 1.3 *The Milky Way*

Object	Mass ($M_\odot$)
Stars	1.8 (11)
Gas	4.5 (9)

Source	Luminosity ($L_\odot$)
Stellar luminosities	
All stars	4.0 (10)
OBA	8.0 (9)
Gas and dust	
[CII] 158 μm	5.0 (7)
FIR	1.7 (10)
Radio	1.5 (8)
γ-rays	3.0 (5)
Mechanical luminosities	
SN	2.0 (8)
WR	2.0 (7)
OBA	1.0 (7)
AGB	1.0 (4)

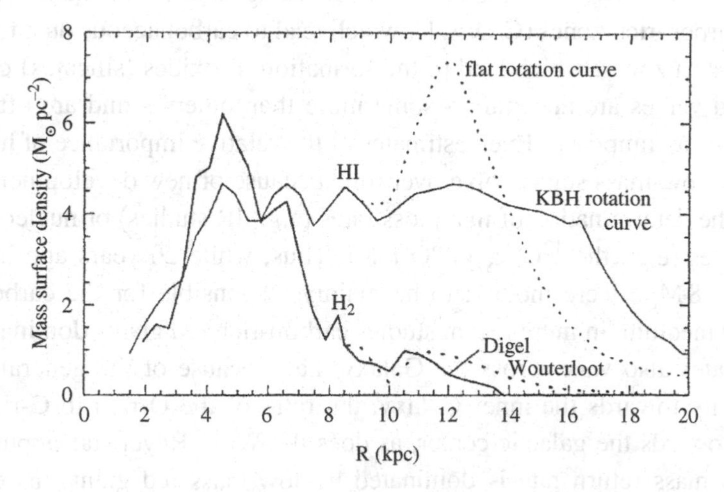

Figure 1.12 The mass surface density of HI and H_2 as a function of galactocentric radius. The distribution of HI in the outer Galaxy is quite sensitive to the adopted galactic rotation curve. Note that the nuclear ring is not shown. Figure reproduced with permission from T. M. Dame, 1993, in *Back to the Galaxy*, ed. S. S. Holt and F. Verter (New York: AIP), p. 267.

limited to the arms. Long structures are present in either tracer and, for example, the Sagittarius–Carina arm is readily recognized. Nevertheless, whether the Milky Way is a two-arm or four-arm spiral has not been settled.

1.4.3 Spiral galaxies

The overall distribution of the ISM can be particularly well studied in other nearby galaxies such as M31. This shows that the bright HII regions, the atomic and molecular gas, and the dust are concentrated in spiral arms. In contrast the interarm regions are much less obscured, although diffuse HI is present throughout the disk as well. Edge-on galaxies such as NGC 891 – an Sb galaxy very similar to our own – show most clearly the narrow disk component; for example, in the dust absorbing background stellar light. The HI distribution of this galaxy shows a component that is somewhat more extended than that of the Milky Way.

1.5 The mass budget of the ISM

Stars of all masses in various stages of their lives indiscriminately pollute their environment with gas, dust, and metals. Except for a few elements produced in the Big Bang, the abundance of all heavy elements reflects this enrichment with stellar nucleosynthetic products. Table 1.4 summarizes gas and dust mass injection rates into the ISM for a variety of stellar objects. The dust budget has been split out into two separate columns according to whether the stellar source contains carbon-rich zones ($C/O > 1$), which lead to carbonaceous dust formation, or oxygen-rich zones, which lead to the formation of oxides (silicates) or metals. The quoted values are uncertain – some more than others – and are often based upon various assumptions. Even estimates of the relative importance of high-mass stars versus low-mass stars evolve over time because of new developments in, for example, the determination of mass loss rates (e.g., IR studies) or nucleosynthetic reaction rates (e.g., the $^{12}C(\alpha, \gamma)^{16}O$ rate). Thus, while 20 years ago high-mass stars ($M > 8\,M_{\odot}$) were thought to be mainly responsible for the carbon in the interstellar medium, in more recent studies carbon-rich red giants dominate. These injection rates also vary across the Galaxy; i.e., because of the general increase in metallicity towards the inner Galaxy, the ratio of the O-rich to C-rich giants increases towards the galactic center, as does the Wolf–Rayet star population.

The gas mass return rate is dominated by low-mass red giants, as expected, since low-mass stars dominate the stellar mass of the Galaxy. However, massive stars are more efficient in producing and injecting heavy elements, such as O and Si, in the ISM (a typical type-II supernova has an enrichment factor of $\sim$10 for such elements). However, on the AGB, low-mass stars inject much C formed

Table 1.4 *Interstellar gas and dust budgets*

Source	$\dot{M}_H^a$ ($M_\odot$ kpc^{-2} Myear^{-1})	$\dot{M}_c^b$ ($M_\odot$ kpc^{-2} Myear^{-1})	$\dot{M}_{sil}^c$ ($M_\odot$ kpc^{-2} Myear^{-1})
C-rich giants	750	3.0	—
O-rich giants	750	—	5.0
Novae	6	0.3	0.03
SN type Ia	—	0.3[d]	2[d]
OB stars	30	—	—
Red supergiants	20	—	0.2
Wolf–Rayet	100[e]	0.06[f]	—
SN type II	100	2[d]	10[d]
Star formation	−3000	—	—
Halo circulation[f]	7000		
Infall[g]	150		

[a] Total gas mass injection rate.
[b] Carbon dust injection rate.
[c] Silicate and metal dust injection rate.
[d] Fraction and composition of dust formed in SN is presently unknown. These values correspond to upper limits.
[e] Dust injection only by carbon-rich WC 8–10 stars.
[f] Mass exchange between the disk and the halo estimated from HI in non-circular orbits and CIV studies.
[g] Estimated infall of material from the intergalactic medium and satellite galaxies.

by the triple-α process. Asymptotic giant branch stars are also the site of the production of the s process (formed through slow neutron capture) elements. Type-Ia supernovae, which also have low mass progenitors, inject a considerable amount of Fe into the ISM.

The time scale for stars to inject or replenish the local interstellar gas mass (8×10^6 M$_\odot$ kpc^{-2}) is $\simeq 5 \times 10^9$ years. The total dust injection rate corresponds to a dust-to-gas ratio in the ejecta of $\sim$1.5%. This is somewhat larger than the average dust-to-gas ratio in the ISM, reflecting the synthesis of heavy elements in type-II supernova. Locally, star formation will convert all molecular gas into stars in only 2×10^8 years; the average star formation rate is heavily weighted towards the inner molecular ring. The different phases of the ISM exchange material at a rapid rate (3×10^7 years) and, considering all the available gas, this time scale to convert gas into stars increases to $\simeq 3 \times 10^9$ years and is more comparable to the stellar injection time scale. Both rates are very short compared with the lifetime of the Galaxy ($\simeq 1.2 \times 10^{10}$ years). The disk of the Galaxy also exchanges material with the lower halo. In particular, the concerted action of supernovae set up a galactic fountain and some 5 M$_\odot$ is exchanged per year. In terms of the mass flux, this circulation pattern dominates the mass budget. Finally, the Galaxy is still accreting primordial material from its environment. The amount of this accretion is controversial. There is a contribution from satellite galaxies such as the Magellanic Clouds, and the Magellanic Stream represents an accretion of some $150\,M_\odot$ kpc^{-2} Myear^{-1}. There are also some indirect arguments that suggest that the Galaxy may have been accreting about 1 $M_\odot$ year^{-1} of metal-poor primordial gas over much of its lifetime or about an order of magnitude more than the Magellanic Stream accretion.

1.6 The lifecycle of the Galaxy

The origin and evolution of galaxies are closely tied to the cyclic processes in which stars eject gas and dust into the ISM, while at the same time gas and dust clouds in the ISM collapse gravitationally to form stars. The ISM is the birthplace of stars, but stars regulate the structure of the gas, and therefore influence the star formation rate. Winds from low-mass stars – and hence, the past star formation rate – control the total mass balance of interstellar gas and contribute substantially to the injection of dust, an important opacity source, and polycyclic aromatic hydrocarbon molecules (PAHs), an important heating agent of interstellar gas. High-mass stars (i.e., the present star formation rate) dominate the mechanical energy injection into the ISM, through stellar winds and supernova explosions, and thus the turbulent pressure that helps support clouds against galactic- and self-gravity. Through the formation of the hot coronal phase, massive stars regulate the thermal pressure as well. Massive stars also control the FUV photon energy budget and the cosmic-ray flux, which are important heating, ionization, and dissociation sources of the interstellar gas, and they are also the source of intermediate-mass elements that play an important role in interstellar dust. Eventually, it is the dust opacity that allows molecule formation and survival. The enhanced cooling by molecules is crucial in the onset of gravitational instability of molecular clouds.

Clearly, therefore, there is a complex feedback between star formation and the ISM. And it is this feedback that determines the structure, composition, chemical evolution, and observational characteristics of the interstellar medium in the Milky Way and in other galaxies all the way back to the first stars and galaxies that formed at redshifts, $z, > 5$. If we want to understand this interaction, we have to understand the fundamental physical processes that link interstellar gas to the mechanical and FUV photon energy inputs from stars.

1.7 Physics and chemistry of the ISM

The key point always to keep in mind when studying the interstellar medium is that the ISM is far from being in thermodynamic equilibrium. In thermodynamic equilibrium, a medium is characterized by a single temperature, which describes the velocity distribution, excitation, ionization, and molecular composition of the gas. While the velocity distribution of the gas can generally be well described by a single temperature, the excitation, ionization, and molecular composition are often very different from thermodynamic equilibrium values at this temperature. This reflects the low pressure of the ISM, so that, for example, collisions cannot keep up with the fast radiative decay rates of atomic and molecular levels. Ionization and chemical composition are also kept from their equilibrium values by the presence of ~100 MeV cosmic-ray particles – clearly a non-Maxwellian component – and a diluted, stellar, EUV-FUV photon field, which is much stronger than a 100 K medium would normally have. Finally, the large-scale velocity field is much influenced by the input of mechanical energy.

Whenever a gas is not in local thermodynamic equilibrium, the level populations, degree of ionization, chemical composition, and of course the temperature are set by balancing the rates of the processes involved. Much of the study of the ISM is thus concerned with identifying the various processes that control the ionization and energy balance, setting up the detailed statistical equilibrium equations and solving them for the conditions appropriate for the medium.

In the remainder of this book, we will thus first study the physical and chemical processes that are of astrophysical relevance (Chapters 2–4). This is followed by a discussion of the physics and chemistry of two important components of the ISM: dust grains (Chapter 5) and large molecules (Chapter 6). With this in hand, we can examine in-depth HII regions, the phases of the ISM, photodissociation regions, and molecular clouds (Chapters 7–10). In each chapter, the ionization and thermal balance are investigated first, setting up the statistical equilibrium equations and deriving the relevant parameters of the medium. This is then followed by a discussion of the observations of these objects and their analysis. In Chapters 11–12, non-equilibrium effects of a different nature are studied. Specifically, the time-dependent effects of strong shock waves on the temperature and chemical composition of clouds are discussed in Chapter 11. Chapter 12 examines then the dynamical effects of expanding HII regions, supernova explosions and stellar winds. Finally, in Chapter 13, some aspects of the lifecycle of interstellar dust – which is a prime example of these types of non-equilibrium effects – are explored.

1.8 Further reading

A general, technical yet accessible, introduction to the interstellar medium is provided by [1].

An introductory-level discussion of the interstellar medium can be found in [2]. A thorough but now dated monograph on physical processes in the interstellar medium is [3], which is an updated version of [4]. These two textbooks are no easy reading matter and the reader should beware that many a promising young scientist was last observed buying these books. A recent monograph on the interstellar medium is [5], which is intended for postgraduate courses. An unsurpassed textbook treating the physics of ionized gas is [6].

A recent overview of the properties of our Galaxy as compared to other galaxies is given by [7].

Panoramic images of the Milky Way at different wavelengths can be found on the web, http://adc.gsfc.nasa.gov/mw/milkyway.html.

Each of these wavelengths traces a different component of the Galaxy. It is instructive to compare these different images.

References

[1] J. B. Kaler, 1997, *Cosmic Clouds: Birth, Death, and Recycling in the Galaxy*, (New York: Freeman Co.)
[2] J. E. Dyson and D. A. Williams, 1980, *Physics of the Interstellar Medium* (Manchester: Manchester University Press)
[3] L. Spitzer, Jr., 1978, *Physical Processes in the Interstellar Medium* (New York: Wiley)
[4] L. Spitzer, Jr., 1968, *Diffuse Matter in Space*, (New York: Interscience)
[5] M. A. Dopita and R. S. Sutherland, 2003, *Astrophysics of the Diffuse Universe*, (Berlin: Springer Verlag)
[6] D. Osterbrock, 1989, *Astrophysics of Gaseous Nebulae and Active Galactic Nuclei* (Mill Valley: University Science Books)
[7] R. C. Kennicutt, 2001, in *Tetons 4: Galactic Structure, Stars and the Interstellar Medium*, ed. C. E. Woodward, M. D. Bicay, and J. M. Shull (San Francisco: ASP), p. 2

2

Gas cooling

In this chapter, we will examine the various line cooling processes of interstellar gas. Together with the heating processes, which are discussed in the next chapter, this will allow us to solve the energy balance for the gas in a wide variety of environments. This will be applied in HII regions (Section 7.3), HI regions (Section 8.2.2), photodissociation regions (Section 9.3), and molecular clouds (Section 10.3). This chapter starts off with a refresher on the concepts of electronic, vibrational, and rotational spectroscopy (Section 2.1). We will then discuss the cooling rate (Section 2.2) with the emphasis on two-level systems (Section 2.3). A two-level system analysis is a powerful tool for understanding the details of gas emission processes and we will encounter this many times in subsequent chapters. We will then consider in detail the cooling processes in ionized (Section 2.4) and neutral (Section 2.5) gas. The last section (Section 2.6) discusses the cooling law.

2.1 Spectroscopy

Table 2.1 summarizes typical properties of transitions. These are, of course, directly related to the binding energies of the species involved. Electronic binding energies for atoms increase from left to right in the periodic system from about 5 eV to some 20 eV. For hydrogen and helium the lowest electronic transitions are fairly high (10.2 and 21.3 eV, respectively); a substantial fraction of the ionization energy. Multi-electron systems have electronic orbitals that are low in energy compared with their ionization potentials. Their electronic transitions thus occur in the visible through extreme UV range. Transitions of bonding electrons in molecules typically also occur in the UV range. However, radicals or ions – which have an unpaired electron in low-lying electronic states – have transitions in the visible range. Molecular vibrational transitions involve motions of the atoms and, because of the larger reduced mass, are shifted by a factor $\sqrt{m_e/M}$ into the mid-infrared, where m_e and M are the masses of the electron and atom,

Table 2.1 *Typical transition strength*[a]

Type of transition	f_{ul}	$A_{ul}(s^{-1})$	Example	λ	$A_{ul}(s^{-1})$
Electric dipole					
UV	1	10^9	Lyα	1216 Å	2.40×10^8
Optical	1	10^7	Hα	6563 Å	6.00×10^6
Vibrational	10^{-5}	10^2	CO	4.67 μm	34.00
Rotational	10^{-6}	3×10^{-6}	CS[b]	6.1 mm	1.70×10^{-6}
Forbidden					
Optical (Electric quadrupole)	10^{-8}	1	[OIII]	4363 Å	1.7
Optical (Magnetic dipole)	2×10^{-5}	2×10^2	[OIII]	5007 Å	2.00×10^{-2}
Far-IR fine structure	$\frac{2\times10^{-7}}{\lambda(\mu m)}$	$\frac{10}{\lambda^3(\mu m)}$	[OIII]	52 μm	9.80×10^{-5}
Hyperfine			HI	21 cm	2.90×10^{-15}

[a] See text for details.
[b] The $J=1\rightarrow0$ transition.

respectively. Molecules can also rotate and the centrifugal energy is proportional to $1/M$. Hence, rotational energies of molecules are smaller by a factor m_e/M compared with electronic energies and occur at (sub)millimeter wavelengths. As an example, the binding energy of CO is 11 eV (1100 Å), the vibrational energy is 0.27 eV (2170 cm^{-1}, 4.67 μm), and the rotational energy is 5×10^{-4} eV (2.6 mm). A final and important point to keep in mind when considering transitions between levels is that they obey certain selection rules depending on whether they are electric dipole, magnetic dipole, electric quadrupole, etc. This directly influences the strength of the transitions (Table 2.1 and Section 2.1.4). We will come back to this later (Table 2.2 and Section 2.1.4).

2.1.1 Electronic spectroscopy

Atoms

Each atomic line can be related to a transition between two atomic states. These states are identified by their quantum numbers. The states are characterized by the principal quantum number, n (1, 2, ...); the orbital angular momentum, ℓ (which can have values between 0 and $n-1$); and the electron spin angular momentum, $s(\pm1/2)$. The orbital angular momenta are designated by s, p, d, ..., corresponding to $\ell=0$, 1, 2, ... The total angular momentum, j, is the vector sum of ℓ and s. For hydrogen, neglecting fine structure, the energies of the atomic states depend only on n. Hydrogen transitions can be grouped into series (downward transitions, $m\rightarrow n$, or upward transitions, $n\rightarrow m$) according to the value of n, with the conventional names for the first few: Lyman ($n=1$), Balmer ($n=2$), Paschen

Table 2.2 *Selection rules*[a]

	Electric dipole "allowed"	Magnetic dipole "forbidden"	Electric quadrupole "forbidden"
1	$\Delta J = 0, \pm1$ $0 \nleftrightarrow 0$	$\Delta J = 0, \pm1$ $0 \nleftrightarrow 0$	$\Delta J = 0, \pm1, \pm2$ $0 \nleftrightarrow 0,\ 1/2 \nleftrightarrow 1/2,\ 0 \nleftrightarrow 1$
2	$\Delta M = 0, \pm1$ $0 \nleftrightarrow 0$ when $\Delta J = 0$	$\Delta M = 0, \pm1$ $0 \nleftrightarrow 0$ when $\Delta J = 0$	$\Delta M = 0, \pm1, \pm2$
3	Parity change	No parity change	No parity change
4[b]	One electron jumping $\Delta l = \pm1$, Δn arbitrary	For all electrons $\Delta l = 0$, $\Delta n = 0$	One electron jumping with $\Delta l = 0, \pm2$, Δn arbitrary or for all electrons $\Delta l = 0$, $\Delta n = 0$
5[c]	$\Delta S = 0$	$\Delta S = 0$	$\Delta S = 0$
6[c]	$\Delta L = 0, \pm1$ $0 \nleftrightarrow 0$	$\Delta L = 0, \Delta J = \pm1$	$\Delta L = 0, \pm1, \pm2$ $0 \nleftrightarrow 0,\ 0 \nleftrightarrow 1$

[a] A short summary of six selection rules describing electric dipole and quadrupole and magnetic dipole transitions.
[b] With negligible configuration interaction.
[c] For LS coupling only.

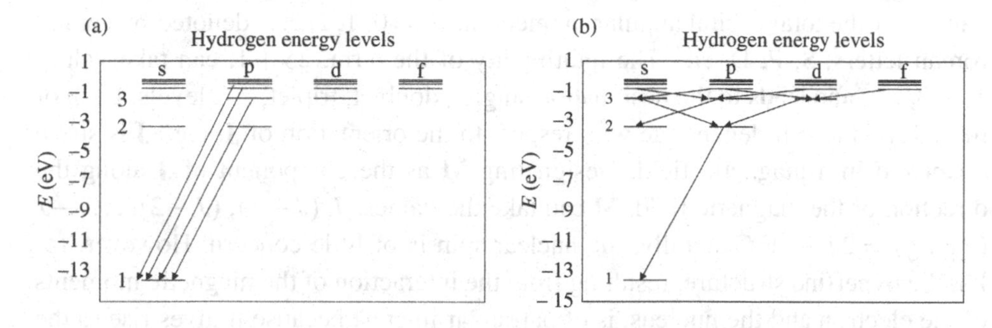

Figure 2.1 The energy level diagram of hydrogen. The principal quantum number, n, increases upwards (for clarity only the lower ones are labeled). The orbital angular momentum number, ℓ, increases rightwards and only the first few are shown. They are labeled s, p, d, f. The arrows indicate some allowed transitions. (a) The Lyman α through δ transitions are indicated. (b) The α transitions between the lowest levels are indicated. Note that the 2s level is metastable.

($n = 3$), Brackett ($n = 4$), Pfund ($n = 5$), and Humphreys ($n = 6$). Successive lines in these series are labeled α, β, γ, etc. Thus, Lyman α emission corresponds to the transition $n = 2 \to 1$. The Balmer transitions are also indicated by the designation, Hα, Hβ, etc.

For heavier species, the electrons are grouped in closed and open shells. The shell energies depend mainly on n. Subshells within each shell have different angular momenta and are labeled by n, ℓ, and the number of electrons in the subshell (to the maximum possible $2(2\ell+1)$). For heavier species, the description of states involved in transitions becomes more complicated than for H because of the interaction between the various electrons present. The energy levels of an atom are characterized by the values of J, the magnitude of the total angular momentum. The total angular momentum, $\mathbf{J}$ is defined by $\mathbf{J}=\mathbf{L}+\mathbf{S}$. Here, $\mathbf{L}$ is the orbital angular momentum: the vector sum of the orbital angular momenta of the individual electrons. The spin of the atom, $\mathbf{S}$, is the sum of the electron spins, where each electron has a spin 1/2 that can be parallel or antiparallel. The total angular momentum, $\mathbf{J}$, is formed by vector addition and can take the integral values, $(L+S)$, $(L+S-1)$, $(L+S-2)$, ..., $|L-S|$. There are thus $2S+1$ possible J values. When relativistic effects are small, a level with a given L and S is split into a number of distinct levels with different values of J, because of spin-orbit coupling. This coupling is a result of the interaction of the magnetic moment of the electron spin with the magnetic moment of the orbital motion. This fine-structure splitting is a small perturbation – of the order of $\alpha^2 Z^4$, with α the fine-structure constant ($=1/137$) and Z the nuclear charge. Transitions among the fine-structure levels occur therefore at near-to far-IR wavelengths.

Energy levels are then designated by term symbols, ${}^{2S+1}L_J$, with the different values for the total orbital angular momentum, $L=0, 1, 2, \ldots$, denoted by capital roman letters, S, P, D, ...[1] The multiplicity of the term, $2S+1$, can take values $1, 2, 3, \ldots$, and leads to the designation singlet, doublet, triplet, ..., levels. Each of these levels is still degenerate with respect to the orientation of $\mathbf{J}$; e.g., $\mathbf{J}$ is space quantized in a magnetic field. Designating $\mathbf{M}$ as the component of $\mathbf{J}$ along the direction of the magnetic field, $\mathbf{M}$ can take the values, J, $(J-1)$, $(J-2)$, ..., $-J$ (e.g., $g_J=2J+1$). Generally, the nuclear spin is of little concern. However, for HI, the hyperfine structure, resulting from the interaction of the magnetic moments of the electron and the nucleus, is of particular interest because it gives rise to the 21 cm line. This interaction is quite weak and the resulting splitting is very small. All this terminology of electronic terms can be interpreted in classical terms of orbital and spin angular momentum and the coupling between them. However, in addition, there is the quantum-mechanical concept of parity, which has no classical analog. Parity describes the behavior of the wave function, $\psi(x, y, z)$, under reflection where even parity corresponds to $\psi(x, y, z) = \psi(-x, -y, -z)$ and odd parity, $\psi(x, y, z) = -\psi(-x, -y, -z)$. Odd parity is indicated by a left superscript o.

Selection rules for electronic transitions are summarized in Table 2.2. These merely reflect conservation principles. Thus, angular momentum has to be

[1] Unfortunately, convention uses very similar notation for $L=0$ (S) and for the electron spin angular momentum (S).

conserved under vector addition and, given that the photon has one unit of angular momentum, this leads to the ΔJ and ΔL rules. The electron spin can only be changed by a magnetic field. The parity rule reflects that the dipole operator in the transition moment integral (proportional to the transition probability) has odd parity and hence only couples states with opposing parity. In the weak spin-orbit coupling regime (LS-coupling, also called Russell–Saunders coupling), the fine-structure energies are small compared to the energy differences of the levels. In the LS-coupling case, both L and S are "good" quantum numbers, which are conserved. When the coupling increases, (e.g., for heavier elements with larger nuclear charge), this rule breaks down and the spin-orbit interactions become as strong as the interactions between individual spins or orbital angular momenta. For LS coupling, conservation of S results in the segregation of transitions into groups associated with singlet, triplet, or quintet states, and transitions between these different S-states are weak. Conversely, this then implies that a species can become "trapped" in the lowest triplet or quintet state, which can represent considerable internal excitation.

As an example, consider neutral carbon with six electrons ($1s^22s^22p^2$). In the ground state, four of these are paired up in 1s and 2s orbitals and two occupy 2p orbitals but with the same spin (e.g., $1s^22s^22p^2$ where the superscript indicates the number of electrons in the shell). These parallel spins give favorable exchange interactions. The ground state has thus $L=1$, $S=1$, and $J=0$, e.g., ${}^3P_0$. Fine-structure interaction then produces two levels, ${}^3P_1$ and ${}^3P_2$ at 16.4 and 43.4 cm^{-1} above the ground state. The state where the electrons have opposing spin has slightly higher energy and has $L=2, S=0$, and $J=2$ with designation, ${}^1D_2$ at about 10 000 cm^{-1} above ground. The next state up is a ${}^1S_0$ state at about 21 600 cm^{-1}. These levels arise from the same electron configuration and are thus only linked through forbidden transitions. In particular, the ^{3}P levels are connected through fine-structure transitions at 609 and 370 μm. The forbidden character of transitions between the low-lying states is a very general point: all the low-lying electronic levels of the more common species arise from the same electronic configuration as their ground state and transitions are forbidden because of the parity selection rule.

Molecules

We will first discuss diatomic molecules, which in many ways we can describe in terms of an equivalent atom. Many of the concepts of atomic spectroscopy carry directly over into molecular spectroscopy. The nuclei in diatomic molecules produce an electric field along the internuclear axis. The component of the angular momentum vector of each electron along the internuclear axis can only take on values $m_l = l, (l-1), (l-2), \ldots, -l$. The variable, λ, is then defined as the absolute value of m_l, where $\lambda = 0, 1, 2, \ldots$, correspond to $\sigma, \pi, \delta, \ldots$, completely analogous to atomic s, p, d, ..., orbitals. For centrosymmetric species (e.g., homonuclear diatomics), parity under inversion is indicated by g for even and u for odd (the wave function does not change sign and changes sign, respectively). The term symbols of electronic levels in diatomic molecules are then characterized by the component of the total angular momentum along the intermolecular axis, Λ equal to the sum of the λs. Analogous to the designation, S, P, D, ..., for atoms, we have for diatomic molecules, $\Sigma, \Pi, \Delta, \ldots$, corresponding to $\Lambda = 0, 1, 2, \ldots$ The multiplicity of the term is indicated by a superscript with the value $2S+1$, reflecting the orientation of the total spin angular momentum, Σ, with respect to the internuclear axis, where Σ can take the values $S, (S-1), (S-2), \ldots, -S$.[2] The total angular momentum Ω is the sum of the orbital and spin angular momenta along the internuclear axis, $\Omega = \Lambda + \Sigma$. We can then designate molecular terms as ${}^{2S+1}\Lambda_{u,g}$, where the subscripts only apply to centrosymmetric species. For Σ terms, a $\pm$ superscript indicates the behavior of the wave function under reflection in the plane containing the two nuclei. As for atoms, coupling of the magnetic moments associated with spinning/orbiting charges is important. In this case, in addition to the orbital and spin angular momenta, we also need to consider the nuclear rotations. Two limiting cases are of interest: Hund's case (a), where the latter is unimportant and spin–orbit coupling is dominant, and Hund's case (b), where the spin couples strongly to the nuclear rotations and weakly to the orbital motions.

As for atoms, the total angular momentum must change by 0 or 1 (and $J = 0 \not\leftrightarrow 0$). The selection rules for allowed electronic transitions are now $\Delta\Lambda = 0, \pm 1, \Delta S = 0, \Delta\Sigma = 0$, and $\Delta\Omega = 0, \pm 1$. The selection rules associated with symmetry are now: for Σ terms, only $\Sigma^+ \leftrightarrow \Sigma^+$ and $\Sigma^- \leftrightarrow \Sigma^-$ are allowed. For centrosymmetric species, only transitions with a change in parity (e.g., $u \leftrightarrow g$) are allowed. As for atoms, the conservation of electronic spin can lead to metastable excited levels.

So, molecular hydrogen has the ground state ${}^1\Sigma_g^+$ with $S = 0$, $\Lambda = 0$, and $\Omega = 0$ and bound excited singlet states, ${}^1\Sigma_u^+$ and ${}^1\Pi_u$ (Fig. 2.2). There is also a repulsive electronic triplet state, ${}^3\Sigma_u^+$, at intermediate energies and bound excited triplet states (${}^3\Pi_u$ and ${}^3\Sigma_g^+$). Molecular oxygen has a ground state ${}^3\Sigma_g^-$ and carbon monoxide has the ground state ${}^1\Sigma^+$. Singlet–triplet transitions are forbidden and the first allowed electronic transitions for molecular hydrogen are the Lyman (${}^1\Sigma_g^+ \rightarrow {}^1\Sigma_u^+$) and Werner (${}^1\Sigma_g^+ \rightarrow {}^1\Pi_u$) bands.

In polyatomic molecules, electronic transitions are often connected to the excitation of specific types of electrons or electrons associated with a small group of

[2] Again, convention uses very similar notations for the $\Lambda = 0$ term (Σ) and the total spin angular momentum (Σ).

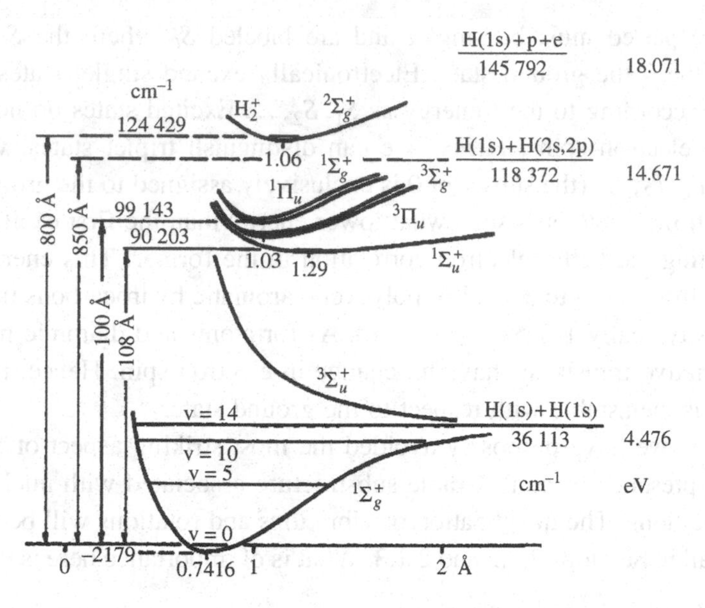

Figure 2.2 Schematic of the low-lying electronic energy levels of H_2 as a function of the separation between the two hydrogen atoms. Some vibrational levels in the ground electronic state are also schematically included. Dissociation and ionization energies are indicated on the right-hand side in eV and cm^{-1}. Some excitation energies are given on the left-hand side in cm^{-1}. All energies are relative to the $v = 0$ level in the ground electronic state. Figure reproduced from G. B. Field, W. B. Somerville, and K. Dressler, and reprinted with permission from *Ann. Rev. Astron. Astrophys.*, **4**, p. 207, ©1966, by *Ann. Rev.* (www.annualreviews.org).

atoms in the species. For example, the $\pi^\star \leftarrow \pi$ transition in a C=C bond occurs around 60 000 cm^{-1} (1600 Å), while in a C=O bond this transition occurs around 36 000 cm^{-1} (2800 Å). In a conjugated chain, this transition will shift to longer wavelengths. It is not possible to review the spectroscopy of polyatomic species here. In short, for a given nuclear geometry, molecular orbitals are constructed from atomic orbitals. The available electrons are then divided over these orbitals to construct electronic configurations, taking electron correlation into account (e.g., the Pauli principle). The lowest electronic configuration is then the ground state. For organic molecules, the ground state possesses a "closed shell" configuration (i.e., all the bonding and non-bonding orbitals are filled with a pair of electrons). Radicals (and often ions) possess molecular orbitals with only one electron.

For our purposes, the relevant electrons are located in the highest filled molecular orbitals. Deeper-lying electrons are not easily excited by the relatively low energies of photons in the interstellar medium. Radicals have transitions at low energies (i.e., in the visible range of the spectrum) associated with the partially occupied molecular orbital. Ground-state configurations for species where the electrons are paired must be singlet and are labeled S_0 where the S indicates singlet and the 0 the ground state. Electronically excited singlet states are then enumerated according to their energy as $S_1, S_2, \ldots$ Excited states do not have to have paired electron spins. Hence, we can distinguish triplet states, which are designated $T_1, T_2, \ldots$ (the subscript 0 is exclusively assigned to the ground state). The lowest triplet state is at somewhat lower energy than the first excited singlet state, reflecting the better electron correlation in the former. This energy difference ranges from $\sim$0.3 to 3 eV. For polycyclic aromatic hydrocarbons this energy difference is typically 1–1.5 eV (Fig. 2.3). As for atoms and diatomic molecules, allowed radiative transitions have no change in electron spin. Hence, the lowest triplet state is metastable with respect to the ground state.

Up to now, we have purposely avoided the most striking aspect of molecular spectra, the presence of considerable substructure associated with nuclear vibrations and rotations. The quantization of vibrations and rotations will be discussed in more detail in Sections 2.1.2 and 2.1.3. What is of importance here is that during

Figure 2.3 Schematic diagram of singlet and triplet electronic levels of an organic molecule. The electronic states are displaced horizontally for clarity but this has no physical meaning. The vertical axis denotes energy on an arbitrary scale. Within each electronic state, vibrational levels are schematically displayed. The rotational sublevels are not shown.

an electronic transition, the molecule may also change its vibrational or rotational state. This may result in vibrational sidebands, which are shifted in energy by one or more quanta of vibrational energy (~1000–3000 cm^{-1}). In addition, these bands also show small-scale structure caused by simultaneous rotational transitions. As for vibrations, these are classified in terms of classical P, Q, and R branches (cf. Section 2.1.2). The general appearance of these bands reflects the difference in rotational constants between the two electronic states involved in the transition.

2.1.2 *Vibrational spectroscopy*

The essence of vibrational spectroscopy is contained within Hooke's law for a harmonic oscillator,

$$\nu = \frac{1}{2\pi c}\sqrt{\frac{\kappa}{\mu}}, \tag{2.1}$$

with ν the fundamental frequency, κ the force constant, and μ the reduced mass of the molecular units vibrating. The factor c is included to transform the unit to wavenumbers (cm^{-1}) for cgs units, commonly used in spectroscopy. Molecular bond strengths are, of course, very similar to binding energies of electrons to an atom. However, the frequencies of the transitions of molecular vibrational levels are, thus, shifted to lower energies by $\sqrt{m_e/M}$, with m_e and M the masses of the electron and atom, respectively, and occur in the near- and mid-IR. Real molecules are not harmonic oscillators and, as a result of anharmonicity of the bonding, hot bands (e.g., 2–1, 3–2) are generally shifted to lower frequencies. All the possible vibrational motions of a molecule can now be described in terms of these fundamental modes. A non-linear molecule containing N atoms will have $3N-6$ normal modes; a linear molecule $3N-5$. Not all of these modes will lead to distinct absorptions. Some of the modes will occur at the same frequency and hence these modes are degenerate. Others will not be infrared-active because the dipole moment does not change during the vibration. So, as an example, methane has nine fundamental modes but only the symmetric and asymmetric stretching and bending vibrations of the C–H bonds about the central C atom show up in the IR absorption spectrum: the ν_4(1306 cm^{-1}), the ν_2(1534 cm^{-1}), and the ν_3(3019 cm^{-1}) modes. Vibrations in homonuclear molecules do not lead to changes in the dipole moment and hence are inactive in the infrared. Mixing of isotopes in such species will lead to infrared absorptions. Likewise, interactions with the environment in a solid will introduce weak infrared activity.

Vibrational transitions are very characteristic for the motions of the atoms in the molecular group directly involved but much less sensitive to the structure of the rest of the molecule. Characteristic band positions of various molecular groups

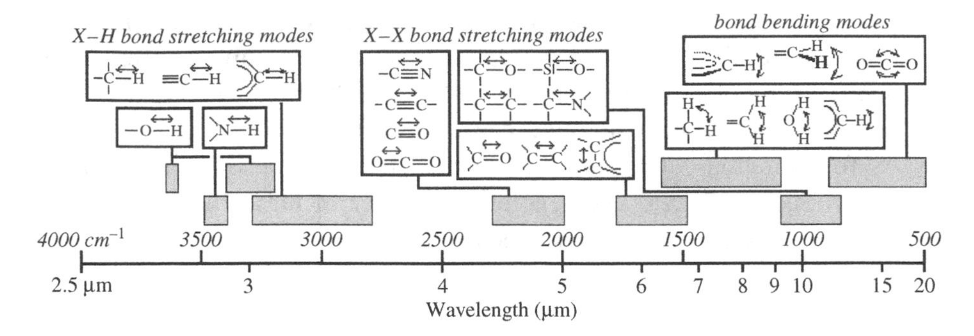

Figure 2.4 Summary of the vibrational frequencies of various molecular groups. The filled boxes indicate the range over which specific molecular groups absorb. The vibrations of these groups are schematically indicated in the linked boxes. Figure kindly provided by D. Hudgins.

are illustrated in Fig. 2.4 and summarized in Table 2.3. Modes involving motions of hydrogen occur at considerably higher frequencies than modes involving similar motions of heavier atoms (Hooke's law again). Thus, H-stretching vibrations occur in the 3 μm region, while stretching motions among (single-bonded) C, N, and O atoms are located around 10 μm. Likewise, when the bond strength increases, the vibration shifts to higher frequencies. Simplistically speaking, single bonds are characterized by two electrons occupying the bonding molecular orbital. Similarly, double bonds and triple bonds correspond to four and six electrons in the bonding molecular orbitals. Because the bond strength increases from single bonds to double bonds to triple bonds, the location of singly, doubly, and triply bonded CC vibrations shifts from about 1000 to 2000 cm^{-1} (Fig. 2.4).

The actual spectrum of a gaseous compound is much more complex because transitions are actually combined rotational–vibrational transitions where both the vibrational and the rotational energy can change. This gives rise to three sets of absorption bands corresponding to $\Delta J = 1$ (R branch), $\Delta J = 0$ (Q branch), and $\Delta J = -1$ (P branch), where $\Delta J = J' - J$ with J' and J the rotational quantum numbers in the upper and lower vibrational state, respectively. For electric quadrupole transitions, the selection rules are $\Delta J = 2$ (S branch), $\Delta J = 0$ (Q branch), and $\Delta J = -2$ (O branch). These sets of bands are, of course, shifted from each other by the difference in rotational energy content of the species. Consider as an example a diatomic molecule, which we represent here as a harmonic oscillator and a rigid rotor. The energy is then given by

$$E(v, J) = \omega \left(v + \frac{1}{2} \right) + B J (J+1), \tag{2.2}$$

with v and J the vibrational and rotational quantum number, ω the vibrational frequency, and B the rotational constant. The R branch ($v = 1 \rightarrow 2,\ J \rightarrow J+1$)

Table 2.3 *Characteristic vibrational band positions*

Group	Mode	Frequency range (cm^{-1})	Note
OH stretch			
	free OH	3610–3645[a]	sharp
	intramolecular[b]	3450–3600	sharp
	intermolecular[c]	3200–3550	broad
	chelated[d]	2500–3200	very broad
NH stretch			
	free NH	3300–3500	sharp
	H-bonded NH	3070–3350	broad
CH stretch			
	≡C–H	3280–3340	
	=C–H	3000–3100	
	CO–CH_3	2900–3000	ketones
	C–CH_3	2865–2885	symmetric
		2950–2975	asymmetric
	O–CH_3	2815–2835	symmetric
		2955–2995	asymmetric
	N–CH_3	2780–2805	aliphatic amines
	N–CH_3	2810–2820	aromatic amines
	CH_2	2840–2870	symmetric
		2915–2940	asymmetric
	CH	2880–2900	
C≡C stretch			
	C≡C	2100–2140	terminal group
	C–C≡C–C	2190–2260	
	C–C≡C–C–C≡C–	2040–2200	
C≡N stretch			
	saturated aliphatic	2240–2260	
	aryl	2215–2240	
C=O stretch			
	non-conjugated	1700–1900	
	conjugated[e]	1590–1750	
	amides	1630–1680	
C=C stretch			
	–HC=C=CH_2	1945–1980	
	–HC=C=CH–	1915–1930	
CH bend			
	CH_3	1370–1390	symmetric
		1440–1465	asymmetric
	CH_2	1440–1480	
	CH	1340	
CO–O–C stretch			
	formates	~1190	
	acetates	~1245	

Table 2.3 (cont.)

Group	Mode	Frequency range (cm^{-1})	Note
C–O–C stretch			
	saturated aliphatic	1060–1150	asymmetric
	alkyl, aryl ethers	1230–1277	
	vinyl ethers	1200–1225	
aromatic modes			
	C–H stretch	~3030	
	C–H deformation	~1160	in plane
	C–H deformation	900–740	out of plane[f]
	C=C stretch	1590–1625, 1280–1315	

[a] The OH frequency of free water occurs at 3756 cm^{-1}.
[b] H-bonded as dimer or polymer.
[c] In a fully H-bonded network such as ice.
[d] The OH is H-bonded to an adjacent C=O group.
[e] Two double bonds separated by a single bond.
[f] Pattern of bonds whose position depends on number of adjacent C-atoms with H.

corresponds to $\nu = \omega + 2B\ (J+1)$ and the P branch ($v = 1 \rightarrow 2,\ J \rightarrow J-1$) to $\nu = \omega - 2BJ$. For all heteronuclear diatomic molecules (and some vibrations of linear polyatomic molecules), the Q branch is absent but its transitions ($v = 1 \rightarrow 2,\ J \rightarrow J$) would occur at $\nu = \omega$. Thus, the spectrum consists of two sets of absorption bands where the individual transitions are separated by the rotational constant. This is illustrated in Fig. 2.5 for CO. In a real molecule, the effects of centrifugal distortion (as the molecule spins faster, the bond stretches, the moment of inertia increases, and B decreases; cf. Section 2.1.3) and the coupling between rotations and vibrations (as the bond stretches, the moment of inertia increases and B decreases; cf. Section 2.1.3) have to be included and this will shift these bands slightly and separate out the individual Q branch transitions; but the principle remains.

In the solid state, rotations are generally suppressed and hence the rotational–vibrational bands collapse to one absorption band at the band origin. This band position will be shifted with respect to the free (gas phase) species because of interactions with neighboring molecules. This shift will be larger if the interaction is stronger. For species isolated in inert matrices (e.g., a noble gas) where only weak van der Waals forces are active, the peak position will be close to the gas phase origin (e.g., the position of the Q branch). For molecules that can form complexes or even stronger H bonds, the shift can be substantial. Water is notorious in this respect and the ν_3 mode of isolated H_2O shifts from the gas phase position, 3756 cm^{-1}, to 3733 cm^{-1} in an Ar matrix, 3727 cm^{-1} in an N_2 matrix and 3707 cm^{-1} in a CO matrix. For the dimer (two H_2O molecules H-bonded to

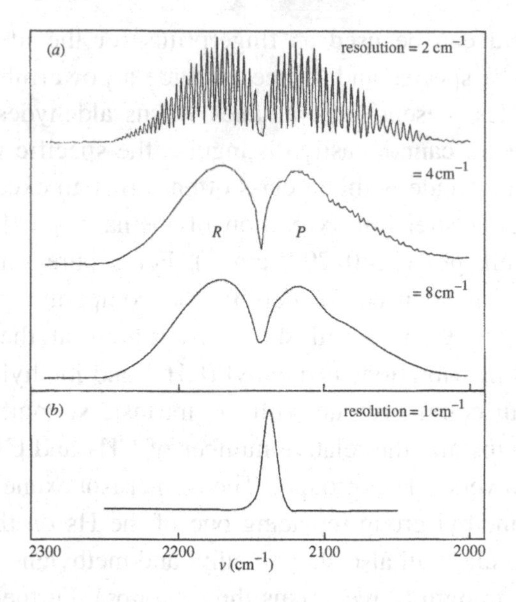

Figure 2.5 The infrared absorption spectrum of CO in the gas phase, at different spectral resolutions (top three curves). To first order, the P and R branches consist of equidistant absorption lines with an absorption intensity that initially increases with *J* owing to the statistical weight but, eventually, Boltzmann statistics take over. The bottom curve shows the absorption spectrum of solid CO. Absorption increases upwards in each curve. Figure reproduced with permission from L. J. Allamandola, 1984, in *Galactic and Extragalactic Infrared Spectroscopy*, ed. M. Kessler and P. Phillips, (Dordrecht: Reidel), p. 5.

each other), the peak positions are 3726 and 3709 cm^{-1} (Ar matrix), 3715 and 3699 cm^{-1} (N_2 matrix), and 3673 and 3493 cm^{-1} (CO matrix). In crystalline H_2O ice, this effect is very strong, on account of H-bonding, and the ν_3 mode is located at 3206 cm^{-1}. In contrast for CO, which forms only weak bonds, these shifts are always small; gas phase 2143 cm^{-1}, CO isolated in N_2 2140 cm^{-1}, CO isolated in Ar 2138.4 cm^{-1}, and solid CO 2138.4 cm^{-1}. Different modes in a molecule are often affected differently. For water, the stretching mode shifts by some 300 cm^{-1} while the bending mode shifts by only 65 cm^{-1}. Finally, besides the peak position, the width of a band is also severely affected by solid-state effects. In the gas phase, the width is set by the Doppler or turbulent broadening of the medium and is, for galactic objects, of the order of 3 km s^{-1} (0.02 cm^{-1}). A species trapped in the solid phase will experience variations in the interaction with the environment from one site to another. This will result in slight shifts in the peak frequency and hence broadening of the line. When the interaction with the environment is small (as for an inert species in a noble gas matrix) this broadening is 1–3 cm^{-1}, but this broadening is very large for H-bonding species in an H-bonding environment (e.g., 300 cm^{-1} for solid H_2O).

Vibrational spectra can be used as fingerprints for the identification of the molecular groups of a species and, hence, provide a powerful tool to determine the class of molecules present (e.g., alkanes versus aldehydes). However, as a rule, vibrational spectra cannot easily distinguish the specific molecule within a class. The smallest molecule within a class often forms an exception to this rule. For example, the C–H stretching vibration of methane ($3019\,cm^{-1}$) is shifted from that of other alkanes (2840–$2975\,cm^{-1}$). For a pure substance, subtleties within the spectra can be used to identify the compound present. Thus, the spectrum of n-hexane (C_6H_{14}) will show absorptions at the positions of the stretching and bending vibrations of methyl (CH_3) and methylene (CH_2) groups in a relative strength commensurate with the intrinsic strength of the modes of these molecular groups and the relative number of CH_3 and CH_2 groups present in the molecule (two versus four groups). The isomer isohexane (2-methylpentane with an additional methyl group replacing one of the Hs on the second C atom of the pentane molecule) will also show methyl and methylene absorptions but in a different relative strength (three versus three groups). Furthermore, the methyl bands will be split because of interaction between the adjacent CH_3 groups. In the gas phase, identification can be considerably aided by resolved P, Q, and R branches. In the solid state, when a mixture of species is present, identification of specific molecular species present within a class is often daunting.

2.1.3 Rotational spectroscopy

Diatomic and linear polyatomic molecules

For a rigidly rotating (i.e., internal nuclear geometry fixed and ignoring electronic motions) diatomic or linear polyatomic molecule, the rotational energy levels are given by

$$\frac{E_{\mathrm{r}}}{hc} = B_{\mathrm{e}}\, J(J+1), \tag{2.3}$$

with B_{e} the rotational constant and J the rotational quantum number (cf. Eq. (2.2)). The rotation constant is equal to

$$B_{\mathrm{e}} = \frac{h}{8\pi^2 cI}, \tag{2.4}$$

with I the moment of inertia,

$$I = \sum_i m_i r_i^2, \tag{2.5}$$

and m_i the mass of atom i at distance r_i from the center of mass. If we allow for centrifugal stretching of the rotating molecule, then we have to add a term $-D_{\mathrm{e}}J^2(J+1)^2$ to Eq. (2.3) with $D_{\mathrm{e}} = 4B_{\mathrm{e}}^3/\omega_{\mathrm{e}}^2$ and ω_{e} the vibrational frequency

Table 2.4 *Characteristics of molecular cooling lines*

Species	Transition	ν_{ul}(GHz)	E_u(K)	A_{ul} (s^{-1})	n_{cr} (cm^{-3})
CO	1–0	115.3	5.5	7.2×10^{-8}	1.1×10^{3}
	2–1	230.8	16.6	6.9×10^{-7}	6.7×10^{3}
	3–2	346.0	33.2	2.5×10^{-6}	2.1×10^{4}
	4–3	461.5	55.4	6.1×10^{-6}	4.4×10^{4}
	5–4	576.9	83.0	1.2×10^{-5}	7.8×10^{4}
	6–5	691.2	116.3	2.1×10^{-5}	1.3×10^{5}
	7–6	806.5	155.0	3.4×10^{-5}	2.0×10^{5}
CS	1–0	49.0	2.4	1.8×10^{-6}	4.6×10^{4}
	2–1	98.0	7.1	1.7×10^{-5}	3.0×10^{5}
	3–2	147.0	14.0	6.6×10^{-5}	1.3×10^{6}
	5–4	244.9	35.0	3.1×10^{-4}	8.8×10^{6}
	7–6	342.9	66.0	1.0×10^{-3}	2.8×10^{7}
	10–9	489.8	129.0	2.6×10^{-3}	1.2×10^{8}
HCO^+	1–0	89.2	4.3	3.0×10^{-5}	1.7×10^{5}
	3–2	267.6	26.0	1.0×10^{-3}	4.2×10^{6}
	4–3	356.7	43.0	2.5×10^{-3}	9.7×10^{6}
HCN	1–0	88.6	4.3	2.4×10^{-5}	2.6×10^{6}
	3–2	265.9	26.0	8.4×10^{-4}	7.8×10^{7}
	4–3	354.5	43.0	2.1×10^{-3}	1.5×10^{8}
H_2CO	2_{12}–1_{11}	140.8	6.8	5.4×10^{-5}	1.1×10^{6}
	3_{13}–2_{12}	211.2	17	2.3×10^{-4}	5.6×10^{6}
	4_{14}–3_{13}	281.5	30	6.0×10^{-4}	9.7×10^{6}
	5_{15}–4_{14}	351.8	47	1.2×10^{-3}	2.6×10^{7}
NH_3	(1,1) inversion	23.7	1.1	1.7×10^{-7}	1.8×10^{3}
	(2,2) inversion	23.7	42	2.3×10^{-7}	2.1×10^{3}
H_2	2–0	1.06×10^{4} [a]	510	2.9×10^{-11}	10
	3–1	1.76×10^{4} [b]	1015	4.8×10^{-10}	300

[a] $\lambda = 28.2\,\mu$m.
[b] $\lambda = 17.0\,\mu$m.

(in cm^{-1}). Radiative rotational transitions are allowed for $\Delta J = \pm 1$, corresponding to

$$E(J+1) - E(J) = 2hcB\,(J+1), \tag{2.6}$$

for a rigid rotor and the spectrum consists of a set of evenly spaced lines in frequency space. Centrifugal distortion will destroy this constant separation. The separation of the lines will now increase in half of the spectrum and decrease in the other half. Table 2.4 summarizes the characteristics of rotational transitions of diatomic molecules. Molecules consisting of heavy atoms have rotational constants corresponding to $\simeq$3 K and the $J = 1 \rightarrow 0$ transition falls at a few millimeters. Because of the lower mass, the rotational energy levels of hydrides are shifted to higher energies; i.e., the submillimeter or far-infrared range. The statistical weight is given by $g_J = 2J + 1$.

Molecular hydrogen is a homonuclear molecule and has no permanent dipole moment and, hence, no allowed dipole transitions. H_2 has weak rotational quadrupole transitions with $\Delta J = \pm 2$. Thus, the energy levels of H_2 divide out into two separate ladders with even or odd J, which are not connected by radiative transitions. Because of symmetry considerations, the even states correspond to para H_2 (i.e., antiparallel nuclear spins) and the odd levels to ortho states (e.g., parallel nuclear spins). Then, owing to nuclear spin statistics ($g_S = 2S + 1$), the ortho levels have three times the statistical weight of the para levels.

Symmetric top molecules

For a non-linear molecule, the rotational energy will depend on the axis of rotation. Two of the moments of inertia are equal – usually by symmetry – in a symmetric top molecule. For a prolate symmetric top, $I_A < I_B = I_C$ (a cigar shape) and for an oblate symmetric top, $I_A = I_B < I_C$ (a disk shape) with I_i the respective moments of inertia. The rotational motion is characterized by two quantum numbers: the total angular momentum, J, and its projection on the symmetry axis, K. For a prolate symmetric top, the energy levels are given by

$$\frac{E_{\rm r}}{hc} = BJ(J+1) + (A - B)K^2, \tag{2.7}$$

where the rotational constants, B and A, are given by

$$B = \frac{h}{8\pi^2 c I_B}, \tag{2.8}$$

$$A = \frac{h}{8\pi^2 c I_A} \tag{2.9}$$

and $A > B$. For an oblate symmetric top, we have

$$\frac{E_{\rm r}}{hc} = BJ(J+1) + (C - B)K^2, \tag{2.10}$$

where the rotational constant, C, is given by

$$C = \frac{h}{8\pi^2 c I_C}. \tag{2.11}$$

For a prolate top, we recover the expression of a linear or diatomic molecule with $I_A = 0$ for $K = 0$. Figure 2.6 illustrates the energy levels for a prolate and an oblate symmetric top. The value of K must not be greater than J (i.e., $|K| = J, J - 1, J - 2, \ldots, 0$). Except for $K = 0$, all K states are doubly degenerate, corresponding to opposite directions of rotation around the figure axis. The energy levels divide thus into separate ladders with $J = K, K+1, K+2, \ldots$ Hence, for a given K, the energy levels of a symmetric top show the same simple structure as linear molecules but the different K ladders are offset by $(A-B)K^2$ in energy. For a given J value, the energy of a level increases with increasing K for a prolate top while it decreases for an oblate top (cf. Fig. 2.6). We recognize that the molecule is rotating around its principal axis when $|K|$ is close to its maximum, J, and rotating end-over-end when $K = 0$.

Radiative transitions are allowed between the J levels within a given K ladder ($\Delta J = 1$, $\Delta K = 0$), which again reflects angular momentum conservation of the oscillating dipole of a rotating polar molecule coupled with the spin of the photon. As for the linear rotor, the spectrum consists of a simple series of equidistant lines, separated by $2B(J+1)$, which, in first order, overlap for all K-values. Thus, the transition $J = 1-0$ for $K = 0$ occurs at $\nu = 2B$, the transitions $J = 2-1$ for $K = 0$, 1 are at $\nu = 4B$, and $J = 3-2$ for $K = 0, 1, 2$ at $\nu = 6B$. However, because of centrifugal distortion, the transitions with different K values separate out. Note that the lowest level in each K ladder is metastable. Each level is, as usual, $2J+1$ fold degenerate.

Asymmetric top molecules

For asymmetric top molecules, all three moments of inertia are different and there is no simple general formula describing the energy levels. When $I_B \simeq I_A \neq I_C$ (oblate) or $I_B \simeq I_C \neq I_A$ (prolate), the levels can be represented by $J_{K^-K^+}$, where K^- and K^+ are approximate quantum numbers. As an example, formaldehyde is a near-prolate molecule. The energy levels show the characteristic pattern of the

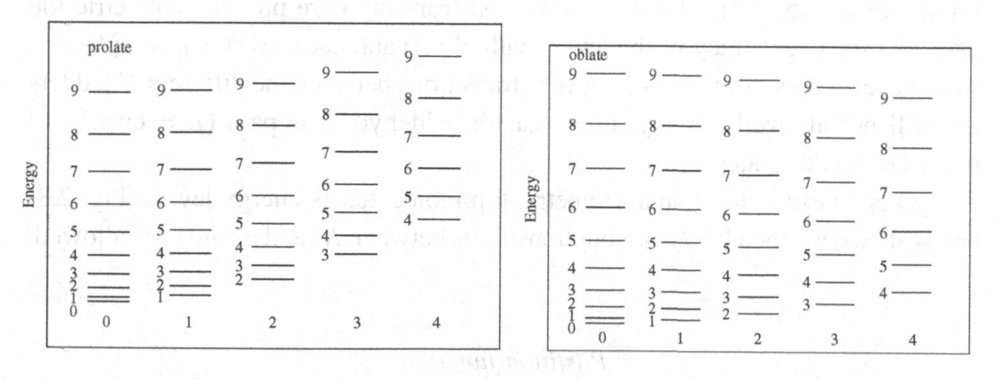

Figure 2.6 Schematic rotational energy level diagram for a prolate (left) and oblate (right) symmetric top molecule. Energy increases upwards. The individual levels have been labeled with their J value. The K ladders have been shifted horizontally. Allowed transitions occur within each K ladder. Because K is the projection of J on the molecular symmetry axis, we have $K \leq J$.

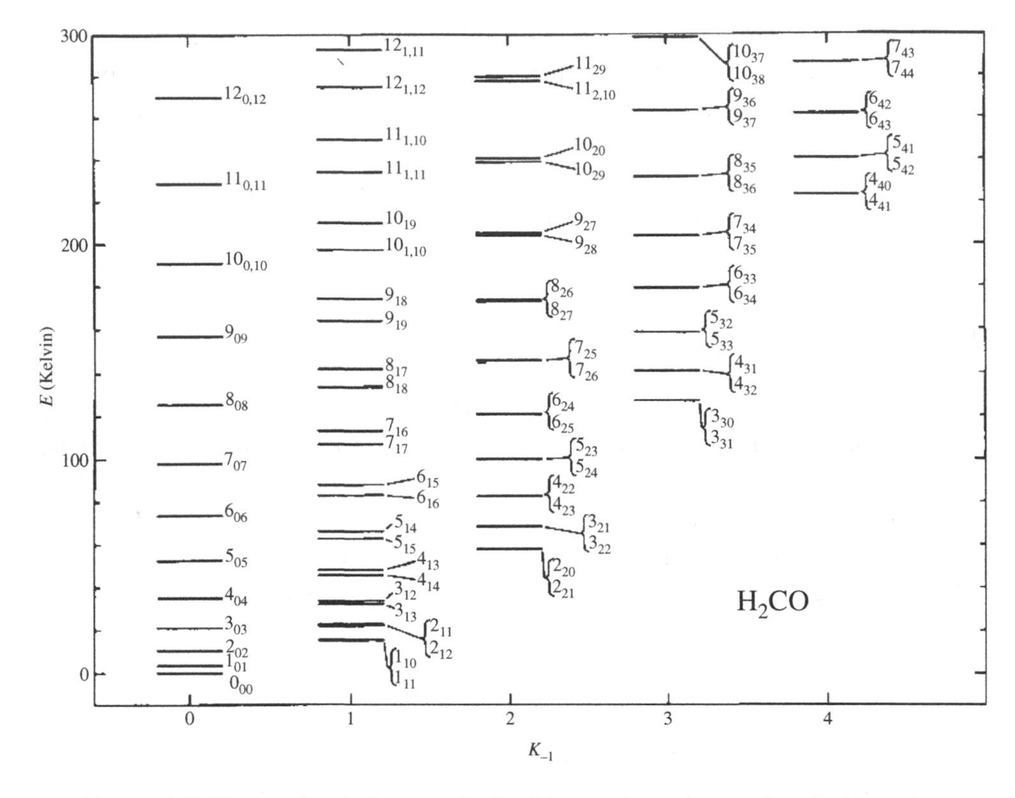

Figure 2.7 Energy level diagram for H_2CO, a near-prolate molecule. Note that formaldehyde has para, K is even, and ortho, K is odd, levels. Figure reproduced with permission from J. G. Mangum and A. Wootten, 1993, *Ap. J. S*, **89**, p. 123.

prolate case (Fig. 2.7). The slight deviation from the pure prolate symmetric top case lifts the degeneracy of the levels with $K > 0$ and each level is now split into two. Nevertheless, for the lowest levels, transitions between the different K ladders are still not allowed. Finally, note that formaldehyde has para ($K =$ even) and ortho ($K =$ odd) states.

H_2O is an example of an asymmetric top molecule. Its energy levels (Fig. 2.8) can still be described by $J_{K^-K^+}$ but transitions between K ladders are now allowed.

Partition function

For rotational transitions, the transition moment is given by

$$\mu^2(J+1, J) = \frac{J+1}{2J+3}\mu_e^2, \tag{2.12}$$

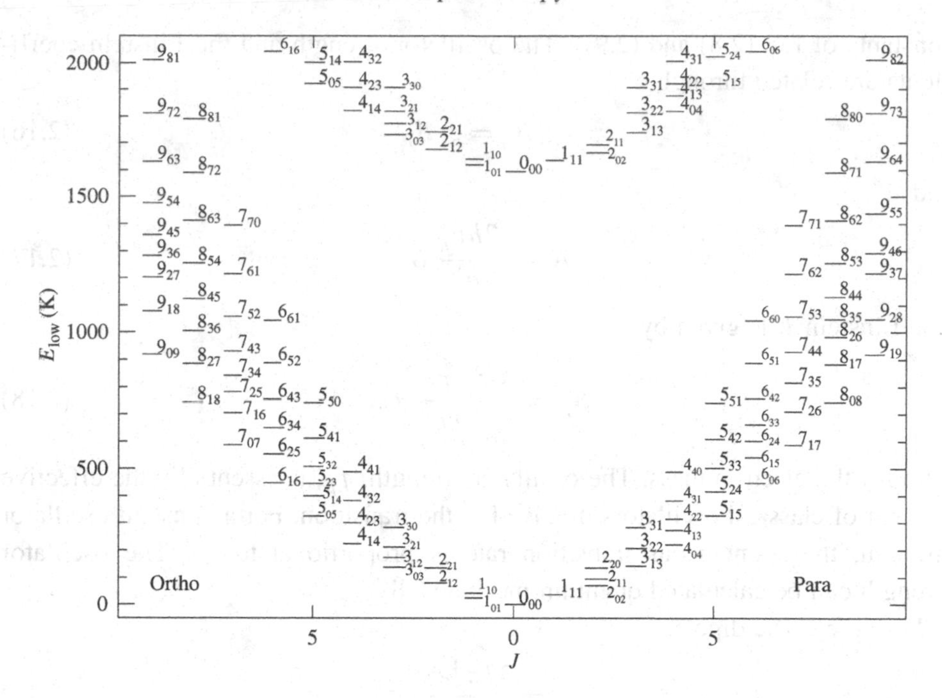

Figure 2.8 Energy level diagram for H_2O. Figure courtesy of F. Helmich.

for diatomic and linear polyatomic molecules and

$$\mu^2(J, K \rightarrow J-1, K) = \frac{J^2 - K^2}{(2J+1)}\, \mu_e^2, \tag{2.13}$$

for symmetric top molecules, with μ_e the permanent electric dipole moment.

For a constant excitation temperature, T_x, the partition function is

$$Q(T_x) = \sum g_i \exp[-E_i/kT_x] \simeq \frac{kT_x}{hcB}, \tag{2.14}$$

for a linear diatomic or polyatomic molecule. For a symmetric top molecule, the partition function is

$$Q(T_x) \simeq 1.8 \left(\frac{kT_x}{hcB}\right) \left(\frac{kT_x}{hcA}\right). \tag{2.15}$$

2.1.4 Transition strength

The transition strength can be expressed in terms of the oscillator strength or through the Einstein A and B coefficients (note that these are not the rotational constants of Eq. (2.8) and (2.9)). The oscillator strength and the Einstein coefficients are related through

$$g_i B_{ij} = g_j B_{ji} \tag{2.16}$$

and

$$A_{ji} = \frac{2h\nu_{ji}^3}{c^2} B_{ji}. \tag{2.17}$$

The Einstein B is given by

$$B_{ji} = \frac{4\pi}{h\nu_{ji}} \frac{\pi e}{m_e c} f_{ji}, \tag{2.18}$$

with m_e the electron mass. The oscillator strength, f_{ji}, is essentially the effective number of classical oscillators involved in the transition. For a constant oscillator strength, the spontaneous transition rate is proportional to ν^3. The oscillator strength can be calculated quantum mechanically.

For an electric dipole,

$$B_{ji} = \frac{32\pi^4 e^2}{3h^2 c} \mu_{ji}^2, \tag{2.19}$$

with μ_{ji} the dipole matrix element. As a rough order of magnitude estimate (see Table 2.1), electric dipole transitions will scale approximately with $\mu_e \simeq ea_0$ with a_0 the radius of the first Bohr orbit ($\mu_e \simeq 2.5$ D). The strongest, electric dipole allowed, ultraviolet transitions have $A \simeq 10^9$ s^{-1} corresponding to $f \sim 1$. Visible electric dipole allowed transitions will have $A \sim 10^7$ s^{-1}. Molecules typically have a permanent dipole of 1–3 D and allowed electronic transitions for molecules have high Einstein As as well. Because the mass involved differs, the oscillator strengths of molecular vibrational transitions are typically a factor $m_e/M \simeq 10^{-4}$ weaker. The selection rule is now $\Delta v = 1$ with v the vibrational quantum number. Molecular rotational transitions have $f \simeq 10^{-6}$.

When a transition is electric dipole forbidden, it can still occur by magnetic dipole, electric quadrupole, or higher-order "poles" interaction. These forbidden lines are indicated by square brackets (e.g., [OI] λ 6300 Å). Electric quadrupole transitions are typically a factor of $e^2 a_0^2 \lambda^2 / e^2 a_0^4 \simeq 10^8$ weaker than electric dipole transitions, or for visible transitions, $A \sim 10^{-1}$ s^{-1}. Magnetic dipole transitions scale with the Bohr magneton ($eh/4\pi mc$) rather than μ_e, resulting in a transition weaker by a factor $\alpha^2 \simeq 5 \times 10^{-5}$ (with α the fine-structure constant). For the optical lines, the magnetic dipole involved is typically much weaker. Far-IR atomic fine structure levels are coupled through magnetic dipole transitions with an oscillator strength of 2×10^{-9} at 100 μm. The HI 21 cm line is a magnetic dipole transition between the two hyperfine levels of the ground state of atomic hydrogen, which are separated by only 6×10^{-6} eV or 0.05 K. Its transition probability is therefore exceedingly small.

2.2 Cooling rate

The radiative cooling rate due to a transition from level i to level j of some species x is given by

$$n^2 \Lambda_x(\nu_{ij}) = n_i A_{ij} h\nu_{ij} \beta(\tau_{ij}) \frac{S_x(\nu_{ij}) - P(\nu_{ij})}{S_x(\nu_{ij})}, \tag{2.20}$$

where n_i is the population density of species x in level i, A_{ij} is the spontaneous transition probability, $h\nu_{ij}$ is the energy difference between levels i and j, τ_{ij} is the optical depth averaged over the line. Here, $\beta(\tau_{ij})$ is the probability that a photon created at optical depth, τ_{ij}, escapes from the cloud. The source function, $S_x(\nu_{ij})$, is given by

$$S_x(\nu_{ij}) = \frac{2h\nu_{ij}^3}{c^2} \left[\frac{g_i n_j}{g_j n_i} - 1 \right]^{-1}, \tag{2.21}$$

where g_i and g_j are the statistical weights of the levels. The background radiation field, $P(\nu_{ij})$, generally contains two terms: the 2.7 K cosmic microwave background and the infrared emission of dust at temperature T_d, and with an emission optical depth, $\tau_d(\nu_{ij})$,

$$P(\nu_{ij}) = B(\nu_{ij}, T = 2.7\,K) + \tau_d(\nu_{ij}) B(\nu_{ij}, T_d). \tag{2.22}$$

The level populations follow from the equation of statistical equilibrium, which for level i reads

$$n_i \sum_{j \neq i}^{k} R_{ij} = \sum_{j \neq i}^{k} n_j R_{ji}, \tag{2.23}$$

with k the total number of levels included and

$$R_{ij} = A_{ij} \beta(\tau_{ij}) (1 + Q_{ij}) + C_{ij}, \qquad i > j \tag{2.24}$$

$$R_{ij} = \frac{g_j}{g_i} A_{ij} \beta(\tau_{ij}) Q_{ij} + C_{ij} \qquad i < j. \tag{2.25}$$

Here, A_{ij} is the Einstein coefficient for spontaneous emission and C_{ij} is the collisional rate from level i to level j. The background radiation is contained in Q_{ij}; viz.

$$Q_{ij} = \frac{c^2}{2h\nu_{ij}^3} P(\nu_{ij}). \tag{2.26}$$

Note that the statistical equilibrium equation (Eq. (2.23) with (2.24)) is deceptively simple. Actually, there is an equation for each point in the cloud and these are linked through the radiation field "hidden" in the escape probability terms. This set of k statistical equilibrium equations is not independent and one equation has to be replaced by the conservation equation,

$$n_x = \sum_{j=0}^{k} n_j, \tag{2.27}$$

with n_x the number density of species x.

2.3 Two-level system

In general, cooling of interstellar gas can be very complex because local physical processes (e.g., collisional excitation) couple with global characteristics of the cloud through radiative transfer. Analysis of a two-level system can provide some insight into the physics of cooling processes.

2.3.1 Optically thin limit

Critical density

Ignoring for the moment radiation trapping ($\beta = 1$ and $P = 0$ in Eq. (2.20)), consider a two-level system (e.g., the fine-structure lines of C^+) connected by collisions with collisional rate coefficients, γ_{ij}, and radiative de-excitation with an Einstein A_{ij} coefficient. The level populations can be found by solving the statistical equilibrium equation,

$$n_{\rm l} n \gamma_{\rm lu} = n_{\rm u} n \gamma_{\rm ul} + n_{\rm u} A_{\rm ul}. \tag{2.28}$$

Because of detailed balance, the collisional rate coefficients of the two levels are related, viz.,

$$\gamma_{\rm lu} = \left(\frac{g_{\rm u}}{g_{\rm l}}\right) \gamma_{\rm ul} \exp[-E_{\rm ul}/kT], \tag{2.29}$$

where $g_{\rm u}$ and $g_{\rm l}$ are the statistical weights of the upper and lower level and $E_{\rm ul}$ is the energy difference between the two levels. This can be rewritten as

$$\frac{n_{\rm u}}{n_{\rm l}} = \frac{g_{\rm u}/g_{\rm l} \exp[-E_{\rm ul}/kT]}{1 + n_{\rm cr}/n}, \tag{2.30}$$

with the critical density, $n_{\rm cr}$, given by

$$n_{\rm cr} = \frac{A_{\rm ul}}{\gamma_{\rm ul}}. \tag{2.31}$$

Thus, when the density is much larger than the critical density, collisions dominate the de-excitation process. We then recover local thermodynamic equilibrium (LTE) and the level populations are given by the Boltzmann expression at the kinetic temperature of the gas.

This analysis can be generalized to multilevel systems where the critical density now compares radiative transitions to all lower levels with the collisional rates to all levels; i.e.,

$$n_{\mathrm{cr}} = \frac{\sum_{\mathrm{l}<\mathrm{u}} A_{\mathrm{ul}}}{\sum_{\mathrm{l}\neq\mathrm{u}} \gamma_{\mathrm{ul}}}. \tag{2.32}$$

The principle is the same: LTE ensues when the density is larger than the critical density.

Collisional de-excitation rates

The collisional de-excitation rate coefficient can be written as

$$\gamma_{\mathrm{ul}} = \frac{4}{\sqrt{\pi}} \left(\frac{\mu}{2kT}\right)^{3/2} \int_0^\infty \sigma_{\mathrm{ul}}(v)\, v^3 \exp\left[-\frac{\mu v^2}{2kT}\right] \mathrm{d}v, \tag{2.33}$$

with μ the reduced mass of the system and $\sigma_{\mathrm{ul}}(v)$ the collisional de-excitation cross section at the relative velocity, v, of the collision partners. The cross section will depend on the interaction potential of the collision partners. Some typical values are summarized in Table 2.5. Neutral–neutral interactions are regulated by short-range van der Waals forces and cross sections are typically some 10–20 Å^2. The resulting de-excitation rate coefficients scale with $T^{1/2}$ because of the temperature dependence of the mean velocity. Electron–neutral excitations of interest here involve a spin change (e.g., exchange of the colliding electron with an atomic electron). These interactions are stronger because of the induced dipole and the rate coefficients are considerably larger. Electrons and ions interact through strong coulomb forces, which favor low collision velocities (i.e., $\sigma \sim v^{-2}$). Because of the coulomb focussing, these interactions have the largest cross sections and rate coefficients at low temperatures (cf. Table 2.5).

Perusing Tables 2.5 and 2.1, it is clear that critical densities for allowed electronic transitions are very high and in the general ISM, these levels will not be in LTE. Because of their relatively low excitation rates, vibrational transitions of molecules also have high critical densities. At the high temperatures of HII regions ($\simeq 10^4$ K), optical forbidden transitions still have lifetimes less than the collisional de-excitation rate. In contrast, far-IR fine-structure lines have critical densities right in the density regime of HII regions. Likewise, low-lying rotational levels of molecules have critical densities that span the range of densities encountered in molecular clouds. The HI hyperfine levels are readily in collisional equilibrium throughout the ISM.

Table 2.5 *Typical de-excitation rate coefficients*

Collision partners	$<\sigma_{\rm ul}>$ (10^{-15} cm^2)	$\gamma_{\rm ul}$ (cm^3 s^{-1})
Neutral–neutral	2	$3\times10^{-11}\,T^{1/2}$
Electron–neutral	30	1×10^{-9}
Electron–ion	$2\times10^4/T$	$1\times10^{-5}\,T^{-1/2}$

Excitation temperature

In general, levels will not be in LTE. It is instructive to define the excitation temperature, $T_{\rm X}$, as

$$T_{\rm X} \equiv \frac{E_{\rm ul}}{k}\left[\ln\left(\frac{n_{\rm l}g_{\rm u}}{n_{\rm u}g_{\rm l}}\right)\right]^{-1}. \tag{2.34}$$

From Eq. (2.30), we find

$$\frac{T}{T_{\rm X}} - 1 = \frac{kT}{E_{\rm ul}}\ln\left(1+\frac{n_{\rm cr}}{n}\right). \tag{2.35}$$

At high densities ($n \gg n_{\rm cr}$), $T_{\rm X} \simeq T$ as expected, while at low densities ($n \ll n_{\rm cr}$), $T_{\rm X}$ will be much less than T. Of course, formally, when $kT \ll E_{\rm ul}$, $T_{\rm X}$ can also be close to T, even if $n \ll n_{\rm cr}$. However, at such low densities and temperatures, excitation of the upper level will be exceedingly low and the transition will be largely irrelevant.

Cooling law

The cooling law (cf. Eq. (2.20)) is now

$$n^2\,\Lambda = n_{\rm u}A_{\rm ul}h\nu_{\rm ul} = \frac{g_{\rm u}/g_{\rm l}\,\exp[-h\nu_{\rm ul}/kT]}{1+n_{\rm cr}/n+g_{\rm u}/g_{\rm l}\,\exp[-h\nu_{\rm ul}/kT]}\,\mathcal{A}_j n\,A_{\rm ul}h\nu_{\rm ul}, \tag{2.36}$$

where we have used the conservation equation for species j

$$n_{\rm l}+n_{\rm u} = \mathcal{A}_j\,n, \tag{2.37}$$

with $\mathcal{A}_j$ the abundance of species j. In the high density limit, we recognize the denominator for the partition function of a two-level system and again recover LTE. In that limit, the cooling scales with n (and T) and the emergent intensity with the column density. In the low density limit, the cooling law simplifies to

$$n^2\,\Lambda \simeq n^2\,\mathcal{A}_j\,\gamma_{\rm lu}\,h\nu_{\rm ul}. \tag{2.38}$$

Every upward collision now results in a cooling photon and the cooling law depends on density squared (and the temperature, through the excitation rate). The emergent intensity then scales with the density times the column density.

2.3.2 Optical depth effects

The escape probability

Consider now the effect of absorption on the level populations. Given a mean photon radiation field, $J_{\rm ul}$, the statistical equilibrium equation for a two-level system can be written as

$$n_{\rm l} n \gamma_{\rm lu} + n_{\rm l} B_{\rm lu} J_{\rm ul} = n_{\rm u} n \gamma_{\rm ul} + n_{\rm u} A_{\rm ul} + n_{\rm u} B_{\rm ul} J_{\rm ul}, \tag{2.39}$$

where the Bs are the Einstein coefficients for absorption and stimulated emission. The level populations constitute a non-linear problem: at any spatial point, the level populations depend on the radiation field while the radiation field depends on the level populations everywhere. Hence, the statistical equilibrium equations couple with the radiative transfer equations and these have to be solved simultaneously for the whole cloud. By introducing the concept of an escape probability – photons produced locally can only be absorbed locally – the local aspect can be recovered. Such a situation can, for example, result from the presence of a strong velocity gradient. Here, we will consider a plane-parallel semi-infinite slab where photons only escape through the front surface. Furthermore, in calculating the optical depth, we will assume that the local-level populations hold globally. Both of these assumptions (locally determined optical depth and escape probability) are generally not correct. However, this decouples the statistical equilibrium equations from the radiative transfer equations and simplifies the problem considerably. In this approximation, we can write that the net absorptions, corrected for stimulated emission, are equal to those photons that do not escape,

$$(n_{\rm l} B_{\rm lu} - n_{\rm u} B_{\rm ul})\, J_{\rm ul} = n_{\rm u}(1 - \beta(\tau_{\rm ul}))A_{\rm ul}, \tag{2.40}$$

where $\beta(\tau)$ is the probability that a photon formed at optical depth τ escapes through the front surface. The level populations are now given by

$$n_{\rm l} n \gamma_{\rm lu} = n_{\rm u} n \gamma_{\rm ul} + n_{\rm u} \beta(\tau_{\rm ul}) A_{\rm ul}. \tag{2.41}$$

Comparing this with Eq. (2.28), we note that when Eq. (2.31) is replaced by

$$n_{\rm cr} = \frac{\beta(\tau_{\rm ul}) A_{\rm ul}}{\gamma_{\rm ul}}, \tag{2.42}$$

Equations (2.30) and (2.36) are recovered. Thus the effect of photon trapping is to lower the density at which LTE is approached; i.e., after each absorption, the gas has a chance to collisionally de-excite the species and return the excitation energy to the thermal energy of the gas.

The optical depth

It now remains to describe the relation between the optical depth and the level populations and between the optical depth and the escape probability. For a turbulent, homogeneous, semi-infinite slab, the line-averaged optical depth is given by

$$\tau_{\rm ul} = \frac{A_{\rm ul} c^3}{8\pi \nu_{\rm ul}^3} \frac{n_{\rm u}}{b/\Delta z} \left[\frac{n_{\rm l} g_{\rm u}}{n_{\rm u} g_{\rm l}} - 1 \right], \tag{2.43}$$

where b is the Doppler broadening parameter and Δz the distance from the surface. The escape probability formalism is also often used in situations with a velocity gradient that is large compared to the thermal motion. In that case, $b/\Delta z$ is replaced by the velocity gradient dv/dz in the expression for the optical depth. The optical depth is then purely local; i.e., once a photon escapes its local environment (has traveled a distance $\Delta v_{\rm D}/(dv/dz)$ with $\Delta v_{\rm D}$ the Doppler width of the line) it will "find" itself in the "wing" of the line profile where the optical depth is small and the photon will escape the cloud.

For various simple geometries, the radiation transfer problem can be solved on an optical depth scale and simple expressions are available. For a plane-parallel slab, the escape probability is given by

$$\beta(\tau) = \frac{1 - \exp(-2.34\tau)}{4.68\tau} \qquad \tau < 7 \tag{2.44}$$

and

$$\beta(\tau) = \left[4\tau \left(\ln\left(\frac{\tau}{\sqrt{\pi}} \right) \right)^{1/2} \right]^{-1} \qquad \tau > 7. \tag{2.45}$$

In the limit of small optical depth, the escape probability goes to 1/2: in a semi-infinite slab, photons escape through half a hemisphere. At large optical depth, β scales with τ^{-1}.

Emergent intensity

The emergent intensity in the line is given, under these approximations (plane-parallel, homogeneous, semi-infinite slab), by

$$I = \frac{1}{2\pi} \int_0^z n^2 \, \Lambda(\tilde{z}) \, d\tilde{z}, \tag{2.46}$$

where the factor 2π takes into account that photons only escape through the front surface. Here, Λ is given by Eq. (2.36) (but remember that now the critical density contains β). In thermodynamic equilibrium, this reduces to

$$I = B(T) \frac{\nu b}{c} f(\tau), \tag{2.47}$$

where we have used the relations between the Einstein coefficients. The optical depth function, $f(\tau)$, is given by

$$f(\tau) = 2\int_0^{\tau} \beta(\tilde{\tau})\, d\tilde{\tau} = 0.428[E_1(2.34\tau) + \ln(2.34\tau) + 0.577\,21], \qquad (2.48)$$

with E_1 the exponential integral. For $\tau \ll 1$, this becomes $f(\tau) = \tau$. The intensity then scales directly with the column density in the upper level, N_u, as

$$I = \frac{A_{ul}\, N_u\, h\nu_{ul}}{4\pi}; \qquad (2.49)$$

i.e., the linear part of the curve of growth. For large optical depth, the logarithm in Eq. (2.48) dominates (i.e., the logarithmic portion of the curve of growth). For $\tau = 3$, $f(\tau)$ is $\simeq 1$, and

$$I \simeq B(T)\frac{\nu b}{c}, \qquad (2.50)$$

i.e., the integrated line intensity is the Planck function multiplied by the width of the line. For even higher optical depth, the damping wings of a line become important. This would give rise to the square root portion of the curve of growth, but – unlike for stellar spectroscopy – this regime is of little importance in ISM studies and its effects are not included in the expression for the escape probability.

If we relax the thermodynamic equilibrium condition ($n \ggg n_{cr}\,\beta$), the intensity is still given by Eq. (2.49) in the optically thin limit. In this case the population of the upper level is not given by the Boltzmann equation but rather by Eq. (2.30). We can then write, for the intensity,

$$I = \frac{\gamma_{lu}\, n \mathcal{A}_j N\, h\nu_{ul}}{4\pi}; \qquad (2.51)$$

where we have used the fact that the population of the upper level is small. In this limit, every upward collision leads to the emission of a photon. For an optically thick line, the level populations will vary through the cloud because of the effects of line trapping. The cooling law is now (cf. Eq. (2.36)),

$$n^2\,\Lambda = \frac{g_u/g_l\, \exp[-h\nu_{ul}/kT]}{1 + n_{cr}\beta(\tau)/n + g_u/g_l\, \exp[-h\nu_{ul}/kT]}\, \mathcal{A}_j n\, A_{ul} h\nu_{ul}\beta(\tau), \qquad (2.52)$$

which simplifies to

$$n^2\,\Lambda = \gamma_{lu}\, \mathcal{A}_j n^2\, h\nu_{ul}. \qquad (2.53)$$

This integrates again to the optically thin limit (cf. Eq. (2.51)). Thus, although the line is optically thick, because collisional de-excitation is unimportant, all upward transitions result in a photon that eventually escapes the cloud. This photon scattering is a random walk process. If we define ϵ as the probability that a photon is converted back into thermal energy of the gas,

$$\epsilon = \frac{n\gamma_{\rm ul}}{n\gamma_{\rm ul} + A_{\rm ul}} \simeq \frac{n}{n_{\rm cr}}, \tag{2.54}$$

then a line is "optically thin" in the subthermal regime when,

$$L < \frac{\lambda}{\sqrt{\epsilon}}, \tag{2.55}$$

with L the size of the cloud and λ the mean free path of a photon which is equal to $1/\kappa$ with κ the line averaged opacity. Thus, when the density is less than

$$n < \frac{n_{\rm cr}}{\tau^2}, \tag{2.56}$$

the cloud is effectively optically thin. For many water lines, with their high critical densities, this can be a very important effect.

2.3.3 *Radiative excitation*

It is also of some interest to consider a gas illuminated by a strong radiation field, either from dust or the 3 K cosmic microwave background. For a two-level system, the level populations are now given by

$$\frac{n_{\rm u}}{n_{\rm l}} = \frac{n\gamma_{\rm lu} + B_{\rm lu}J_{\rm lu}}{A_{\rm ul} + n\gamma_{\rm ul} + B_{\rm ul}J_{\rm ul}}. \tag{2.57}$$

If we ignore collisional excitation, this simplifies to

$$\frac{n_{\rm u}}{n_{\rm l}} = \frac{g_{\rm u}}{g_{\rm l}} \frac{J_{\rm lu}}{\frac{2h\nu_{\rm ul}^3}{c^2} + J_{\rm ul}}. \tag{2.58}$$

For a gas immersed in a black body with radiation temperature $T_{\rm R}$, we find

$$\frac{n_{\rm u}}{n_{\rm l}} = \frac{g_{\rm u}}{g_{\rm l}} \exp[-h\nu_{\rm ul}/kT_{\rm R}]. \tag{2.59}$$

The gas is now thermalized by the radiation field at the radiation field temperature (of course, inside a black body, the kinetic temperature should be equal to the radiation field temperature). More generally, for a radiation field characterized by a (dust) temperature $T_{\rm d}$, a (dust) optical depth $\tau_{\rm d}$, with $\tau_{\rm d} \ll 1$, and a dilution factor W, we have

$$\frac{n_{\rm u}}{n_{\rm l}} = \frac{g_{\rm u}}{g_{\rm l}} \frac{W\,\tau_{\rm d}}{\exp[h\nu_{\rm ul}/kT_{\rm d}] - 1 + W\,\tau_{\rm d}}. \tag{2.60}$$

A radiation field will tend to drive the gas away from the kinetic temperature to the radiative temperature. Ignoring the $W\tau_d$ term in the denominator, the radiation field will dominate over collisions when

$$\left(\frac{n_{cr}}{n}\right)\frac{W\tau_d}{\exp[h\nu/kT_R]-1} \gg 1. \tag{2.61}$$

Thus, when the density is low compared with the critical density or when $h\nu/kT_R \ll 1$, the radiation field becomes important. Obviously, the radiation field has a harder time to keep up with collisions when W is small. In practice, vibrational transitions of molecules have a high critical density and hence are easily influenced by a strong IR radiation field.

2.4 Gas cooling in ionized regions

The lowest excited level of atomic H is 10.5 eV (~120 000 K) above ground and hence, for temperatures well below this energy, the excitation rates of atomic H will be very small,

$$n^2\,\Lambda(\mathrm{Ly}\alpha) = 7.3\times10^{-19}\,n_e\,n_{HI}\,\exp[-118\,400/T]. \tag{2.62}$$

An ionized gas can also cool through free–bound transitions where an electron recombines radiatively with a proton. The kinetic energy of the electron is then effectively removed from the thermal energy of the gas. The energy loss is then given by

$$n^2\,\Lambda_{fb} = n_e\,n_p\,kT_e\,\beta(T_e), \tag{2.63}$$

where n_e and n_p are the electron and proton density, T_e is the electron temperature, and β is the recombination cooling coefficient (i.e., the kinetic-energy-averaged recombination coefficient; cf. Chapter 3), which is weakly dependent on temperature. Generally, free–bound cooling is not very important. Finally, hydrogen can cool through free–free transitions, where an electron is decelerated in the electric field of a proton. The cooling rate integrated over frequency is then

$$n^2\Lambda_{ff} = n_e\,n_p\,\frac{2^5\,\pi\,e^6\,Z^2}{3^{3/2}\,h\,mc^3}\left(\frac{2\pi kT_e}{m}\right)^{1/2}. \tag{2.64}$$

This reduces to

$$n^2\,\Lambda_{ff} \simeq 1.4\times10^{-27}Z^2\,T^{1/2}\,n_e\,n_p\,\mathrm{erg\,cm^{-3}\,s^{-1}}. \tag{2.65}$$

Still, except for a pure hydrogen nebula, these cooling processes are not very important.

Despite their low abundance, the dominant cooling process for ionized gas is excitation of low lying electronic states ($\lesssim$10 000 K) of trace species, such as O^{++} or N^{+}. Electrons are the dominant collision partners and the collisional de-excitation rate from level u to level l can be written as

$$\gamma_{\rm ul} = \left(\frac{2\pi}{kT}\right)^{1/2} \frac{\hbar^2}{m_{\rm e}^{3/2}} \frac{\Omega({\rm u,l})}{g_{\rm u}} \simeq \frac{8.6\times10^{-6}}{T^{1/2}} \frac{\Omega({\rm u,l})}{g_{\rm u}} \,{\rm cm^3\,s^{-1}}, \tag{2.66}$$

where $\Omega(\rm u, l)$ is the collision strength, which is typically of order unity, and $g_{\rm u}$ is the statistical weight of the upper level. Values for some important transitions are summarized in Table 2.6. All these transitions arise between states with the same electron configuration and are thus dipole-forbidden transitions. However, the densities in the ISM are generally low compared to the critical densities (cf. Table 2.6). Hence, collisional de-excitation is not very important and these lines can be effective coolants. Energy level diagrams for O^{++} and N^{+} are shown in Fig. 2.9. These transitions separate out into two groups: (1) Those involving fine-structure levels of the ground state that are separated by less than 1000 K: the populations of these levels are therefore not very sensitive to the temperature of the gas when $T_e \simeq 10^4$ K. (2) Transitions involving levels separated by 1–3 eV: these are sensitive to the electron temperature and hence their excitation acts as a thermostat regulating the temperature of the ionized gas to $\simeq 10^4$ K. Or, to phrase it more to the point: because typically the fine-structure transitions do not deliver enough cooling, the gas temperature has to rise sufficiently ($\sim$10 000 K) to excite these higher levels. Because these are generally forbidden transitions and regions of ionized gas are typically limited in size, optical depth effects are not very important.

2.5 Gas cooling in neutral atomic regions

As for regions of ionized gas, the dominant cooling occurs through excitation of low-lying electronic states. Because temperatures are often lower, fine-structure levels with $E < 1000$ K are generally most important. Regions where atomic H is neutral can still contain some electrons through ionization of trace species with ionization potentials below the ionization potential of hydrogen, 13.6 eV. These include atomic C, S, and Si. Also, cosmic rays and X-rays can maintain a small fraction of ionized hydrogen and helium and, for a moderate degree of ionization (e.g., at low densities), excitation by electrons can be very important. At high densities, excitation by atomic H or molecular H_2 dominates. Table 2.7 summarizes the cooling parameters for the most important transitions in neutral gas. Cooling by the C^{+} line at 158 μm is often very important. In the low

Table 2.6 *Cooling lines of ionized gas*

Ion	u	l	λ_{ul}(Å)	E_{ul}(K)	A_{ul}(s)	Ω_{ul} (cm^{-3})	n_{cr}(u)
O^+	$^2D_{5/2}$	$^4S_{3/2}$	3728.8	3.9×10^4	3.8×10^{-5}	0.82	1.3×10^3
O^+	$^2P_{1/2}$	$^2D_{5/2}$	7318.8	2.0×10^4	6.1×10^{-2}	0.30	4.3×10^6
O^+	$^2P_{1/2}$	$^2D_{3/2}$	7329.6	2.0×10^4	1.0×10^{-1}	0.28	4.3×10^6
O^+	$^2P_{1/2}$	$^4S_{3/2}$	2470.2	5.8×10^4	2.3×10^{-2}	0.14	4.3×10^6
O^+	$^2P_{3/2}$	$^2D_{5/2}$	7319.9	2.0×10^4	1.2×10^{-1}	0.74	6.3×10^6
O^+	$^2P_{3/2}$	$^2D_{3/2}$	7330.7	2.0×10^4	6.1×10^{-2}	0.41	6.3×10^6
O^+	$^2P_{3/2}$	$^4S_{3/2}$	2470.3	5.8×10^4	5.6×10^{-2}	0.28	6.3×10^6
N^+	$^3P_1$	$^3P_0$	2040000.0	7.0×10^1	2.1×10^{-6}	0.41	4.5×10^1
N^+	$^3P_2$	$^3P_1$	1220000.0	1.2×10^2	7.5×10^{-6}	1.12	2.8×10^2
N^+	$^1D_2$	$^3P_2$	6583.4	2.2×10^4	3.0×10^{-3}	1.47	8.6×10^4
N^+	$^1D_2$	$^3P_1$	6548.1	2.2×10^4	1.0×10^{-3}	0.88	8.6×10^4
N^+	$^1S_0$	$^1D_2$	5754.6	2.5×10^4	1.0	0.83	1.1×10^7
N^+	$^1S_0$	$^3P_1$	3062.8	4.7×10^4	3.3×10^{-2}	0.10	1.1×10^7
S^+	$^2D_{3/2}$	$^4S_{3/2}$	6730.8	2.1×10^4	8.8×10^{-4}	2.76	3.6×10^3
S^+	$^2D_{5/2}$	$^4S_{3/2}$	6716.4	2.1×10^4	2.6×10^{-4}	4.14	1.3×10^3
S^+	$^2P_{1/2}$	$^2D_{5/2}$	10370.5	1.4×10^4	7.8×10^{-2}	2.20	9.8×10^5
S^+	$^2P_{1/2}$	$^2D_{3/2}$	10336.4	1.4×10^4	1.6×10^{-1}	1.79	9.8×10^5
S^+	$^2P_{1/2}$	$^4S_{3/2}$	4076.4	3.5×10^4	9.1×10^{-2}	1.17	9.8×10^5
S^+	$^2P_{3/2}$	$^2D_{5/2}$	10320.5	1.4×10^4	1.8×10^{-1}	4.99	5.7×10^6
S^+	$^2P_{3/2}$	$^2D_{3/2}$	10286.7	1.4×10^4	1.3×10^{-1}	3.00	5.7×10^6
S^+	$^2P_{3/2}$	$^4S_{3/2}$	4068.6	3.5×10^4	2.2×10^{-1}	2.35	5.7×10^6
O^{++}	$^3P_1$	$^3P_0$	884000.0	1.6×10^2	2.7×10^{-5}	0.54	5.0×10^2
O^{++}	$^3P_2$	$^3P_1$	518000.0	2.8×10^2	9.8×10^{-5}	1.29	3.4×10^3
O^{++}	$^1D_2$	$^3P_2$	5006.9	2.9×10^4	2.0×10^{-2}	1.27	6.9×10^5
O^{++}	$^1D_2$	$^3P_1$	4958.9	2.9×10^4	7.0×10^{-3}	0.76	6.9×10^5
O^{++}	$^1S_0$	$^1D_2$	4363.2	3.3×10^4	1.7	0.58	2.4×10^7
O^{++}	$^1S_0$	$^3P_1$	2321.4	6.2×10^4	2.3×10^{-1}	0.10	2.4×10^7
S^{++}	$^3P_1$	$^3P_0$	335000.0	4.3×10^2	4.7×10^{-4}	3.98	1.4×10^3
S^{++}	$^3P_2$	$^3P_1$	187000.0	7.7×10^2	2.1×10^{-3}	7.87	1.2×10^4
S^{++}	$^1D_2$	$^3P_2$	9530.9	1.5×10^4	5.5×10^{-2}	3.86	6.2×10^5
S^{++}	$^1D_2$	$^3P_1$	9068.9	1.6×10^4	2.1×10^{-2}	2.32	6.2×10^5
S^{++}	$^1S_0$	$^1D_2$	6312.1	2.3×10^4	2.3	1.38	1.4×10^7
S^{++}	$^1S_0$	$^3P_1$	3721.7	3.9×10^4	8.4×10^{-1}	0.39	1.4×10^7

density limit ($n < 3\times10^3$ cm^{-3}), the [CII] 158 μm cooling rate is approximately given by

$$n^2\,\Lambda_{[\mathrm{CII}]} \simeq 3\times10^{-27}\,n^2\left(\frac{\mathcal{A}_\mathrm{C}}{1.4\times10^{-4}}\right) \times\left(1+0.42\left(\frac{x}{10^{-3}}\right)\right)\exp[-92/T]\ \mathrm{erg\,cm^{-3}\,s^{-1}}, \quad (2.67)$$

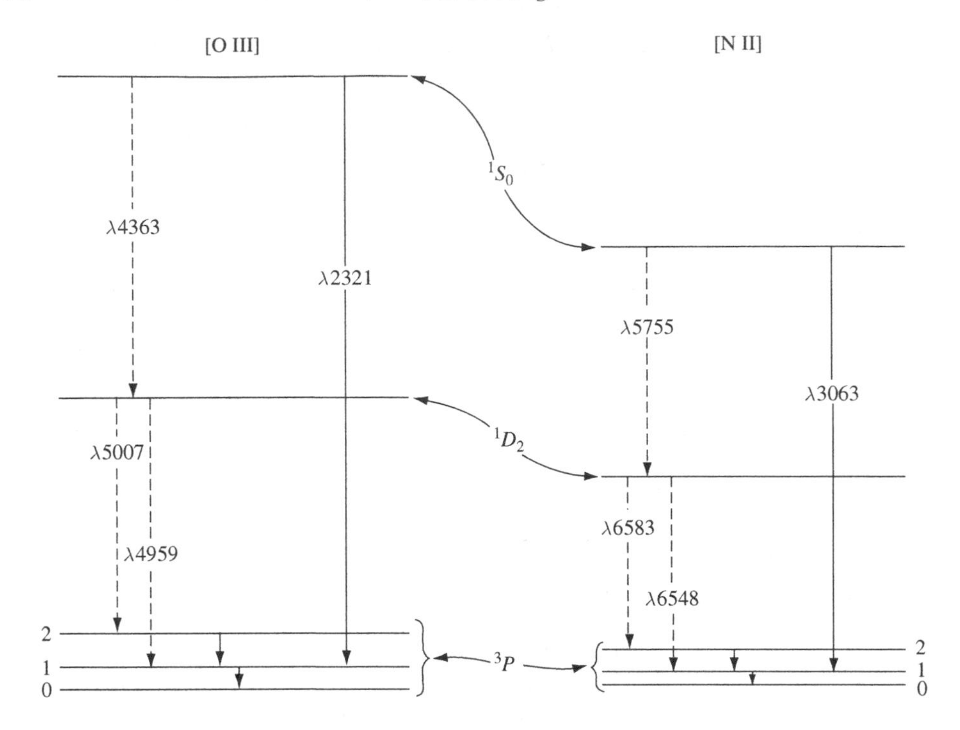

Figure 2.9 A schematic energy level diagram of the lowest energy levels of the isoelectronic O^{++} and N^+ ions. The strongest transitions are indicated. Figure reproduced with permission from D. Osterbrock, 1989, *Astrophysics of Gaseous Nebulae and Active Galactic Nuclei* (Mill Valley: University Science Books).

with $\mathcal{A}_C$ the carbon abundance and x the degree of ionization. Here, we have assumed that all C is singly ionized and evaluated the de-excitation rates at 100 K.

Optical depth effects can become important for some of these transitions. Table 2.7 includes the equivalent H column density, N_τ, for which a transition reaches a line-averaged optical depth of unity assuming cosmic abundances and that all of the element is in the respective ionic form and in the lower level. A line width of $1\,\mathrm{km\,s^{-1}}$ has been adopted in this calculation.

$$N_\tau = \frac{8\pi b}{A_{\rm ul}\lambda_{\rm ul}^3}\frac{g_j}{g_i}\mathcal{A}_j^{-1}$$

$$\simeq 6.8\times10^{21}\left(\frac{10^{-5}}{A_{\rm ul}}\right)\left(\frac{10^{-4}}{\mathcal{A}_j}\right)\times\left(\frac{100\,\mu\mathrm{m}}{\lambda_{\rm ul}}\right)^3\left(\frac{b}{1\,\mathrm{km\,s^{-1}}}\right). \quad (2.68)$$

Typically, these equivalent H column densities correspond to a FUV dust optical depth of unity – which happens to be approximately equal to the atomic-to-molecular transition size scale – and these lines achieve therefore moderate optical depths ($\tau \simeq 1$) in astrophysically relevant situations.

Table 2.7 *Cooling lines of neutral gas*[a]

Species	u	l	λ_{ul}(Å)	E_{ul}(K)	A_{ul}(s)	p, e collisions[a] γ^b_{ul} (cm^3 s^{-1})	p, e collisions[a] n^b_{cr} (cm^{-3})	H collisions α^b (cm^3 s^{-1})	H collisions β^b (cm^{-3})	H collisions n^b_{cr} (cm^{-2})	N_τ
C^+	$^2P_{3/2}$	$^2P_{1/2}$	1 577 000	92	2.4 (−6)	3.8 (−7)	6.3 (0)	8.9 (−10)	0.02	2.7 (3)	1.2 (21)
C	$^3P_1$	$^3P_0$	6 092 000	23.6	7.9 (−8)	3.0 (−9)	2.6 (1)	1.6 (−10)	0.14	4.9 (2)	4.2 (20)
C	$^3P_2$	$^3P_1$	3 690 000	38.9	2.7 (−7)	1.5 (−8)	1.8 (1)	2.9 (−10)	0.26	9.3 (2)	1.0 (21)
C	$^3P_2$	$^3P_0$	230	62.5	2.0 (−14)	5.0 (−9)	—	9.2 (−11)	0.26	—	1.9 (28)
O	$^3P_1$	$^3P_2$	632 000	228	8.95 (−5)	1.4 (−8)	6.4 (3)	9.2 (−11)	0.67	9.7 (5)	1.0 (21)
O	$^3P_0$	$^3P_1$	1 456 000	98	1.7 (−5)	5.0 (−9)	3.4 (3)	1.1 (−10)	0.44	1.5 (5)	7.8 (20)
O	$^3P_0$	$^3P_2$	440 000	326	1.0 (−10)	1.4 (−8)	—	4.4 (−11)	0.80	—	8.0 (27)
O	$^1D_2$	$^3P_2$	6 300.3	2.3(4)	6.3 (−3)	2.5 (−9)[c]	2.4 (6)[c]	—	—	—	—
O	$^1D_2$	$^3P_1$	6 363.8	2.3(4)	2.1 (−3)	1.5 (−9)[c]	1.4 (6)[c]	—	—	—	—
Si^+	$^2P_{3/2}$	$^2P_{1/2}$	348 000	414	2.2 (−4)	4.3 (−7)	5.1 (2)	6.5 (−10)	—	3.4 (5)	4.6 (23)
S	$^3P_2$	$^3P_1$	252 000	571	1.4 (−3)	3.3 (−8)	4.2 (4)	7.5 (−10)	0.17	1.8 (6)	6.4 (22)
S	$^3P_0$	$^3P_1$	566 000	255	3.0 (−4)	1.2 (−8)	2.5 (4)	4.2 (−10)	0.17	7.1 (5)	4.7 (22)
S[d]	$^3P_2$	$^3P_0$	174 000	826	6.7 (−8)	3.3 (−8)	—	7.1 (−10)	0.17	—	1.2 (28)
Fe	$^5D_3$	$^5D_4$	242 000	595	2.5 (−3)	1.2 (−7)	2.1 (4)	8.0 (−10)	0.17	3.1 (6)	9.9 (23)
Fe	$^5D_2$	$^5D_3$	347 000	415	1.6 (−3)	9.3 (−8)	1.7 (4)	5.3 (−10)	0.17	3.0 (6)	3.7 (21)
Fe	$^5D_4$	$^5D_2$	143 000	1010	1.0 (−9)	1.2 (−7)	—	6.9 (−10)	0.17	—	1.7 (32)
Fe^+	$^6D_{7/2}$	$^6D_{9/2}$	260 000	554	2.1 (−3)	1.8 (−6)	6.0 (2)	9.5 (−10)	—	2.2 (6)	9.1 (23)
Fe^+	$^6D_{5/2}$	$^6D_{7/2}$	354 000	407	1.6 (−3)	8.7 (−7)	1.8 (3)	4.7 (−10)	—	3.3 (6)	5.2 (23)
Fe^+	$^6D_{5/2}$	$^6D_{9/2}$	150 000	961	1.5 (−9)	1.8 (−6)	—	5.7 (−10)	—	—	9.0 (30)

[a] Electron collision rates for ions (these have a $T^{-0.5}$ dependence), proton collision rates for neutrals.
[b] De-excitation collisional rate coefficient and critical density evaluated at 100 K, unless otherwise noted.
[c] Evaluated at 10 000 K. Scales with $T^{0.6}$.
[d] The cooling parameters of S^+ are given in Table 2.6.

When fine-structure lines cannot keep up with the heating, low-lying electronic states of trace species take over and the temperature has to rise above 5000 K; e.g., for the [OI] λ 6300 Å transition

$$n^2 \Lambda = 1.8 \times 10^{-24} \mathcal{A}_{\rm O} n^2 \exp[-22,800/T]. \qquad (2.69)$$

If these lines do not suffice, the HI lines have to be excited to initiate the cooling. Lyman α cooling is given by

$$n^2 \Lambda_{\rm Ly\alpha} \simeq 7.3 \times 10^{-19} n_{\rm e} n_{\rm HI} \exp[-118,400/T] \, {\rm erg\,cm^{-3}\,s^{-1}}. \qquad (2.70)$$

Note that when Lyman α excitation becomes important, collisional ionization will also be energetically feasible (e.g., $E(n=2) \simeq 75\%$ of the HI ionization potential) and the gas will be easily ionized.

2.6 Cooling law

The various processes that contribute to or influence the cooling of the gas have now been discussed. In this section the resulting cooling laws for atomic and molecular gas are summarized.

2.6.1 Atomic gas

The cooling curve for low-density atomic gas is shown in Fig. 2.10 for the optically thin limit – not a bad assumption, except for dense photodissociation regions. At low temperatures, cooling is dominated by the excitation of fine-structure levels of trace elements, in particular CII. As the temperature increases, fine-structure excitation of Si^+ and Fe^+ becomes more important. As indicated, their excitation is sensitive to the degree of ionization. The cooling curve starts to level off when the temperature reaches the energy difference of these levels (a few hundred K). Above 6000 K, excitation of low-lying levels of O, N, and other species becomes more important. When the temperature increases above $\simeq$10 000 K, cooling is largely due to collisional excitation of neutral H (Lyman α). The steep increase at this temperature is also partly due to a sharp increase in (collisional) ionization, which results in more efficient excitation. For higher temperatures, cooling through highly ionized trace elements (C, O, and Fe) becomes important. Above $\simeq 10^5$ K, most species (except for Fe) retain few bound electrons and the cooling drops. H is fully ionized for $T \gtrsim 10^4$ K and does not play a role in the cooling any more until temperatures of about 10^7 K where free–free radiation becomes important.

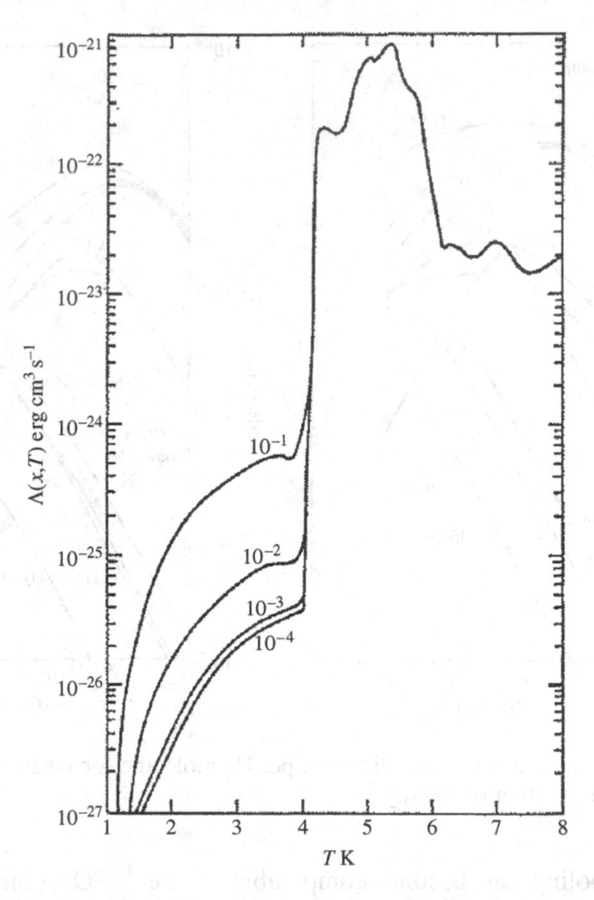

Figure 2.10 The (low-density) cooling rate for interstellar gas as a function of temperature. At low temperatures, the cooling rate depends, as indicated, on the degree of ionization. Figure reproduced from A. Dalgarno and R. A. McCray: reprinted with permission from the *Ann. Rev. Astron. Astrophys.*, **10**, p. 375, ©1972, by Ann. Rev. (www.annualreviews.org).

2.6.2 Molecular gas

Because there are more levels available and because the energies involved are much smaller, rotational transitions of molecules provide a more effective way of cooling gas than atomic transitions. Molecular clouds are therefore generally cold. Figure 2.11 summarizes calculated cooling rates for interstellar molecular clouds. These cooling rates depend on assumed abundances or, more precisely, on the ratio of the abundance to the velocity gradient, which sets the optical depth of the transitions involved (cf. Eq. (2.43)) and reasonable assumptions have been made. At low densities, CO dominates the cooling because of its high abundance in molecular clouds. Around a density of 10^3 cm^{-3}, ^{13}CO starts to contribute to the cooling. Despite the low isotope ratio of ^{13}C/^{12}C (= 1/65 in the

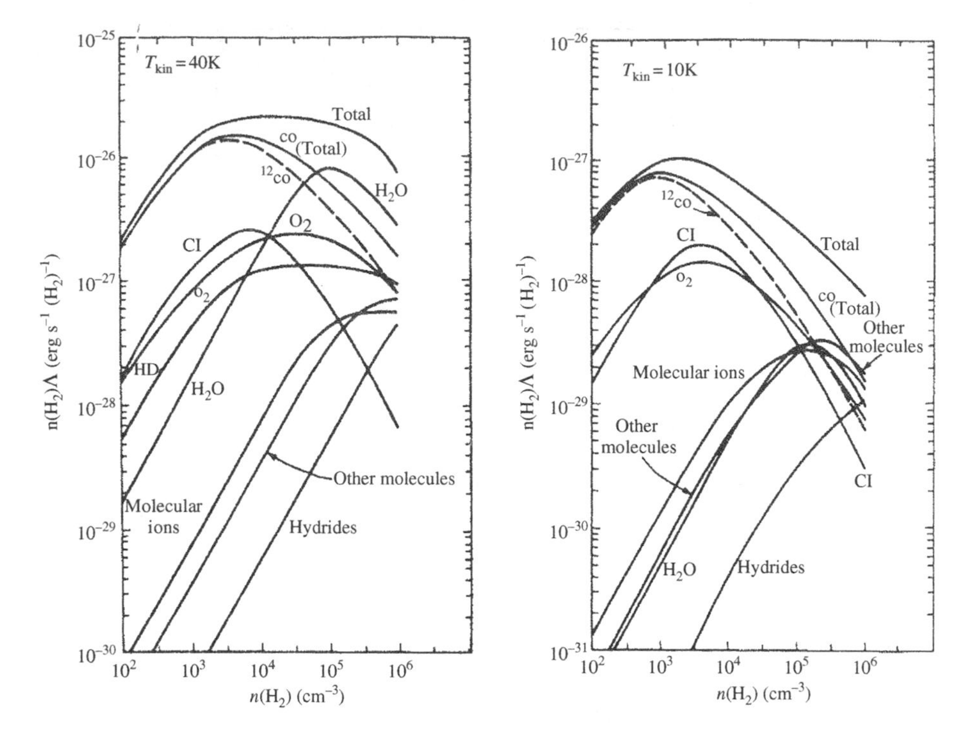

Figure 2.11 The calculated cooling rate per H_2 molecule for 10 and 40 K molecular gas as a function of density.

ISM), ^{13}CO cooling can become comparable to the ^{12}CO cooling because the latter is optically thick. Because of the low critical density of the low-J levels of CO, its importance diminishes at higher densities where other species – first C and O_2 and then H_2O – start to contribute or even take over. Water and other hydrides have strong transitions at higher frequencies and become progressively more important at higher temperatures. Despite their low abundance, molecular ions are also important coolants, on account of their large dipole moments.

2.7 Further reading

Electronic spectroscopy of atoms and molecules is discussed in [1], [2], and [3]. There are various excellent textbooks on vibrational spectroscopy of molecules, e.g., [4], [5], and [6]. I prefer [7]. Rotational spectroscopy of molecules is discussed in [8]. A general textbook on spectroscopy is [9]. Atomic and molecular collision processes are described in detail in the monograph [10].

I strongly recommend [11], which describes the ins and outs of cooling of ionized gas in great detail. No comparable textbook exists describing cooling of neutral atomic and molecular gas. An early review on cooling of atomic gas is [12]. The appendixes in [13] and [14] contain much of the required detail on atomic cooling. Cooling of molecular gas has been discussed in [15].

Information on spectroscopy and excitation of atoms and molecules can be found at the website of the NIST:

http://www.nist.gov/srd/atomic.htm

and the website maintained by D. Verner,

http://www.pa.uky.edu/~verner/atom.html.

Rotational spectra of molecules can be accessed through the websites maintained by JPL and by the Cologne Database for Molecular Spectroscopy:

http://spec.jpl.nasa.gov/,
http://www.ph1.uni-koeln.de/vorhersagen/.

Dipole moments of astrophysically relevant species have also been compiled in [16]. A database of infrared spectra of polycyclic aromatic hydrocarbons is:

http://www.astrochem.org/pahdata/index.html.

The HITRAN and GEISA database – compilations of spectroscopic parameters for simulation of transmission and emission of light in the atmosphere – can be accessed through:

http://cfa-www.harvard.edu/hitran//welcometop.html and
http://ara.lmd.polytechnique.fr/htdocs-public/products/GEISA/access.html.

The BASECOL database provides collisional excitation rates for rotational–vibrational excitation of molecules by various collision partners:

http://basecol.obs-besancon.fr/BASECOL/REF/basecolpub.html.

The website of Sheldon Green reports collisional de-excitation cross sections of various molecules but has not been updated since his death,

http://www.giss.nasa.gov/data/mcrates/.

Optical properties of astrophysically relevant solid materials are available at:

http://www.astro.uni-jena.de/Laboratory/Database/databases.html.

A database of laboratory spectra of interstellar ice analogs is available at:

http://www.strw.leidenuniv.nl/~lab/databases/.

References

[1] G. H. Herzberg, 1944, *Atomic Spectra and Atomic Structure*, (New York: Dover)
[2] G. H. Herzberg, 1991, *Electronic Spectra and Electronic Structure of Polyatomic Molecules*, (New York: Krieger)

[3] J.M. Hollas, 1996, *Modern Spectroscopy*, (New York: Wiley and Sons)
[4] G.H. Herzberg, 1991, *Infrared and Raman Spectra of Polyatomic Molecules*, (New York: Krieger)
[5] L.J. Bellamy, 1958, *The Infrared Spectra of Complex Organic Molecules*, 2nd edn. (New York: Wiley)
[6] K. Nakamoto, 1997, *Infrared and Raman Spectra of Inorganic and Co-ordination Compounds*, Parts I and II (New York: Wiley)
[7] G. Socrates, 1994, *Infrared Characteristic Group Frequencies*, (New York: Wiley)
[8] C.H. Townes and A.L. Schawlow, 1975, *Microwave Spectroscopy*, (New York: Dover)
[9] P.F. Bernath, 1995, *Spectra of Atoms and Molecules*, (Oxford: Oxford University Press)
[10] D. Flower, *Molecular Collisions in the Interstellar Medium*, (Cambridge: Cambridge University Press)
[11] D. Osterbrock, 1989, *Astrophysics of Gaseous Nebulae and Active Galactic Nuclei*, (Mill Valley: University Science Books)
[12] A. Dalgarno and R.A. McCray, 1972, *Ann. Rev. Astron. Astrophys.*, **10**, p. 375
[13] A.G.G.M. Tielens and D. Hollenbach, 1985, *Ap. J.*, **291**, p. 722
[14] D. Hollenbach and C.F. McKee, 1989, *Ap. J.*, **342**, p. 306
[15] P.F. Goldsmith and W.D. Langer, 1978, *Ap. J.*, **222**, p. 881
[16] Y.H. Le Teuff, T.J. Millar, and A.J. Markwick, 2000, *A. & A.*, **146**, p. 157

3
Gas heating

3.1 Overview

This chapter presents the other instalment required for the energy balance: the processes that heat the gas. While there are a number of energy sources with comparable energy densities in the ISM (see Table 1.2), the processes that couple the gas to the radiation field generally dominate. This coupling mainly goes through photo-ionization where the electron takes away some of the photon energy ($\simeq$1–20 eV) in the form of kinetic energy. This kinetic energy is quickly shared with the other atoms and electrons in the gas. The typical elastic electron–electron and electron–atom energy transfer time scales are

$$t_{\mathrm{e,e}} \simeq 10^4 \left(\frac{E_{\mathrm{e}}}{1\,\mathrm{eV}}\right)^{3/2} \frac{1}{n_{\mathrm{e}}}\ \mathrm{s} \tag{3.1}$$

and

$$t_{\mathrm{e,H}} \simeq 2 \times 10^7 \left(\frac{E_{\mathrm{e}}}{1\,\mathrm{eV}}\right) \frac{1}{n_{\mathrm{H}}}\ \mathrm{s}, \tag{3.2}$$

where E_{e} is the energy of the electron. In an HII region, the former time scale is the important one and this time scale is much shorter than the recombination time scale. HI and molecular clouds have typical electron fractions of 10^{-4} and 10^{-7}, respectively. In that case, the second process is somewhat more important. In any case, the energy-transfer time scale is much shorter than other time scales and a Maxwellian equilibrium is quickly established.

In discussing photo-ionization, we have to distinguish between HII regions where very energetic photons are available and (neutral) HI regions where no photons more energetic than 13.6 eV – the ionization potential of hydrogen – are present (cf. Chapters 7 and 8). In the former, H ionization dominates the heating of the gas, while in the latter, photo-ionization of large (stable) molecules and small dust grains takes over. In molecular regions, photodissociation of (small) molecules can play a similar role to photo-ionization. In practice that is only important in dense regions illuminated by strong FUV fields (photodissociation regions, see Chapter 9) where photodissociation of H_2 can be of importance. The gas can also tap the FUV radiation field indirectly. Dust grains are heated through absorption of these photons (see Chapter 5) and collisions will then couple the gas thermodynamically to the dust. Gas species can also be vibrationally or rotationally excited by the IR dust continuum radiation field. Collisional de-excitation then heats the gas. Of course, this collisional de-excitation process competes with radiative decay of the excited level and this process is only important at high densities.

Besides the EUV and FUV radiation fields, the ISM is also pervaded by very energetic photons and particles and X-ray heating and cosmic-ray heating can be important, particularly when the UV fields have been strongly attenuated (e.g., inside dense molecular clouds) or close to bright sources of energetic photons or particles (e.g., supernova remnants). Finally, gas can also couple energetically to macroscopic fluid motion. The decay of (subsonic) turbulence is treated in this chapter. Shock waves convert supersonic motion into heat and this is treated in Chapter 11.

3.2 Photo-ionization of atoms

The heating, $n\Gamma_i$, due to photo-ionization of element i can be written as

$$n\Gamma_i = n_i \int_{\nu_i}^{\infty} 4\pi \mathcal{N}(\nu)\alpha_i(\nu)h(\nu - \nu_i)\mathrm{d}\nu, \tag{3.3}$$

where n_i is the density of species i, $\alpha_i(\nu)$ is the photo-ionization cross section, and ν_i is the photo-ionization threshold. The mean photon-intensity, $\mathcal{N}(\nu)$, is given by

$$\mathcal{N}(\nu) = \frac{J(\nu)}{h\nu}, \tag{3.4}$$

with $J(\nu)$ the mean intensity of ionizing photons. The average energy per photo-ionization, $3/2\,kT_i$, carried away by the photo-electron can then be defined as

$$\frac{3}{2}kT_i \int_{\nu_i}^{\infty} \mathcal{N}(\nu)\alpha_i(\nu)\mathrm{d}\nu \equiv \int_{\nu_i}^{\infty} \mathcal{N}(\nu)\alpha_i(\nu)h(\nu - \nu_i)\mathrm{d}\nu. \tag{3.5}$$

This average energy depends, therefore, on the spectrum of the ionizing radiation field but not on the absolute value of the mean intensity. Likewise, the average energy does not depend on the absolute value of the cross section.

3.2.1 Photo-ionization heating in ionized gas

In an HII region, photo-ionization of HI is the dominant heating term and this average energy, $3/2\,kT_{\rm H}$, is given by

$$\frac{3}{2}kT_{\rm H}\int_{\nu_{\rm H}}^{\infty}\mathcal{N}(\nu)\alpha_{\rm H}(\nu)\mathrm{d}\nu=\int_{\nu_{\rm H}}^{\infty}\mathcal{N}(\nu)\alpha_{\rm H}(\nu)h(\nu-\nu_{\rm H})\mathrm{d}\nu, \tag{3.6}$$

with $h\nu_{\rm H}$ the energy of the hydrogen ionization edge (13.6 eV). Figure 3.1 shows this average energy calculated as a function of stellar effective temperature for a black body. Typically, $T_{\rm H}$ is $\simeq 0.5-0.7\,T_{\rm eff}$ or some 25 000 K. Because of the factor $\nu-\nu_{\rm H}$, a harder radiation field is somewhat more effective than a softer radiation field. For a black body, this effect is small but not negligible in calculating the temperature of a nebula.

3.2.2 Photo-ionization heating in neutral gas

There are no H-ionizing photons in an HI region (see Chapter 8) and photo-ionization of trace elements takes over as heating source. Of these, carbon, because of its abundance, is the most important. The average energy per ionization is then

$$\frac{3}{2}\,kT_{\rm C}\int_{\nu_{\rm C}}^{\nu_{\rm H}}\mathcal{N}(\nu)\,\alpha_{\rm C}(\nu)\,\mathrm{d}\nu=\int_{\nu_{\rm C}}^{\nu_{\rm H}}\mathcal{N}(\nu)\,\alpha_{\rm C}(\nu)\,h(\nu-\nu_{\rm C})\mathrm{d}\nu, \tag{3.7}$$

where the integration runs from the carbon ionization edge (11.3 eV) to the hydrogen ionization edge (13.6 eV). Adopting, for the moment, a frequency-independent

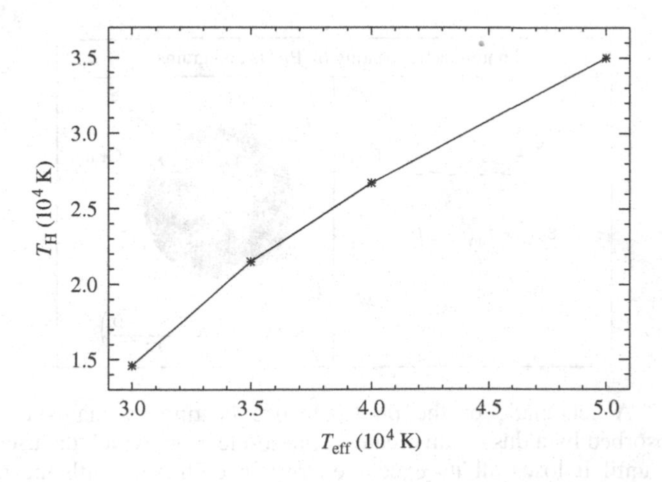

Figure 3.1 The average excess energy carried away by the photo-electron upon ionization of hydrogen calculated as a function of stellar effective temperature. The ionizing spectrum is assumed to be given by a black body at temperature $T_{\rm eff}$.

photon field and a constant ionization cross section, the mean energy per ionization translates into $T_C \simeq 9000$ K. Thus, because of the limited energy range over which heating can take place, the average energy ($\sim$1 eV) is much less than for H ionization.

Adopting the spectral energy distribution of the interstellar radiation field, the heating rate is given by

$$n\Gamma_{CI} = 2.2 \times 10^{-22} f(\mathrm{CI}) \mathcal{A}_C n G_0 \exp[-2.6A_v] \,\mathrm{erg\,cm^{-3}\,s^{-1}}, \qquad (3.8)$$

where $\mathcal{A}_C$ is the atomic carbon abundance in the gas phase, $f(\mathrm{CI})$ is the neutral fraction of carbon, G_0 is the intensity of the radiation field in units of the average interstellar radiation field, and the exponential factor is the dust attenuation in the FUV at a depth corresponding to a visual extinction of A_v into a cloud.

3.3 Photo-electric heating

The most important heating process in the neutral ISM is the photo-electric effect working on large PAH molecules and small dust grains. Figure 3.2 schematically shows the physics associated with the photo-electric effect on interstellar grains and PAHs. Far-ultraviolet photons absorbed by a grain will create energetic (several eV) electrons. While these electrons diffuse in the grain, they will lose energy through collisions. However, if, during this diffusion process, they reach the surface with enough energy to overcome the work function W of the grain and the coulomb potential, ϕ_c (if the grain is positively charged), they can be

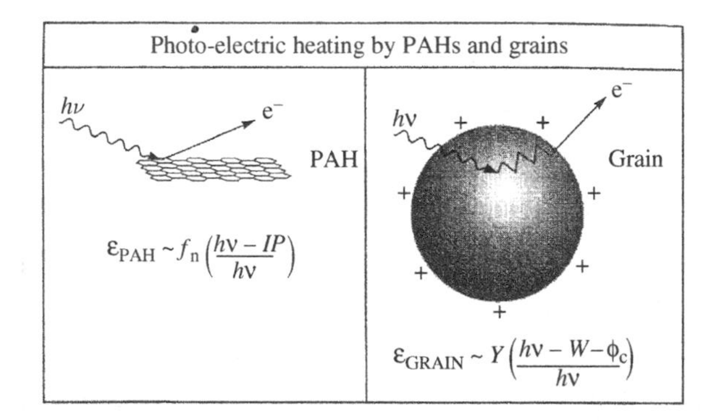

Figure 3.2 A schematic of the photo-electric heating mechanism. An FUV photon absorbed by a dust grain creates a photo-electron, which diffuses through the grain until it loses all its excess energy in collisions with the matrix or finds the surface and escapes. For PAH molecules, the diffusion plays no role. Simple expressions for the heating efficiencies, ϵ, of these different processes are indicated. Note that f_n is the neutral fraction (e.g., $f\,(Z=0)$) in the text.

injected into the gas phase with excess kinetic energy. The average energy per ionization is then

$$\frac{3}{2}kT_{\rm grain}\int_{\nu_i}^{\nu_{\rm H}}\mathcal{N}(\nu)\alpha_{\rm grain}(\nu){\rm d}\nu=\int_{\nu_i}^{\nu_{\rm H}}\mathcal{N}(\nu)\alpha_{\rm grain}(\nu)E_{\rm kin}{\rm d}\nu, \tag{3.9}$$

with $\alpha_{\rm grain}$ the ionization cross section of the grains, $E_{\rm kin}$ the kinetic energy of the photo-electron, and $h\nu_i = W + \phi_{\rm c}$ the photon energy required for ionization. The expression for PAH heating is analogous. With a typical ionization potential of 5 eV and adopting an ionization cross section and ionizing radiation field independent of energy, the average energy per ionization of a neutral grain becomes, $\overline{E}_{\rm kin}/k \simeq 50\,000$ K. While this is an appreciable amount of energy, the actual heating rate will depend on the ionization rate as well as the grain charge.

3.3.1 *The photo-electric effect on grains*

For a mono-energetic radiation field, the efficiency $\epsilon_{\rm grain}$ of the photo-electric effect on a grain per absorbed FUV photon – or equivalently, the ratio of gas heating to the grain FUV absorption rate – is given by the yield Y, which measures the probability that the electron escapes multiplied by the fraction of the photon energy carried away as kinetic energy by the electron;

$$\epsilon_{\rm grain} \sim Y\left(\frac{h\nu - W - \phi_{\rm c}}{h\nu}\right), \tag{3.10}$$

where, for simplicity, we have assumed that no energy is lost within the grain. The yield is a complex function of the grain size, a, the collision length scale for low-energy electrons in solids, $\ell_{\rm e}$ ($\simeq$10 Å), the FUV absorption length scale inside a grain, $\ell_{\rm a}$ ($\simeq$100 Å), and the photon energy, $h\nu$. For large grains and photon energies well above threshold, the photons are absorbed $\sim$ 100 Å inside the grain and the photo-electrons rarely escape ($Y \sim \ell_{\rm e}/\ell_{\rm a} \simeq 0.1$). For very small grains, $a \sim \ell_{\rm e}$, the yield can approach unity. With a typical FUV photon energy of 10 eV and a work function of 5 eV, the maximum efficiency is only 0.05 for large grains. Generally, because of positive charging of the grains, the efficiency will be much less, because of two effects. First, because of the coulomb interaction, the ejected electron will carry away less kinetic energy. Second, only a smaller fraction of all the available photons is able to ionize the grain. Grain charging will be discussed in Section 5.2.3. When photo-electric heating is important, the grain charge is a balance between photo-ionization and electron recombination and thus controlled by the parameter, $\gamma = G_0 T^{1/2}/n_{\rm e}$. The electrostatic grain potential is then $\phi_{\rm c} = Z_{\rm d}e^2/a \sim 7\times10^{-4}\gamma$ eV, with $Z_{\rm d}$ the grain charge. For the diffuse ISM ($\gamma \simeq 1500$ cm^3 K$^{1/2}$), this reduction factor is not very important. However, for dense photodissociation regions ($\gamma \simeq 10^5$ cm^3 K$^{1/2}$), charging severely reduces the photo-electric heating rate by grains.

3.3.2 The photo-electric effect on large molecules

Large molecules are much more efficient in photo-electrically heating the gas than grains. We will focus here on PAHs. While a fraction of the photon energy (generally taken to be $\sim$0.5) may remain behind as electronic excitation energy, the yield is much higher for planar PAH molecules. The limiting factor now is that the ionization potential, *IP*, of a charged PAH is readily larger than 13.6 eV. Far-ultraviolet photons absorbed by such a PAH cation do not lead to the creation of a photo-electron or, therefore, to any gas heating. The ionization potential of a charged PAH is given by (cf. Section 6.3.1)

$$IP = W + \phi_c = W + (Z + 0.5)e^2/C = W + (Z + 0.5)\pi e^2/2a, \qquad (3.11)$$

where W is the work function of bulk graphite, ϕ_c is the coulomb potential, Z is the PAH charge, a is the PAH radius, and C is the capacitance of the PAH, which in the left-hand side has been approximated as a thin disk. In the limit of large PAHs, the ionization potential goes to the bulk work function, but for small sizes and high numbers of charges, *IP* can exceed 13.6 eV. For example, the second ionization potential of pyrene, $C_{16}H_{10}$, is 16.6 eV, well above the hydrogen ionization limit. The photo-electric heating efficiency ϵ_{PAH} for small PAHs is then reduced by the ("neutral") fraction, $f(Z = 0)$, of PAHs that can still be ionized by FUV photons,

$$\epsilon_{PAH} = \frac{1}{2} f(Z = 0) \left(\frac{h\nu - IP}{h\nu} \right). \qquad (3.12)$$

Thus, with a typical photon energy of 10 eV and an ionization potential of 7 eV, the maximum efficiency is 0.15.

The PAH neutral fraction can be found from the PAH ionization balance (see Chapter 6). Consider a PAH with two ionization stages, 0 and +1. The neutral fraction is then given by

$$f(Z = 0) = (1 + \gamma_0)^{-1}, \qquad (3.13)$$

where γ_0 is the ratio of the ionization rate over the recombination rate, which for small PAHs is given by (see Section 6.3.7)

$$\gamma_0 = 3.5 \times 10^{-6} N_c^{1/2} \gamma, \qquad (3.14)$$

with $\gamma = G_0 T^{1/2}/n_e$. For a 50 C atom PAH, $\gamma_0 \simeq 0.05$ and the neutral fraction is $f(Z = 0) \simeq 0.95$ in a diffuse HI cloud. In a dense photodissociation region, PAH charging is slightly more important with $\gamma_0 \simeq 3.5$ and $f(Z = 0) \simeq 0.2$.

Nevertheless, compared with grains, PAHs are more efficient in heating the gas photo-electrically and the effects of charging are smaller.

3.3.3 The photo-electric heating rate

The photo-electric heating rate is given by

$$n\Gamma_{\rm pe} = \int_{a_-}^{a_+} \mathrm{d}a n(a) \sum_i \int_{\nu_i(a)}^{\nu_{\rm H}} \mathcal{N}(\nu)\alpha_{\rm grain}(\nu)E_{\rm kin}(a,i)\mathrm{d}\nu, \tag{3.15}$$

where the summation extends over all accessible charge states of species with size a and ionization energy $h\nu_i(a)$. Figure 3.3 shows the contribution to the photo-electric heating by grains of different sizes, adopting a power law size distribution extending into the molecular PAH domain (cf. Section 6.7.3). This figure includes the effects of grain charging for typical conditions in a dense photodissociation region (see Chapter 9). About half the gas heating is due to grains with sizes less than 15 Å ($N_{\rm c} \simeq 1500$ C atoms). The other half originates in grains with sizes between 15 and 100 Å ($1500 < N_{\rm c} < 5 \times 10^5$ C atoms). Grains larger than 100 Å contribute negligibly to the photo-electric heating of

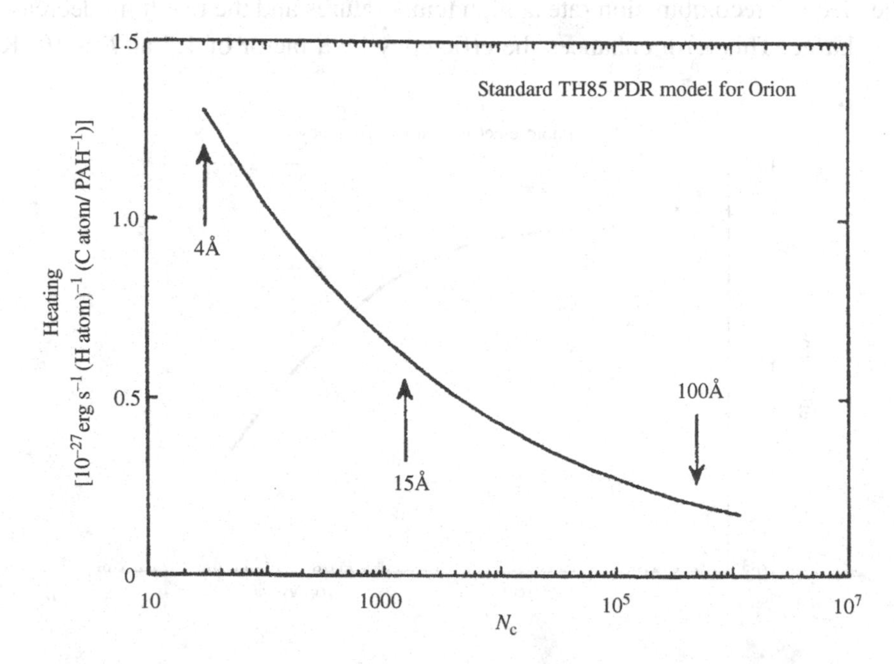

Figure 3.3 The contribution to the photo-electric heating of interstellar gas by PAHs and grains containing different numbers of carbon atoms, $N_{\rm c}$. The results of these calculations are presented in such a way that equal areas under the curve correspond to equal contributions to the heating. Typical PAH and grain sizes are indicated.

the interstellar gas, partly because small grains dominate the FUV absorption extinction and partly because photo-electrons escape more easily from PAHs and small grains than from large grains.

The photo-electric heating efficiency, ϵ, depends on the grain or PAH charge and thus on the ratio of the photo-ionization rate over the recombination rate of electrons with grains and PAHs. Thus, when γ is small, grains or PAHs are predominantly neutral and the photo-electric heating has the highest efficiency. When γ increases, grains or PAHs will charge up and the photo-electric efficiency will drop. Theoretical calculations on the heating efficiency, ϵ, by a grain size distribution of PAHs and small grains, including the effects of charge, are shown in Fig. 3.4 and these have been fitted by a simple analytical formula,

$$\epsilon = \frac{4.87 \times 10^{-2}}{1 + 4 \times 10^{-3} \gamma^{0.73}} + \frac{3.65 \times 10^{-2} (T/10^4)^{0.7}}{1 + 2 \times 10^{-4} \gamma}. \tag{3.16}$$

The first term takes the ionization balance into account and is the equivalent of Eq. (3.12) with Eq. (3.13) for a single species. The dependence of γ is slightly less steep because larger PAHs have more charge states available. The second term introduces an additional temperature dependence, which reflects an increase in the electron recombination rate at high temperatures and the resulting decreased grain charge. This term enhances the efficiency by a factor of 1.7 at $T \sim 10^4$ K.

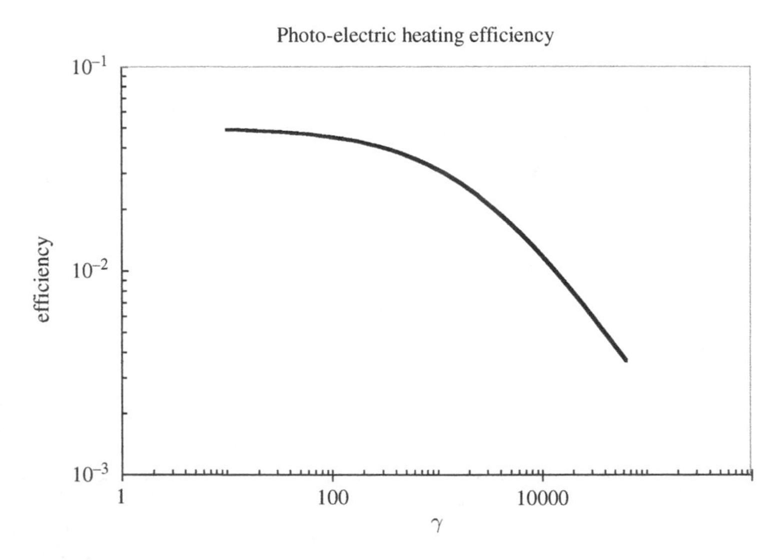

Figure 3.4 The photo-electric heating efficiency as a function of the charging parameter $\gamma \equiv G_0 T^{1/2}/n_e$, which is proportional to the ionization rate over the recombination rate). For low γ, PAHs and grains are neutral and the photo-electric heating is at maximum efficiency. For increasing γ, grains and PAHs charge up and the overall heating efficiency decreases.

In terms of the efficiency (Eq. 3.16), the total photo-electric heating rate is given by

$$n\Gamma_{\rm pe} = 10^{-24}\epsilon n G_0 \,{\rm erg\,cm^{-3}\,s^{-1}}. \tag{3.17}$$

We note that for a single species, in the small γ limit, the photo-electric heating is proportional to $G_0 n$ (e.g., for predominantly neutral species) and ionization scales with G_0, while in the large γ limit, the heating rate is independent of the intensity of the FUV field and proportional to nn_e through the recombination rate (cf. Eqs. (3.13) and (3.14)). For a size distribution, the latter is only approximately correct, because the transition from small to large γ occurs at different γs for different sized grains. For neutral PAHs and grains ($\gamma \ll 10^3$ K$^{1/2}$ cm^3), a maximum efficiency of $\sim$0.05 is reached and the photo-electric heating rate is then $\simeq 5\times 10^{-26}\, G_0$ erg (H atom)$^{-1}$ s^{-1}. In the cold neutral medium, $\gamma \simeq 10^3$ K$^{1/2}$ cm^3 resulting in $\epsilon \simeq 0.03$ and near-maximum heating. In the warm neutral medium, $\gamma \simeq 2\times 10^4$ K$^{1/2}$ cm^3 and ϵ is only 0.01. In a dense PDR, $\gamma \simeq 10^5$ K$^{1/2}$ cm^3, and the efficiency is calculated to be 0.003.

For comparison, the heating due to carbon photo-ionization is $\sim 3\times 10^{-26}$ f(CI) G_0 erg (H atom)$^{-1}$ s^{-1} (cf. Eq. (3.8)), which in the cold neutral medium is $\simeq 10^{-29}\, G_0\, n$ erg cm^{-3} s^{-1}. Polycyclic aromatic hydrocarbon molecules (and small dust grains) are much more efficient in heating the gas than carbon atoms largely because of the much larger neutral fraction of these species. Essentially, PAH molecules (and small dust grains) have so many internal degrees of freedom that can take up the incoming kinetic energy of a recombining electron that their recombination rate coefficient ($\sim 10^{-5}$ cm^3 s^{-1}) is much higher than the (radiative) recombination rate coefficient of atomic ions ($\sim 10^{-11}$ cm^3 s^{-1}). Even if we include the additional neutralization of C^+ due to recombination with large PAH anions (cf. Section 3.2.1), the difference in the heating rate due to C^0 ionization and PAH ionization is still a factor of $\sim 10^3$.

3.4 Photon heating by H_2

After photodissociation of a molecule, the fragments will carry away some of the photon energy as kinetic energy, heating the gas. After (re)formation of a molecule, the newly formed species may be left in a vibrationally excited state. Infrared photons can also vibrationally excite molecules directly. In both cases, collisional de-excitation can then heat the gas.

Because of abundance considerations, photodissociation of H_2 will dominate this heating term. The photochemistry of H_2 is discussed extensively in Section 8.7.1. In short, H_2 photodissociation proceeds through FUV line absorption from the ground electronic state to an excited electronic state. A small fraction

($\simeq$10%) of the radiative decays back to the electronic ground state occur in the vibrational continuum (see Section 8.7.1), delivering about 0.25 eV to the gas. The heating by this process, $n\Gamma_{pd}$, is given by

$$n\Gamma_{pd} = 4 \times 10^{-14} n(H_2) k_{pump} \, \text{erg cm}^{-3} \, \text{s}^{-1}. \tag{3.18}$$

The pump rate of molecular hydrogen is given by

$$k_{pump} = 3.4 \times 10^{-10} \beta(\tau) G_0 \exp[-2.6 \, A_v] \, \text{s}^{-1}, \tag{3.19}$$

where the numerical factor is the pumping rate in the average interstellar radiation field. The two additional factors take attenuation into account: $\beta(\tau)$ is the reduction of the FUV pumping radiation field due to self-shielding by molecular hydrogen molecules that provide an optical depth τ to the cloud surface (see Section 8.7.1 for details) and the exponential factor takes dust absorption into account.

Most of the FUV absorption is by H_2 decay back to a bound vibrational excited state of the electronic ground state. The molecule will then slowly decay radiatively through the emission of IR photons. At high densities ($n \gtrsim 10^4$ cm^{-3}, depending on T), the vibrationally excited molecule can also be collisionally de-excited, thereby heating the gas. The heating efficiency of this process is then approximately

$$\epsilon(H_2) \simeq \left(\frac{E_{vib}}{h\nu} \right) f_{H_2} \simeq 0.17 \, f_{H_2}, \tag{3.20}$$

where E_{vib} is the vibrational energy converted into heat and $h\nu$ is the energy of the pumping FUV photon. The fraction of the FUV photon flux pumping H_2, f_{H_2}, reflects the competition between H_2 and the dust for these FUV photons. This depends on the location of the HI/H_2 transition zone. For dense regions ($G_0/n \lesssim 4 \times 10^{-2}$ cm^3), H_2 self-shielding is important, the H_2 transition is near the surface, and most of the photons that can pump H_2 are absorbed by H_2 rather than dust (Section 8.7.1). Under these conditions, $f_{H_2} \simeq 0.25$, and this process provides an efficient coupling to the FUV photon flux of the star. Indeed, this efficiency is comparable to that of the photo-electric effect. In low-density regions ($G_0/n \gtrsim 4 \times 10^{-2}$ cm^3), dust rather than H_2 dominates the absorption of the pumping photons and this heating process is unimportant.

The heating rate by this process depends, thus, on the density of vibrationally excited species. Here, we will concentrate on the edge of dense photodissociation regions where atomic H is the dominant collision partner. Then, if we approximate the H_2 molecule with one pseudo-vibrational level at an energy of 2 eV, the heating rate is given by

$$n\Gamma_{H_2} = 3.2 \times 10^{-12} n \gamma n_2^{\star}, \tag{3.21}$$

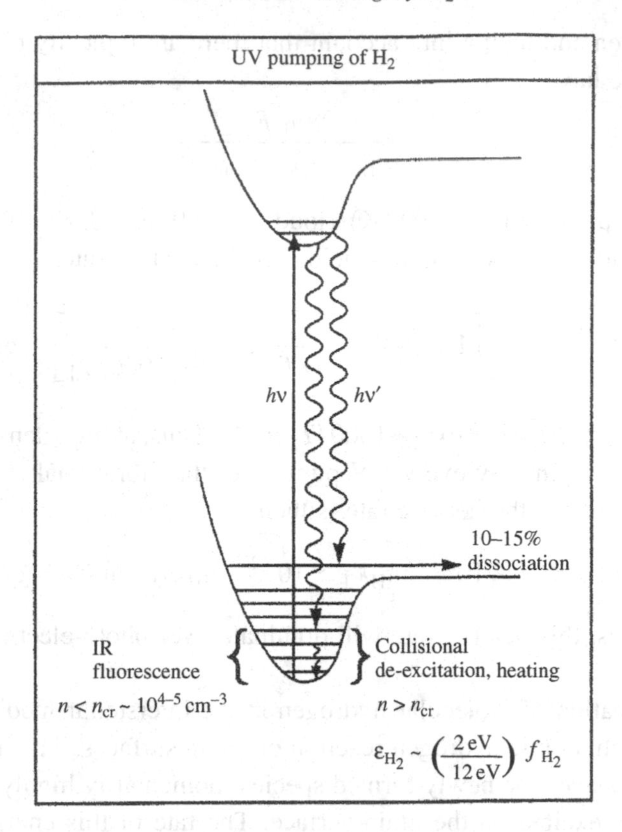

Figure 3.5 A schematic diagram of the H_2 pumping, fluorescence, dissociation, and heating through the absorption of FUV photons. Absorption of an FUV photon followed by FUV radiative decay can leave H_2 vibrationally excited in the ground electronic state. The excess vibrational energy can be emitted as a near-IR photon or the molecule can be de-excited through collisions, thereby heating the gas. In about 10–15% of the FUV pumps, H_2 decays to the vibrational continuum of the ground electronic state and the molecule dissociates.

with γ the de-excitation rate ($\gamma \simeq 10^{-12}\, T^{0.5} \exp[-1000/T]$ cm^3 s^{-1} for atomic H). The population of excited H_2, $n_2^\star$, can be approximated by balancing collisional de-excitation ($n_2^\star \gamma n_H$) and radiative decay ($A n_2^\star$) with the UV pump rate ($n\,(H_2)\, k_{pump}$),

$$n_2^\star = \frac{n(H_2) k_{pump}}{\gamma n_H + A + k_{pump}}, \tag{3.22}$$

where we have ignored direct photodissociation out of the vibrationally excited state of H_2. Now, assuming that the formation of H_2 from atomic H on grain surfaces ($n\, n_H k_d$, with $k_d \simeq 3 \times 10^{-17}$, see Section 8.7.1) is balanced by photodissociation and taking into account that there are typically nine pumps per dissociation, we have

$$n_2^\star = \frac{9 n n_{\rm H} R}{n_{\rm H}\gamma + A + k_{\rm pump}}. \tag{3.23}$$

At the edge of a warm ($T \sim 1000$ K) cloud, $A_{\rm v} = 0$ and $\beta(\tau = 0) = 1$ (i.e., all photons are available to pump), this yields for the heating rate

$$n\Gamma_{\rm H_2} \simeq 2.9\times 10^{-11} n n_{\rm H} k_{\rm d} \left[1 + \left(\frac{n_{\rm cr}}{n}\right) + \frac{4.4\times 10^2\, G_0}{n\, T^{1/2} \exp[-1000/T]}\right]^{-1} \text{erg cm}^{-3}\,\text{s}^{-1}, \tag{3.24}$$

where $n_{\rm cr} = A/\gamma \simeq 10^5/T^{0.5} \exp[-1000/T]\,\text{cm}^{-3}$. Thus, at high densities ($n \gg n_{\rm cr}$ and $n/G_0 \gg 1$), essentially every UV pump into the vibrational states results in heating of the gas and the heating rate is then

$$n\Gamma_{\rm H_2} \simeq 3\times 10^{-11} n n_{\rm H} k_{\rm d} \simeq 10^{-27}\, n n_{\rm H}\, \text{erg cm}^{-3}\,\text{s}^{-1}. \tag{3.25}$$

At high densities, this heating rate will dominate over photo-electric heating (cf. Eq. (3.17)).

Finally, formation of molecular hydrogen in the interstellar medium generally proceeds through atomic hydrogen reaction on grain surfaces. The heat of formation ($\simeq$4.5 eV) leaves the newly formed species momentarily highly vibrationally and rotationally excited on the grain surface. The fate of this energy is unclear. Energy transfer to the grain through multipolar–multipolar interaction may be efficient, particularly for the higher vibrational states of H_2, which are near resonant with vibrational modes of surface functional groups such as C–H and O–H. However, if the newly formed species is ejected quickly from the grain, a fraction, $\epsilon_{\rm chem}$, of the chemical energy may be available for gas heating; $\epsilon_{\rm chem}$ might be as small as 0.2. The heating rate is then given by

$$n\Gamma_{\rm chem} = 7.2\times 10^{-12} \frac{n_{\rm H} n k_{\rm d}}{1 + n_{\rm cr}/n} \epsilon_{\rm chem}\, \text{erg cm}^{-3}\,\text{s}^{-1}. \tag{3.26}$$

Thus, this chemical energy heating term is typically small ($0.24\,\epsilon_{\rm chem}$) compared to the heating term due to UV pumping. Heating due to molecular hydrogen formation may be important in dissociative shocks inside dense clouds, where UV photons play little part (see Section 11.2.2).

3.5 Dust-gas heating

In the interstellar medium, gas and dust are not in thermodynamical equilibrium and often have quite different temperatures. If the dust is warmer than the gas (say, in the envelope of a protostar), gas atoms bouncing off a grain can be an important gas heating source. The heating rate is then given by

$$n\Gamma_{\rm g-d} = n n_{\rm d}\sigma_{\rm d}\left(\frac{8kT}{\pi m}\right)^{1/2}(2kT_{\rm d} - 2kT)\alpha_{\rm a}, \tag{3.27}$$

where $n_{\rm d}\sigma_{\rm d}$ is the dust density and geometric cross section ($10^{-21}\,n$ cm^{-1}), and $2kT$ is the average kinetic energy of a gas atom striking the grain surface. The accommodation coefficient, $\alpha_{\rm a}$, measures how well the gas atom accommodates to the grain; i.e., an $\alpha_{\rm a}$ of unity corresponds to a bouncing particle that has completely thermalized and leaves with $2kT_{\rm d}$. Experimental and theoretical studies suggest that $\alpha_{\rm a} \simeq 0.15$. The heating rate is then

$$n\Gamma_{\rm g-d} \simeq 10^{-33} n^2 T^{1/2}(T_{\rm d} - T)\,\text{erg cm}^{-3}\,\text{s}^{-1}. \tag{3.28}$$

Of course, if the gas is warmer than the dust (e.g., in the diffuse ISM or in a photodissociation region), this process is really a cooling process. If the gas and grains are charged, the cross section can be enhanced by coulomb focussing by a factor of 3 (see Section 5.2.3). This heating rate can also be substantially enhanced if the dust moves relative to the gas at high velocities. This may occur near a bright star where radiation pressure accelerates the dust or in shock fronts where the kinetic energy of the gas is quickly thermalized but dust grains keep moving because of their inertia.

3.6 Cosmic-ray heating

Low-energy cosmic rays (~1–10 MeV) are most efficient in ionizing and heating the gas. A high-energy proton can ionize a gas atom. The substantial kinetic energy (~35 eV) of the resulting primary electron can be lost through (elastic) collisions with other electrons or through ionization or excitation of gas atoms or molecules. The total ionization rate, $\xi_{\rm CR}$, including secondary ionizations, depends then on the energy of the primary and the electron fraction, $x_{\rm e}$:

$$n\xi_{\rm CR} = n\zeta_{\rm CR}[1 + \phi^{\rm H}(E, x_{\rm e}) + \phi^{\rm He}(E, x_{\rm e})], \tag{3.29}$$

where $\zeta_{\rm CR}$ is the primary ionization rate, and the $\phi^i(E, x_{\rm e})$s are the average number of secondary ionizations of H and He per primary ionization. For diffuse interstellar clouds where the electron abundance is low, the number of secondaries is about 0.8. Above $x_{\rm e} \simeq 10^{-3}$, an increasing fraction of the energy of the primary electron is lost through coulomb interaction with thermal electrons of the gas rather than through further ionization.

Because their propagation is also most affected by the interstellar and Solar System magnetic field, the flux of low-energy cosmic rays is not well constrained by direct observations (see Section 1.3.3) but can be inferred from observations of the degree of ionization in HI and molecular clouds (see Section 8.8). With a primary ionization rate of 2×10^{-16} s^{-1}, the total cosmic-ray ionization rate is $\xi_{\mathrm{CR}} \simeq 3 \times 10^{-16}$ s^{-1}. The heating rate is then given by

$$n\Gamma_{\mathrm{CR}} = n\zeta_{\mathrm{CR}} E_{\mathrm{h}}(E, x_{\mathrm{e}}), \tag{3.30}$$

with $E_{\mathrm{h}}(E, x_{\mathrm{e}})$ the average heat deposited per primary ionization. For low degrees of ionization, $E_{\mathrm{h}}(E, x_{\mathrm{e}}) \simeq 7\,\mathrm{eV}$. The cosmic-ray heating rate is then

$$n\Gamma_{\mathrm{CR}} = 3 \times 10^{-27} n \left[\frac{\zeta_{\mathrm{CR}}}{2 \times 10^{-16}}\right] \mathrm{erg\,cm^{-3}\,s^{-1}}. \tag{3.31}$$

3.7 X-ray heating

As for cosmic rays, the primary electron created by X-ray absorption can be energetic enough to lead to secondary ionization. The primary ionization rate is given by

$$n\zeta_{\mathrm{XR}} = 4\pi n \int \mathcal{N}_{\mathrm{XR}}(\nu) e^{-\sigma(\nu)N} \sigma(\nu) \mathrm{d}\nu, \tag{3.32}$$

with $\mathcal{N}_{\mathrm{XR}}$ the X-ray mean photon intensity and σ the X–ray ionization cross section. The observed diffuse X-ray background contains extragalactic components as well as contributions from a number of galactic components (the local bubble, galactic disk, and halo). The X-ray mean intensity is included in Fig. 1.8. The cross section for X-ray absorption for Solar System abundances is shown in Fig. 3.6. As for cosmic rays, because of coulomb losses, the total ionization rate and total heating rate – including the effects of secondary ionization – depend on the electron fraction in the gas. Absorption by intervening material is very important for soft X-rays and is explicitly included in Eq. (3.32).

Figure 3.7 shows the calculated ionization and heating rates in the local solar neighborhood as a function of absorbing H column density for different electron fractions. The ionization rate and heating rate decrease with increasing depth into a cloud because of the attenuation. Because of the presence of more than one spectral component in the diffuse X-ray background coupled with the increased mean free path for harder X-rays, the ionization and heating rates drop more slowly with column density than a simple exponential. When the degree of ionization increases, more energy of the primary electron is lost to coulomb losses and hence the heating rate increases while the ionization rate decreases. The X-ray heating rate is substantially less than the photo-electric heating rate due to FUV photons, which merely reflects the low photon flux (Fig. 1.9) and the low ionization cross section of atoms (Fig. 3.6) in the X-ray range. For small attenuating columns, X-ray ionization can be important compared to cosmic rays.

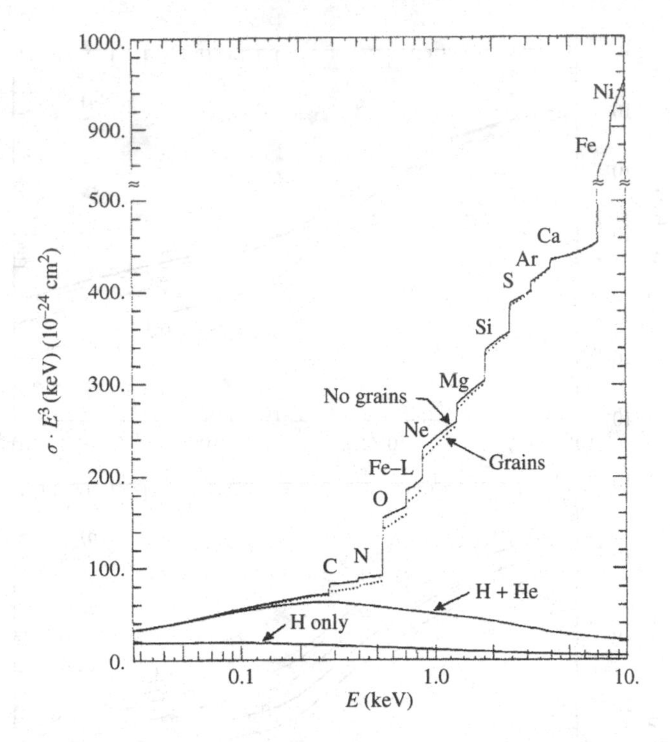

Figure 3.6 The photo-electric absorption cross section per H atom, scaled by E^3(keV). The solid line assumes that all the elements are in neutral form in the gas phase while the dotted line assumes typical dust abundances. The absorption edges due to various elements are indicated. Figure reproduced with permission from R. Morrison and D. McCammon, 1983, *Ap. J.*, **270**, p. 119.

3.8 Turbulent heating

Incompressible fluids tend to undergo a transition from a smooth laminar flow to a state characterized by large and unpredictable fluctuations in velocity or pressure when the Reynolds number is large. The Reynolds number describes the behavior of a flow with substantial velocity gradients perpendicular to the flow (shear). It measures the ratio of the inertial forces to the viscous forces in a flow,

$$Re = \frac{vL}{\nu}, \tag{3.33}$$

with v the mean velocity, L a characteristic length scale, and ν the kinematic viscosity. In a gas, the viscosity reflects the diffusion of momentum through the system. When $Re \geq 3000$, turbulent flow will ensue. For the ISM, the Reynolds number is 10^6–10^8 and hence the ISM is prone to develop turbulence.

We will consider here fully developed, incompressible turbulence fed by energy input at the largest scale. At the smallest scale, this energy is eventually dissipated through viscous interaction. The turbulent velocity field consists of many eddies

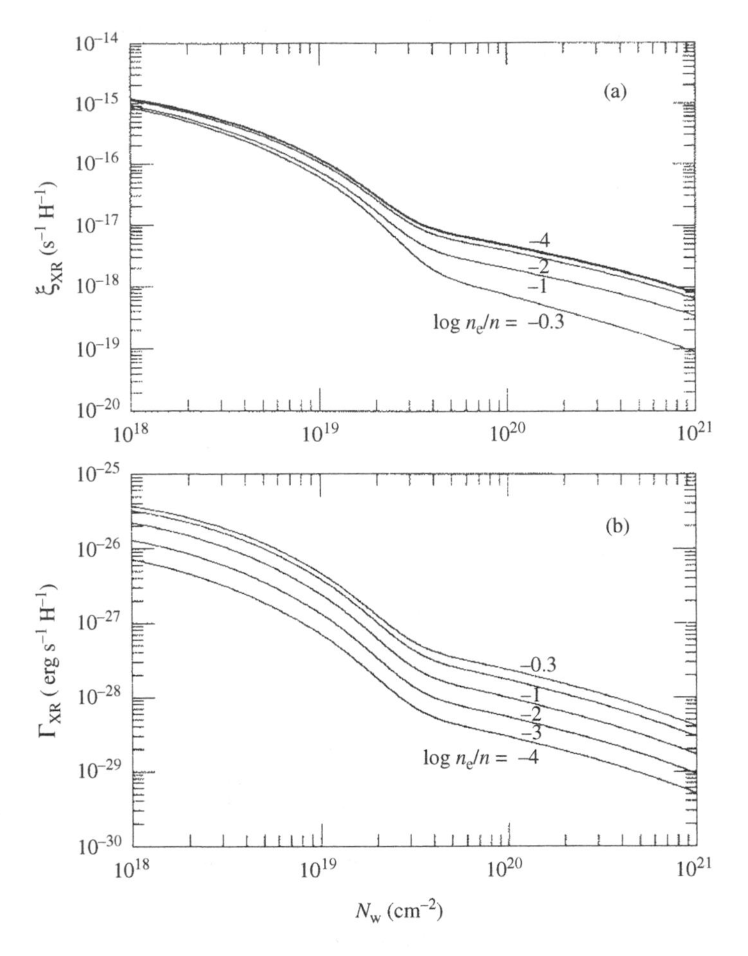

Figure 3.7 The ionization rate (top) and heating rate (bottom) due to the soft X-ray background as a function of absorbing H column density. The different curves are labeled by the adopted electron fraction.

of different sizes, where eddies of a given size, ℓ, will have a characteristic velocity, $v(\ell)$. The energy fed into the system at the largest scales cascades into smaller scale eddies and these in turn feed their energy into even smaller eddies, all the way down to the molecular dissipation level where the energy is converted into heat. The smallest scale size will be set by the viscosity, $\ell_s v(\ell_s) \simeq \nu$. The viscosity, ν, is given by

$$\nu = v_{\text{therm}}/n\sigma \simeq 10^{20}/n \, \text{cm}^2 \, \text{s}^{-1}, \tag{3.34}$$

where v_{therm} is the thermal velocity ($\sim$1 km s^{-1}), and σ is a typical cross section (10^{-15} cm^2). Thus, the power spectrum should reflect that the "energy flow" through the system is constant (e.g., $\dot{\epsilon} = v^2/(\ell/v) = constant$), which then leads to

$$v = (\dot{\epsilon}\ell)^{1/3}. \tag{3.35}$$

The smallest length scale, ℓ_s, is then

$$\ell_s \sim \left(\frac{\nu^3}{\dot{\epsilon}}\right)^{1/4}. \tag{3.36}$$

This length scale is typically very small, 10^{16} cm for the WNM and 10^{13} cm for molecular clouds (see below). The kinetic energy density of the turbulent fluid at scale $k = 2\pi/\ell$ is $\mathrm{d}(v^2)/\mathrm{d}k$. This results in

$$E(k)\mathrm{d}k = \frac{2}{3}(2\pi\dot{\epsilon})^{2/3}k^{-5/3}\mathrm{d}k, \tag{3.37}$$

which is known as the Kolmogorov energy spectrum.

Thus, by making the energy flow assumption, we can estimate the energy transfer rate – and thus the gas heating rate – if we know the characteristics of the turbulent fluid at one scale size,

$$n\Gamma_{\mathrm{turbulence}} = \frac{1/2nm_{\mathrm{H}}v^2}{\ell/v}. \tag{3.38}$$

In the warm neutral medium, $v = 10\,\mathrm{km\,s^{-1}}$ on a scale of 200 pc and the turbulent heating rate is

$$n\Gamma_{\mathrm{turbulence}} = 2.8 \times 10^{-30} n\,\mathrm{erg\,cm^{-3}\,s^{-1}}. \tag{3.39}$$

For molecular cloud cores, observations give $v \simeq 1\,\mathrm{km\,s^{-1}}$ at $n = 10^4\,\mathrm{cm^{-3}}$ and $\ell = 1$ pc and this results in a heating rate of

$$n\Gamma_{\mathrm{turbulence}} = 3 \times 10^{-28} n\,\mathrm{erg\,cm^{-3}\,s^{-1}} \tag{3.40}$$

for Kolmogorov turbulence.

3.9 Heating due to ambipolar diffusion

In a partially ionized gas, such as a molecular cloud core, ions and neutrals may develop a small (drift) velocity difference, for example, through the counterplay of magnetic fields and gravity. The friction between the ions and the neutrals heats the gas at a rate

$$n\Gamma_{\mathrm{ad}} = n_{\mathrm{e}}n_{\mathrm{n}}m < \sigma v > v_{\mathrm{d}}^2, \tag{3.41}$$

with $< \sigma v >$ the average collision rate (2×10^{-9} cm^3 s^{-1}), and v_d the neutral–ion drift velocity. In molecular cloud cores, calculated drift velocities are $\simeq 5 \times 10^3$ cm s^{-1}, which corresponds to a drift of 0.5 pc on a typical ambipolar diffusion time scale of 10^7 years. The degree of ionization is $\simeq 10^{-5}/n^{1/2}$ (see Section 10.2). The heating rate is then

$$n\Gamma_{\rm ad} \simeq 1.6 \times 10^{-30} n^{3/2} \, {\rm erg\,cm^{-3}\,s^{-1}}. \tag{3.42}$$

Ambipolar diffusion heating plays a role in the slow leakage of magnetic fields from pre-stellar molecular cloud cores and the accompanying settling into an isothermal sphere structure. This process will not be discussed in this book. Ambipolar diffusion also plays an important role in a type of shocks in molecular clouds, called C shocks. This will be discussed extensively in Section 11.3.

3.10 Gravitational heating

Gravitational collapse will heat the gas by compression. Adopting the free-fall solution, we find

$$n\Gamma_{\rm gravity} = \frac{5}{2} kTv \left| \frac{{\rm d}n}{{\rm d}r} \right| = \frac{15}{4} kT \frac{n}{\tau_{\rm ff}}, \tag{3.43}$$

with the free-fall time scale given by

$$\tau_{\rm ff} = \left(\frac{3\pi}{32 G n m_{\rm H}} \right)^{1/2} \simeq \frac{4 \times 10^7}{n^{1/2}} \, {\rm years}. \tag{3.44}$$

Thus, in a dense cloud core, the gravitational heating rate is given by

$$n\Gamma_{\rm gravity} \simeq 4.3 \times 10^{-31} n^{3/2} T \, {\rm erg\,cm^{-3}\,s^{-1}}. \tag{3.45}$$

Obviously, this heating process is important during the collapse phase of molecular cloud cores and we will not discuss it further in this book.

3.11 The heating of the interstellar medium

3.11.1 Heating of diffuse clouds

Figure 3.8 compares the different heating processes for conditions relevant for the diffuse clouds: a gas temperature of 100 K, the average interstellar radiation field ($G_0 = 1$), and an electron fraction of 1.4×10^{-4}. The photo-electric effect dominates over the full range. Coupling between the gas and cosmic rays is of some importance at the lowest densities. Photo-ionization of neutral carbon was calculated assuming a balance between photo-ionization and radiative electron

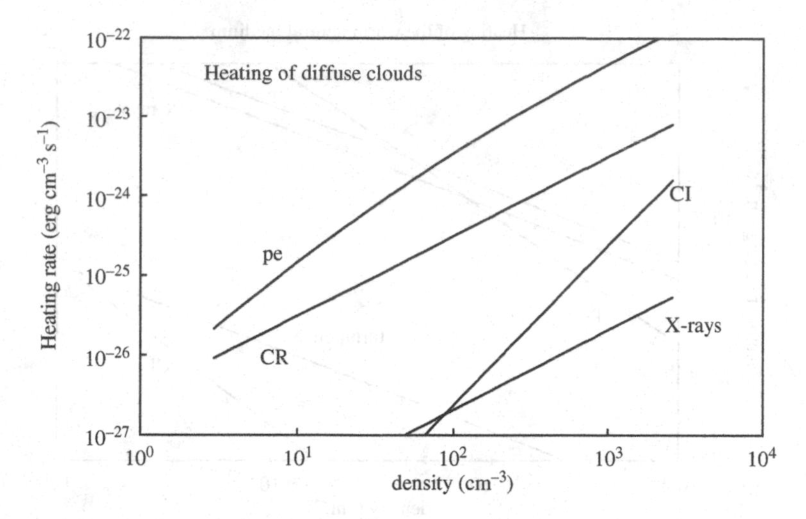

Figure 3.8 Heating processes in diffuse interstellar clouds as a function of the density. The processes displayed are the photo-electric effect (pe), photo-ionization of neutral carbon (CI), cosmic-ray ionization (CR), and X-ray ionization (X-rays).

recombination of C^+ (see Section 8.2.1). Coupling between the gas and X-rays was calculated assuming a shielding column of 10^{20} cm^{-2}. Neither of these two processes is ever important.

3.11.2 Heating of the warm neutral medium

Figure 3.9 compares the different heating processes for conditions relevant for the warm neutral medium: a gas temperature of 8000 K, the average interstellar radiation field ($G_0 = 1$), and an electron fraction of 3×10^{-3}. The photo-electric effect dominates at high densities, while cosmic rays (with $\zeta_{CR} = 2 \times 10^{-16}$ s^{-1}) and X-rays (assuming a column typical for the WNM, 10^{19} cm^{-2}) are important at low densities. Carbon I ionization and decay of turbulence are unimportant throughout.

3.11.3 Heating of molecular cloud cores

Figure 3.10 compares the different heating processes for conditions relevant for the molecular clouds: a gas temperature of 10 K, no FUV radiation, and an electron fraction of 10^{-7}. A number of different processes now contribute. In the absence of FUV (and X-ray) photons, cosmic rays are now the most important external heating source, which, with a primary ionization rate of 2×10^{-16}, dominates over the full density range. In order of importance, decay of turbulence is the next

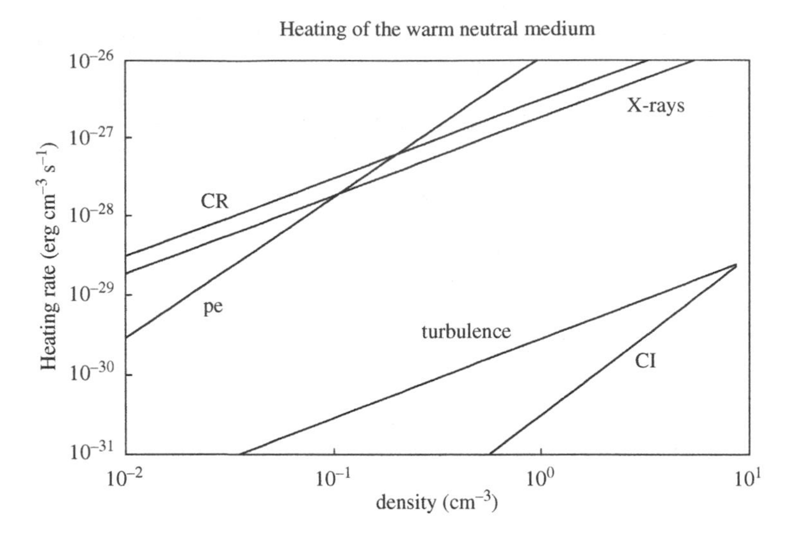

Figure 3.9 Heating processes in the warm neutral medium as a function of the density. The processes displayed are the photo-electric effect (pe), photo-ionization of neutral carbon (CI), cosmic-ray ionization (CR), decay of turbulence (turbulence), and X-ray ionization (X-rays).

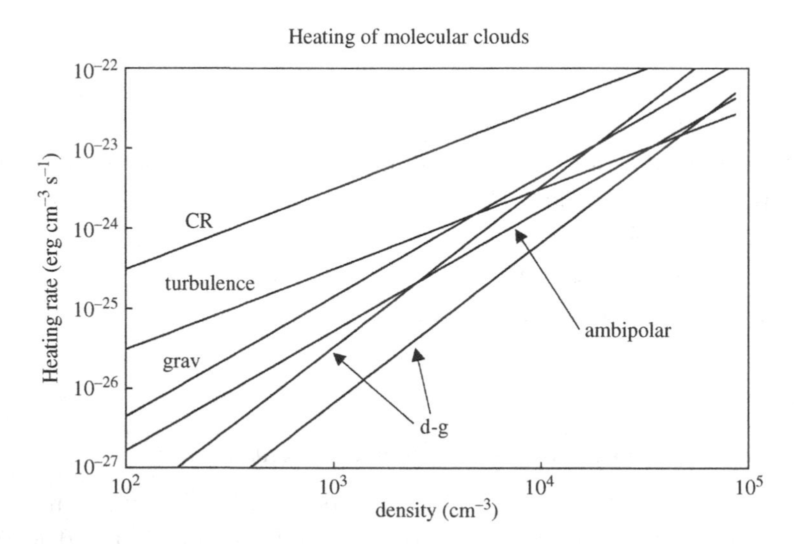

Figure 3.10 Heating processes in molecular clouds as a function of the density. The processes displayed are turbulence, cosmic-ray ionization (CR), gravitational heating (grav), heating due to ambipolar diffusion and two curves for heating through collisions of gas with warm dust grains (d-g) at temperature differences of 2 and 10 K, respectively.

process at low densities. Heating by ambipolar diffusion (during the precollapse phase) or gravitational heating (during the actual collapse phase) are next in line. Once a protostar has formed and heated the dust around it, dust and gas are well coupled to $\Delta T \simeq 10\,\mathrm{K}$, in a dense core.

3.12 Further reading

Heating of ionized gas is extensively discussed in the excellent monograph [1].

Various authors have worked on the photo-electric effect on interstellar dust, starting with [2] and followed by [3] and [4]. The importance of the photo-electric effect on interstellar molecules has been discussed in [5] and [6]. The discussion in this chapter, including the simple analytical approximations, has been taken from [7].

Heating by collisional de-excitation of UV-pumped vibrational excited H_2 has been discussed in detail in [8]. The analytical approximation in this chapter has been taken from [9].

Heating due to gas grain collisions, including an evaluation of experimental and theoretical estimates of the thermal accommodation coefficients, has been discussed in [10].

The coupling of cosmic rays to the gas and its dependence on the degree of ionization have been discussed in, among others, [11] and [12].

The diffuse X-ray background has been estimated in [13]. The gas photo-ionization cross sections are provided in a convenient analytical form in [14]. The effects of secondaries are again provided in [12].

Heating of the diffuse interstellar medium has a long history dating back to [15]. A thorough discussion is provided by [16]. Heating of molecular clouds has been discussed in [11] and [17].

References

[1] D.E. Osterbrock, 1989, *Astrophysics of Gaseous Nebulae and Active Galactic Nuclei* (Mill Valley: University Science Books)
[2] W.D. Watson, 1972, *Ap. J.*, **176**, p. 103
[3] T. de Jong, 1977, *A. & A.*, **55**, p. 137
[4] B.T. Draine, 1978, *Ap. J. S.*, **36**, p. 595
[5] L. Verstraete, A. Léger, L. d'Hendecourt, O. Dutuit, and D. Défourneau, 1990, *A. & A.*, **237**, p. 436
[6] S. Lepp and A. Dalgarno, 1988, *Ap. J.*, **335**, p. 769
[7] E.L.O. Bakes and A.G.G.M. Tielens, 1994, *Ap. J.*, **427**, p. 822
[8] A. Sternberg and A. Dalgarno, 1989, *Ap. J.*, **338**, p. 199
[9] M. Burton, D. Hollenbach, and A.G.G.M. Tielens, 1990, *Ap. J.*, **365**, p. 620
[10] J.R. Burke and D. Hollenbach, 1983, *Ap. J.*, **265**, p. 223

[11] A. E. Glassgold and W. D. Langer, 1973, *Ap. J.*, **179**, p. L147
[12] J. M. Shull and M. E. van Steenberg, 1985, *Ap. J.*, **298**, p. 268
[13] G. P. Garmire, J. A. Nousek, K. M. V. Apparao, *et al.*, 1992, *Ap. J.*, **399**, p. 694
[14] M. Balucinska-Church and D. McCammon, 1992, *Ap. J.*, **400**, p. 699
[15] G. B. Field, D. W. Goldsmith, and H. J. Habing, 1968, *Ap. J.*, **155**, p. L149
[16] M. G. Wolfire, D. Hollenbach, C. F. McKee, A. G. G. M. Tielens, and E. L. O. Bakes, 1995, *Ap. J.*, **443**, p. 152
[17] J. Black, 1987, in *Interstellar Processes*, ed. D. Hollenbach and H. Thronson, (Dordrecht: Reidel), p. 731

4
Chemical processes

In a way, astrochemistry describes a cosmic dance of the elements in which atoms are constantly reshuffled from one species to another. This molecular rearrangement may be effected by gas phase binary collisions where atoms change partner or through recombination on grain surfaces. This "dance" is driven by the action of various energy sources, including photons and cosmic rays. In order to appreciate astrochemistry properly, we first have to get a basic understanding of the "dance" steps involved. This chapter will focus on the basic chemical processes that are of importance. In later chapters, we will then overview the resulting chemical reaction schemes that drive molecular complexity in the Universe.

There is a variety of processes that can lead to the formation of molecules in the interstellar medium, but these can be separated into two broad classes: reactions that occur in the gas phase and reactions that occur on the surfaces of small grains prevalent throughout the interstellar medium. These two classes of reactions are discussed in turn in the two sections of this chapter. The focus is on the formation of relatively simple species. The chemistry of large and complex interstellar molecules is discussed in Chapter 6.

4.1 Gas-phase chemical reactions

Gas phase reactions can be divided into different categories depending on their general effects. There are the bond-formation processes, including radiative association (cf. Section 4.2), which link atoms into simple or more complex species. Reactions such as photodissociation, dissociative recombination, and collisional dissociation are bond-destruction processes, which fragment species into smaller species. Finally, there are the bond-rearrangement reactions – ion–molecule exchange reactions, charge-transfer reactions, and neutral–neutral reactions – which transfer parts of one coreactant to another one. Here we will discuss these reactions in some detail. Generic examples of these reactions and typical reaction

Table 4.1 *Generic gas phase reactions and their rates*

	reaction	rate	unit	note
Photodissociation	$AB + h\nu \rightarrow A + B$	10^{-9}	s^{-1}	(*a*)
Neutral–neutral	$A + B \rightarrow C + D$	4×10^{-11}	$cm^3\,s^{-1}$	(*b*)
Ion–molecule	$A^+ + B \rightarrow C^+ + D$	2×10^{-9}	$cm^3\,s^{-1}$	(*c*)
Charge-transfer	$A^+ + B \rightarrow A + B^+$	10^{-9}	$cm^3\,s^{-1}$	(*c*)
Radiative association	$A + B \rightarrow AB + h\nu$			(*d*)
Dissociative recombination	$A^+ + e \rightarrow C + D$	10^{-7}	$cm^3\,s^{-1}$	
Collisional association	$A + B + M \rightarrow AB + M$	10^{-32}	$cm^6\,s^{-1}$	(*c*)
Associative detachment	$A^- + B \rightarrow AB + e$	10^{-9}	$cm^3\,s^{-1}$	(*c*)

(*a*) Rate in the unshielded radiation field.
(*b*) Rate in the exothermic direction and assuming no activation barrier (i.e., radical–radical reaction).
(*c*) Rate in the exothermic direction.
(*d*) Rate highly reaction specific.

rates are summarized in Table 4.1. For a bimolecular reaction, the corresponding rates of formation or destruction of a species are

$$\frac{dn(A)}{dt} = -k\,n(A)\,n(B) = -\frac{dn(C)}{dt}. \tag{4.1}$$

For a unimolecular reaction, we have

$$\frac{dn(A)}{dt} = -k\,n(A) = -\frac{dn(C)}{dt}. \tag{4.2}$$

Similarly, for a termolecular reaction, we have

$$\frac{dn(A)}{dt} = -k\,n(A)\,n(B)\,n(M) = -\frac{dn(AB)}{dt}. \tag{4.3}$$

4.1.1 Photochemistry

The FUV photons permeating the diffuse ISM are a dominant destruction agent for small molecules. Typical bonding energies of molecules are in the range 5–10 eV, corresponding to wavelengths of $\simeq$3000 Å and shorter. However, direct absorption into the dissociating continuum of the ground state is generally negligible. Rather, dissociation occurs through a transition to the continuum of an electronic excited state. Alternatively, dissociation can occur because the excited electronic state mixes with a crossing, dissociative state (predissociation). There is a third route, which happens to be the only one available for molecular hydrogen, whereby dissociation occurs after the UV excited electronic state decays radiatively to the continuum of the ground electronic state. This is discussed in more detail in Section 8.7.1.

In the diffuse interstellar medium, the photodissociation rate is given by

$$k_{pd} = \int_{\nu_i}^{\nu_H} 4\pi\, \mathcal{N}_{ISRF}(\nu)\, \alpha_{pd}(\nu)\, d\nu, \tag{4.4}$$

where $\mathcal{N}_{ISRF}(\nu)$ is the mean photon intensity of the interstellar radiation field, $\alpha_{pd}(\nu)$ the photodissociation cross section, and the integration runs from the dissociation limit to the hydrogen photo-ionization limit. If there are allowed transitions to dissociative channels available, the cross section is typically 0.01–0.1 Å^2 for dipole-allowed transitions in the wavelength region of interest. In the interstellar radiation field ($\sim 10^8$ FUV photons cm^{-2} s^{-1} sr^{-1}), this translates into a rate of 10^{-9}–10^{-10} s^{-1} or a lifetime of 10^9–10^{10} s. Inside a cloud, the radiation field will be attenuated by dust. Because dust absorption and scattering is not gray, the frequency distribution of the FUV radiation field will depend on the depth into the cloud. Assuming standard dust properties (Chapter 5), extensive radiative transfer studies have been done to determine the dependence of the photodissociation reaction rates on cloud depth. Table 4.2 lists some of the important ones appropriate for infinite homogeneous slabs, tabulated in the form

$$k_{pd} = a \exp[-bA_v], \tag{4.5}$$

with a the unshielded rate and A_v the visual extinction due to dust.

Table 4.2 *Some important photo reactions and their rates*[a]

Reaction	*a*	*b*
$CH \rightarrow C + H$	2.7 (−10)	1.3
$CH_2 \rightarrow CH + H$	5.0 (−11)	1.7
$CH \rightarrow C + H$	2.7 (−10)	1.3
$O_2 \rightarrow O + O$	3.3 (−10)	1.4
$OH \rightarrow O + H$	7.6 (−10)	2.0
$CO \rightarrow C + O$	1.7 (−10)	3.2[b]
$H_2O \rightarrow OH + H$	5.1 (−10)	1.8
$CH \rightarrow CH^+ + e$	3.2 (−10)	3.0
$CH_2 \rightarrow CH_2^+ + e$	1.0 (−9)	2.3
$CH^+ \rightarrow C^+ + H$	1.8 (−10)	2.8
$CH_2^+ \rightarrow CH^+ + H$	1.7 (−9)	1.7
$CH_3^+ \rightarrow CH_2^+ + H$	1.0 (−9)	1.7
$CH_3^+ \rightarrow CH^+ + H_2$	1.0 (−9)	1.7

[a] Rates are $k_{pd} = a \exp[-bA_v]$.
[b] Self-shielding can be important for CO dissociation.

These rates do depend on the adopted extinction and particularly scattering properties of the dust – which generally dominates the opacity. These dust properties are generally measured in the diffuse ISM (cf. Chapter 5) and the properties of dust in dense clouds may have been modified considerably because of the effects of, for example, coagulation. Moreover, these rates have been evaluated for homogeneous clouds – generally slabs or spheres. In reality, clouds are very inhomogeneous and, except for dense cloud cores, any spot within a cloud may have some sight-lines with high optical depth while others are practically optically thin. FUV photons can then penetrate much deeper than naively expected from the physical depth. In some cases, absorption of FUV photons by (abundant) atoms or molecules can also be important. This will lead to an extra reduction factor $\beta(\tau)$ in the expression for the photodissociation rate, with β the probability that a UV photon will not be absorbed by these species along the way to the (optical) depth, τ, from the surface.[1] As for the dust attenuation term, this β will also depend on the geometry and the clumpiness of the cloud. Absorption by the molecule under consideration is a special case of this. Such self-shielding is of particular importance for H_2 and for CO, which are dissociated through line absorption. For H_2, this is discussed in some detail in Section 8.7.1.

Some FUV photons will be present even deep within dense cloud cores. In particular, primary electrons produced by cosmic rays can still have enough energy to excite H_2. The electronically excited H_2 will decay radiatively. Because cosmic rays have a long penetration depth into a molecular cloud, this leads to a constant source function and FUV photon intensity (per H atom) in the cloud. The photodissociation rate of species m due to this source of photons is

$$k_{\mathrm{pd}}(\mathrm{CR}, m) = \frac{p_m}{1-\omega}\,\xi_{\mathrm{CR}}, \tag{4.6}$$

with ω the FUV dust albedo ($\simeq$0.5) and p_m the efficiency for the dissociation of molecule m. These efficiencies really express the competition between dust absorption and molecular photodissociation for these locally created FUV photons,

$$p_m\,\xi_{\mathrm{CR}} = 4\pi \int \frac{\sigma_{\mathrm{pd},m}(\nu)}{\sigma_{\mathrm{d}}(\nu)}\,\mathcal{N}_{\mathrm{CR}}(\nu)\,\mathrm{d}\nu, \tag{4.7}$$

where $\sigma_{\mathrm{pd},m}(\nu)$ is the photodissociation cross section of species m and $\mathcal{N}_{\mathrm{CR}}(\nu)$ is the mean photon intensity produced by the cosmic rays, which is $\simeq$400 photons cm^{-2} s^{-1} sr^{-1} or about 5×10^{-4} of the average interstellar radiation field. Table 4.3 provides some typical values for these efficiencies.

[1] We encountered the escape probability in Section 2.2 as the probability that a photon created at depth τ makes it to the surface and escapes. This is the reverse situation.

Table 4.3 *Cosmic-ray driven photo reaction rates*[a]

species	p_m
OH	509
H_2O	971
CO	7[b]
CO_2	1708
H_2CO	2660
C_2H_2	5155
HCN	3114

[a] Photodissociation rate given as $k_{pd} = \xi_{CR}\, p_m/(1-\omega)$.
[b] For CO, self-shielding rather than dust is the limiting factor. The photodissociation rate is $\xi_{CR}\, p_m$.

4.1.2 Neutral–neutral reactions

Neutral–neutral reactions are reactions of the type

$$A + B \rightleftharpoons C + D + \Delta E, \qquad (4.8)$$

where it is assumed that the reaction is exothermic to the right. The thermochemistry of a reaction, describing the heat flow, can be evaluated from the enthalpy (H) difference between products and reactants

$$\Delta H(\text{reaction}) = H(\text{products}) - H(\text{reactants}), \qquad (4.9)$$

where a negative enthalpy corresponds to release of energy by a reaction (e.g., $\Delta E = -\Delta H$; for an exothermic reaction $\Delta H < 0$; for an endothermic reaction $\Delta H > 0$). The enthalpy of a species is not just the bond energy but also involves the internal energy and the work done on the environment. Enthalpy is always measured relative to a standard state. In evaluating the enthalpy change of a reaction, the stoichiometry of the reaction, the phases of the reactants and products, and the temperature of the reaction have to be kept in mind. Finally, the enthalpy change does not depend on the reaction pathway and hence can be determined from the heat of formation and the thermodynamic properties of the species. For stable molecules, heats of formation are generally negative quantities, which implies that the formation of a compound from its elements is usually an exothermic process. For atoms, the heat of formation can be derived from the bond strength if the compounds are diatomic gases in their standard state (e.g., H_2, O_2, N_2). For those elements that are solid in their standard state (e.g., C), the heat of formation can

Table 4.4 *Neutral–neutral reactions* [a]

reaction	α	β	γ
$H_2+O \rightarrow OH+H$	9.0(−12)	1.0	4.5(3)
$H+OH \rightarrow O+H_2$	4.2(−12)	1.0	3.5(3)
$H_2+OH \rightarrow H_2O+H$	3.6(−11)		2.1(3)
$H+H_2O \rightarrow OH+H_2$	1.5(−10)		1.0(4)
$H+O_2 \rightarrow OH+O$	3.7(−10)		8.5(3)
$OH+O \rightarrow O_2+H$	4.0(−10)		6.0(2)
$H_2+C \rightarrow CH+H$	1.2(−9)	0.5	1.4(4)
$H+CH \rightarrow C+H_2$	1.2(−9)	0.5	2.2(3)
$C^++H_2 \rightarrow CH^++H$	9.4(−12)	1.25	4.7(3)

[a] Reaction rates of the form $k = \alpha\,(T/300)^{\beta}\,\exp[-\gamma/kT]$.

be evaluated from vapor pressure data. For ions the enthalpy can be derived from their ionization potentials. The heats of formation at 0 K for some astrophysically relevant species are listed in Table 4.5. These have to be transferred to the temperature of the gas under consideration using the specific heat of the species but that correction is generally small for astrophysical conditions.

Neutral–neutral reactions often possess appreciable activation barriers because of the necessary bond breaking associated with the molecular rearrangement. The forward and backward reaction rates, $k_{\rm f}$, and $k_{\rm b}$, are related through the equilibrium constant, K,

$$K = \frac{k_{\rm f}}{k_{\rm b}} \exp[\Delta E/kT]. \tag{4.10}$$

Ignoring, for the moment, the activation energy involved, the forward rate can be estimated from the classical "hard-collision" cross section; i.e., the collision where the impact parameter is small enough that the attractive interactive forces can overcome the angular momentum of the collision. For neutral–neutral collisions, the attractive interaction is due to van der Waals forces,

$$V(r) = -\frac{\epsilon_1 \epsilon_2}{r^6}\, I, \tag{4.11}$$

with ϵ_i the polarizability of the species involved and I the harmonic mean of the ionization potentials. The rate coefficient is then

$$k_{\rm f} = 13.5\,\pi \left(\frac{\epsilon_1 \epsilon_2 I}{\mu}\right)^{1/3} < v^{1/3} > \simeq 4\times 10^{-11} {\rm cm}^3\,{\rm s}^{-1}, \tag{4.12}$$

with μ the reduced mass, and we have adopted typical values ($\epsilon = 10^{-24}$ cm^3, $\mu = 3\times 10^{-24}$ g, $I = 13.6$ eV, and $T = 100$ K). If there is an activation barrier involved,

Table 4.5 *Heat of formation at 0 K*

species	H_0 [a] (kJ mole^{-1})
H_2	0.0
H	216.0
H^+	1528.0
H^-	143.2
H_3^+	1107.0
O_2	0.0
O	246.8
OH	38.4
H_2O	−238.9
C	711.2
CH	592.5
CH_2	390.0
CH_3	149.0
CH_4	−66.8
C_2H_2	228.6
CO	−113.8
CO_2	−393.1
N_2	0.0
NH	376.5
NH_2	191.6
NH_3	−38.9
HCN	135.5
HCO	44.8
H_2CO	−104.7
CH_3OH	−190.7
HCO^+	825.6

[a] The molar heat of formation of a compound is equal to the enthalpy change when one mole of compound is formed at 0 K and 1 atm from elements in their stable form.

this reaction rate has to be multiplied by the Boltzmann factor, $\exp[-E_a/kT]$. Even a modest barrier of 1000 K makes a reaction prohibitive at the 100 K typical for the diffuse ISM, let alone at the 10 K characteristic of molecular clouds. Such neutral–neutral reactions can be of importance when the gas is warm; e.g., in stellar ejecta, in hot cores associated with protostars, in dense photodissociation regions associated with luminous stars, or in shocks. Table 4.4 provides some examples of astrophysically relevant reactions. In general, the only neutral–neutral reactions that are of consequence for the cold conditions of dark clouds are those involving atoms or radicals, often with non-singlet electronic ground states that do not have activation barriers. Some examples are listed in Table 4.6.

Table 4.6 *Radical reactions*[a]

reaction	α	β
$C+OH \rightarrow CO+H$	1.1 (−10)	0.5
$C+O_2 \rightarrow CO+O$	3.3 (−11)	0.5
$O+CH \rightarrow CO+H$	4.0 (−11)	0.5
$O+CH \rightarrow HCO^+ + e$	2.0 (−11)	0.44
$O+CH_2 \rightarrow CO+H+H$	2.0 (−11)	0.5

[a] Reaction rates are of the form $k = \alpha\,(T/300)^{\beta}$.

4.1.3 Ion–molecule reactions

Exothermic ion–molecule reactions, as a rule, occur rather rapidly because the strong polarization-induced interaction potential can be used to overcome any activation energy involved. For an ion and a neutral, the induced-dipole interaction potential is given by

$$V(r) = -\frac{\alpha\, e^2}{2r^4}, \tag{4.13}$$

with α the polarizability of the neutral species ($\sim 10^{-24}$). The "hard" collision reaction rate coefficient – the so-called Langevin rate – is then

$$k_{\rm L} = 2\pi\, e \left(\frac{\alpha}{\mu}\right)^{1/2}, \tag{4.14}$$

with μ the reduced mass. Typically, $k_{\rm L} \simeq 2\times 10^{-9}\,{\rm cm^3\,s^{-1}}$ with no temperature dependence (cf. Table 4.7). This is some two orders of magnitude faster than neutral–neutral reactions, even disregarding the activation barriers generally involved in the latter. It is clear, then, that a small amount of ionization can be very effective in driving interstellar chemistry. If the ion–molecule reaction involves a neutral species possessing a permanent dipole, then the reaction rate coefficient can be much larger than the Langevin rate coefficient. If we assume that the dipole becomes locked in a close collision with an ion, the reaction rate coefficient is given by average dipole orientation theory as

$$k_{\rm dip} = 2\pi e \left(\left(\frac{\alpha}{\mu}\right)^{1/2} + \mu_{\rm D} \left(\frac{2}{\pi \mu k T}\right)^{1/2} \right), \tag{4.15}$$

with $\mu_{\rm D}$ the dipole moment of the neutral molecule and α its polarizability. This leads to an enhanced rate compared with the Langevin rate, particularly at low temperatures with values at 10 K of up to $10^{-7}\,{\rm cm^3\,s^{-1}}$ for a strong dipole.

Proton transfer reactions from one species to another are of particular relevance. These will occur if the proton affinity of the "receptor"-species is larger than that

Table 4.7 *Ion–molecule reactions*

reaction	α
$H_2^+ + H_2 \rightarrow H_3^+ + H$	2.1 (–9)
$H_3^+ + O \rightarrow OH^+ + H_2$	8.0(–10)
$H_3^+ + CO \rightarrow HCO^+ + H_2$	1.7 (–9)
$H_3^+ + H_2O \rightarrow H_3O^+ + H_2$	5.9 (–9)
$OH^+ + H_2 \rightarrow H_2O^+ + H$	1.1 (–9)
$H_2O^+ + H_2 \rightarrow H_3O^+ + H$	6.1(–10)
$C^+ + OH \rightarrow CO^+ + H$	7.7(–10)
$C^+ + H_2O \rightarrow HCO^+ + H$	2.7 (–9)
$CO^+ + H_2 \rightarrow HCO^+ + H$	2.0 (–9)
$He^+ + CO \rightarrow C^+ + O + He$	1.6 (–9)
$He^+ + O_2 \rightarrow O^+ + O + He$	1.0 (–9)
$He^+ + H_2O \rightarrow OH^+ + H + He$	3.7(–10)
$He^+ + H_2O \rightarrow H_2O^+ + He$	7.0(–11)
$He^+ + OH \rightarrow O^+ + H + He$	1.1 (–9)

[a] Reaction rates are of the form $k = \alpha$.

of the "donor"-species. Table 4.8 provides a compilation of proton affinities for some astrophysically relevant species taken from the NIST database. The rates are then well reproduced by the Langevin or average dipole orientation theory.

Ion–molecule reactions are ultimately driven by cosmic-ray ionization. Some relevant cosmic-ray ionization rates are summarized in Table 4.9.

4.1.4 Charge transfer reactions

Charge transfer involving an atom can be important in setting the ionization balance. Rates can be of the order of 10^{-9} cm^3 s^{-1} in the exothermic direction when there is an energy level in the product ion that is resonant (within ~ 0.1 eV) with the recombination energy of the incoming ion and the Franck–Condon factor connecting the upper and lower states is large.[2] Reactions involving only atoms are important for the ionization balance of trace species in HII regions (cf. Section 7.2.3). The charge exchange reaction between O and H^+ is of particular importance in this regard. This reaction is also important in driving interstellar chemistry because it transfers ionization to oxygen which can then readily participate in the chemistry. Because the ionization potentials are close, this reaction has a large rate coefficient. Charge transfer reactions involving other neutral atoms are

[2] The Franck–Condon principle states that a transition is favored when there is a large overlap between the vibrational wave functions of the initial and final state; e.g., this assumes that the transition occurs while the nuclei are stationary.

Table 4.8 *Proton affinities*

Species	PA[a] (kJ mol^{-1})[b]
H_2	422.3
C	623.0
N	405.0
O	485.0
Si	836.0
S	656.0
CH	731.0
CH_2	836.0
CH_3	673.0
CH_4	534.5
C_2	694.0
C_2H	753.0
C_2H_2	641.4
C_6H_6	750.4
$C_{10}H_8$	1802.9
$C_{24}H_{12}$	861.3
CO	594.0
CO_2	540.5
H_2CO	712.9
CH_3OH	754.3
CH_3CN	779.2
HCOOH	742.0
CH_3OCH_3	792.0
NH	589.0
NH_2	769.0
NH_3	853.6
N_2	493.8
HCN	712.9
HNC	772.3
HC_3N	751.2
OH	593.2
H_2O	691.0
O_2	421.0
SH	690.0
H_2S	705.0

[a] The value of the proton affinity is the negative of the enthalpy change in the proton transfer reaction involving the given neutral species.

[b] Conversion factor to eV molecule^{-1} and to K molecule^{-1}: multiply by 1.04×10^{-2} and 1.2×10^{2}, respectively.

rarely important. Charge transfer reactions between atoms and molecular species will have larger cross sections because of the larger number of electronic states available. Of course, for large molecules, other reaction channels (e.g., proton transfer) may also open up.

Table 4.9 *Cosmic-ray ionization rate*

Species	ξ_0[a]
H_2	2.3
H	1.1
He	1.2
C	4.4
N	5.2
O	6.5
CO	7.5

[a] Reaction rates are of the form $k = \xi_0 \zeta_{CR}$, with ζ_{CR} the primary cosmic-ray ionization rate.

4.1.5 Radiative association reactions

In radiative association reactions, the collision product is stabilized through the emission of a photon. With a radiative lifetime of $\tau_{rad} \simeq 10^{-7}$ s for an allowed transition, and a collision time scale of $\tau_{col} \simeq a/v \simeq 10^{-13}$ s (a the distance over which the Einstein A is large and v the collision velocity), the efficiency of the radiative association process is 10^{-6}. The efficiency can be greatly enhanced if the collision partners form a long-lived activated complex that then stabilizes through the emission of a photon. Of course it is most probable that the activated complex still redissociates. In laboratory settings, collisions with other species, rather than radiation, dominate stabilization. Hence, the rates of these reactions are generally estimated through theoretical calculations.

The reaction can be represented by

$$\mathrm{A} + \mathrm{B} \underset{k_d}{\overset{k_a}{\rightleftarrows}} \mathrm{AB}^* \xrightarrow{k_r} \mathrm{AB}, \tag{4.16}$$

with A and B the reactants and AB* the activated complex. The overall radiative association rate, k_{ra}, for the reaction is then given by

$$k_{ra} = k_a \left(\frac{k_r}{k_r + k_d} \right) \simeq k_r \left(\frac{k_a}{k_d} \right), \tag{4.17}$$

with k_a, k_d, and k_r the association, dissociation, and radiative stabilization rate of the activated complex. The collision complex will temporarily store the kinetic energy and the interaction energy in vibrational energy. Stabilization then occurs through emission in a vibrational transition with a radiative rate of typically 10^3 s^{-1}. The forward and the backward rates are related through detailed balance,

$$k_a \Phi_A \Phi_B = k_d \Phi_{AB^*}, \tag{4.18}$$

where the Φs are the partition functions per unit volume of the colliding species and the excited complex. This corresponds to the ergodic assumption; i.e., the energy is divided over all available modes of the collision complex. At the temperatures we are concerned with, the partition functions of the colliding species are proportional to the rotational partition functions. The partition function of the activated complex depends on the density of states of the activated complex at the dissociation energy of the complex. Consider a collision complex consisting of s harmonic oscillators of frequency ν and an internal energy equivalent to n quanta. Now, assuming that dissociation will occur when all these quanta are in any one mode, we can write for the probability that this occurs

$$P = \frac{n!(s-1)!}{(n+s-1)!}. \tag{4.19}$$

Using the frequency of the oscillator, this can be converted into a lifetime for the activated complex,

$$k_{pd}^{-1} = \frac{(n+s-1)!}{n!\,(s-1)!}\,\nu^{-1}. \tag{4.20}$$

With $s = 9, n = 5$, and $\nu = 5 \times 10^{13}\,\mathrm{s}^{-1}$, we find $k_{pd}^{-1} \simeq 2 \times 10^{-11}$ s, which is considerably longer than the direct collision time scale (10^{-13} s). Hence association can be greatly enhanced. For ion–molecule reactions, k_a is $2 \times 10^{-9}\,\mathrm{cm^3\,s^{-1}}$ and the overall rate is 10^{-16}–$10^{-17}\,\mathrm{cm^3\,s^{-1}}$. The density of states is a rapidly rising function of the internal energy of the species and of the number of atoms in the species (i.e., the number of modes over which this energy can be divided). Thus, a more tightly bound species or a larger species will lead to a longer lived complex and hence a higher radiative association rate. Some important radiative association reactions are summarized in Table 4.10. These types of radiative association reactions, and particularly their inverse, unimolecular photodissociation reactions, are of great importance for the photochemistry of large polycyclic aromatic hydrocarbon molecules in the interstellar medium and we will come back to this in Section 6.6.

4.1.6 Dissociative electron recombination reactions

Dissociative recombination involves the capture of an electron by an ion to form a neutral in an excited electronic state that can dissociate. Rate coefficients are

Table 4.10 *Radiative association reactions*

reaction	α [a]
$H + C \rightarrow CH$	1.0 (−17)
$C^+ + H \rightarrow CH^+$	1.7 (−17)
$C^+ + H_2 \rightarrow CH_2^+$	6.0 (−16)

[a] Reaction rates are of the form $k = \alpha$.

Table 4.11 *Electron recombination reactions*[a]

reaction	α	β
$OH^+ + e \rightarrow O + H$	3.8 (−8)	−0.5
$CO^+ + e \rightarrow C + O$	2.0 (−7)	−0.5
$H_2O^+ + e \rightarrow O + H + H$	2.0 (−7)	−0.5
$H_2O^+ + e \rightarrow OH + H$	6.3 (−8)	−0.5
$H_2O^+ + e \rightarrow O + H_2$	3.3 (−8)	−0.5
$H_3O^+ + e \rightarrow H_2O + H$	3.3 (−7)	−0.3
$H_3O^+ + e \rightarrow OH + H + H$	4.8 (−7)	−0.3
$H_3O^+ + e \rightarrow OH + H_2$	1.8 (−7)	−0.3
$H_3^+ + e \rightarrow H_2 + H$	3.8 (−8)	−0.45
$H_3^+ + e \rightarrow H + H + H$	3.8 (−8)	−0.45
$HCO^+ + e \rightarrow CO + H$	1.1 (−7)	−1.0
$CH^+ + e \rightarrow C + H$	1.5 (−7)	−0.4
$CH_2^+ + e \rightarrow CH + H$	1.4 (−7)	−0.55
$CH_2^+ + e \rightarrow C + H + H$	4.0 (−7)	−0.6
$CH_2^+ + e \rightarrow C + H_2$	1.0 (−7)	−0.55
$CH_3^+ + e \rightarrow CH_2 + H$	7.8 (−8)	−0.5
$CH_3^+ + e \rightarrow CH + H + H$	2.0 (−7)	−0.4
$CH_3^+ + e \rightarrow CH + H_2$	2.0 (−7)	−0.5

[a] Electron recombination rate coefficients are given as $k_{rec} = \alpha(T/300)^{\beta}$.

typically $10^{-7}\,cm^3\,s^{-1}$. For polyatomic molecules, the neutral products and branching ratios are generally not well known. Some typical electron recombination rates are listed in Table 4.11. The temperature dependence ranges from $T^{-1/3}$ to T^{-1}. For large molecular species such as PAHs, the electronic excitation energy of the neutral product will quickly transfer to the vibrational manifold. This vibrational energy will then be radiated away through IR vibrational emission and their electron recombination rates are much higher.

4.1.7 Collisional association and dissociation reactions

In laboratory settings, three-body reactions generally dominate chemistry,

$$A + B + M \longleftrightarrow AB + M, \tag{4.21}$$

with rates $\simeq 10^{-32}\,\mathrm{cm^6\,s^{-1}}$. These reactions are generally of little importance in astrophysical environments except for dense gas near stellar photospheres or in dense ($\simeq 10^{11}\,\mathrm{cm^{-3}}$) circumstellar disks.

4.1.8 Associative detachment reactions

In associative attachment, an anion and an atom collide and the neutral product stabilizes through electron emission,

$$A + B^- \longrightarrow AB + e. \tag{4.22}$$

Generally, anions play little role in the chemistry of the ISM. The reaction

$$H + H^- \longrightarrow H_2 + e, \tag{4.23}$$

is an exception, with a rate of $k = 1.3 \times 10^{-9}\,\mathrm{cm^3\,s^{-1}}$. This may well be the only way molecular hydrogen was made in the early Universe (cf. Section 8.7.1).

4.1.9 Non-LTE effects

In this discussion, we have always tacitly assumed that the various degrees of freedom are in equilibrium and can be described by a single temperature. However, in general, the interstellar medium is far out of equilibrium. With regard to the chemical processes, this can be of importance if a reaction with a high activation barrier is promoted by internal vibrational excitation of one of the species. For example, in photodissociation regions, molecular hydrogen can be highly vibrationally excited due to FUV pumping in the Lyman–Werner bands. The reactions $O + H_2 \rightarrow OH + H$ and $C^+ + H_2 \rightarrow CH^+ + H$ are notable examples that may well then go at the collision rate of these species with vibrationally excited molecular hydrogen. In contrast, in molecular clouds, the excitation of molecules may be much less than the kinetic temperature of the gas would predict if the density is below the critical density. In that case, reaction rate coefficients may be much less than determined in the laboratory. This should be kept in mind when evaluating the effects of neutral–neutral reactions in the warm post-shock gas.

Reactions may also be promoted by electronic excitation. Fine-structure excitations of O, C, and Si are of interest. The interaction of these species with neutral radicals is dominated by long-range forces involving the atomic quadrupole. These interactions can suppress barriers in the potential energy surfaces that result from chemical interactions at intermediate separations. The reactivity is then determined by the quadrupole moment of the lowest spin-orbit states and, hence, is sensitive to the population of the fine-structure levels. The rate coefficients can then be enhanced by an order of magnitude compared with the usual extrapolations of laboratory studies (at 300 K) to low T using a $T^{1/2}$ dependence of neutral–neutral reactions.

The ortho–para ratio of molecular hydrogen is another example where non-LTE effects are often important. Various processes influence the abundance ratio of ortho (parallel nuclear spins) to para (antiparallel nuclear spins) H_2. These include chemical reactions in the gas phase and physical spin flipping reactions on grain surfaces. Because the gas and grain temperatures are generally out of equilibrium, the ortho-to-para ratio is also not well described by the kinetic temperature of the gas. A notable reaction for which this may be of importance is the reaction of N^+ with H_2, which is the bottleneck in the formation of ammonia. This reaction has a 170 K barrier for H_2 in its lowest vibrational and rotational state. For para H_2, this barrier is important but it is of no consequence for ortho H_2.

Finally, some reactions are promoted by translational energy. Now, generally, gas species have a Maxwellian velocity distribution, which is rapidly established through collisions with the gas; e.g., H_2 molecules. The excess energy associated with exothermic reactions may leave the reaction products translationally hot. If the reaction between this reaction product and H_2 is promoted by excess translational energy then this reaction may occur at a much faster rate than the kinetic temperature would indicate. As an example, the reaction $He^+ + N_2 \rightarrow N^+ + N + 0.29\,eV$ provides N^+ with 0.15 eV of kinetic energy and this helps the reaction $N^+ + H_2 \rightarrow NH^+ + H$, which has a small energy barrier (170 K). All of these effects have to be carefully evaluated for interstellar chemistry.

4.1.10 Gas-phase chemistry networks

Based upon these chemical considerations, a few general rules governing gas phase chemical networks can be formulated. If an ion can react with H_2, those reactions will take precedence over all other reactions because of abundance considerations. If not, electron recombination generally dominates the loss channel for ions. If electron reactions are also inhibited, loss through reactions with neutrals will take over. For neutrals, if they can occur, reactions with ions will dominate. Because of abundance considerations, the important ions in diffuse clouds are C^+, H^+, and He^+, while in molecular clouds H_3^+, HCO^+, H_3O^+, and He^+ take over. If the neutral does not react with these species, neutral–neutral reactions with small radicals (atomic or diatomic) become important. The formation of neutrals is often dominated by electron recombination of an ion. Note that, in the general scheme of things, this will only be important if that ion does not react with H_2.

Detailed gas phase chemical networks for various astrophysical environments will be discussed in Sections 8.7, 9.5, 10.4–6, and 11.2.2. Here, we emphasize that involving ions in the reaction network carries a huge premium in terms of driving molecular complexity. For that reason, the small amount of ionization in molecular clouds caused by penetrating cosmic-ray ions takes on great significance for astrochemistry. Furthermore, molecular build-up in the interstellar medium starts off with an atomic gas. Given the low rates of radiative association reactions, the first step combining two atoms into a diatomic species forms a bottleneck in the gas phase. Here, chemistry on grain surfaces comes to the rescue. The grains act as an extra body, readily accepting excess energy and momentum. This is particularly true for molecular hydrogen formation. Once molecular hydrogen has formed on grain surfaces, gas-phase chemistry can really take off. The physical and chemical principles involved in grain-surface chemistry are discussed in the next section.

Once a set of chemical species and the reactions among them have been selected, differential equations for the abundance of each of the species can be written,

$$\frac{dn(i)}{dt} = -n(i)\sum_j n(j)k_{ij} + \sum_{j,k} n(j)n(k)k_{jk} + \sum_j n(j)k_j - n(i)\sum_j k_j, \tag{4.24}$$

where the first sum includes all coreactants for species i while the second sum pertains to all bimolecular reactions resulting in the formation of species i, and the last two sums refer to unimolecular reactions (e.g., photodissociation, photoionization, or cosmic-ray ionization) forming or destroying species i. In the steady state, this results in a set of quadratic equations,

$$\mathcal{F}(\mathbf{n}) = 0, \tag{4.25}$$

where the vector $\mathbf{n}$ contains the abundances of the species. Because in each reaction, formation and destruction balance, this set of equations is dependent (e.g., each term appears as often with a plus sign as with a minus sign in this set of equations). However, for each element, we can replace one equation by a conservation equation. In addition, we can replace one equation by charge conservation. The resulting set of equations can be solved iteratively using standard Newton–Rhapson methods,

$$\mathbf{n}^{j+1} = \mathbf{n}^j - \mathcal{J}^{-1}\mathcal{F}, \tag{4.26}$$

where $\mathcal{J}$ is the Jacobian of this set of equations and $\mathbf{n}^j$ and $\mathbf{n}^{j+1}$ contain the current and an improved estimate of the abundances. Adopting initial guesses for the abundances, this set of equations is readily iterated upon for a given temperature, density, UV radiation field, cosmic-ray ionization rate, and set of elemental abundances. Convergence depends, of course, on the initial guesses, but is generally rapid. Also, in chemistry, sometimes more than one (physical) solution is possible. Of course, many unphysical solutions are also available – e.g., where one or more abundances are negative – and care should be taken to avoid those in numerical schemes.

More generally, the time-dependent evolution of the chemical abundances is followed. In that case, the set of differential equations (4.24) is integrated starting from some initial value – often appropriate for diffuse atomic clouds – and adopting values for the physical conditions. Of course, the evolution of the physical conditions must be followed simultaneously. It should be recognized that the time scales for the reactions in a realistic chemical network can be very different. In particular, cosmic-ray ionization reactions and H_2 formation on grain surfaces are very slow compared with ion–neutral reactions. This results in a set of stiff differential equations, which have to be solved using special numerical methods such as the Gears method. Routines for this are available in numerical libraries. Alternatively, and sometimes more practically, the chemical set of species can be divided into species whose abundances evolve slowly and those that are in rapid "equilibrium" with their environment. Cases in point are species whose abundances are controlled by proton transfer, reaction with H_2, or dissociative electron recombination reactions. In that case, the abundances of slowly evolving species can be explicitly followed by solving the relevant differential equations while the abundance of rapidly evolving species can be determined solving a set of non-linear equations. Numerical routines also exist for solving a set of differential equations and non-linear equations simultaneously. In either case, the size of time-steps should be carefully chosen to keep the propagation of errors under control. Conservation equations provide helpful checks on the accuracy.

As mentioned above, sometimes more than one physical solution is available in steady state. Time history provides, then, the guide to which solution will be realized starting from given initial conditions. When more than one solution is available, the system may start to oscillate between the possible states or, even, develop chaotic behavior. This aspect will not be covered in this book, but the reader should beware that, in general, chemistry can be prone to chaotic behavior.

4.2 Grain-surface chemistry

Interstellar grains provide a surface on which accreted species can meet and react and to which they can donate excess reaction energy (Fig. 4.1). Grain surface chemistry is thus governed by the accretion rate – which sets the overall time

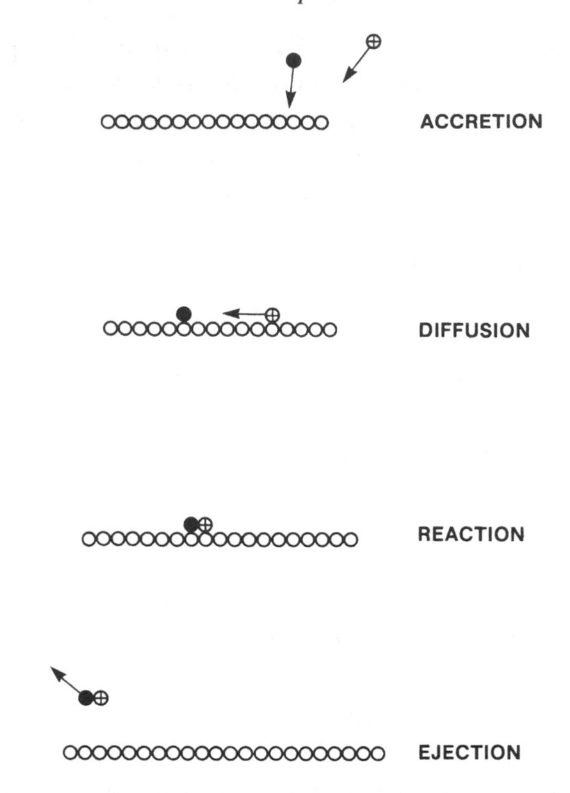

Figure 4.1 A schematic of the formation of molecules on grain surfaces. Gas phase species accrete, diffuse, and react on an interstellar grain surface.

scale for the process – and the surface migration rate – which governs the reaction network. We will briefly discuss these "steps" and the parameters that control them.

4.2.1 Accretion

The accretion rate of species on grains is given by

$$k_{\rm ac} = n_{\rm d}\sigma_{\rm d} v S(T, T_{\rm d}) \simeq 10^{-17}\left(\frac{T}{10\,\rm K}\right)^{1/2} n\,\rm s^{-1}, \qquad (4.27)$$

where T and $T_{\rm d}$ are the gas and dust temperature and a mean mass appropriate for CO has been assumed. The sticking coefficient depends on the species accreting, the thermal velocity of the gas, and the excitation of the phonon spectrum of the grain, as well as the interaction energy of the gas phase species and the surface. Except for atomic H, this sticking coefficient is expected to be close to unity at the low temperatures of interstellar molecular clouds. Based upon experiments and theoretical models, the sticking coefficient for H on grain surfaces has been evaluated to be given by

$$S(T, T_d) = \left[1 + 4\times 10^{-2}(T + T_d)^{1/2} + 2\times 10^{-3}T + 8\times 10^{-6}T^2\right]^{-1}, \quad (4.28)$$

and is 0.8 at 10 K, dropping to 0.5 at 100 K. The time scale at which gas phase species deplete out onto grains is $4\times 10^9/n$ years or as short as 4×10^5 years in a dense core. This is much shorter than the dynamical evolution time scale of molecular clouds ($\simeq 3\times 10^7$ years for giant molecular clouds and $\sim 10^6$ years for dense cores the [free-fall time scale]). This depletion time scale is quite comparable to the chemical evolution time scale of dense cores. We can also evaluate the time scale for species to arrive on a particular grain,

$$\tau_{\rm ar} = (n_i \sigma_d v)^{-1} \simeq 3 \left[\frac{10^4\,{\rm cm}^{-3}}{n}\right]\left[\frac{1000\,{\rm Å}}{a}\right]^2 \text{ days}, \quad (4.29)$$

where this equation has again been evaluated for CO and a gas temperature of 10 K. So, in a dense core, this time scale is typically a few days.

4.2.2 Binding energies

A gas phase species approaching a surface will feel at a large distance a weak attraction due to van der Waals forces. These are due to mutually induced dipole moments in the electron shells of the gas phase species and the atoms in the surface. At short range, forces associated with the overlap of the wave functions of the approaching species and the surface atoms lead to much stronger binding. Thus, we can recognize weak physisorbed sites with well depths in the range 0.01–0.2 eV (100–2000 K) at a distance of a few angstroms and chemisorbed sites with a strength of $\sim$1 eV (10 000 K) close to the surface (Fig. 4.2). The interaction potential depends not only on the distance to the surface but also on the location on the surface. A perfect crystal will have a regular variation of the potential energy across the surface with an array of wells evenly spaced in between the surface atoms. In a disordered material, this regularity in the location is lost. Moreover, the wells may show a spread in their binding properties. In either case, the topology resembles a mountainous terrain with peaks associated with the surface atoms and valleys representing the physisorbed (at high altitudes) and the chemisorbed (deepest) wells. These valleys are separated by saddle points, which present activation barriers to the motion of the adsorbate across the surface as well as from physisorbed to chemisorbed wells (see Fig. 4.2).

Typical interaction energies for atomic H on astrophysically relevant surfaces are summarized in Table 4.12. Somewhat larger binding energies can be expected

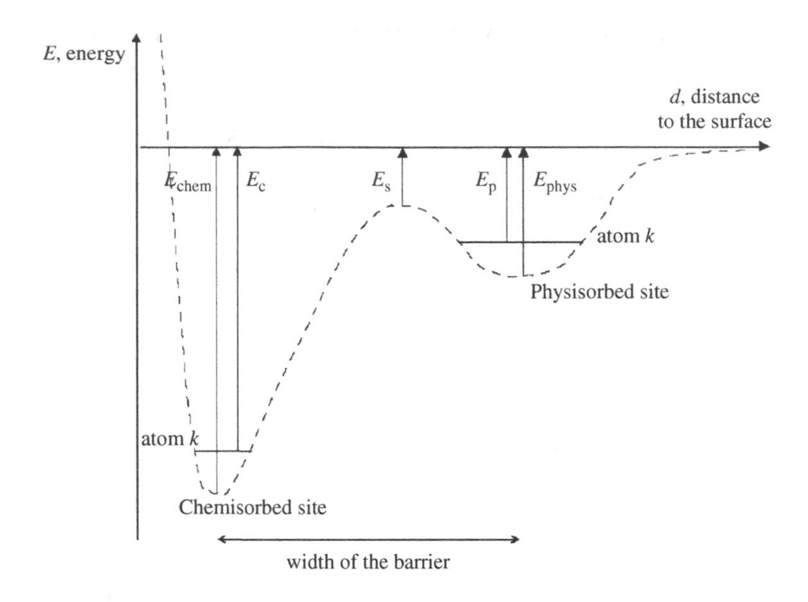

Figure 4.2 The interaction between an adsorbate and a surface as a function of the distance to the surface. Two types of sites can be recognized: physisorbed sites due to van der Waals interaction (binding energy, E_{phys}) and chemisorbed sites involving shared electrons (binding energy, E_{chem}). The actual binding energies of a species, E_p and E_c, take the zero-point energy into account. The two types of sites are separated by a saddle point with energy, E_s.

Table 4.12 *H–surface interaction energies*

surface	Energy(eV)	
	physisorbed	chemisorbed
graphite	0.06	~ 2.5
silicate	0.05	~ 2.5

for H physisorbed in defect sites such as kinks, ledges, and corners. When the surface is partly covered by chemisorbed species, the chemisorption energy may be either somewhat weakened or strengthened, depending on adsorbate and surface, due to the electron redistribution associated with the presence of adsorbates. In molecular clouds, chemisorbed sites on grain surfaces will be quickly covered with a (strongly bonded) layer of "ice" on which species will only be physically adsorbed. Thus, for ice mantle formation, only physisorption is relevant. Table 4.13 summarizes binding energies for some astrophysically relevant species on ices.

Table 4.13 *The interaction of atoms and molecules with an H_2O surface*

species	E_b^a (K)	10 K		30 K	
		τ_{ev}^b (s)	τ_m^c (s)	τ_{ev}^b (s)	τ_m^c (s)
H	350	1.6 (3)	1 (−12)	1 (−7)	1 (−12)
H_2	450	3 (7)	4 (−12)	3 (−6)	4 (−12)
C	800		1 (−2)	4 (−1)	2 (−9)
N	800		1 (−2)	4 (−1)	2 (−9)
O	800		1 (−2)	4 (−1)	2 (−9)
S	1100		1 (2)	8 (13)	4 (−8)
CO	1900		6 (12)		2 (−4)
N_2	1700		1 (10)		2 (−5)
O_2	1600		7 (8)		9 (−6)
CH_4	2600				2 (−1)
H_2O	4000				2 (5)

[a] Estimated binding energy on an H_2O surface.
[b] Evaporation time scale from an H_2O surface. No value given when it exceeds the molecular cloud lifetime.
[c] Migration time scale on an H_2O surface. No value given when it exceeds the molecular cloud lifetime.

The binding energy of one H_2 molecule to a clean H_2O surface is about 860 K, but decreases rapidly with H_2 surface coverage. As a result, an interstellar ice grain will rapidly accrete a (partial) H_2 layer until the H_2 accretion rate balances the evaporation rate. Experiments suggest that this occurs for a surface coverage of about 0.2 and a corresponding binding energy of 450 K. Because of its lower mass and polarizability, the binding energy of an H atom will be somewhat less than that of H_2 or about 350 K. Note that, as a result, the thermal evaporation time scale of H is coupled to the accretion rate of H_2. The binding energy of D will, likewise, be coupled to that of H_2. However, heavier species are unlikely to notice the presence of such a partial H_2 layer.

Surface reactions may now occur because surface species are mobile and "collisions" between them lead to a reaction – the so-called Langmuir–Hinshelwood mechanism – or because impinging gas phase species react directly with already adsorbed species – the Eley–Rideal mechanism. The former is of particular relevance and hence mobility between surface sites is at the center of grain-surface chemistry.

4.2.3 *Evaporation*

The residence time of a species on a grain surface is given by

$$\tau_{ev} = \nu_0^{-1} \exp(E_b/kT_d) \tag{4.30}$$

where ν_0 is the vibrational frequency of the adsorbed species on the grain surface. This is essentially the characteristic time scale for a species to acquire sufficient energy through thermal fluctuations to evaporate. Consider a species with mass m, adsorbed with an energy E_b, on a surface with a number density of surface sites N_s ($N_s \simeq 2\times10^{15}$ sites cm^{-2}). For a symmetric harmonic potential, the frequency of the adsorbed species in the surface site is

$$\nu_z = \left(\frac{2N_s E_b}{\pi^2 m}\right)^{1/2}. \tag{4.31}$$

For a binding energy of 1000 K, ν_z is about 5×10^{12} and 10^{12} s^{-1} for an H atom and an O atom, respectively. In a chemisorbed site with $E_b \simeq 20\,000$ K, these frequencies are 3×10^{13} and 7×10^{12} s^{-1}. An adsorbed species has replaced (at least) one degree of translational freedom by a vibrational degree of freedom of the bond with the surface. For a monatomic gas that is completely mobile on the surface, we can find ν_0 from thermodynamic analysis,

$$\nu_0 = \frac{kT}{h} f_z^{-1}, \tag{4.32}$$

where f_z is the vibrational partition function of the adsorbate in the mode perpendicular to the surface. When this mode is fully excited (e.g., high temperatures), f_z is just $kT/h\nu_z$ and $\nu_0 = \nu_z$. At low temperatures, when the surface bonding mode is not excited at all, $f_z = 1$ and ν_0 is $kT/h = 2\times10^{11}(T/10\,\text{K})$ s^{-1}. When the species is not mobile on the surface, Eq. (4.32) has to be modified to

$$\nu_0 = \frac{kT}{h}\frac{1}{f_x f_y f_z}\frac{2\pi m kT}{N_s h^2}, \tag{4.33}$$

where f_x, f_y, and f_z are the partition functions in the x, y, and z direction and the last factor takes into account the number of possible ways in which adsorbates can be distributed over the surface sites. This is a situation that is appropriate for CO on an ice surface. For an immobile species with $m = 30$ atomic mass units, we find at 10K, $\nu_0 = 1.4\times10^{12}$ s^{-1}. Thus, we will adopt here: ν_0 is 2×10^{13} s^{-1} for an H atom, ν_0 is equal to $kT/h \simeq 2\times10^{11}$ s^{-1} for C, N, and O, and 1.4×10^{12} s^{-1} for other species. The evaporation time scales are summarized in Table 4.13. Except for atomic H, accreted species will not evaporate over the lifetime of a molecular cloud at the low dust temperatures of molecular clouds ($T_d \simeq 10$ K). However, this is very sensitive to the dust temperature and at 30 K accreted atoms do evaporate quickly.

As emphasized in Section 4.2.2, the atomic H evaporation rate is coupled to the accretion rate of molecular hydrogen through the coverage dependent binding energy of H_2 and H. The evaporation time scale for atomic H is thus,

$$\tau_{\rm ev} = 3 \times 10^2 \frac{n}{10^4\,{\rm cm}^{-3}}\ {\rm s}, \quad (4.34)$$

independent of T for $T < 19$ K, corresponding to the H_2 binding energy on a clean H_2O surface.

4.2.4 Surface migration

Of importance for the reaction network is the time scale for migration of accreted species on a grain surface. At low temperatures, hydrogen (and deuterium) will migrate through quantum mechanical tunneling. The tunneling time through a rectangular barrier with height $E_{\rm m}$, and width a, is equal to

$$\tau_{\rm t} = \nu_0^{-1} \exp\left[\frac{2a}{\hbar}(2mE_{\rm m})^{1/2}\right], \quad (4.35)$$

with ν_0 the vibrational frequency of the species in the well and m the mass of the species. For H on an H_2O surface partially covered with H_2, the migration barrier is not well known. However, it should be less than 350 K. With $\nu_0 = 2 \times 10^{13}$ s^{-1} and $a = 1$ Å, the migration time scale is estimated to be 1.5×10^{-10} s for H. We can contrast this with the migration time scale of an H atom on a clean silicate surface (in the diffuse ISM), which is estimated from experiments to be some 10^{-2} s, reflecting the much larger binding energy.

For heavier species, tunneling is unimportant and the limiting step is, again, the "acquisition" of sufficient thermal energy. The thermal hopping time scale is given by

$$\tau_{\rm m} = \nu_{\rm m}^{-1} \exp[E_{\rm m}/kT_{\rm d}] \quad (4.36)$$

where $\nu_{\rm m}$ is the frequency for vibrational motion parallel to the surface, which is comparable to ν_0. The barrier against migration on the surface, $E_{\rm m}$, is typically $E_{\rm b}/3$. Migration time scales are also listed in Table 4.13.

This is the time scale to move to an adjacent site. Surface migration is a random walk process in two dimensions. After N steps, a species will have visited $N/\ln N$ distinct sites. The average time scale to visit all sites at least once on a surface with N_t sites is equal to $(N_t \ln N_t)\tau_{\rm m}$. With a typical density of sites of 2×10^{15} sites cm^{-2}, a 1000 Å grain has $\sim 2 \times 10^6$ sites. Thus, before evaporating, an H atom can be expected to visit all sites many times. In the diffuse ISM, the migration time scale of H on a clean silicate surface is 10^{-2} s, and an H atom will visit some 10^5 sites before the next H atom lands (H atom evaporation is unimportant at 10 K). For a C, N, or O atom, this surface scanning time scale is some 100 hours on a 1000 Å icy grain. Here too, the surface scanning time scale should be compared to the time scale to accrete the next species on the grain ($\sim$75 h in a molecular cloud).

4.2.5 Reactions involving radicals

Reactions involving species with unpaired electrons can be expected to occur upon "collision" on the surface. Of course, this presumes that one of the reactants is mobile on the surface. Thus, specifically, this includes reactions involving H, C, N, or O atoms with themselves or with radicals such as OH, CH, etc. A number of relevant reactions are listed in Table 4.14. This list is not complete, as longer and longer carbon backbones can be made through the sequential addition of carbon atoms. Also, D-isotopic reactions can be included analogously to those of H atoms.

Table 4.14
Radical–radical surface reactions

reactants		products
$H + O$	$\longrightarrow$	OH
$H + OH$	$\longrightarrow$	H_2O
$H + C$	$\longrightarrow$	CH
$H + CH$	$\longrightarrow$	CH_2
$H + CH_2$	$\longrightarrow$	CH_3
$H + CH_3$	$\longrightarrow$	CH_4
$H + N$	$\longrightarrow$	NH
$H + NH$	$\longrightarrow$	NH_2
$H + NH_2$	$\longrightarrow$	NH_3
$H + O_2H$	$\longrightarrow$	H_2O_2
$H + NO$	$\longrightarrow$	HNO
$H + CN$	$\longrightarrow$	HCN
$H + CNO$	$\longrightarrow$	$HCNO$
$H + HCO$	$\longrightarrow$	H_2CO
$H + HCOO$	$\longrightarrow$	$HCOOH$
$H + CH_3O$	$\longrightarrow$	CH_3OH
$H + NCHO$	$\longrightarrow$	$NHCHO$
$H + NHCHO$	$\longrightarrow$	NH_2CHO
$H + CCHO$	$\longrightarrow$	$CHCHO$
$H + CHCHO$	$\longrightarrow$	CH_2CHO
$H + CH_2CHO$	$\longrightarrow$	CH_3CHO
$H + N_2H$	$\longrightarrow$	N_2H_2
$O + O$	$\longrightarrow$	O_2
$O + N$	$\longrightarrow$	NO
$O + C$	$\longrightarrow$	CO
$O + CN$	$\longrightarrow$	OCN
$O + HCO$	$\longrightarrow$	$HCOO$
$C + N$	$\longrightarrow$	CN
$C + HCO$	$\longrightarrow$	$CCHO$
$N + N$	$\longrightarrow$	N_2
$N + NH$	$\longrightarrow$	N_2H
$N + HCO$	$\longrightarrow$	$NCHO$

4.2.6 Reactions of H atoms possessing activation barriers

Grain surfaces are the "watering holes" of astrochemistry where species come to meet and mate. Indeed the prolonged residence time on grain surfaces allows reactions that are inhibited in the gas phase to proceed owing to strong activation barriers and this is one of the distinguishing characteristic of grain surface reaction networks. Specifically, consider an H atom in a site where it can react with a neighboring species with a barrier of height E_a, and width a. The probability for reaction is now given by the competition between migration and the penetration of the reaction barrier,

$$p = \frac{p_0}{p_0 + p_m}, \tag{4.37}$$

where p_m is the inverse of the site-to-site migration time scale and p_0 is given by

$$p_0 = \nu_0 \exp\left[-\frac{2a}{\hbar}(2m\, E_a)^{1/2} \right]. \tag{4.38}$$

Typically, p_0 will be much less than p_m (unless the reaction activation barrier is smaller than the migration barrier) and the H atom will generally migrate to the next site. Now, reactions may still occur because H atoms scan the surface very quickly and revisit the site many times or because the surface coverage of coreactants is high. For a surface coverage of these coreactants for H atoms of θ_r, the number of times that a site allowing such a reaction to occur is visited is given by

$$n_r = \frac{\tau_{ev}}{\tau_m} \theta_r. \tag{4.39}$$

The probability that a reaction occurs the kth time such a site is visited is given by

$$p_k = (1-p)^{k-1}\, p. \tag{4.40}$$

The probability for a reaction before evaporation is, then,

$$p_r = \sum_{k=1}^{n_r} p_k = 1 - (1-p)^{n_r} \simeq n_r\, p = \tau_{ev} \theta_r\, p_0, \tag{4.41}$$

where the right-hand side assumes that $p \ll 1$. Because this probability scales with the overall time an atom spends in sites where it can react, this time scale is independent of the value of the migration time scale. Inserting values, $\tau_{ev} = 3 \times 10^2$ s, $\theta_r = 10^{-6}$, an H atom has a 50% probability of reacting with a species with an activation barrier of 3400 K. If the surface concentration of coreactants is 0.1, this limiting barrier is equal to 7500 K.

Table 4.15 *Hydrogen reactions with activation barriers*

reactants		products	E_a(K)
$H + CO$	$\longrightarrow$	HCO	1000
$H + H_2CO$	$\longrightarrow$	CH_3O[a]	1000
$H + O_2$	$\longrightarrow$	HO_2	1200
$H + H_2O_2$	$\longrightarrow$	$H_2O + OH$	1400
$H + O_3$	$\longrightarrow$	$O_2 + OH$	450
$H + C_2H_2$	$\longrightarrow$	C_2H_3	1250
$H + C_2H_4$	$\longrightarrow$	C_2H_5	1100
$H + H_2S$	$\longrightarrow$	$SH + H_2$	860
$H + N_2H_2$	$\longrightarrow$	$N_2H + H_2$	650
$H + N_2H_4$	$\longrightarrow$	$N_2H_3 + H_2$	650

[a] Product species could also be CH_2OH.

In general, an accreted H atom may be able to pick from among many possible coreactants. The probability for reaction of atomic H with species i is then given by

$$\phi_i = \frac{\theta_i \, p_0(i)}{\sum_j \theta(j) \, p_0(j)}, \tag{4.42}$$

where p_0 is given by Eq. (4.38) and the summation is over all possible coreactant species. Table 4.15 lists some relevant reactions. The activation barriers involved have been taken from gas phase studies and their applicability to low temperature grain-surface chemistry is unclear.

4.2.7 *Reactions involving H_2*

At low temperatures, there can be an appreciable surface coverage factor of H_2 on a grain surface ($\simeq 0.2$; cf. Section 4.2.2). Any reactions involving molecular hydrogen will have appreciable activation barriers. However, as for atomic H, these reactions may still occur because of tunneling. We have to approach the problem from a slightly different point of view than for atomic H reactions. In this case, consider a coreactant (newly accreted or made on the surface). We are now interested in the probability that this species will react with an H_2 molecule on the surface before another migrating radical (e.g., H or O) arrives. The probability for reaction is now

$$p_r = \tau_{ar} \theta(H_2) \, p_0. \tag{4.43}$$

For an arrival time scale of $\tau_{ar} = 10^5$ s, an H_2 coverage of 0.2, and assuming that the activation barrier is in the entrance channel, the limiting activation barrier is $\simeq$4700 K. This is only logarithmically sensitive to the accretion time scale and the H_2 surface coverage. If the barrier is in the exit channel, an H atom has to tunnel out and this limiting barrier is about 10 500 K. Thus, reactions involving H_2 that are relevant for grain-surface chemistry include the important reaction of OH with H_2.

4.2.8 *Reactions with activation barriers involving O, N, or C*

The atoms C, N, and O are also mobile and hence reactions with activation barriers have some probability of occurring. The probability for reaction can be evaluated analogously to that for H_2 and the arrival time is the limiting factor. For these atoms, tunneling is of little consequence and the thermal hopping rate has to be used. As a result, the limiting barrier is much lower ($\simeq$360 K). Few reactions are in this category. Possibly, $O + O_2 \longrightarrow O_3$, $O + CS \longrightarrow OCS$, $O + SO \longrightarrow SO_2$, and $O + CO \longrightarrow CO_2$ are relevant, but the laboratory evidence for some of these reactions (e.g., $O + CO \longrightarrow CO_2$) is controversial.

4.2.9 *Grain surface chemical networks*

The chemical network will depend on the chemical composition of the accreting gas, their accretion rates, and the migration time scale of accreted species on the grain surface. The accretion time scale is much longer than in typically terrestrial laboratory settings. Interstellar grain-surface chemistry is, therefore, in the "diffusion" limit, where the reaction rate is essentially limited by the rate at which species are transported to the surface. At any time, at most two species that can migrate appreciably within an accretion or evaporation time scale (whichever is shorter) are present on any grain surface (compare Eqs. 4.29 and 4.36). Now, an accreted H will scan the whole surface very rapidly. C, N, and O atoms can also scan an appreciable fraction of a grain surface, searching for a reaction partner, before a next radical lands. Heavier species, on the other hand, are trapped in the site where they originally land and are forced to wait for a reaction partner to pass by. Thus, the reactions that need to be considered are reactions of migrating atoms (H, C, N, and O) with themselves and with other non-mobile species.

There is an important point buried in this discussion. Normally, the chemical consequences of the Langmuir–Hinshelwood mechanism are evaluated using rate equations where the reaction rate scales with the product of the surface coverages of the species involved and their collision rate on the surface (i.e., their mobility). However, this formalism implicitly assumes an "infinite" surface where even small θs are meaningful. As emphasized above, this is not the case in the interstellar medium. A typical reaction sequence may involve the accretion of a migrating atom on a coreactant-free surface; e.g., at that point, the surface coverage of the accreted species is $1/4\pi a^2 N_s$ while the surface coverage of coreactants is zero. The first coreactant that lands will react, setting all surface concentrations to zero until the next species lands. If a number of different coreactants could accrete, the reaction rate would then scale with the probability that the next accreting species is the specific one under consideration. For mobile species (e.g., H, D, C, N, and O), the surface collision rate does not enter at all in this. Now, using rate equations to calculate the reaction rate may lead to erroneous results, particularly when several species with very different surface mobilities accrete from the gas phase with very similar rates.

It is instructive to consider the accretion of species A and B at rates r_A and r_B (per site), which can react to form species A_2, AB, and B_2. In the diffusion limit where transport to the surface is the limiting factor, the formation rates of these species are, $r_A^2/2(r_A + r_B)$, $r_B^2/2(r_A + r_B)$, and $r_A r_B/(r_A + r_B)$. For the rate equation approach, we have in steady state,

$$r_A - 2k_A\theta_A^2 - k_{AB}\theta_A\theta_B = 0, \tag{4.44}$$

and

$$r_B - 2k_B\theta_B^2 - k_{AB}\theta_A\theta_B = 0, \tag{4.45}$$

where the k_is are the collision frequencies. These two equations can be manipulated to yield

$$\frac{\theta_B}{\theta_A} = \frac{1}{4}\left(\left(\left(\frac{k_{AB}}{k_B}\right)^2\left(1 - \frac{r_B}{r_A}\right)^2 + 16\frac{k_A}{k_B}\frac{r_B}{r_A}\right)^{1/2} - \frac{k_{AB}}{k_B}\left(1 - \frac{r_B}{r_A}\right)\right), \tag{4.46}$$

and

$$\theta_A = \left(\frac{k_{AB}}{r_A}\frac{\theta_B}{\theta_A} + 2\frac{k_A}{r_A}\right)^{-1/2}, \tag{4.47}$$

and from these we can calculate the formation rates, $k_A\theta_A^2$, $k_B\theta_B^2$, and $k_{AB}\theta_A\theta_B$. If A and B have the same mobility (i.e., $k_{AB} = 2k_A = 2k_B$), these formation rates are equivalent to those calculated from the stochastic method. However, if the two species have very different mobilities, the rate equation method can lead to very erroneous results. Essentially, in these calculations, the surface coverage of A is always $<10^{-7}$ and, hence, meaningless for grains smaller than 3000 Å (cf. Fig. 4.3).

When the surface collision rates are very different, the surface coverage develops a sharp transition at $r_A = r_B$ where the surface goes from A-dominated to B-dominated. The formation rates of the various species follow this flip-flop. This

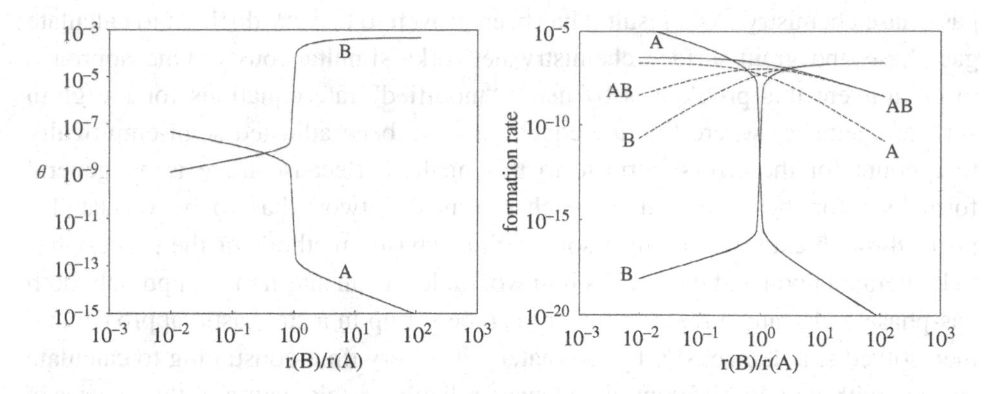

Figure 4.3 Left: The surface coverage of species A and B as a function of the ratio of the accretion rates calculated using the rate equation approach. Note that an interstellar grain has approximately $N = 2.5 \times 10^6 (a/1000\,\text{Å})^2$ sites (e.g., $\Theta \geq 1/N$). Right: The formation rates of A_2 (labeled A), B_2 (labeled B), and AB using rate equations (solid) and stochastic equations (dashed) in the diffusion limit ($r = 4 \times 10^{-7}$, $k_A = 10^{10}\,s^{-1}$, $k_B = 1\,s^{-1}$, and $k_{AB} = k_A + k_B\,s^{-1}$).

switch-over becomes more pronounced when the ratio of the accretion rate to the surface collision rates decreases.

This simple example can be solved analytically in the stochastic approach. But generally this is not the case because some of the accreted species or the intermediate reaction products are immobile on the surface. In that case, the probability that more than one (immobile) radical is present on the surface as a possible reaction partner for a newly accreted, mobile species has to be accounted for. Thus, one has to evaluate the joint probability that there are n_A species A, n_B species B, n_C species C, … , present as reaction partners for the mobile radical. Such systems have been calculated using Monte Carlo techniques where the different reaction sequences possible are followed explicitly. In this approach, random number generators are used to evaluate the accretion history of the different species and their reaction sequences with the various reaction partners present. Using a large number of drawings, the system will converge to the steady state results (within statistical uncertainties). Alternatively, this system can be solved using a Markov chain approach. The state of the surface is then represented by a probability vector, which contains the surface concentration of the various species. A transition matrix links these different states (e.g., this matrix contains the probability that the surface state will transit from i to j). Of course, with each accretion step, the size of this matrix will grow and one has to impose cutoffs on the values of the n_is. With modern computer techniques, both these approaches will work.

However, either stochastic method for computing the grain-surface chemistry effectively decouples it from the rate equation method used to calculate the gas-phase chemistry. As a result it has been proven to be very difficult to calculate gas-phase and grain-surface chemistry networks simultaneously. One approach to circumvent this problem is by using "modified" rate equations for the grain surface chemistry where the rate equations have been adjusted semi-empirically to account for the errors intrinsic to this method. Because there is no general formalism for these corrections, each chemical network has to be validated a priori through extensive comparisons with stochastic methods or the process has to be iterated upon and this makes it unworkable. As an alternative approach, both gas-phase and grain-surface chemistry can be set up in a stochastic approach and then solved simultaneously. Unfortunately, it is very time-consuming to calculate species with widely different abundances reliably in this way and this approach has not been actively pursued. Calculating abundances for trace species is, of course, a problem for all stochastic calculations but very clever Monte Carlo techniques have been devised to overcome this problem, albeit not yet applied to the specific problem of grain-surface chemistry.

Finally, there are some astrophysical environments where the gas–grain interaction is only one way. The chemistry of hot cores (cf. Section 10.6.3) is a case in point, where ice mantles rapidly evaporate, following the formation of a protostar, and flood the gas phase with high abundances of grain surface products. Such situations can be reliably evaluated using gas phase chemistry alone, starting from "evaporated" initial conditions. Likewise, formation of H_2 in a largely atomic environment will be handled well by the rate-equation method (essentially because H flows predominantly to H_2, in any case). In a general sense, the interplay of gas-phase and grain-surface chemistry is a key factor in all of astrochemistry and this area of research is not yet handled well by theoretical models.

4.3 Further reading

The textbooks on interstellar chemistry, [1] and [2], provide an excellent introduction into the field. There is also a series of symposia on astrochemistry and their proceedings, which generally include a session on new reaction mechanisms and new rate measurements. The last one is [3].

The molecular physics of interstellar gas-phase reactions has been discussed in detail in [4] and [5]. Both of these are highly recommended. The importance of ion–molecule reactions in interstellar molecular clouds is described in [6] and [7]. The radiative transfer involved in photochemistry and a compilation of reaction rates is provided in [8] and [9]. The importance of cosmic-ray-induced internal radiation fields inside dense molecular clouds was realized in [10]. An extensive study of cosmic-ray-driven photochemical reaction rates is provided by [11].

Various non-thermal effects are discussed in [12], [13], [14], [15], and [16]. The effects of chaotic behavior in gas-phase models is discussed in [17].

The discussion on the physics of interstellar grain-surface chemistry has been taken from [18] and [19]. Sticking coefficients have been discussed in [20]. A very limited number of experiments have been made of astrophysically relevant surface reactions, including [21], [22], and [23].

Two compilations of gas-phase reaction rates are accessible on the web. The UMIST data base

http://www.rate99.co.uk,

is maintained by Tom Millar and described in [24]. The new standard model database,

http://www.physics.ohio-state.edu/~eric/research.html,

is maintained by Eric Herbst. Java applets are provided to query these databases, extracting reactions given reactants or products, and including references to laboratory studies and estimates of the uncertainty involved. [24] provides heats of formation for many astrophysically relevant species. These can also be obtained from the NIST website,

http://webbook.nist.gov/chemistry/.

The discussion around the inherent difficulties associated with the rate equation approach to grain-surface chemistry in the diffusion limit dates back to an astrochemistry workshop at the Lorentz Center in Leiden. A recent summary of the various approaches to interstellar grain-surface chemistry is provided by [25].

References

[1] W. W. Duley and D. A. Williams, 1984, *Interstellar Chemistry* (London: Academic Press)
[2] E. L. O. Bakes, *Astrochemical Evolution of the Interstellar Medium* (Vledder: Twin Press)
[3] Y. C. Minh and E. F. van Dishoeck, ed. 2000, *Astrochemistry: From Molecular Clouds to Planetary Systems* (San Francisco: Astronomical Society of the Pacific)
[4] A. Dalgarno, 1975, *Molecular Processes in Interstellar Clouds, Saas Fee Advanced Course*, ed. M. C. E. Huber and H. Nussbaumer (Sauverny: Observatoire de Genève)
[5] W. D. Watson, 1976, *Rev. Mod. Phys.*, **48**, p. 513
[6] W. D. Watson, 1973, *Ap. J.*, **182**, p. L69
[7] E. Herbst and W. Klemperer, 1973, *Ap. J.*, **185**, p. 505
[8] W. G. Roberge, A. Dalgarno, and B. P. Flannery, 1981, *Ap. J.*, **243**, p. 817
[9] W. G. Roberge, D. Jones, S. Lepp, and A. Dalgarno, 1991, *Ap. J. S.*, **77**, p. 287
[10] S. S. Prasad and S. P. Tarafdar, 1983, *Ap. J.*, **267**, p. 603
[11] R. Gredel, S. Lepp, A. Dalgarno, and E. Herbst, 1989, *Ap. J.*, **347**, p. 289

[12] W. D. Langer and A. E. Glassgold, 1990, *Ap. J.*, **352**, p. 123
[13] J. le Bourlot, 1991, *A. & A.*, **242**, p. 235
[14] M. M. Graff, 1989, *Ap. J.*, **339**, p. 239
[15] M. M. Graff and A. Dalgarno, 1987, *Ap. J.*, **317**, p. 432
[16] A. F. Wagner and M. M. Graff, 1983, *Ap. J.*, **267**, p. 603
[17] J. le Bourlot, G. Pineau des Forets, E. Roueff, and P. Schilke, 1993, *Ap. J.*, **416**, p. L87
[18] A. G. G. M. Tielens and W. Hagen, 1982, *A. & A.*, **114**, p. 245
[19] A. G. G. M. Tielens and L. J. Allamandola, 1987, in *Interstellar Processes*, ed. D. Hollenbach and H. Thronson, (Dordrecht: Reidel), p. 397
[20] J. R. Burke and D. Hollenbach, 1983, *Ap. J.*, **265**, p. 223
[21] V. Pirronello, C. Liu, J. Roser, and G. Vidali, 1999, *A. & A.*, **344**, 681; *Ap. J.*, **483**, p. L131
[22] K. Hiraoka, T. Sato, S. Sato *et al.*, 2002, *Ap. J.*, **577**, p. 265
[23] N. Watanabe and A. Kouchi, 2002, *Ap. J.*, **571**, p. L173
[24] Y. H. Le Teuff, T. J. Millar, and A. J. Markwick, 2000, *A. & A. Suppl. Ser.*, **146**, p. 157
[25] E. Herbst and V. I. Shematovitch, 2003, *Ap. & S. S.*, **285**, p. 725

5
Interstellar dust

5.1 Introduction

Interstellar dust is an important component of the interstellar medium. Dust provides the dominant opacity source in the interstellar medium for non-ionizing photons and therefore controls the spectral energy distribution of the ISM at all wavelengths longer than 912 Å. Dust grains also lock up a substantial fraction of all heavy elements. Grains provide a surface on which species can accrete, meet, and react – giving rise to an interesting and complex chemistry. This chapter will look at the physical processes involving dust, including their interaction with light – in particular their energy balance and the resulting temperature – and their charge balance. Dust also regulates the gas phase abundances of the elements through accretion and destruction processes. This chapter discusses the physical processes involved in dust destruction. The chemical processes that control accretion and ice mantle formation are described in Section 4.2. The growth and characteristics of interstellar ice mantles are discussed in Sections 10.6 and 10.7.4. We will return to the lifecycle of interstellar dust and the depletion of the elements in Chapter 13. The composition of interstellar dust has been widely debated and silicates and graphite are generally considered the most important interstellar dust components. In this chapter, we have therefore focussed on these compounds. However, the discussion is very general and one might often substitute minerals for silicates and amorphous carbon for graphite in the discussion of the physical processes.

Section 5.2.1 considers the absorption of light by small particles, which is key in determining the dust temperature as well as the analysis of much of the observational data. The radiative energy balance and the resulting temperature of a grain are discussed in Section 5.2.2. The grain charge balance is examined in Section 5.2.3. The processes involved in dust destruction (Section 5.2.4) round off this discussion on physical processes. Section 5.3 discusses the various observations of interstellar dust and their implications for dust characteristics, notably abundances, sizes, and composition.

5.2 Physical processes

5.2.1 Absorption and scattering by small particles

The dust optical depth, τ, is given by

$$\tau(\lambda) = n_{\rm d} C_{\rm ext}(\lambda) L, \tag{5.1}$$

where $C_{\rm ext}(\lambda)$ is the extinction cross section, $n_{\rm d}$ is the dust density, and L the pathlength. The relation between extinction in optical depth units, $\tau(\lambda)$, and in magnitudes, A_λ, is given by

$$\tau(\lambda) = 1.086 A_\lambda. \tag{5.2}$$

The extinction cross section of a grain is often expressed in terms of the extinction efficiency, $Q_{\rm ext}$, and the geometric cross section, $\sigma_{\rm d}$,

$$Q_{\rm ext}(\lambda) = \frac{C_{\rm ext}(\lambda)}{\sigma_{\rm d}}. \tag{5.3}$$

Extinction is the sum of absorption and scattering processes; viz.,

$$Q_{\rm ext}(\lambda) = Q_{\rm abs}(\lambda) + Q_{\rm sca}(\lambda). \tag{5.4}$$

The albedo, $\tilde{\omega}$, is then

$$\tilde{\omega}(\lambda) = \frac{Q_{\rm sca}(\lambda)}{Q_{\rm ext}(\lambda)}. \tag{5.5}$$

The extinction, absorption, and scattering cross sections and efficiencies are, in general, size dependent. As an example, for a grain size distribution $n_{\rm d}(a)$, the extinction optical depth is given by

$$\tau(\lambda) = L \int_{a_-}^{a_+} n_{\rm d}(a) C_{\rm ext}(a, \lambda) {\rm d}a, \tag{5.6}$$

where a_- and a_+ are the grain size limits involved.

In general, scattering is also a function of the scattering angle, θ,

$$C_{\rm sca}(\lambda) = \int_0^{2\pi} \int_0^{\pi} \frac{{\rm d}\sigma(\lambda)}{{\rm d}\Omega} \sin\theta {\rm d}\theta {\rm d}\phi. \tag{5.7}$$

In the astronomical and the light-scattering communities, the differential scattering cross section, ${\rm d}\sigma/{\rm d}\Omega$, is often indicated by the symbol $S(\theta)$. The scattering phase function asymmetry, g, is given by

$$g \equiv <\cos\theta> = \frac{1}{C_{\rm sca}} \int_0^{2\pi} \int_0^{\pi} \frac{{\rm d}\sigma(\lambda)}{{\rm d}\Omega} \cos\theta \sin\theta {\rm d}\theta {\rm d}\phi. \tag{5.8}$$

Isotropic scattering is characterized by $g = 0$. In the small particle limit ($2\pi a \ll \lambda$, with a the size), scattering is described by the Rayleigh formalism (see below) and the differential scattering cross section is given by

$$\frac{1}{C_{\rm sca}}\frac{{\rm d}\sigma}{{\rm d}\Omega} = \frac{3}{16\pi}(1+\cos^2\theta). \tag{5.9}$$

This also yields $g = 0$; however Rayleigh scattering is not isotropic. Before the advent of computers, the differential scattering cross section was often approximated by the Henyey–Greenstein phase function. However, this function was only introduced for computational convenience and has no deeper physical significance.

The asymmetry factor is related to the momentum transferred to the grain; e.g., the radiation pressure force, $F_{\rm rp}$, exerted by a beam of intensity I, is

$$F_{\rm rp}(\lambda) = C_{\rm rp}(\lambda)\frac{I(\lambda)}{c}, \tag{5.10}$$

with c the speed of light. The radiation pressure cross section, $C_{\rm rp}$, is given by

$$C_{\rm rp} = C_{\rm abs} + (1-g)C_{\rm sca}. \tag{5.11}$$

Besides wavelength, size (and shape), the extinction cross section depends on the optical properties of the material. These enter into the theory through the complex index of refraction, $m = n + {\rm i}k$, where the real and imaginary part are functions of the wavelength. Alternatively, the optical properties of a material can be expressed in terms of the dielectric constant, $\epsilon = \epsilon_1 + {\rm i}\epsilon_2$. The dielectric constant and the complex index of refraction are related through

$$\epsilon = m^2, \tag{5.12}$$

or, written out,

$$\epsilon_1 = n^2 - k^2 \tag{5.13}$$

and

$$\epsilon_2 = 2nk. \tag{5.14}$$

Both m and ϵ are referred to as optical constants, although neither of them is actually constant with wavelength. Some further insight in their significance can be obtained by considering a plane wave traveling in the z direction,

$$E = E_0 \exp[{\rm i}\kappa z - {\rm i}\omega t]. \tag{5.15}$$

In free space, the wave vector κ is given by $\omega/c = 2\pi/\lambda$, with λ the wavelength in vacuo. In a material with complex index of refraction m, $\kappa = \omega m/c$. Thus,

$$E = E_0 \exp\left[-\frac{\omega}{c}kz\right]\exp\left[-{\rm i}\omega\left(t-\frac{nz}{c}\right)\right]. \tag{5.16}$$

Thus, the real part, n, of the complex index of refraction introduces a phase shift (i.e., a phase velocity $v = c/n$) while the imaginary part, k, results in damping.

The electric displacement and the electric field at a frequency ω are related through the complex dielectric constant. For any such linear system, there exists a dispersion relation between the real and imaginary part; i.e., the phase lag and the damping. Mathematically, this relationship follows directly from Fourier analysis. Physically, this relationship expresses that a cause precedes a response. These so-called Kramers–Kronig relations read for the dielectric constant

$$\epsilon_1(\omega_0) - 1 = \frac{2}{\pi}\mathcal{P}\int_0^\infty \frac{\omega\epsilon_2(\omega)}{\omega^2 - \omega_0^2}\,d\omega \tag{5.17}$$

and

$$\epsilon_2(\omega_0) = \frac{2}{\pi}\mathcal{P}\int_0^\infty \frac{\epsilon_1(\omega) - 1}{\omega^2 - \omega_0^2}\,d\omega, \tag{5.18}$$

where $\mathcal{P}$ indicates the principal value of the integral. From these relations, we can immediately conclude that any system that produces absorption will also produce dispersion. Conversely, any system that does not produce absorption also produces no dispersion and is therefore indistinguishable from free space. Furthermore, these relations impose limits on the frequency dependence of the real and imaginary part of the dielectric function at low and high frequencies. Specifically, for high enough frequencies, ϵ_1 approaches 1; physically, the polarization of the material cannot respond to the electric field on such short time scales. Finally, in the limit ω_0 tends to 0, these relations are known as sum rules,

$$\epsilon_1(\omega_0) - 1 = \frac{2}{\pi}\int_0^\infty \frac{\epsilon_2(\omega)}{\omega}\,d\omega. \tag{5.19}$$

Once the dielectric (or optical) constants, the size, and shape of the grain have been specified, the extinction and scattering efficiencies can be calculated. Illustrative extinction and scattering efficiency factors for dielectric spheres are shown in Fig. 5.1 as a function of the size parameter, $x = 2\pi a/\lambda$. For small x, the extinction rises slowly with x and reaches a maximum around $x = 6$. The slowly modulating maxima and minima apparent in these curves are, in essence, interference fringes between the unperturbed incident wave and the forward scattered light (i.e., the phase lag through the particle is given by $\rho = 2x(m-1)$ and constructive and destructive interference occurs when this is equal to $2\pi(\ell + 3/4)$ and $2\pi(\ell + 1/4)$, respectively [with ℓ an integer]). The higher frequency ripple also apparent in these calculated extinction efficiencies does not have such a simple physical interpretation. In an average over a size distribution, these ripples would wash out, however. They are also much less prominent with increasing k. For particles that are large compared with the wavelength, the

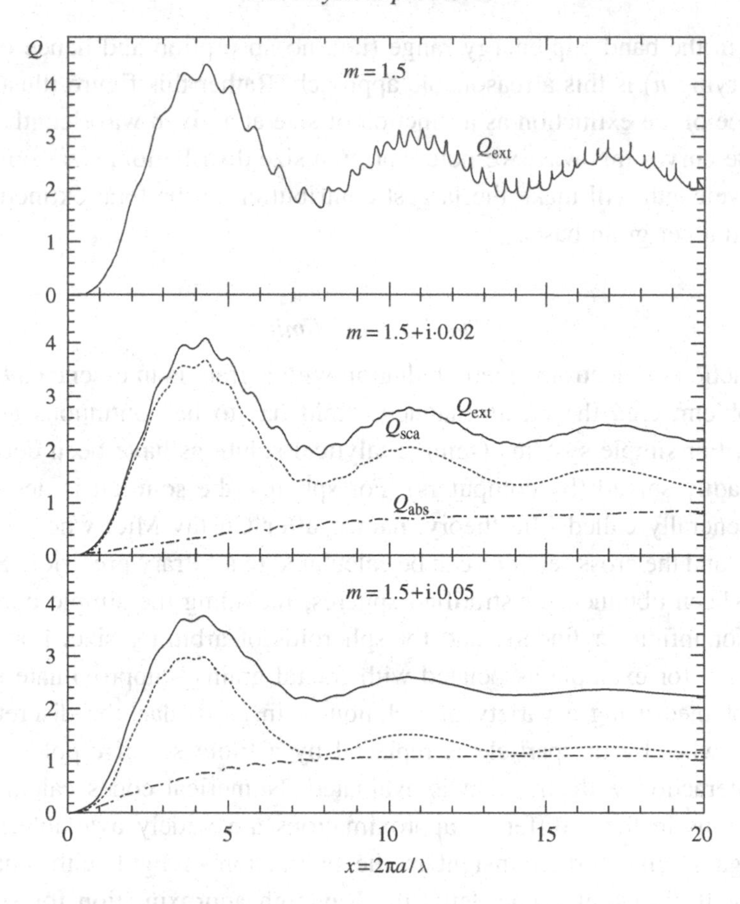

Figure 5.1 The extinction and scattering efficiency calculated using Mie theory for spherical grains plotted as a function of the size parameter, $x = 2\pi a/\lambda$. The adopted optical constants are indicated in the panels. Figure adapted from H.C. van der Hulst, 1981, *Light Scattering by Small Particles*, (New York: Dover).

scattering and the absorption efficiency both approach unity and the extinction efficiency goes to 2. This result in the geometric optics limit may seem, at first sight, somewhat paradoxical. However, all the light falling within the geometric cross section of the grain is either absorbed or scattered. Moreover, the beam of light is also diffracted (scattered at small angles) at the edges of the grain. This diffraction also removes exactly an amount of light given by the geometric cross section. Of course, for objects which are near, this diffraction loss is imperceptible but at "interstellar" distances this light is lost from the beam.

Sometimes, results such as presented in Fig. 5.1 are used to infer the wavelength dependence of the extinction and scattering efficiency. However, generally, the optical constants are not constant with wavelength. Only for truly dielectric materials in the band gap energy range (i.e., no absorption and hence only very slowly varying n) is this a reasonable approach. Rather this figure illustrates the dependence of the extinction as a function of size at a given wavelength. Considering these curves this way, we note that, in a size distribution, sizes comparable to the wavelength will make the largest contribution to the total extinction cross section, on a per grain basis.

The Rayleigh limit

The interaction of electromagnetic radiation with a grain is in essence a boundary value problem; e.g., the electromagnetic field has to be continuous across the boundary. For simple systems (semi)-analytical solutions have been derived that can be readily solved (by computers). For spheres, the solution is described by what is generally called Mie theory, named after Gustav Mie, who derived the equations, and the cross sections can be calculated to arbitrary precision. Solutions have also been obtained for stratified spheres, including the simple core-mantle spheres, for infinite cylinders, and for spheroids of arbitrary size. For complex geometries – for example associated with fractal grains – approximate solutions can be obtained using a variety of techniques; in particular, the discrete dipole technique, whereby the particle is replaced by a finite set of dipoles and their mutual interaction with the light is evaluated. Numerical codes calculating the cross sections in these different approximations are widely available. Here, in order to gain some further insight in the interaction of light with small particles, we will discuss in some detail the Rayleigh approximation for spheroidal particles.

In Mie theory for spheres, the absorption and scattering efficiencies are written as a series expansion in the size parameter, $x = 2\pi a/\lambda$. In the Rayleigh limit, when x is small and mx is small – a size much less than the wavelength and a negligible phase shift in the particle – only the leading terms need to be retained; i.e.

$$Q_{\rm ext} = 4x{\rm Im}\left[\frac{\epsilon-1}{\epsilon+2}\right] \tag{5.20}$$

and

$$Q_{\rm sca} = \frac{8}{3}x^4\left|\left(\frac{\epsilon-1}{\epsilon+2}\right)\right|^2, \tag{5.21}$$

where Im denotes the imaginary part of the complex expression. It is obvious now that, for constant ϵ, absorption will dominate over scattering when x is small (i.e., in the infrared).

Table 5.1 *Geometric factors for spheroids*[a]

shape	L_1	L_2	L_3
rod ($b=0$)	0.0	0.5	0.5
prolate spheroid ($a/b=5$)	0.056	0.472	0.472
prolate spheroid ($a/b=2.5$)	0.134	0.433	0.433
prolate spheroid ($a/b=1.25$)	0.276	0.362	0.362
sphere	1/3	1/3	1/3
oblate spheroid ($a/b=0.8$)	0.396	0.302	0.302
oblate spheroid ($a/b=0.4$)	0.588	0.206	0.206
oblate spheroid ($a/b=0.2$)	0.75	0.125	0.125
disk ($a=0$)	1.0	0.0	0.0

[a] Spheroids with semi-axis a and $b=c$ (oblate) or b and $a=c$ (prolate).

Actually, in the Rayleigh approximation, these results can be generalized to spheroids. It is then convenient to start with the polarizability, α_j, for light with the electric vector along one of the three major axes of the spheroid, j,

$$\alpha_j = \frac{V}{4\pi}\frac{\epsilon-1}{L_j(\epsilon-1)+1}, \tag{5.22}$$

with V the particle volume. The geometric factors, L_j, (sometimes called the depolarization factors) contain the dependence on particle shape. The L_js lie in the range 0–1 and for any shape fulfill the relation

$$\Sigma_j L_j = 1. \tag{5.23}$$

Table 5.1 summarizes some typical values for these geometric factors.

The extinction and scattering cross sections are now given by

$$C^j_{\rm ext} = 4\pi\frac{2\pi}{\lambda}{\rm Im}(\alpha_j) \tag{5.24}$$

and

$$C^j_{\rm sca} = \frac{8\pi}{3}\left(\frac{2\pi}{\lambda}\right)^4 |\alpha_j|^2. \tag{5.25}$$

For spheres, these expressions reduce to the ones derived above from Mie theory. These cross sections can be written as

$$C^j_{\rm ext} = \frac{2\pi}{\lambda}V\frac{\epsilon_2}{\left(L_j(\epsilon_1-1)+1\right)^2+\left(L_j\epsilon_2\right)^2} \tag{5.26}$$

and

$$C^j_{\rm sca} = \frac{8\pi^3}{3}\frac{V^2}{\lambda^4}\frac{(\epsilon_1-1)^2+\epsilon_2^2}{\left(L_j(\epsilon_1-1)+1\right)^2+\left(L_j\epsilon_2\right)^2}. \tag{5.27}$$

For constant optical constants, the extinction cross section is proportional to λ^{-1} while scattering is proportional to λ^{-4}. As an example, for air, the absorptivity in the visible is very small and only scattering contributes to the extinction. This familiar result for Rayleigh scattering gives rise to blue skies and red sunsets. Actually, for most dielectric materials far from resonances, ϵ_2 (and k) will scale with λ^{-1} while ϵ_1 (and n) is approximately constant. The extinction cross section, then, scales with λ^{-2} and absorption dominates over scattering at long wavelengths. It can also be recognized that, in the Rayleigh limit, absorption is proportional to the total volume of the grain. In principle, the infrared thus provides a good handle on the total dust volume if the optical constants are known. Also, in a size distribution, on a per volume basis, "Rayleigh" grains contribute equally to the total absorption. Scattering, on the other hand, is dominated by the largest grains. Of course, only as long as $2\pi a < \lambda$; for grains large compared to the wavelength, absorption and scattering "saturate" at the geometrical cross section and on a per volume basis, the contribution by large grains decreases with increasing size (cf. Fig. 5.1).

For randomly oriented spheroids, the average cross section is given by

$$\overline{C}_{\text{ext}} = \frac{1}{3} \Sigma_j C^j_{\text{ext}} \tag{5.28}$$

and similarly for the scattering cross section.

Resonances

Examination of the expression for the extinction cross section for a spheroidal grain in the Rayleigh limit shows that strong resonances will occur when ϵ_1 becomes negative; specifically when

$$\epsilon_1 = 1 - \frac{1}{L_j} \tag{5.29}$$

and the denominator in Eq. (5.26) becomes small. Thus, the peak frequency of the resonance depends on the shape of the particle through the geometry factor L_j. For spheres (Table 5.1), this condition reads $\epsilon_1 = -2$. The extinction cross section is, then,

$$C^j_{\text{ext}} = \frac{2\pi}{\lambda} V \frac{9}{\epsilon_2}. \tag{5.30}$$

The real part of the dielectric constant can become negative near a resonance. As an example, the dielectric constant for a Lorentz oscillator near resonance at frequency ω_0 is given by

$$\epsilon(\omega) = \epsilon_\infty + \frac{S\omega_0^2}{\omega_0^2 - \omega^2 + i\omega\Gamma_0}, \tag{5.31}$$

with ϵ_∞ the high frequency dielectric constant, Γ_0 the width, and S the strength. Dielectric constants for different values of S are presented in Fig. 5.2. The imaginary part, ϵ_2, shows the resonance of a damped oscillator, while the real part, ϵ_1, shows a region of anomalous dispersion (normally, dispersion increases with increasing frequency; e.g., blue is more diffracted than red in a rainbow). As expected, the variations in the dielectric constants become more pronounced as the strength of the oscillator increases. Note also the marked Lorentz wings on the ϵ_2 profile. For strong oscillators, the real part of the dielectric constant does become negative and small particles will show strong resonance effects. Figure 5.2 shows calculated extinction cross sections for spheres adopting the Lorentz oscillator dielectric constants. For low strength, the variations in the dielectric constant are small and the extinction cross section follows the variation in ϵ_2. However, when the transition becomes very strong, ϵ_1 becomes small and even negative. As a

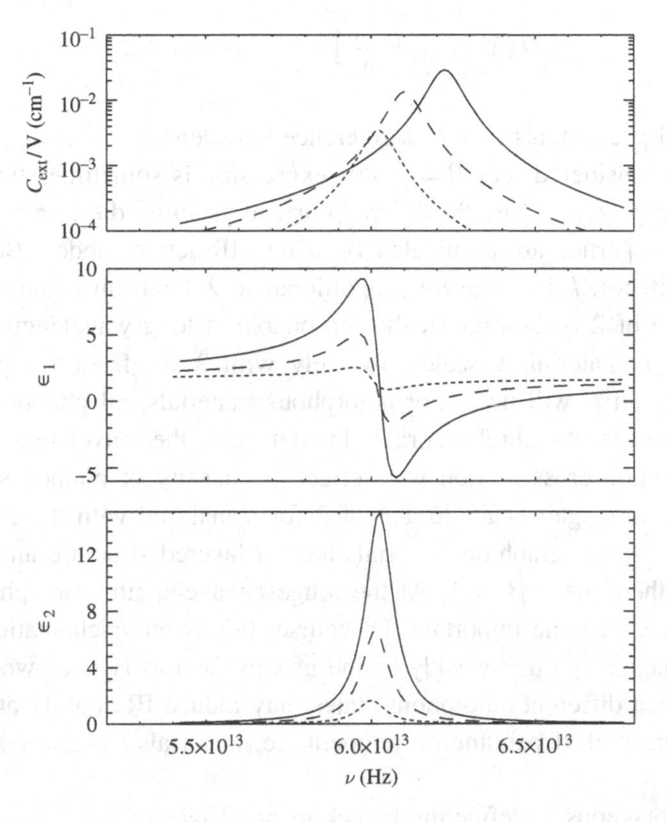

Figure 5.2 The real and imaginary parts of the dielectric constant due to a Lorentz oscillator for different values of the strength. Top panel: the extinction cross section per unit volume calculated for spheres using the dielectric constants given in the bottom two panels.

result the peak in the extinction cross section shifts towards higher frequencies. The location of the peak then becomes sensitive to the shape of the particle. Also, the absolute value of the cross section can become very large.

Such resonance effects are quite common in astronomical settings. They occur whenever the integrated strength of a mode is high or the band is very narrow. Examples are the $\pi \longrightarrow \pi^\star$ transition in graphite (cf. Section 5.5.2), the vibrational stretching mode in solid CO, the vibrational stretching and bending modes in solid CO_2, and the various vibrational modes of crystalline silicates.

IR absorption cross sections

The absorption efficiency of dust in the infrared plays an important part in the radiative energy balance of dust. Often the absorption efficiency is approximated by a simple power law; viz.

$$Q(\lambda) = Q_0 \left(\frac{\lambda_0}{\lambda}\right)^\beta \qquad \text{for } \lambda > \lambda_0, \tag{5.32}$$

with Q_0 and β constants and λ_0 a reference wavelength. Typically, βs between 1 and 2 are considered. For $\beta = 1$, this expression is sometimes further simplified by setting $Q_0 = 1$ and $\lambda_0 = 2\pi a$. For a crystalline dielectric material, the absorption properties are dominated by a few IR-active modes. Because for a Lorentz oscillator, k is inversely proportional to λ far from resonance and n is constant, a β of 2 is then physically reasonable at long wavelengths. Likewise, for a metallic material, k scales inversely with λ far from the plasma resonance and again β will be 2. For amorphous materials, all phonons will be IR active due to the structural disorder. In that case, the wavelength dependence of the absorption cross section will reflect the density of phonon states. In the Debye limit, this again leads to a β of 2 for a material with three-dimensional structure. However, graphitic materials have a layered structure and amorphous carbon has therefore a β of 1. At the longest wavelengths, two-phonon difference processes become important. Of course, this is only schematic. Moreover, it must be recognized that weakly bound groups such as H_2O as well as "transitions" between different amorphous states may induce IR activity at long wavelengths. In general, depending on temperature, materials can show a range of β values.

It is advantageous to define the Planck mean efficiency,

$$< Q_P(T_d) >= \frac{\pi}{\sigma T_d^4} \int_0^\infty Q(\lambda) B(T_d, \lambda) d\lambda, \tag{5.33}$$

where σ is the Stefan–Boltzmann constant. For an emission efficiency given by Eq. (5.32) and assuming that only emission at wavelengths longer than λ_0 is relevant, this results in

$$< Q_P(T_d) >= \frac{15}{\pi^4}(\beta+3)!\zeta(\beta+4)Q_0\left(\frac{\lambda_0 k T_d}{hc}\right)^\beta , \quad (5.34)$$

where ζ is the Riemann zeta function. For a gray body ($\beta = 0$), this reduces to $< Q_P(T_d) >= Q_0$.

The IR absorption properties of dielectric materials are dominated by resonances. Moreover, the absorption efficiency of grains generally depend on their size. Nevertheless, the following approximations hold well for silicate grains with sizes less than 1 μm,

$$\begin{aligned} < Q_P(T_d) > &\simeq 1.25\times10^{-5}T_d^2\left(\frac{a}{1\,\mu\text{m}}\right) && \text{for } 10 < T_d < 250\,\text{K} \\ &\simeq 1.25\times10^{-3}\left(\frac{a}{1\,\mu\text{m}}\right) && \text{for } 250 < T_d < 1000\,\text{K}. \end{aligned} \quad (5.35)$$

The low temperature behavior merely reflects the λ^{-2} behavior of the efficiency of dielectric materials longwards of the strongest resonances. At $\lambda \simeq 10$ and 20 μm, silicates show strong resonances. Likewise, for typical interstellar grain sizes, the efficiency increases rapidly towards even shorter wavelengths. Nevertheless, over a black body spectrum, these resonances average out and silicates show gray behavior at high temperatures. The Planck mean of graphite grains shows more dependence on size and temperature. The following are approximations for $a < 0.3\,\mu$m,

$$\begin{aligned} < Q_P(T_d) > &\simeq 3.2\times10^{-5}T_d^{1.8}\left(\frac{a}{1\,\mu\text{m}}\right) && \text{for } 10 < T_d < 65\,\text{K} \\ &\simeq 6\times10^{-2}\left(\frac{a}{1\,\mu\text{m}}\right) && \text{for } 65 < T_d < 250\,\text{K} \\ &\simeq 1\times10^{-4}T_d\left(\frac{a}{1\,\mu\text{m}}\right) && \text{for } 250 < T_d < 1000\,\text{K}. \end{aligned} \quad (5.36)$$

Graphite is a semimetal and the atomic vibrational modes are of little importance compared with the electronic absorption. At very long wavelengths ($\lambda > 100\,\mu$m), the absorption is due to free electron intraband transitions that lead to a λ^{-2} behavior. Around 75 μm, graphite will show spectral structure due to electronic interband transitions that give rise to a broad maximum in the absorption. For large grains ($a > 0.1\,\mu$m), magnetic dipole effects become important, which results in a strongly size-dependent absorption cross section. The Planck-mean absorptivities reflect these spectral characteristics of graphite grains. Particularly, at the higher temperatures, the approximation is not as good as for silicates.

5.2.2 Grain temperature

The temperature of a dust grain is set by the radiative energy balance. The energy absorbed by a dust grain is given by

$$\Gamma_{\rm abs} = 4\pi\sigma_{\rm d}\int_0^\infty Q(\lambda)J(\lambda){\rm d}\lambda, \tag{5.37}$$

with $J(\lambda)$ the mean intensity of the radiation field at wavelength λ. The energy emitted by a dust grain is

$$\Gamma_{\rm em} = 4\pi\sigma_{\rm d}\int_0^\infty Q(\lambda)B(T_{\rm d},\lambda){\rm d}\lambda, \tag{5.38}$$

with $T_{\rm d}$ the radiative equilibrium dust temperature and B the Planck function. Using the Planck mean absorption efficiency Eq. (5.32), we can write, for the energy emitted by a grain,

$$\Gamma_{\rm em} = 4\sigma_{\rm d} < Q_{\rm P}(T_{\rm d}) > \sigma T_{\rm d}^4. \tag{5.39}$$

The radiative dust temperature can now be found by equating absorption and emission of radiation. For the diffuse interstellar medium, $4\pi J = cU$ with U the energy density, which is about 7×10^{-13} erg cm^{-3} for stellar light and 4×10^{-13} erg cm^{-3} for the 3K background radiation. For a black body, the radiative temperature is then $\simeq$3.5 K. Small grains are not black bodies and assuming an IR emission efficiency given by Eq. (5.32) with $\beta = 1$ and an absorption efficiency of unity for $\lambda < \lambda_0$, and adopting $Q_0 = 1$ and $\lambda_0 = 2\pi a$, we find

$$T_{\rm d} = \left(\frac{hc}{k}\right)\left(\frac{4\pi J}{384\pi^2 ahc^2\,\zeta\,(5)}\right)^{1/5}, \tag{5.40}$$

with J the integrated mean intensity of the radiation field. With $\zeta(5) = 1.037$, such a grain heated by stellar radiation in the diffuse ISM is thus $T_{\rm d} \simeq 9\,(1\,\mu{\rm m}/a)^{0.2}$ K or about 14 K for a 1000 Å grain. Using the Planck mean efficiencies (Eq. 5.35), the temperature of silicate grains heated by stellar photons in the diffuse ISM is $T_{\rm sil} \simeq 14\,(1\,\mu{\rm m/a})^{0.16}$ K, or about 20 and 30K for a 1000 Å and a 100 Å grain, respectively. Actually, calculations result in somewhat lower temperatures and shallower size dependencies because the absorption efficiencies throughout the visible and UV are not unity and, moreover, size dependent. A reasonable approximation is

$$T_{\rm sil} = 13.6\left(\frac{1\mu{\rm m}}{a}\right)^{0.06}\ {\rm K}. \tag{5.41}$$

Graphite grains will reach very similar temperatures (cf. Eq. (5.36)), $T_{gra} = 13\,(1\ \mu\text{m}/a)^{0.17}$ K, or about 19 and 29 K for a 1000 Å and a 100 Å grain, respectively. Again, the actual graphite grain temperature, allowing for the size and wavelength dependence of the absorption efficiency, is slightly different; i.e.

$$T_{gra} = 15.8 \left(\frac{1\ \mu\text{m}}{a}\right)^{0.06} \text{K}. \tag{5.42}$$

Because of the intense radiation field, dust reaches much higher temperatures near luminous stars. Expressing the radiation field in terms of a one-dimensional Habing field (1.6×10^{-3} erg cm^{-2} s^{-1}; Section 1.3.1), we have, for a star with luminosity $L_\star$, and a distance d,

$$G_0 = 2.1 \times 10^4 \left(\frac{L_\star}{10^4 \text{L}_\odot}\right)\left(\frac{0.1\ \text{pc}}{d}\right)^2. \tag{5.43}$$

We should further note that an extra factor of four arises in the radiative energy balance because, in this anisotropic radiation field, absorption is proportional to the geometric cross section while emission occurs with the actual surface area. We find, then, for the dust temperature,

$$\begin{aligned}
T_d &\simeq 33.5 \left(\frac{1\mu\text{m}}{a}\right)^{0.2} \left(\frac{G_0}{10^4}\right)^{0.2} && \text{K for } \beta = 1 \\
T_{sil} &\simeq 50 \left(\frac{1\mu\text{m}}{a}\right)^{0.06} \left(\frac{G_0}{10^4}\right)^{1/6} && \text{K for } T_{sil} < 250\,\text{K} \\
T_{gra} &\simeq 61 \left(\frac{1\mu\text{m}}{a}\right)^{0.06} \left(\frac{G_0}{10^4}\right)^{1/5.8} && \text{K for } T_{gra} < 65\text{K}.
\end{aligned} \tag{5.44}$$

Inside dense clouds, where the FUV–visible radiation field is severely attenuated, only the far-IR dust continuum and the cosmic microwave background radiation will heat the dust and the temperature drops to some 6 K. Finally, inside HII regions, trapped Lyman alpha photons can become an important heating source for dust and the resulting temperature can be much higher than estimated here (cf. Section 7.2.4).

Ignoring extinction, the IR intensity is given by

$$I(\lambda) = \int n_d(r) \pi a^2 Q(\lambda) B(T_d, \lambda) d\ell, \tag{5.45}$$

where n_d is the dust density and the integral is evaluated along the line of sight, ℓ. For a homogeneous cloud of size L, this becomes

$$I(\lambda) = B(T_d, \lambda) \tau_d(\lambda), \tag{5.46}$$

with the optical depth given by $\tau(\lambda) = n_d \pi a^2 Q(\lambda) L$. For a grain size distribution, $n_d(a)da$, the intensity equation is

$$I(\lambda) = \int \int_{a_-}^{a_+} n_d(r, a) \pi a^2 Q(\lambda, a) B(T_d, \lambda) da d\ell, \tag{5.47}$$

where the integral over the sizes runs from the maximum size, a_+, to the minimum size, a_-.

5.2.3 Grain charge

The charge of dust grains is set by the balance between positive ion recombination and photo-ionization on the one hand and electron recombination on the other hand. Considering only collisions of electrons and ions of charge $Z = 1$, the probability of finding a grain at charge $Z_d e$, $f(Z_d)$, can be found from detailed balance as

$$f(Z_d)[J_{pe}(Z_d) + J_{ion}(Z_d)] = f(Z_d + 1) J_e(Z_d + 1), \tag{5.48}$$

where J_{pe} is the rate of photo-electron emission, and J_e and J_{ion} are the accretion rates of electrons and ions, respectively. The charge distribution functions can then be found by successively applying the following two sets of equations,

$$\begin{aligned} f(Z_d) &= f(0) \prod_{Z'_d=1}^{Z_d} \left[\frac{J_{pe}(Z'_d - 1) + J_{ion}(Z'_d - 1)}{J_e(Z'_d)} \right] & Z_d > 0 \\ f(Z_d) &= f(0) \prod_{Z'_d=Z_d}^{-1} \left[\frac{J_e(Z'_d + 1)}{J_{pe}(Z'_d) + J_{ion}(Z'_d)} \right] & Z_d < 0. \end{aligned} \tag{5.49}$$

This set of equations is closed by

$$\sum_{Z_d=-\infty}^{\infty} f(Z_d) = 1. \tag{5.50}$$

So, in order to solve the charge balance of the dust, we have to specify the photo-electric emission rate and the electron and ion collision rates.

Collisional rates

The collisional rate, $J_i(Z_d)$, of gas particles i, with charge q_i, number density n_i, mass m_i, with a grain of charge $Z_d e$ is given by

$$J_i(Z_d) = n_i s_i \left(\frac{8kT}{\pi m_i} \right)^{1/2} \pi a^2 \tilde{J}(\tau, \nu), \tag{5.51}$$

where we have introduced the reduced temperature, $\tau = akT/q_i^2$, and the charge ratio, $\nu = Z_{\rm d}e/q_i$. Here, s_i is an effective sticking coefficient and $\widetilde{J}(\tau, \nu)$ is the reduced rate, which contains the coulomb interaction aspects. At low kinetic energies of the impacting species, the sticking coefficient is probably close to unity, even for an electron. For higher energies, secondary electron emission has to be taken into account. Moreover, an electron or ion, impacting at very high velocity, may actually be able to traverse a small grain and exit on the other side. Both these effects will affect the sticking coefficient, particularly for electrons (see below).

Consider a grain, moving at a speed $v_{\rm gr}$, relative to the gas. The reduced rate is given by

$$\widetilde{J}(\tau, \nu) = \left(\frac{\pi^3 m_i}{2kT}\right)^{1/2} \int_0^\infty {\rm d}v\, \widetilde{\sigma}(v) v^3 \int_0^\pi f(v, v_{\rm gr}, \theta) \sin\theta {\rm d}\theta, \tag{5.52}$$

where the reduced cross section, $\widetilde{\sigma}$, is the ratio of the cross section to the geometric cross section and it should be noted that, for repulsive potentials, the cross section is zero for velocities below a minimum value corresponding to the coulomb barrier (modified by the image charge). The velocity distribution function, $f(v, v_{\rm gr}, \theta)$, of species with asymptotic velocity v, relative to the grain moving at an angle θ to $v_{\rm gr}$ is given by

$$f(v, \theta) = \left(\frac{m_i}{2\pi kT}\right)^{3/2} \exp\left[-\frac{m_i}{2kT}(v_{\rm gr}^2 + v^2 - 2v_{\rm gr} v \cos\theta)\right]. \tag{5.53}$$

For a Maxwellian velocity distribution, the reduced rate, $\widetilde{J}$, is given by,

$$\widetilde{J}(\tau, \nu) = \int_0^\infty \widetilde{\sigma}(x, \nu) x \exp[-x] {\rm d}x. \tag{5.54}$$

For $\nu = 0$ (neutral grains), the charge interacts through the image potential and

$$\widetilde{J}(\tau, \nu = 0) = 1 + \left(\frac{\pi}{2\tau}\right)^{1/2}. \tag{5.55}$$

For $\nu < 0$, the attractive interaction enhances the cross section by electrostatic focussing. The reduced rate is then, to a good approximation,

$$\widetilde{J}(\tau, \nu < 0) \simeq \left(1 + \frac{|\nu|}{\tau}\right)\left(1 + \left(\frac{2}{\tau + 2|\nu|}\right)^{1/2}\right), \tag{5.56}$$

where the first term on the right-hand side is the familiar coulomb focussing factor $(1 + |Z_{\rm d}eq_i|/akT)$ and the second term represents the image polarization interaction. For $\nu > 0$, the interaction is repulsive and a good approximation is provided by

$$\widetilde{J}(\tau, \nu > 0) \simeq (1 + (4\tau + 3\nu)^{-1/2})^2 \exp[-\theta_\nu/\tau], \tag{5.57}$$

with

$$\theta_\nu \simeq \frac{\nu}{1+\nu^{-1/2}} \qquad (\nu > 0). \tag{5.58}$$

The exponential term in this expression reflects that only those collisions that have enough initial kinetic energy to overcome the repulsive interaction potential will contribute.

Photo-electric rates

The photo-electric ejection rate is given by

$$J_{\rm pe}(Z_{\rm d}) = 4\pi \int_{\nu_{Z_{\rm d}}}^{\nu_{\rm H}} \frac{J(\nu)}{h\nu} \sigma_{\rm d}(\nu) Y_{\rm ion}(Z_{\rm d}, \nu) {\rm d}\nu. \tag{5.59}$$

The photo-ionization yield, $Y_{\rm ion}$, initially rises rapidly with photon energy above the ionization potential and then levels off at a constant yield for higher energies. A semi-empirical relation for this yield is given by

$$Y_{\rm ion}(Z_{\rm d}, \nu) = Y_\infty \left(1 - \frac{IP(Z_{\rm d})}{h\nu}\right) f_{\rm y}(a), \tag{5.60}$$

where Y_∞ is the photo-ionization yield for bulk materials and $f_{\rm y}(a)$ is the yield enhancement factor for small grains. While for bulk materials the ionization potential, IP, is equal to the work function, W (4–6 eV for relevant materials), this is not the case for very small grains. For small grains, the ionization potential increases because of the electrostatic work required to charge up the particle,

$$IP(Z_{\rm d}) - W = \left(Z_{\rm d} + \frac{1}{2}\right) \frac{e^2}{C} = \left(Z_{\rm d} + \frac{1}{2}\right) \frac{e^2}{a}, \tag{5.61}$$

where the capacitance of a spherical grain, $C = a$, has been inserted. This becomes important for sizes, $a \simeq 3|Z_{\rm d}|$ Å. Thus, for large grains, the ionization potential is equal to the work function of the bulk. For small grains, the ionization potential can become rapidly very large, severely limiting the total charging up such a grain may experience.

The yield enhancement factor, $f_{\rm y}(a)$, takes into account that for bulk materials the photon attenuation depth, $l_{\rm a}$, in the material is larger than the electron mean free path, $l_{\rm e}$. The photo-electron created deep within bulk material never makes it out to the surface and the photo-electric effect is suppressed. Without derivation, this factor is approximately given by

$$f_{\rm y}(a) = \left(\frac{\zeta}{\alpha}\right)^2 \frac{(\alpha^2 - 2\alpha + 2 - 2\exp[-\alpha])}{(\zeta^2 - 2\zeta + 2 - 2\exp[-\zeta])}, \tag{5.62}$$

with $\alpha = a(l_{\rm e} + l_{\rm a})/l_{\rm e} l_{\rm a}$ and $\zeta = a/l_{\rm a}$. For small grains, the yield enhancement factor is given by $(l_{\rm e} + l_{\rm a})/l_{\rm e}$, while for large grains, $f_{\rm y}$ goes to unity (e.g., within this formalism, the photo-electric yield is normalized on that measured for bulk materials and is enhanced for small grains). Typical values for Y_∞, and l_e and l_a are 0.15, and 10 and 100 Å, respectively.

Assuming absorption cross sections for very small graphitic grains and the diffuse interstellar radiation field, the photo-electric ionization rate can now be found from Eq. (5.59):

$$J_{pe}(Z_d) \simeq 2.5\times10^{-13}(13.6 - IP\,(Z_d))^2\, N_c f_y(N_c)\, G_0 \,\text{electrons s}^{-1}, \quad (5.63)$$

where it is assumed that the total FUV absorption cross section is proportional to the total number of carbon atoms, N_c, in the grain. For grains larger than $a > 100$ Å, the FUV absorption cross section will scale with the surface area;

$$\begin{aligned} J_{pe}(Z_d) &\simeq 1.6\times10^{-7}(13.6 - IP(Z_d))^2\left(\frac{a}{100\,\text{Å}}\right)^2 f_y(N_c) G_0 \,\text{electrons s}^{-1} \\ &\simeq 2.6\times10^{-11}(13.6 - IP(Z_d))^2 N_c^{2/3} f_y(N_c) G_0 \,\text{electrons s}^{-1}. \end{aligned} \quad (5.64)$$

Ion field emission

When a grain becomes highly positively charged, the strong electric field near the surface will lead to the ejection of positive ions. The critical field strength is typically $\simeq 2-5$ VÅ^{-1}. Because ion field emission is governed by an exponential law, this critical field is insensitive to the characteristics of the material and the temperature. Adopting a value of 3 VÅ^{-1}, the critical grain charge required for field emission to start is

$$Z_{d(cr)} \simeq 2100\left(\frac{a}{100\,\text{Å}}\right)^2. \quad (5.65)$$

All other things being equal, field emission will lead to a rapid destruction of the grain since, as the grain size diminishes, the electric field near the surface increases in strength, leading to further ion loss. However, this runaway process may be counteracted by a concomitant increase in the ionization potential of the grain; viz.

$$IP(Z_{d(cr)}) \simeq 300\frac{a}{100\,\text{Å}}\ \text{eV}. \quad (5.66)$$

It is clear that field emission is only important for very small grains subjected to an intense radiation field of highly energetic photons.

Electron field emission

When a grain is highly negatively charged, the strong electric field near the surface will lead to the ejection of electrons. Again, the critical field strength is not very sensitive to the material parameters or the temperature. Because of the small mass of the electron – which promotes tunneling – the critical field strength for electron emission is much less than for ions, $-0.3\ \text{V Å}^{-1}$. The critical grain charge for electron field emission is then

$$Z_{\text{d(cr)}} \simeq -210 \left(\frac{a}{100\,\text{Å}} \right)^2. \tag{5.67}$$

This process can be important in environments where collisional charging operates and this leads to the following constraint on the temperature of the plasma (cf. Eq. 5.74),

$$T_{\text{cr}} \simeq 1.4 \times 10^5 \frac{a}{100\,\text{Å}}\ \text{K}. \tag{5.68}$$

The ionization potential that corresponds to this critical grain charge is

$$IP(Z_{\text{d(cr)}}) \simeq 30 \frac{a}{100\,\text{Å}}\ \text{eV}. \tag{5.69}$$

For very small grains, highly energetic electrons are needed; i.e., plasma temperatures in excess of 3×10^5 K, and it seems that this process never really dominates.

Sticking coefficients

Highly energetic particles incident on a grain can lead to electron emission. In particular, primary electrons with energies in excess of some 50 eV will create secondary electrons, which may diffuse through the grain, find the surface and – if they have enough energy to overcome the work function and coulomb potential – escape, thereby limiting the build-up of negative charge. This can become particularly important for incident electron energies of 200–400 eV where the yield of secondaries per incident electron is of the order of unity. The typical escape length of low energy electrons in solid materials is some 10 Å. Hence, for more energetic primary electrons, the yield will drop with E^{-2} because they deposit their energy too deeply in the grain and the resulting secondary electrons do not make it out. Really energetic electrons impinging on small grains may penetrate through the whole grain and leave on the other end. This will reduce the electron sticking coefficient. The range of an incident electron with energy, E_{e}, is $R \simeq 150(E_{\text{e}}/1\,\text{keV})^{1.5}$ Å. Of course, when the penetration depth of a primary electron becomes comparable to the size of a grain, secondaries can more easily escape as well.

Ions impacting on a grain can also lead to secondary electron emission. However, in this case, the energies involved are much higher. For protons and helium ions impacting a solid, the secondary electron yield reaches about unity for energies of some 5 keV ($T \simeq 5 \times 10^7$ K). The yield keeps on rising to a maximum of a few electrons around 100 keV for H and 1 MeV for He. Thus, such ion impacts are very effective in positively charging a grain. Of course, the range for these ions becomes important as well. Typically, the stopping range for a proton with energy, E_H, is $\simeq 100(E_H/1\,\text{keV})$ Å, or about 1 μm for a 100 keV proton. For helium ions, the range is about 0.6 that of H.

All these effects have to be folded in when evaluating the grain charge in high temperature plasmas ($T > 10^6$ K) or when grains are moving hypersonically ($v_{\text{drift}} > 100\,\text{km s}^{-1}$) through a medium.

Collisional charging

Consider now the simplified case of a low temperature plasma where collisional charging dominates and the sticking coefficients are unity. In that case, the grains will be negatively charged because electrons move faster than ions. The ratio of the abundances of adjacent ionization stages is then given by

$$\frac{f(Z_d+1)}{f(Z_d)} = \frac{J_{\text{ion}}(Z_d)}{J_e(Z_d+1)} = \left(\frac{m_e}{m_{\text{ion}}}\right)^{1/2} \frac{\tilde{J}(\tau, \nu = Z_d)}{\tilde{J}(\tau, \nu = -Z_d - 1)}, \tag{5.70}$$

where Z_d is of course negative and we have assumed charge conservation ($n_{\text{ion}} = n_e$) and sticking coefficients of unity. For large $\tau\,(= akT/e^2)$, the charge distribution function approaches a Gaussian,

$$f(Z_d) = \frac{1}{2\pi\sigma_Z} \exp\left[-\frac{(Z_d - Z_p)^2}{2\sigma_Z^2}\right]. \tag{5.71}$$

The peak of the distribution is now found from the condition that the colliding fluxes of ions and electrons balances;

$$\left(\frac{m_e}{m_{\text{ion}}}\right)^{1/2} \tilde{J}(\tau, \nu = Z_p) = \tilde{J}(\tau, \nu = -Z_p), \tag{5.72}$$

with Z_p the grain charge at the peak of the distribution. More negatively charged grains will have a higher ion-to-electron collision ratio while more positively charged grains will have the opposite. Using the relevant equations (Eqs. (5.56) and (5.57)) in the limit τ and $\nu \gg 1$, this leads to the familiar transcendental equation for the grain charge,

$$\exp\left[-\frac{|Z_p|e^2}{akT}\right] = \left(\frac{m_e}{m_{\text{ion}}}\right)^{1/2} \left(1 + \frac{|Z_p|e^2}{akT}\right). \tag{5.73}$$

For electron and proton collisions, this results in

$$\frac{Z_p e^2}{akT} = -2.5. \tag{5.74}$$

Thus, for the "peak" charge, the coulomb enhancement factor for ion–grain collisions is a factor of 3.5 while the reduction of the electron collision rate is a factor of 12.3. The grain charge is then

$$Z_{\rm p} \simeq -1.5\left(\frac{a}{1000\,\text{Å}}\right)\left(\frac{T}{100\,\text{K}}\right). \tag{5.75}$$

The dispersion of the distribution is given by

$$\sigma_Z \simeq 0.9\left(\frac{akT}{e^2}\right)^{1/2} \simeq 1.2\left(\frac{a}{1000\,\text{Å}}\right)^{1/2}\left(\frac{T}{100\,\text{K}}\right)^{1/2}. \tag{5.76}$$

Collisional charging is rarely dominant in the interstellar medium except for very dense environments well shielded from UV radiation (e.g., dense molecular clouds, protoplanetary disks, or stellar outflows). For dense molecular clouds, τ is small and, in that limit, only two charge states are of importance, -1 and 0. As long as the grains are not the main charge carriers (but see Section 10.5), the ionization fractions are approximately given by

$$f(0) \simeq \left[1+\left(\frac{\tau}{\tau_0}\right)^{1/2}\right]^{-1}$$

$$f(-1) \simeq \left(\frac{\tau}{\tau_0}\right)^{1/2}\left[1+\left(\frac{\tau}{\tau_0}\right)^{1/2}\right]^{-1}, \tag{5.77}$$

where τ_0 is,

$$\tau_0 = \frac{8m_{\rm e}}{\pi m_{\rm ion}}. \tag{5.78}$$

In a molecular cloud with a temperature of 10 K and a typical ion mass of 30 amu, we have $\tau \simeq 0.06$ for a 1000 Å grain and $\tau_0 \sim 5\times10^{-5}$. In that case, $f(-1)\simeq 0.97$ and $f(0)\simeq 0.03$. For comparison, the fraction of singly positively charged grains is $\simeq 2\times10^{-5}$ in that case.

Photo-electric charging

In most of the diffuse interstellar medium, the grain charge is dominated by the photo-electric effect balanced by electron recombination. The abundance ratio of adjacent ionization stages is then given by

$$\frac{f(Z_{\rm d}+1)}{f(Z_{\rm d})} = \frac{J_{\rm pe}(Z_{\rm d})}{J_{\rm e}(Z_{\rm d}+1)}, \tag{5.79}$$

where $Z_{\rm d}$ is of course positive and we have neglected ion–grain recombination. Evaluating this expression for large, multiply charged grains in the diffuse ISM with an ionization potential equal to the work function (which we set equal to 5 eV), we find

$$\frac{f(Z_{\rm d}+1)}{f(Z_{\rm d})} \simeq 6.1 \frac{G_0}{n_{\rm e} T^{1/2}} \left[1 + \frac{Z_{\rm d}+1}{\tau}\right]^{-1}. \tag{5.80}$$

When $(Z_{\rm d}+1)/\tau$ is large, coulomb focussing is important and the natural parameter describing the ionization is $G_0 T^{1/2}/n_{\rm e}$ (the coulomb interaction prefers the slowest electrons). For small $(Z_{\rm d}+1)/\tau$, the important interaction parameter is $G_0/n_{\rm e}T^{1/2}$.

As for the pure collisional interaction, the charge distribution will resemble a Gaussian (cf. Eq. 5.71) with a peak charge given by the condition that electron ejection and electron recombination balance,

$$Z_{\rm p} \simeq 36 \left(\frac{\gamma}{1000\,{\rm cm}^3{\rm K}^{1/2}}\right)\left(\frac{a}{1000\,\text{Å}}\right), \tag{5.81}$$

with $\gamma = G_0 T^{1/2}/n_{\rm e}$, which is of the order of 1000 cm ^{3} K$^{1/2}$ in the diffuse ISM. Thus, the photo-electric effect can charge up grains considerably in the diffuse ISM with potentials of $Z_{\rm p}e^2/akT \simeq 6\gamma/T \simeq 60$. Ignoring the dependence of the photo-electric rate on the grain charge through the ionization potential (e.g., the grains are not too highly charged) and assuming that $Z_{\rm p}e^2/akT \gg 1$, the width of this Gaussian distribution is

$$\sigma_Z = \sqrt{Z_{\rm p}}. \tag{5.82}$$

The charge distribution function

Some representative results for the grain charge distribution function are shown in Fig. 5.3. The charge distribution functions are reasonably well represented by Gaussians. However, as a result of the various approximations made, the peak positions are somewhat displaced from the simple analytical estimates (Eqs. (5.75) and (5.81)). As expected, larger grains are more negatively charged when collisions dominate and more positively charged when the photo-electric effect dominates. In either case, the distribution is also broader for larger grains. The electrostatic grain potential is given by

$$\phi = \frac{Z_{\rm d}e^2}{a}, \tag{5.83}$$

which is typically independent of grain size. In particular, in a photo-electric charging environment, we have $\phi \simeq 6.9 \times 10^{-4}\gamma$ eV.

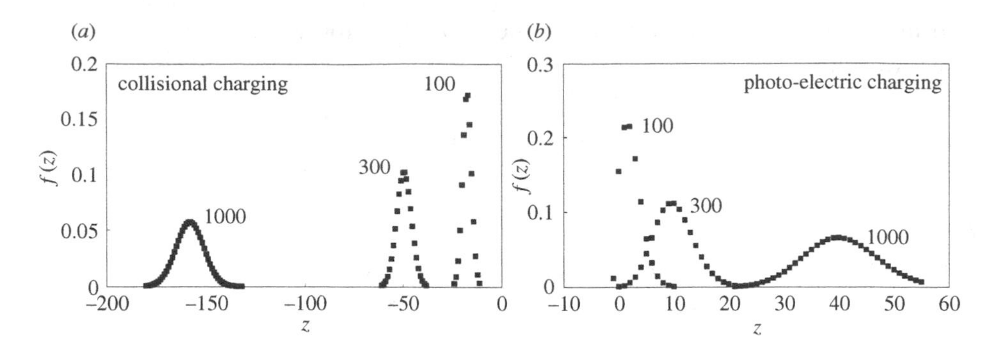

Figure 5.3 Typical charge distribution function for interstellar grains. (*a*) Collisional charging calculated for a temperature of 10^4 K. (*b*) Photo-electric charging calculated for $\gamma = 1500\,\mathrm{cm}^3\,\mathrm{K}^{1/2}$.

5.2.4 Dust destruction processes

Dust is mainly destroyed in the ISM by strong shock waves in the warm phases of the interstellar medium. Sputtering by impacting gas species and vaporization by high velocity grain–grain collisions return grain material to the gas phase. At the same time, shattering by grain–grain collisions considerably modifies the grain size distribution. Here, we will discuss the physical processes involved. Dust destruction in interstellar shocks is discussed further in Chapter 13.

Sputtering

The interaction of energetic ions with solid materials can lead to the ejection of atoms into the gas phase. At "low" energy (<100 eV amu^{-1}) collisions, nuclear interaction dominates the energy deposition, while at higher (>1 keV amu^{-1}) energies electronic excitations by the fast moving ions are more important. For shocks, the former process controls the sputtering process. At higher energies, recoil atoms are energetic enough to create a cascade, which ejects surface atoms. The sputtering yield depends on the ratio of the energy deposited near the surface to the binding energy of the solid. At low energies, the energy deposited will depend on the projectile-to-target-mass ratio (lighter projectiles are less efficient in transferring energy) and on the square root of the ion energy. The nuclear stopping power decreases again after $\sim$100 eV amu^{-1}. At low energies and for light projectiles, sputtering is due to the reflection of the penetrating ion at a deeper layer, and during the return journey a surface atom is knocked out. For light projectiles, the threshold energy for sputtering is then $E_{\rm th} = U_0/(g(1-g))$, with U_0 the binding energy and g the maximum fractional energy transfer possible in a head-on collision, $g = 4m_1m_2/(m_1 + m_2)^2$, with m_i the projectile and target mass. Threshold energies for astrophysically relevant materials have been collected in Table 5.2. Typically, for light ions, the ion energy has to exceed this threshold value by a factor of 3 for appreciable yields.

Table 5.2 *Sputtering threshold energies*

Material	Sputtering				Grain–grain collisions	
	U_0 (eV)	m_2 (amu)	E_{th}(H) (eV)	E_{th}(He) (eV)	$v_{th}(vapor)$ (km^{-1})	$v_{th}(shatter)$ (km^{-1})
Silicate	5.7	23	42	23	19	2.7
Amorphous carbon[a]	4.0	12	20	22	23	1.2
SiC	6.3	20	43	26	25	8.8
Iron	4.1	56	64	16	11	2.2
Ice	0.53	18	2.7	2.2	6.5	1.8

[a] This ignores chemical sputtering, which is very important for amorphous carbon and graphite at low energies.

Sputtering of materials is of great importance in many plasma environments – ranging from the silicon industry to fusion in tokomaks – and a multitude of experimental studies have been performed. In addition, various theoretical models have been perfected. In particular, the TRIM numerical code – which calculates the interaction of high energy ions with materials – is widely available to evaluate sputtering yields of materials. These studies have revealed the existence of a "universal" sputtering law, which relates the sputtering rate to the energy of the impacting ion through an appropriate scaling of target and ion parameters. Experimental data show reasonable agreement with this relationship, particularly at moderate energies (10 keV to MeV range). For low ion masses and near the threshold energy – and this is of most interest to astrophysics – a material-dependent "fudge" factor has to be introduced, which has to be determined semi-empirically by comparison to experiments.

This "universal" relation is compared to available laboratory data for the astrophysically relevant materials, graphite, SiC, Fe, SiO_2, and H_2O-ice, in Fig. 5.4. Overall, this relationship compares well to the experimental data. For astrophysics, sputtering near the threshold energy is particularly relevant and, unfortunately, there is a general lack of experimental data near the threshold energy for many relevant materials. Theoretical studies using the TRIM code are of little use at low energies. The TRIM code relies on binary collisions to evaluate the ion trajectory and energy transfer. However, near the threshold energy, the ion will experience many-body interactions, reflecting the increased coulomb cross section at low energies. Therefore, the "success" of the universal sputtering relation has to be judged from the general good agreement with sputtering data in this energy range, which is actually very impressive. For carbonaceous materials, a caveat should be made here. Near the threshold energy, sputtering of carbonaceous materials is dominated by chemical processes and hence the sputtering yield near the threshold energy is much enhanced compared with the values derived from this universal sputtering relation. For silicates, few data are available in general and the available data apply to quartz rather than Mg or Fe silicates. For H_2O-ice, the sharp rise in the sputtering yield above 1 keV results from electronic stopping losses, which have not been included in the theoretical relationship. In astrophysics, sputtering of ice is dominated by low energy impacts and this discrepancy is of little consequence but, for planetary studies, this energy range is of great importance. One key parameter entering in these fits is the cohesive binding energy (i.e., the energy per atom necessary to disperse a solid into its constituent atoms). For most materials these can be taken directly from tabulations of heats of formation. However, graphite is actually amorphized under the influence of high doses of energetic ions. This actually reduces the cohesive binding

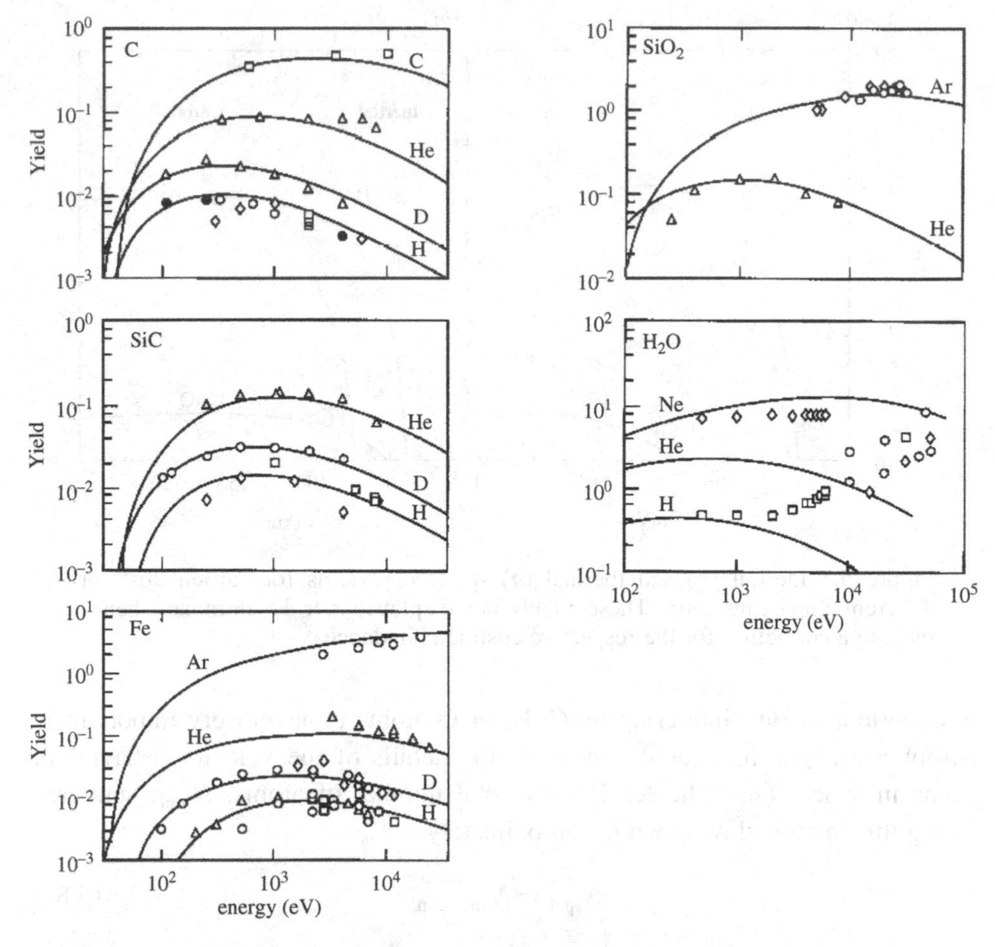

Figure 5.4 Comparison of experimental sputtering yields with the universal sputtering relation for some astrophysically relevant materials. Figure reproduced with permission from A. G. G. M. Tielens, C. F. McKee, C. G. Seab, and D. Hollenbach, 1994, *Ap. J.*, **431**, p. 321.

energy from 7.5 eV C atom^{-1} to 4 eV C atom^{-1} – typical for amorphous carbon materials – and this value has been used in the fit shown in Fig. 5.4.

These sputtering yields have to be convolved with gas phase abundances to derive relative sputtering yields. For low shock velocities (<250 km s^{-1}; radiative shocks; Section 12.3), grains are moving with respect to the gas owing to their own inertia until they are dragged to a halt (Chapter 13). Because of the mass difference, an impacting He ion will have a higher energy than an H atom and hence will have a higher yield. This difference in yield more than compensates for the difference in abundance and He sputtering dominates at low velocities (Fig. 5.5). At higher velocities (i.e., well above threshold), H sputtering becomes

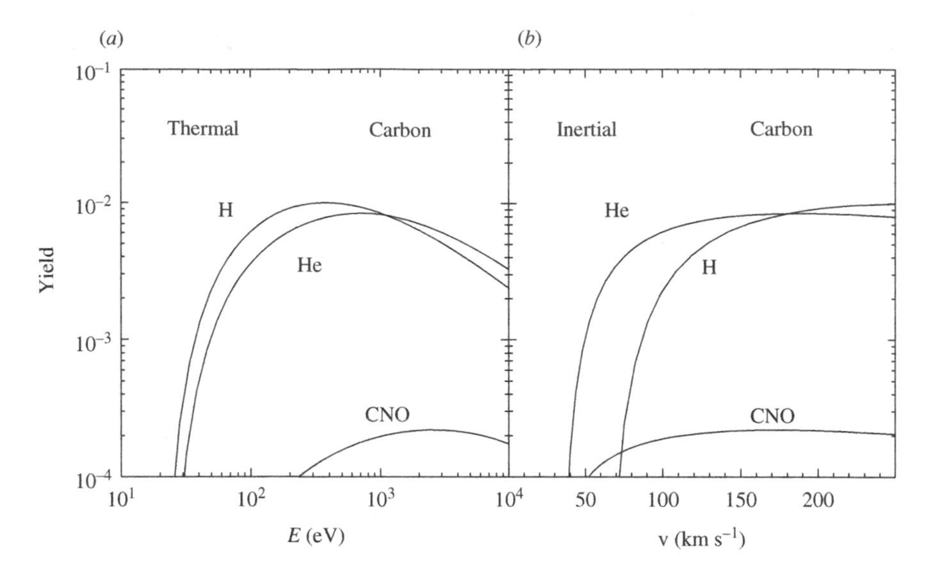

Figure 5.5 Inertial (b) and thermal (a) sputtering yields for carbon dust for different impacting ions. These yields are displayed per H atom and hence include a correction for the respective cosmic abundances.

as important as He. Sputtering by C, N, or O atoms is never very important in astrophysics. Ignoring, for the moment, the details of the velocity behavior of grains in shocks (see Chapter 13), the total number of atoms, $N_{\rm sput}$, sputtered during this inertial slow down is approximately

$$N_{\rm sput} = N_{\rm ion}\overline{Y}_{\rm ion} \tag{5.84}$$

where $N_{\rm ion}$ is the total number of ions the grain collides with and $\overline{Y}_{\rm ion}$ the average yield. For a grain to be slowed down by a factor e, it has to collide with its own mass in gas phase atoms. Thus, taking a carbonaceous grain as an example, $N_{\rm H} \simeq 8.6N_{\rm C}$, with $N_{\rm C}$ the total number of C atoms in the grain, and the numerical factor also allows for collisions with He. At $100\,{\rm km\,s^{-1}}$, the sputtering yield by H atoms and He atoms is $\simeq 10^{-2}$ and $\simeq 10^{-1}$, respectively. The fraction of atoms sputtered is then $\simeq 0.17$, independent of grain size. This corresponds to only a 6% reduction in size. For higher velocities, the yield does not increase much but several e factors in reduction are involved before the velocity drops below the threshold for appreciable sputtering (cf. Fig. 5.5) and the sputtered fraction increases accordingly. More importantly, at high velocities ($>250\,{\rm km\,s^{-1}}$), sputtering can be more appreciable because, while grains are quickly dragged to a halt, the gas stays hot for a long time (i.e., the shocks are adiabatic). In sputtering by thermal gas, impacting ions will have the same energy distribution and, then, H and He sputtering are quite comparable over the full range (cf. Fig. 5.5). The total sputtered fraction now depends on the total time spent in the hot gas (e.g., the lifetime of the supernova remnant; cf. Chapter 13.2.2).

Grain–grain collisions

Collisions between grains at moderate to high velocities can lead to vaporization of grain material as well as to shattering or crater formation. Much experimental and theoretical research has focussed on this subject because of the micrometeorite hazard on space craft surfaces, because of the importance of cratering in the early Solar System, and because of armory protection and (nuclear) weapons considerations.

When a projectile impacts a target at high velocity, shock waves will propagate from the impact area into the target and the projectile. When the shock wave reaches the back of the projectile or the "free" surfaces of the target, rarefaction waves will travel into the shocked regions relieving the high pressures. At the end of this (early-time) phase, the coupling of energy and momentum to the target is essentially complete. In the second phase, the excavation phase, the shock has departed from the immediate vicinity of the projectile and the transient crater. There are now two separate flows in the target: 1) The detached shock penetrating deeper and deeper into the target while compressing the material and decaying in strength. 2) The excavation flow that will eventually carve out the final crater. These two phases are very similar to the first two phases of the evolution of a supernova remnant in the interstellar medium (Section 12.3). The third phase of an SNR evolution, the radiative phase, is of no concern here because the relevant processes in a solid will be over by then.

The decay of the pressure into the target will depend on the coupling of energy and momentum to the target. As for the size-energy relation of supernova remnants discussed in Section 12.3, extensive numerical and experimental studies have revealed a relatively simple relationship between the mass and velocity of the impactor and the target mass shocked to a given pressure: typically, but not precisely, $E \sim PV$ (because some of the energy is lost while extra momentum is deposited by the hot plasma "escaping" from the impact site). The "damage" (e.g., vaporized mass, crater volume) done reflects the target volume shocked to some critical pressure. For vaporization, this critical pressure corresponds to a critical energy that is twice the binding energy. Appreciable vaporization of the target volume requires an internal energy that is about five times the binding energy. For shattering (and crater formation), this critical pressure is the ultimate tensile strength ($G/6$, with G the shear strength of the material) for submicron-sized targets. If we translate this critical pressure to a critical velocity of the impact,

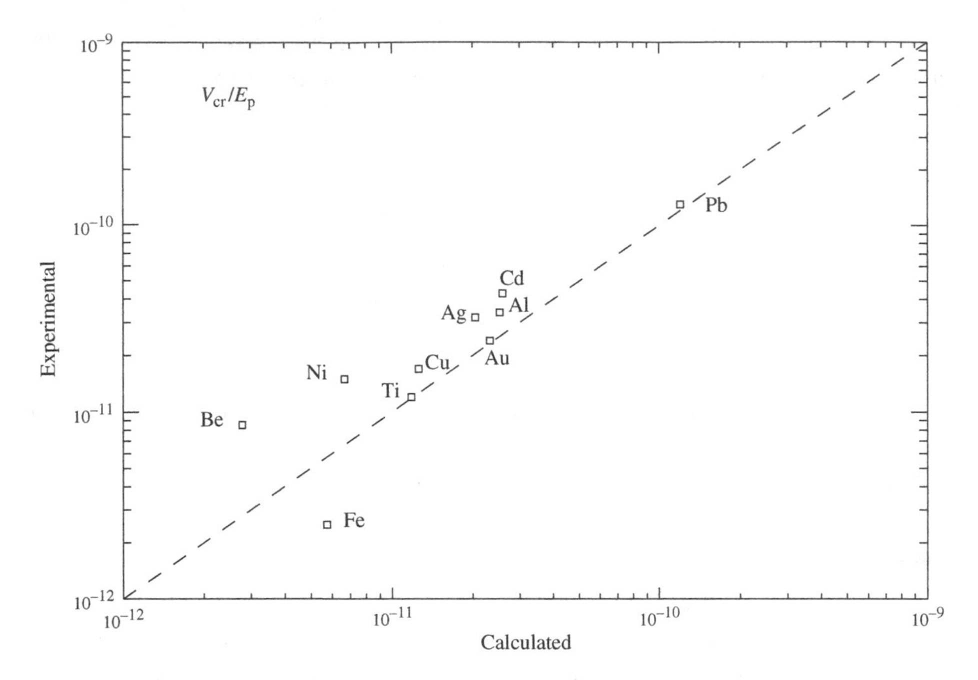

Figure 5.6 The calculated ratio of crater volume to projectile kinetic energy is compared to the measured ratio for micron-sized iron projectiles impacting various surfaces at $\simeq 5\ \mathrm{km\,s^{-1}}$. Figure reproduced with permission from A. P. Jones, A. G. G. M. Tielens, and D. Hollenbach, 1996, *Ap. J.*, **469**, p. 740.

then the ratio of the mass of target material processed to the projectile mass scales with $(v/v_{cr})^{16/9}$.

Figure 5.6 compares calculated ratios of the crater volume to the projectile kinetic energy with experimental data for impacts of iron projectiles on various metallic surfaces. Shock-wave data have been taken from experiments. In all cases, the shear modulus was adopted for the critical pressure. The results adopting the simple scaling relation agree reasonably well with the data. But note that the impact velocities in these experimental studies are much less than relative grain collision velocities in an interstellar shock ($\simeq 100\ \mathrm{km\ s^{-1}}$). The size distribution of the fragments produced by the cratering flow is not well known experimentally. Theoretically, a size distribution with a power law $\sim a^{-\alpha}$, with α near 3.5, is expected. Essentially, the fragment size produced at any depth depends on the pressure experienced, which drops almost linearly with the volume affected.

Critical relative grain velocities for vaporization and shattering are summarized in Table 5.2. Typically, the critical velocity at which cratering ensues, v_{crit}, is 1–10 $\mathrm{km\,s^{-1}}$. Consistent with the relative ease with which graphite flakes, its critical velocity ($1.2\ \mathrm{km\,s^{-1}}$) is at the low end of this range. Likewise, ice shatters readily ($1.8\ \mathrm{km\,s^{-1}}$). However, silicon carbide ($8.8\ \mathrm{km\,s^{-1}}$) and (single crystal) diamond ($16.1\,\mathrm{km\,s^{-1}}$) are rather robust materials that withstand shattering better than most other materials. The ratio of the vaporized mass to total mass affected by the collision (vaporized plus shattered) is well represented by $M_v/M_T = 0.76\,(G/P_v)^{1.07}$, where G is the shear strength of the material and P_v is the critical pressure for vaporization. For graphite and silicates, this ratio is 0.005 and 0.023, respectively. Obviously, shattering will be the dominant process affecting the interstellar grain size distribution.

Most of the shattered volume in interstellar shock waves results from high velocity impacts of small projectiles on large targets. For high enough velocity, complete destruction of the target will ensue (catastrophic destruction). As an example, consider a $100\,\mathrm{km\,s^{-1}}$ impact of a 50 Å grain on a 1000 Å target. The (minimum) velocity at which catastrophic destruction of the target occurs, v_{cat}, is 75–$600\,\mathrm{km\,s^{-1}}$ depending on the material parameters. Graphite ($v_{\mathrm{cat}} = 75\,\mathrm{km\,s^{-1}}$) and ice ($115\,\mathrm{km\,s^{-1}}$) are the most easily destroyed this way, somewhat more easily than silicates ($175\,\mathrm{km\,s^{-1}}$). SiC ($560\,\mathrm{km\,s^{-1}}$) and diamond ($1000\,\mathrm{km\,s^{-1}}$) are the sturdiest. For a 100 Å impact, these catastrophic velocities are reduced to 23–$175\,\mathrm{km\,s^{-1}}$. Of course, a 1000 Å impact only needs the critical shattering velocity for complete disruption of the target.

5.3 Observations

5.3.1 Interstellar extinction

The presence of dust in the interstellar medium was first recognized by its reddening effect on the light from distant stars. The apparent magnitude, m, of a star is given by

$$m(\lambda) = M(\lambda) + 5\log[d] + A_\lambda, \tag{5.85}$$

with M the absolute magnitude, d the distance, and A_λ the extinction due to dust. Now, the extinction can be derived by comparing a reddened star with a nearby star with the same spectral type, resulting in a magnitude difference given by

$$\Delta m(\lambda) = 5\log\left[\frac{d_1}{d_2}\right] + A_\lambda. \tag{5.86}$$

The color excess between two wavelengths, λ_1 and λ_2, can then be defined as

$$E(\lambda_1 - \lambda_2) = \Delta m(\lambda_1) - \Delta m(\lambda_2) = A_{\lambda_1} - A_{\lambda_2}. \tag{5.87}$$

Because extinction generally increases with decreasing wavelength through the visible, extinction is often called reddening. Color differences between different stars can readily be compared after normalization on a common color difference. Often, the B–V color in the Johnson color system is used for this; viz.

$$\frac{E(\lambda - V)}{E(B - V)} = \frac{A_\lambda - A_V}{A_B - A_V}. \tag{5.88}$$

Measurement of extinction in terms of reddening is, thus, rather straightforward. However, it is physically of more interest to consider the normalized extinction ratio, A_λ/A_V. Define the total-to-selective extinction ratio, R_V, as

$$R_V = \frac{A_V}{E(B - V)}. \tag{5.89}$$

The extinction ratio, A_λ/A_V can then be expressed in terms of the color excess as,

$$\frac{A_\lambda}{A_V} = \frac{1}{R_V}\frac{E(\lambda - V)}{E(B - V)} + 1. \tag{5.90}$$

Because extinction rapidly decreases with increasing wavelength into the infrared, we can write,

$$R_V = -\lim_{\lambda \to \infty} \frac{E(\lambda - V)}{E(B - V)}. \tag{5.91}$$

Thus, the total-to-selective extinction ratio can be determined by extrapolating measured color excesses into the infrared and hence the extinction ratio A_λ/A_V can be determined.

While R_V is introduced here to connect reddening to extinction, this parameter is often used in a different capacity as well. In particular, extensive observations have shown that the observed interstellar extinction – from the mid-IR through the visible and near- and far-UV – can be characterized by one free parameter, which is then chosen to be the total-to-selective extinction ratio. Figure 5.7 shows the mean extinction curve in the solar neighborhood for three different values of R_V. In each case, the extinction is characterized by a gradual increase from the infrared through the visible to the near-ultraviolet. In the near-infrared part of the spectrum the "continuum" extinction is approximately proportional to $\lambda^{-1.7}$, while the visible extinction is somewhat less steep ($A \sim \lambda^{-1}$). The near-ultraviolet is characterized by a distinct knee around $\lambda^{-1} = 2.5\,\mu\text{m}^{-1}$. The ultraviolet portion of the extinction curve is dominated by a pronounced bump centered at 2175 Å (or $4.6\,\mu\text{m}^{-1}$) and a steep rise to short wavelengths.

Thus, the interstellar extinction curve separates out into four distinct parts – the infrared, the visible, the 2175 Å bump, and the far-UV rise. Extinction curves vary from region to region but they can (almost) all be parameterized by the total-to-selective extinction ratio, R_V, over the full wavelength range 0.1–2 μm. The value of R_V depends on the environment traversed by the line of sight. The diffuse ISM is characterized by $R_V = 3.1$, while dense molecular clouds have values

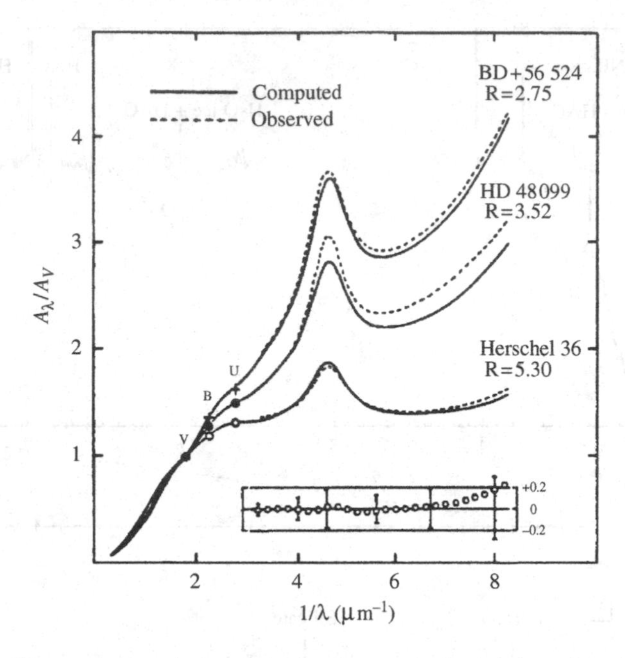

Figure 5.7 Three observed extinction curves are shown as a function of λ^{-1}. These curves show the range in wavelength behavior of the extinction laws in the interstellar medium. The solid lines show, for comparison, the computed parameterized extinction. The insert shows the deviations. Figure courtesy of J. S. Mathis; reprinted with permission from *Ann. Rev. Astron. Astrophys.*, **28**, p. 37, ©1990 by *Ann. Rev.* (www.annualreviews.org).

in the range 4–6. Figure 5.7 compares observed extinction curves in different environments with the R_V-parameterized curve.

Structure in the extinction curve can provide important clues to the nature of the absorbing materials. The ultraviolet bump is the most prominent feature of the interstellar extinction curve. Its peak position is very constant ($\lambda_p = 2175$ Å with a mean deviation of 9 Å). The width of the bump varies much more widely with a typical value of 480 Å but extremes of 360 and 770 Å. After subtraction of an underlying linear continuum extinction, the 2175 Å feature is well represented by a so-called Drude profile, characteristic for absorption associated with a resonance in a conductor.

The visible extinction curve shows weak fine structures, called the diffuse interstellar bands (DIBs), with typical widths of 1–20 Å and strengths $< 0.01\,A_V$. Some 200 bands have been discovered, spanning the wavelength range of the near-UV (4300 Å) to the far-red (~1 μm). These bands are now generally thought to be due to absorption by large molecules rather than dust (cf. Section 6.7.4).

Several absorption features are present in the infrared (cf. Fig. 5.8). The 9.7 μm feature is the strongest one – $\tau_{9.7}/A_V = 18.5$ in the solar neighborhood. This band

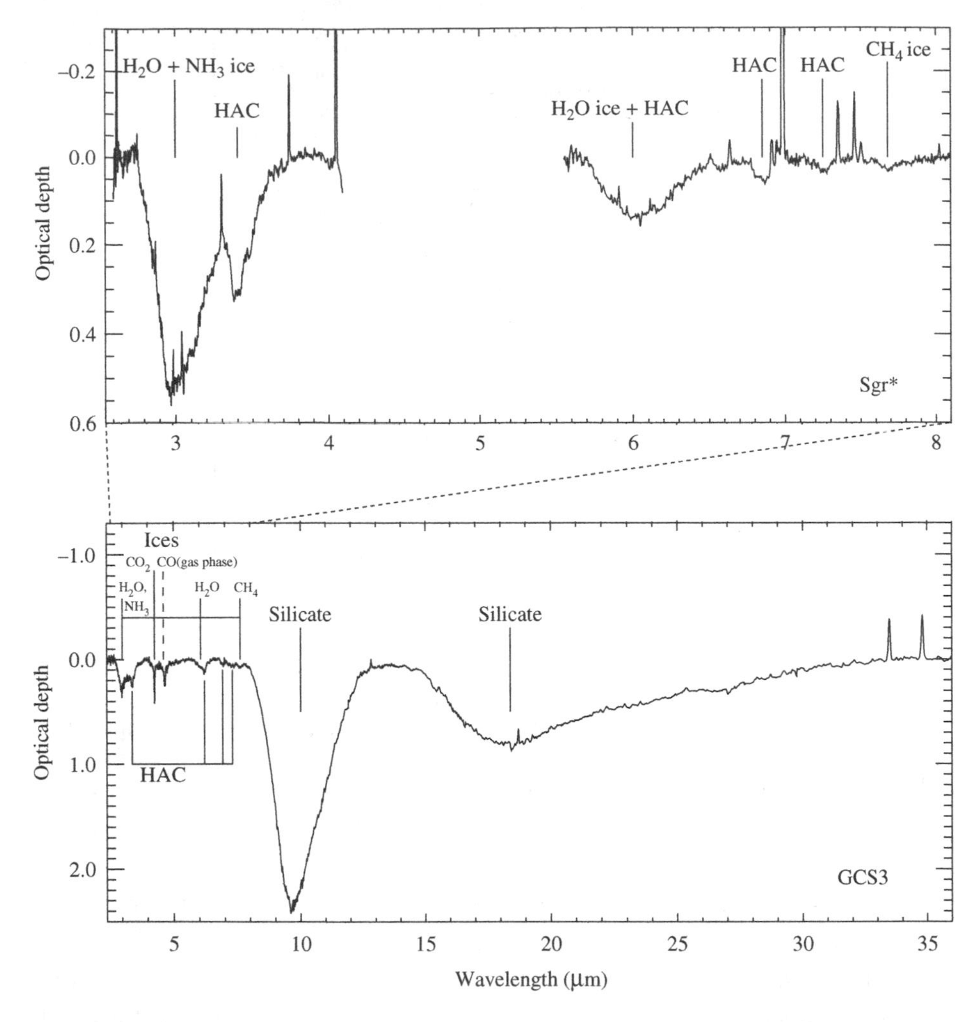

Figure 5.8 The spectrum of sources in the galactic center show strong absorption features due to dust grains along the line of sight. These are shown here on an optical depth scale. Identifications are indicated on top. Some of these features originate in the diffuse interstellar medium (e.g., the hydrocarbon [HAC] bands), while others are due to material in molecular clouds (e.g., H_2O, NH_3, and CH_4 ice). Some bands have probable contributions from both media. Figure courtesy of J. E. Chiar. Data from J. E. Chiar, *et al.*, 2000, *Ap. J.*, **537**, p. 749.

is very broad and structureless, $\Delta\lambda \simeq 2.5\,\mu$m. It is accompanied by a slightly weaker – $\tau_{18}/\tau_{9.7} = 0.6$ – and even broader ($\sim 5\,\mu$m) feature at about $18\,\mu$m. There is also a weak feature at $3.4\,\mu$m, which shows characteristic substructure. For sources in or behind dense molecular clouds, a plethora of absorption features appear. These are attributed to interstellar ice mantles growing on the dust grains in the shielded environment of interstellar clouds and these are discussed in detail in Section 10.7.5.

5.3.2 Scattered light

As reflection nebulae attest, interstellar dust grains are efficient scatterers of radiation. Adopting a geometry for the nebula and star, the scattering properties of the dust can be derived. Because of the uncertainties in geometry, derived properties are limited to the scattering cross section, or the albedo, and the mean value of the cosine of the scattering angle. The albedo is shown in Fig. 5.9. The albedo is quite high, $\simeq 0.6$, and relatively constant in the visible region of the spectrum. Thus, because the extinction rises throughout the visible, the scattering cross section has to increase rapidly towards shorter wavelengths as well. This gives reflection nebulae their characteristic blue color. The broad dip in the albedo curve occurs at the position of the 2175 Å bump in the extinction curve and demonstrates that this is an absorption feature. Likewise, the rapid drop in the albedo at the shortest wavelengths implies that the far-UV rise is an absorption phenomenon. The mean value of the cosine of the scattering angle is about 0.6 in

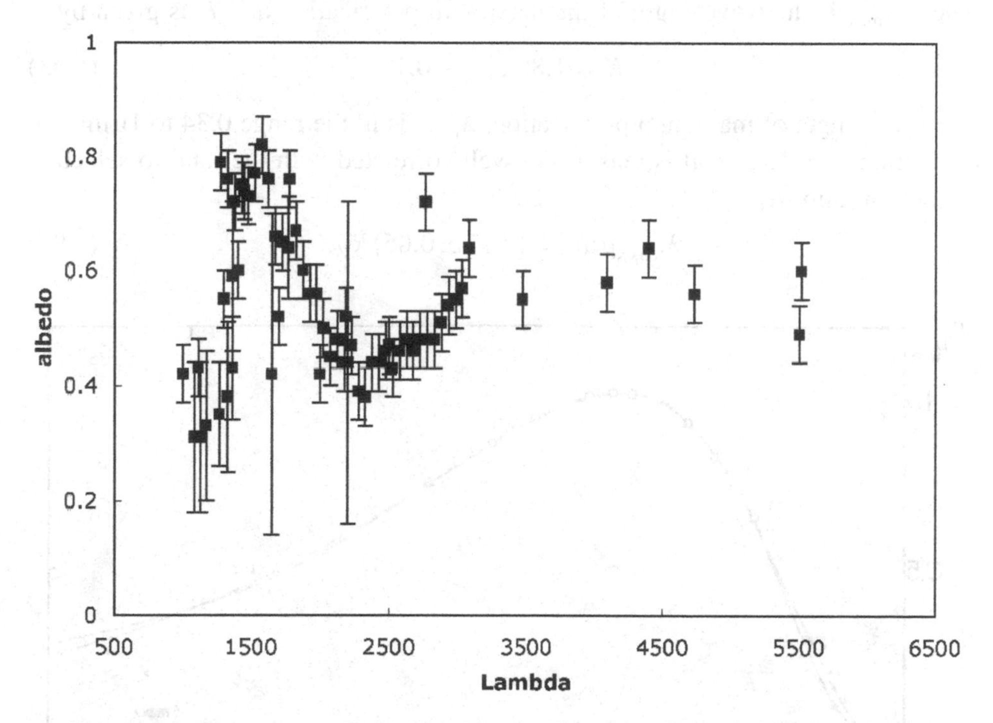

Figure 5.9 The measured albedo of interstellar dust. Note the fairly constant albedo throughout the visible, the presence of the "dip" at the location of the 2175 Å extinction feature, and the rapid drop towards shorter wavelengths. Figure adapted from K. Gordon, 2004, in *Astrophysics of Dust*, ASP Conference Series Vol **309**, ed. Adolf N. Witt, Geoffrey C. Clayton, and Bruce T. Draine, (San Francisco: Astronomical Society of the Pacific), pp. 77–91.

the visible (i.e., strong forward scattering) and decreases to shorter wavelengths (i.e., more isotropic scattering).

5.3.3 Interstellar polarization

Linear polarization

Polarization of starlight is due to propagation of light in a medium where the grains are elongated (the cross section is larger in one direction than in the other) and (partially) aligned. Linear polarization curves rise slowly with decreasing wavelength until they reach a maximum, typically in the visible, and then fall more rapidly (cf. Fig. 5.10). Clearly, the polarization behavior is quite different from the extinction behavior of interstellar dust. Empirically, the linear polarization curve in the visible through near-UV range is given by the so-called Serkowski law,

$$p(\lambda) = p(\lambda_{max}) \exp[-K \ln^2(\lambda/\lambda_{max})], \tag{5.92}$$

where λ_{max} is the wavelength of the maximum polarization and K is given by

$$K = 1.86\lambda_{max} - 0.1. \tag{5.93}$$

The wavelength of maximum polarization, λ_{max}, is in the range 0.34 to 1 μm with an average of 0.55 μm. It is reasonably well correlated with the total-to-selective extinction ratio, R_V,

$$\lambda_{max}(\mu m) = (0.17 \pm 0.05) R_V. \tag{5.94}$$

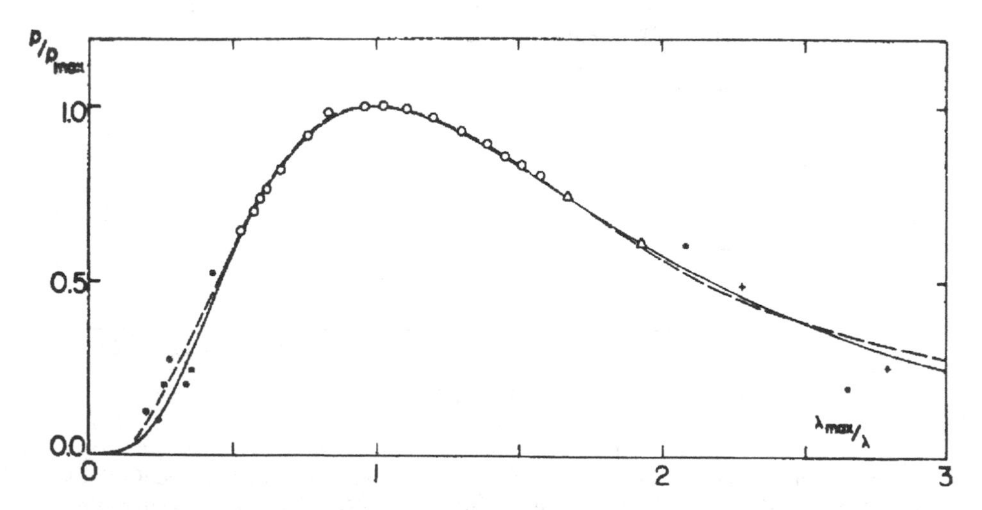

Figure 5.10 The normalized interstellar polarization curve as a function of the normalized wavelength (actually frequency) with $p(\lambda_{max})$ the maximum polarization at the wavelength, λ_{max}.

The FUV polarization often shows excess polarization relative to the Serkowski law but the 2175 Å feature is not polarized. In the near-infrared, the polarization varies approximately with $\lambda^{-1.6}$, independent of λ_{max}. The 9.7 μm feature is polarized, but for the one sight-line measured – towards the galactic center – the 3.4 μm feature is not. The linear polarization-to-extinction ratio, $p(\lambda_{max})/A(\lambda_{max})$, is very small (<0.03 magnitude^{-1}).

Finally, radio synchroton emission due to electrons spiraling around a magnetic field will be polarized orthogonal to the magnetic field. In galactic regions of highly ordered polarization, the optical polarization due to dichroic extinction by dust grains and the radio continuum polarization due to relativistic electrons are highly correlated but perpendicular to each other. Hence, interstellar dust grains are aligned with their long axis perpendicular to the magnetic field.

Circular polarization

Light traversing through a medium of aligned grains whose alignment angle changes position will become circularly polarized. This interstellar birefringe is necessarily weak, at the 0.1% level. Observations show that the circular polarization typically decreases throughout the visible and changes sign around the maximum of linear polarization, λ_{max}, (cf. Fig. 5.11). This behavior is typical for dielectric materials ($k < 0.05$ in the visible).

Polarized infrared emission

The far-IR emission from aligned dust grains will also be polarized, this time with a direction along the long axis of the grains. Far-infrared polarization has been observed for a large number of molecular clouds that have 'high' dust emission optical depths at long wavelengths ($\simeq$0.1–1 mm).

Polarization due to scattering

Scattering of light by dust grains generally also leads to polarization. For single scattering, the polarization vector is perpendicular to the direction between the light source and the scattering grain. Figure 5.12 shows the measured polarization in the K band superimposed on a map of the IR scattering nebula around the young stellar object, BN. The degree of polarization and its distribution provide information on the characteristics of the scattering grains and the geometry of the nebula.

5.3.4 Infrared emission

Dust grains are heated by the interstellar FUV and visible radiation field and emit their absorbed energy in the infrared. As an example, Fig. 5.13 shows the

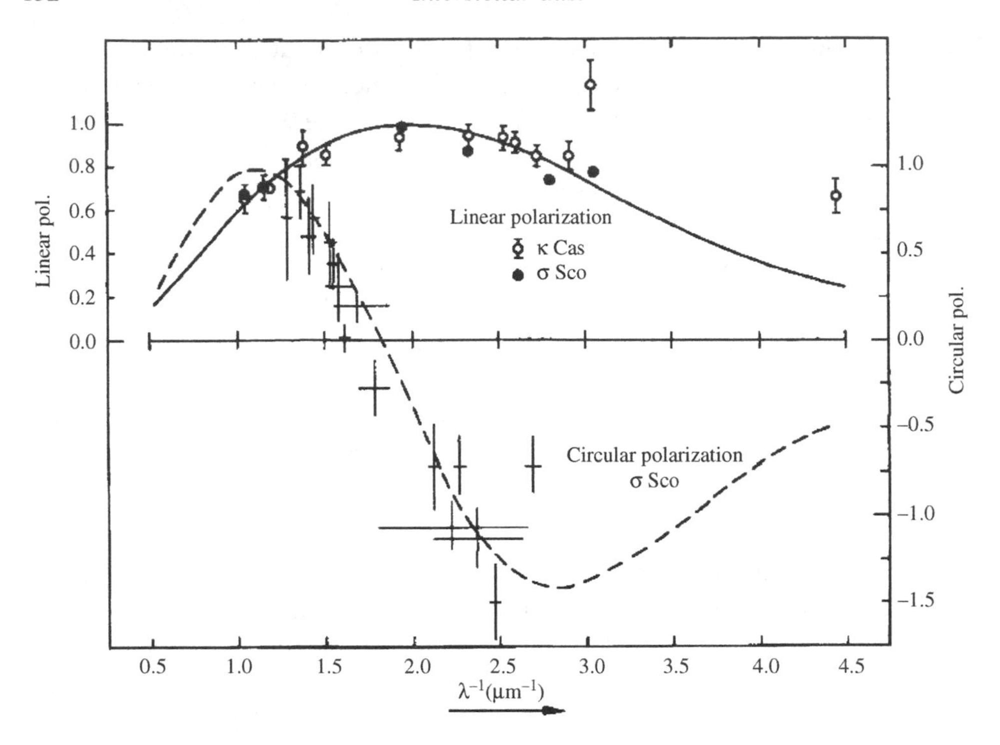

Figure 5.11 The linear and circular polarization curves measured towards the stars σ Sco and κ Cas. The solid and dashed curves present theoretical model calculations for a dielectric grain model. Figure reproduced with permission from J. M. Greenberg and S. S. Hong, 1980, *A. & A.*, **88**, p. 194.

observed IR emission spectrum of the diffuse ISM. Two components are obvious: a cold ($T \simeq 15-20$ K) component emitting mainly at long wavelengths and a hot ($T \sim 500$ K) component dominating the near- and mid-IR emission. The former is due to large dust grains in radiative equilibrium with the interstellar radiation field and the temperatures agree well with those calculated (cf. Chapter 5.2.2). The latter is due to PAH species that are heated by a single FUV photon to temperatures of $\simeq$1000 K and cool rapidly in the near- and mid-IR. These species and their emission process are discussed in Section 6.4.

5.3.5 Dust-to-gas ratio

Dust and gas are well mixed within the ISM. The color excess, $E(B-V)$, is well correlated with the column density of total hydrogen nuclei, $N(\text{HI}) + 2N(\text{H}_2)$,

$$\frac{N\,(\text{HI}) + 2N\,(\text{H}_2)}{E(B-V)} = 5.8 \times 10^{21} \text{atoms cm}^{-2} \text{ magnitude}^{-1}. \tag{5.95}$$

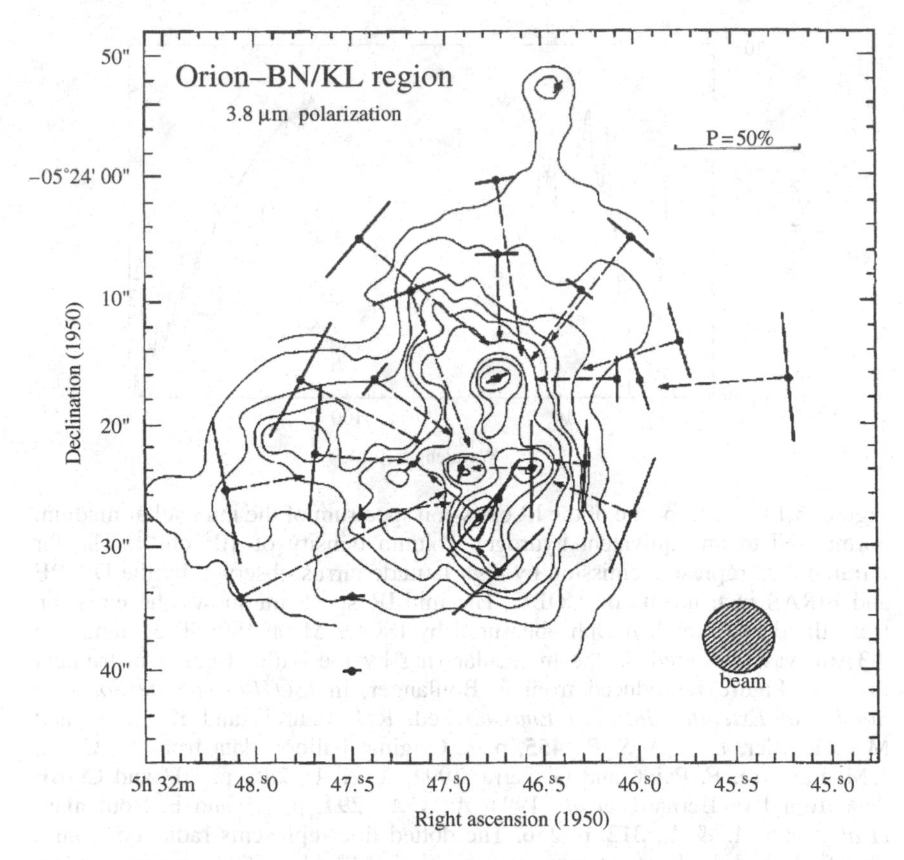

Figure 5.12 The linear polarization of the infrared reflection nebula associated with the region of massive star formation in Orion. The linear polarization vectors measured at K are superimposed on a map of the scattered light intensity. The arrows point back towards the illuminating source. Figure reproduced with permission from M. W. Werner, R. W. Capps, and H. L. Dinerstein, 1983, *Ap. J.*, **265**, p. L13.

With an R_V of 3.1, this corresponds to

$$\frac{N\,(\mathrm{HI}) + 2N\,(\mathrm{H_2})}{A_V} = 1.9 \times 10^{21} \mathrm{atoms\,cm^{-2}\ magnitude^{-1}}. \qquad (5.96)$$

If we assume that the visual extinction is due to 1000 Å grains (e.g., $2\pi a \simeq \lambda$) with $Q_{\mathrm{ext}}(V) \simeq 1$ and a typical specific density of $2.5\,\mathrm{g\,cm^{-3}}$, we arrive at a dust-to-gas mass ratio of 1.1×10^{-2}. There are notable variations. For the star ρ Oph, located behind a dense cloud, the hydrogen column density to color excess ratio is 1.5×10^{22} atoms $\mathrm{cm^{-2}}$ magnitude^{-1}. Even after taking the increased R_V (5.1) into account, this still implies a decreased visual extinction per H nucleus in this dense environment.

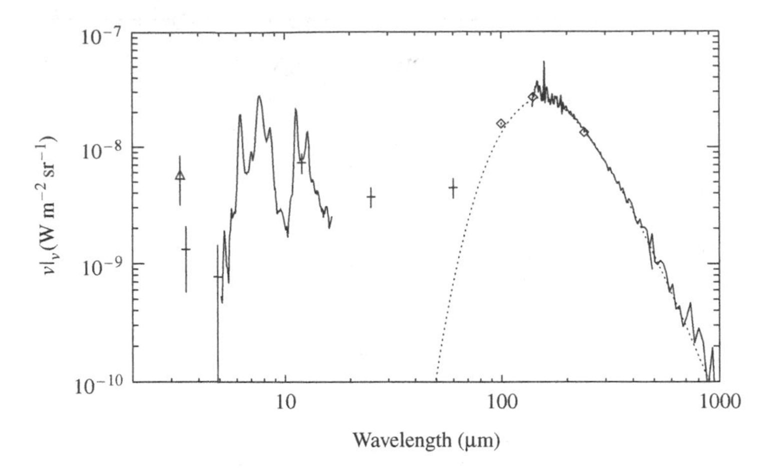

Figure 5.13 A composite of the IR emission spectrum of the interstellar medium normalized to an equivalent hydrogen column density of $10^{20}\,cm^{-2}$. The far infrared data represent emission by high latitude cirrus observed by the DIRBE and FIRAS instruments on COBE. The mid-IR spectrum shows the emission from the dense cloud, ρ Oph, measured by ISOCAM on ISO. The triangle at 3.3 μm was measured on the molecular ring by the balloon-borne instrument Pronaos. Figure reproduced from F. Boulanger, in *ISO Beyond the Sources: Studies of Extended Infrared Emission*, ed. R. J. Laureis and K. Leech and M. F. Kessler, *E. S. A.-S. P.*, **455**, p. 3. Original balloon data from M. Giard, J. M. Lamarre, F. Pajot, and G. Serra, 1994, *A. & A.*, **286**, p. 203 and COBE data from J. P. Bernard, *et al.*, 1994, *A. & A.*, **291**, p. L5 and F. Boulanger, *et al.*, 1996, *A. & A.*, **312**, p. 256. The dotted line represents radiation from a modified black body at a temperature of $T_{dust} = 17.5\,K$ with an emissivity law proportional to ν^2.

The dust distribution in the sky is very patchy; as is very obvious from optical photographs of the Milky Way (Fig. 1.1). In the Solar neighborhood, per kpc, there are about six "Spitzer-type" clouds with a typical visual extinction of 0.2 magnitude and a hydrogen column density of $4 \times 10^{20}\,cm^{-2}$ and about 0.8 large clouds with a typical visual extinction of 0.6 magnitude ($1.2 \times 10^{21}\,cm^{-2}$). Of course, in reality there is a distribution of cloud sizes and masses measured in the HI 21 cm line and the CO $J = 1 - 0$ transition (cf. Chapter 10). On average, then, there is about 2 magnitudes of visual extinction per kpc in the Solar neighborhood, but there are many lines of sight, particularly near dense clouds, that show much higher extinctions.

5.3.6 Interstellar depletion

Atomic absorption lines can be used to determine abundances by comparison with the atomic and molecular hydrogen column densities measured through

Table 5.3 *Elemental depletions*[a]

Element	Solar abundances	Dust core abundances[b]	Dust abundances[c]
C	391	251	259
N	85	−8	9
O	490	170	170
Mg	35	12	32
Si	34	15	33
Fe	28	21	28

[a] Per million H atoms.
[b] The difference between solar abundances and halo abundances (i.e., where only resilient dust cores are expected to survive).
[c] The difference between Solar abundances and abundances measured along the heavily depleted line-of-sight towards ζ Oph.

Lyman alpha and Lyman–Werner transitions. The abundance of many elements is much less in the ISM than that measured in the Sun, the Solar System (e.g., mainly meteorites), or in nearby stars. The difference is generally presumed to be locked up in the solid phase. The question of which reference abundance to take is of some importance and has been contentious in recent years. Young F and G stars have abundances very similar to those of the Sun. B stars, on the other hand, have abundances that are consistently lower. This difference may reflect grain–gas separation during formation of these types of stars. Table 5.3 summarizes measured interstellar gas phase abundances for some of the major elements, relevant reference abundances, and derived elemental fractions locked up in dust. The total dust mass is then $7.6 - 9.2 \times 10^{-3}$ times that of hydrogen.

Figure 5.14 shows measured depletions graphically as a function of dust condensation temperature. The observed good correlation may be taken to reflect the condensation history of the dust in circumstellar environments. This suggests that dust is mainly formed at high temperatures in stellar ejecta and this heritage is preserved through most of the grain's lifetime in the interstellar medium. However, in a general sense, the condensation temperature is a measure for the strength with which an element is bonded in the solid form and, hence, this correlation may also result from selective dust destruction, or accretion of gaseous species onto dust grains in the ISM.

Some elements are observed to be hardly depleted at all. Atomic oxygen is a point in case, presumably because, even if all other heavy elements (except for C) are locked up as oxides, only a minor fraction ($\simeq$20%) of the elemental oxygen is involved. The noble gases, nitrogen, zinc, and sulfur are other examples of species that generally form only rather volatile compounds. In fact, when total H column densities are not available, the column density of one of these species

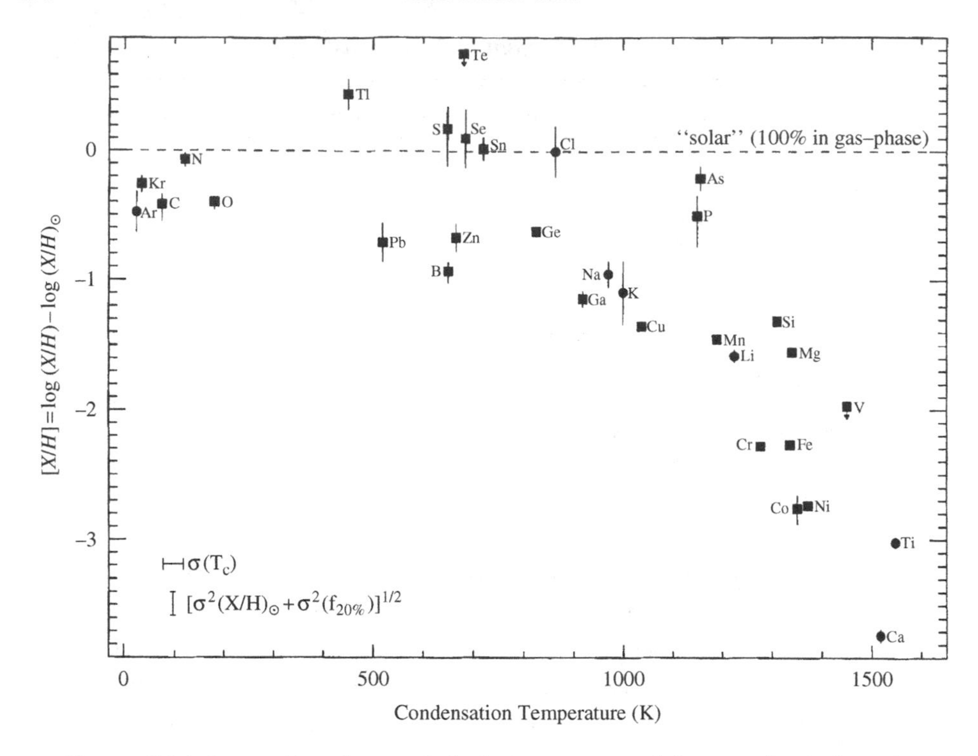

Figure 5.14 Interstellar elemental depletions measured for a variety of species are shown as a function of the condensation temperature (i.e., the temperature at which 50% of this element condenses out in solid to form a cooling gas). Figure courtesy of B. D. Savage and K. R. Sembach; reprinted with permission from the *Ann. Rev. Astron. Astrophys.*, **34**, p. 279, ©1996 by Ann. Rev. (www.annualreviews.org).

is often substituted to estimate H column densities and thereby depletions of other elements. Other species are depleted by large factors; calcium and aluminum by some three orders of magnitude! Clearly, whatever process regulates interstellar abundances of these species, it must be very efficient and this cannot just reflect dust condensation efficiency in stardust birthsites but instead indicates active chemical interchange in the ISM itself.

Elemental depletions are observed to vary from region to region in the ISM. In general, depletions increase with the average density along the line of sight (i.e., with $<n> = N/d$ with N the column density of H nuclei and d the stellar distance). Probably this reflects much higher gas phase abundances in the warm neutral medium ($<n> \simeq 0.5\,\text{cm}^{-3}$) than in the cold neutral medium ($[<n> \simeq 50\,\text{cm}^{-3}]$; cf. Section 8.3). Variable mixing of these two phases along lines of sight can then lead to a smooth correlation between mean density and elemental depletions. The observed decrease in depletion with height above the plane may likewise result from an increased importance of the warm neutral medium at high galactic latitudes. The different dust abundances in these two phases is generally taken to reflect the importance of shock processing in the low density phase and accretion in the high density phase (cf. Chapter 13). Interstellar depletions are also observed to decrease with increasing velocity of the gas probed. The high velocity gas is presumably recently shocked and this correlation points again to the importance of shock processing of grains in the ISM.

5.3.7 *Luminescence*

Many dusty objects – including the diffuse ISM, dark clouds, reflection nebulae, post-AGB objects, planetary nebulae, HII regions, and starburst galaxies – show a broad (600–1000 Å), featureless emission band in the red part of the spectrum (6000–8000 Å). This so-called extended red emission (ERE) was first discovered in the spectrum of the post-asymptotic-giant-branch object, the Red Rectangle, and indeed provides the red color to this nebula. The peak wavelength of the ERE varies from source to source, dependent on the radiation density or hardness of the environment.

The ERE cannot be due to thermal emission but rather represents a luminescence process whereby an absorbed UV or visible photon is down-converted to the ERE range. The observed efficiency of this process is very high: assuming that all the FUV and visible photons are absorbed by the ERE carrier, the photon production efficiency is 0.1 in the diffuse ISM. This is a strict lower limit and, probably, the actual efficiency is much higher. The energy conversion efficiency is of the order of 0.04.

Strong luminescence is a characteristic of many semiconductors with moderate band gaps. The absorbed photon creates an electron-hole pair and the resulting electron loses its excess kinetic energy through interaction with the matrix. Eventually, recombination will lead to emission of a red photon. The efficiency of this process is highest if the electron is strongly localized. Of course, this reflects back on the discussion concerning charging (Section 5.2.3) and the photo-electric effect (Section 3.3).

5.4 The sizes of interstellar grains

There are various independent handles on the interstellar grain size distribution. First, the extinction keeps rising throughout the infrared, visible and near- and far-UV wavelength range (Fig. 5.7). Since extinction "saturates" when the size becomes comparable to the wavelength ($a \sim \lambda$; Section 5.2.1), interstellar grains have to span a range of sizes. The visible extinction is then due to grains with a typical size of 2000 Å. The ultraviolet extinction, on the other hand, reflects the presence of grains with sizes in the range 50–200 Å. Then, ascribing the observed visual extinction per H atom to $\simeq$2000 Å grains with an extinction efficiency $Q_v \simeq 1$, implies an abundance of such grains of $\simeq 4\times10^{-13}$ (cf. Eq. (5.1)). Similarly, the 10 times higher FUV extinction per H atom results in an abundance of $\simeq$100 Å grains of 2×10^{-9}.

On a more quantitative footing, once the dominant grain components have been identified, their optical properties measured, and their shapes characterized, the observed extinction curve can be inverted to give the interstellar grain size distribution. Of course, none of these factors is well known and, hence, there is considerable ambiguity in the results – often reflecting personal taste of the theoreticians involved. All of these agree – of course – in the global range of sizes present but not in the specifics. The most widely used model is the Mathis–Rumpl–Nordsieck model (or MRN model), named after the scientists who first derived this particular grain size distribution. This model consists of graphite and silicate grains with a power law size distribution with an exponent of about -3.5 in the range 50–2500 Å. A convenient representation is given by

$$n_i(a)da = A_i n_{\rm H} a^{-3.5} da, \tag{5.97}$$

with a the grain size, and where the constants A_i for silicate and graphite are given by $A_{\rm sil} = 7.8\times10^{-26}$ and $A_{\rm gra} = 6.9\times10^{-26}$ cm$^{2.5}$ (H atom)$^{-1}$, respectively. Figure 5.15 shows a more detailed size distribution, but the caveats – on composition, optical properties, shapes – mentioned above should be kept in mind. Also, there is little constraint on the grain size distribution at the small end, because for small grains ($<$200 Å) extinction is in the Rayleigh limit and independent of grain size (cf. Section 5.2.1). The presence of very small grains and even large molecules is not derived from extinction measurements but rather inferred from studies of the mid-IR emission of the Galaxy (cf. Section 6.4.8). For very large grains ($\simeq$1 μm), extinction in the visible is gray and dust abundances are mainly derived from abundance constraints on the elements making up these grains. Despite these caveats, all dust models agree that the total dust volume is dominated by the large grains while the number density and surface area are dominated by the small grains, which follows directly from the simple analysis described above.

Polarization studies provide an independent handle on the grain size distribution. For dielectric grains, the wavelength of maximum polarization is related to the grain size through

$$\lambda_{\rm max} \simeq 2\pi(n-1)a. \tag{5.98}$$

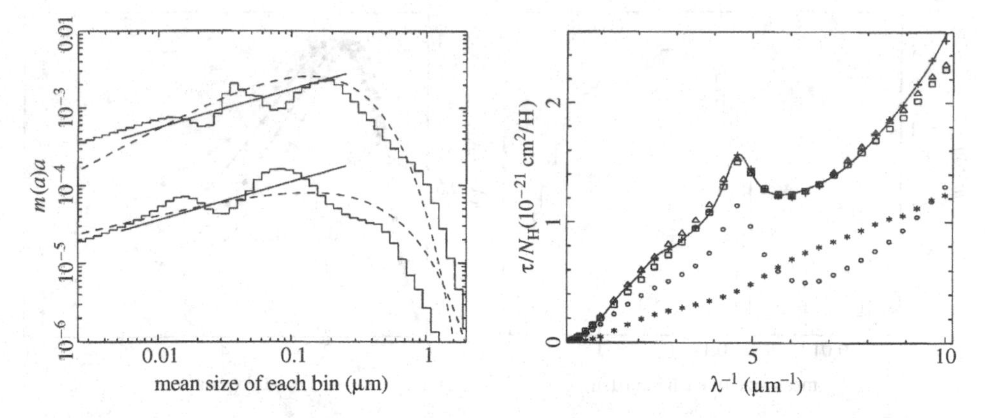

Figure 5.15 The interstellar grain size distribution – plotted as mass fractions – derived from different types of fits to the observed extinction curve (left panel). The top histogram and dashed line show silicates; the bottom histogram and dashed line shows graphite, displaced downwards by factor 10. The solid lines show for comparison the simple MRN power law size distributions. The calculated extinction curve is compared to the observations in the right panel. The contributions due to silicates (∗) and graphite (∘) are shown separately. Figure reprinted with permission from S.-H. Kim, P. G. Martin and P. D. Henry, 1994, *Ap. J.*, **422**, p. 164.

Thus, for silicates with $n = 1.5$, an observed λ_{max} in the visible implies a grain size of $\simeq$1800 Å. Besides the size distribution, the observed wavelength dependence of polarization depends also on the shape and alignment distribution of the grains. The peak in the polarization curve is a confluence of these factors. The observed decrease of the polarization towards the infrared reflects the decrease in the size distribution towards larger sizes. The decline towards the UV, on the other hand, reflects that small grains apparently do not contribute appreciably to the polarization, either because they are more spherical or because they are not well aligned. Figure 5.16 shows the derived size distribution for perfectly aligned, spheroidal silicate grains. The size distribution for spheroidal grains derived from extinction is displaced slightly upwards from that for spherical grains. It is clear that small silicate grains do not contribute to the polarization at all. It should also be emphasized that – in view of the observed circular polarization and the absence of polarization of the 2175 Å feature – graphite grains do not contribute at all to interstellar polarization.

5.4.1 Interstellar dust mass

The interstellar medium can be considered as a very dilute medium containing small scattering and absorbing grains. Such a medium can be characterized by a

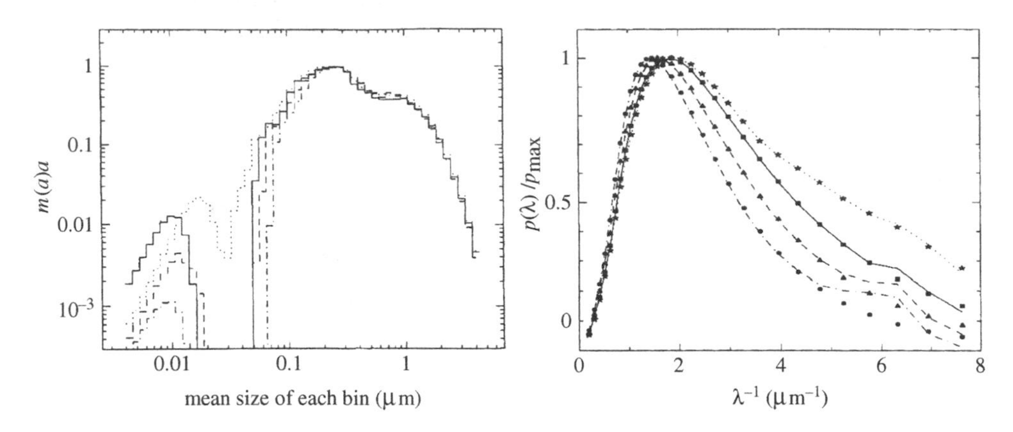

Figure 5.16 Four different polarization curves (right) represented by Serkowski laws with different $\lambda_{\rm max}$ and the size distribution (left) of perfectly aligned silicate grains derived from these polarization data assuming 2:1 oblate spheroid grains. $\lambda_{\rm max} = 0.55\,\mu$m (squares, solid), $\lambda_{\rm max} = 0.615\,\mu$m (triangles, dashed), $\lambda_{\rm max} = 0.68\,\mu$m (circles, dot-dashed), $\lambda_{\rm max} = 0.52\,\mu$m (stars, dotted). For the last, the data for HD 7252 have been substituted for $\lambda^{-1} > 3\,\mu$m^{-1}. Figure reprinted with permission from S.-H. Kim and P.G. Martin, 1994, *Ap. J.*, **431**, p. 783.

complex dielectric constant, $\epsilon_{\rm ISM}$, where the real part is associated with a phase lag through the medium and the complex part represents a decrease in intensity due to absorption and scattering by the individual grains. Now, the real and imaginary part of the dielectric constant of this medium are connected through the Kramers–Kronig relations and this provides a probe of the total interstellar dust volume from the observed extinction curve. Specifically, we can evaluate the Kramers–Kronig relation at zero frequency; i.e.

$$\epsilon_{1,\rm ISM}(0) - 1 = \frac{2}{\pi}\int_0^\infty \frac{\epsilon_{2,\rm ISM}(\omega)}{\omega}\,{\rm d}\omega. \tag{5.99}$$

The right-hand side of this equation can be written in terms of an integral of the observed extinction curve, while the left-hand side can be evaluated on the basis of the individual grains.

The attenuation can be described by (cf. Eq. (5.14))

$$\epsilon_{2,\rm ISM} = \frac{\lambda}{2\pi} n_{\rm d} C_{\rm ext}, \tag{5.100}$$

which can be written as (cf. Eq. (5.1))

$$\epsilon_{2,\rm ISM} = 0.92\frac{\lambda}{2\pi} A_\lambda, \tag{5.101}$$

where A_λ is the extinction at wavelength λ in magnitudes cm^{-1} and λ the wavelength in cm. For a very dilute system, the real part of the index of refraction at very low frequency can be calculated assuming that each grain is exposed to the same electric field. For spherical grains, small compared to the wavelength, this results in

$$\epsilon_{1,\mathrm{ISM}}(0) - 1 = n_{\mathrm{d}} V_{\mathrm{d}} f, \tag{5.102}$$

with n_{d} and V_{d} the number density of dust grains and the volume of an individual dust grain, and

$$f = 3\frac{\epsilon_{\mathrm{d}} - 1}{\epsilon_{\mathrm{d}} + 2}, \tag{5.103}$$

with ϵ_{d} the static dielectric constant of the grain material. Actually, the expression for f can be evaluated for arbitrary shapes and, for dielectric grains, the result is not very sensitive to the adopted shape. Combining the expressions for the real and imaginary parts yields then for the total dust-to-gas ratio by mass

$$\frac{\rho_{\mathrm{d}}}{\rho_{\mathrm{H}}} = 2.2 \times 10^{-3} \frac{\rho_{\mathrm{s}}}{f} \int_0^\infty \frac{A_\lambda}{A_{\mathrm{v}}} \lambda^2 \mathrm{d}(\lambda^{-1}), \tag{5.104}$$

where ρ_{s} is the specific density of the grain material, λ is in the astronomically customary units of $\mu\mathrm{m}^{-1}$, and we have used the observed average extinction per H atom (cf. Section 5.3.5). The optical region contributes about 0.8 to the integral while the UV gives 0.3 (which merely restates that the dust volume is dominated by the largest grains, which produce the optical extinction). If we use the dielectric constant at near-IR wavelengths, typically 2, the factor f is 0.75. For large ϵ_{d}, f does become very sensitive to shape and for metallic needles this factor can become very large at low frequencies. But, then, such grains would give a large contribution to the IR extinction, which we do not observe (and which, if present, should of course be included in the evaluation of the integral on the right-hand side as well). With a density of $3\,\mathrm{g\,cm^{-3}}$, we find then a dust to gas ratio of 7.3×10^{-3} by mass. There is of course some uncertainty associated with the choice of ϵ_{d} and ρ_{s} but these factors tend to work in opposite directions. Finally, we note that this discussion is related to the Clausius–Mossotti relation and the determination of number density of scatterers and Avogadro's number from optical studies of gases.

5.5 The composition of interstellar dust

A number of different grain components have been identified in interstellar dust by various means and this section provides a summary of this.

5.5.1 Silicates

The broad and structureless 9.7 and 18 μm features (Fig. 5.8) are generally attributed to the Si–O stretching and bending modes in amorphous silicate grains. In the Solar neighborhood, the peak strength of this feature has been measured relative to A_V ($A_V/\tau(9.7) = 18.5$) towards a number of bright background sources. Together with a normal dust-to-gas ratio ($N_H/A_V = 1.9 \times 10^{21}$ cm^{-2} magnitude^{-1}), an observed width of 220 cm^{-1}, and the typical integrated absorption strength of 1.2×10^{-16} cm (Si atom)$^{-1}$, this corresponds to an abundance of silicon in solid form of 5.2×10^{-5} per H atom. This is almost 50% more than the Solar elemental abundance of silicon and the origin of this discrepancy is not well understood. With an adopted silicate composition of $(Mg,Fe)_2SiO_4$, the total silicate dust mass is 9×10^{-3} per H atom. The silicate abundance is known to vary in the Galaxy. Towards the Galactic center, the measured $A_V/\tau(9.7)$ is about a factor of two less than in the Solar neighborhood (A_V is actually scaled from the observed near-IR color excess).

Silicates are dielectrics and hence are prime candidates for the carriers of the observed visible polarization. This is supported by presence of a dichroically polarized 9.7 μm absorption feature towards some sources. If silicates indeed cause the observed visible polarization, they have to make an important contribution to the visible extinction and hence a major fraction of their mass must have sizes in the range 0.1–0.3 μm.

While silicates in the interstellar medium are amorphous to a high degree – with a crystalline fraction less than 5×10^{-3} – crystalline silicates can be abundant in many circumstellar environments, particularly when the system is characterized by a long-lived circumstellar disk. This includes both stars in the later stages of their evolution, when they inject much of their mass back into the interstellar medium, as well as newly formed stars surrounded by their protoplanetary disks. A particularly striking example of these crystalline features is shown in Fig. 5.17. Most of the spectral structure in this spectrum is due to crystalline forsterite (Mg_2SiO_4), the magnesium-rich end member of the olivine family. Many sources also show IR emission features owing to enstatite ($MgSiO_3$), the magnesium-rich end member of the pyroxenes. These compounds are very magnesium rich and iron poor. This is best illustrated on the basis of the 69 μm band, whose peak position is very sensitive to the cations present (Fig. 5.18). This band is also sensitive to the temperature of the emitting grains. The shift in peak position and width of this band with temperature is a general characteristic of vibrational phonon modes of crystalline materials and reflects the shrinking of the lattice with decreasing temperature – coupled with the anharmonicity of the bonds – and the lower phonon density at low temperature.

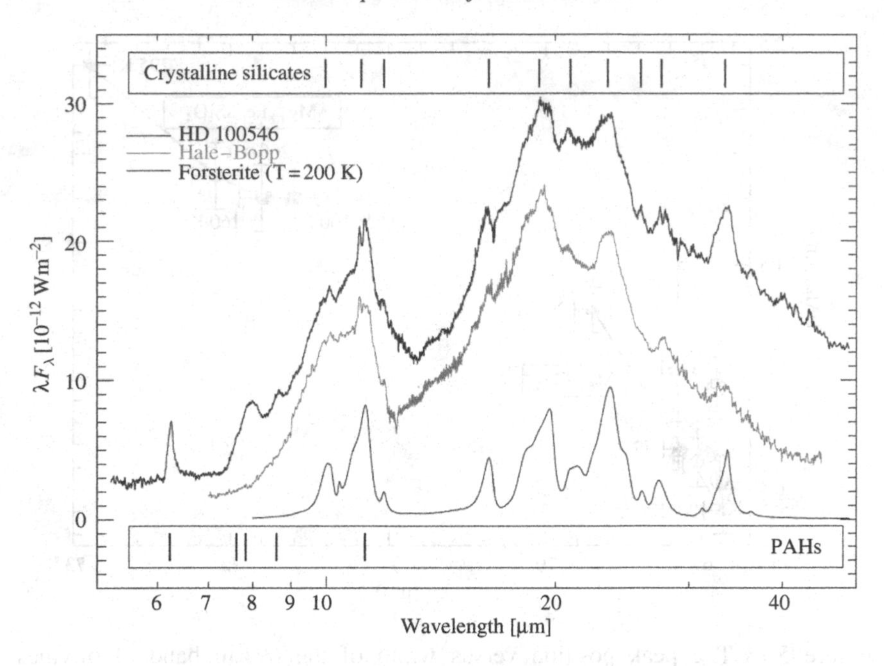

Figure 5.17 The mid-IR spectrum of the Herbig AeBe star, HD 100546 (dark top trace), is characterized by strong emission features due to forsterite (Mg_2SiO_4; laboratory spectrum bottom trace). There is a good resemblance to the spectrum of the comet, Hale–Bopp (middle trace). Note that this young stellar object shows PAH emission features at shorter wavelength. Figure courtesy of F. Molster and S. Hony. The spectrum of HD 100546 is taken from K. Malfait, C. Waelkens and L. B. F. M. Waters, 1998, *A. & A.*, **332**, p. L25. The spectrum of Hale–Bopp is taken from J. Crovisier, *et al.*, 1997, *Science*, **275**, p. 1904. The laboratory spectrum of forsterite is taken from C. Koike, H. Shibai, and A. Tuchiyama, 1993, *M. N. R. A. S.*, **264**, p. 654.

5.5.2 Graphite

The presence of graphite in the ISM has been deduced from the strong 2175 Å feature that dominates the observed interstellar extinction curve (Fig. 5.7). Graphite has a strong resonance at about 2000 Å due to $\pi \longrightarrow \pi^\star$ transitions associated with its aromatic bonds (e.g., sp^2 bonded carbon). Because this is an inherently strong resonance, the exact peak position and profile are very sensitive to grain size, shape, the presence of coatings, and exact optical constants. The observed profile of the interstellar 2175 Å feature is well fitted by theoretically calculated extinction cross sections for either 200 Å graphite spheres or for graphite prolate spheroids with a size of 30 Å and an axial ratio of 1.6. Good fits to the observed spectra can also be obtained with laboratory measured extinction spectra of small ($\simeq$100 Å) hydrogenated amorphous carbon grains – containing 33% H atoms by number – which share the aromatic character of their bonding

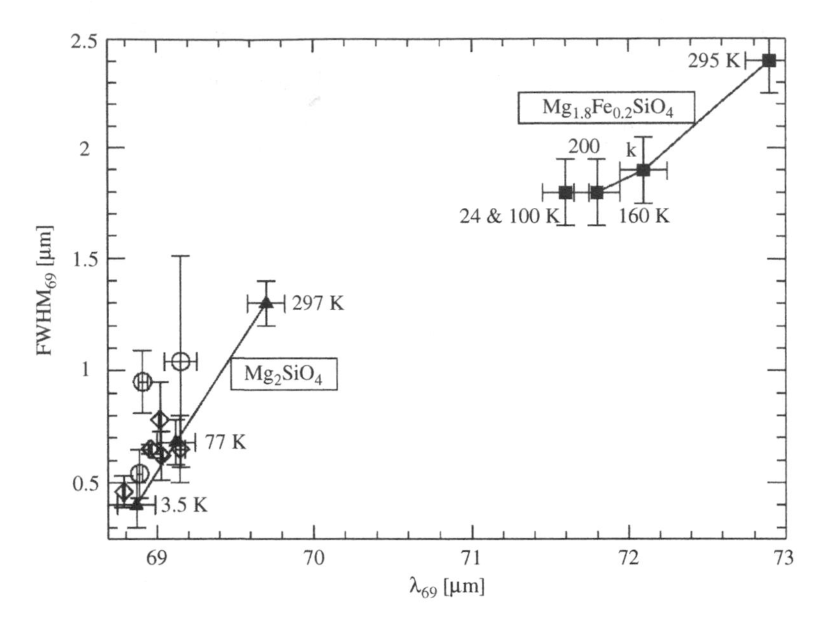

Figure 5.18 The peak position versus width of the 69 μm band of olivines ($Mg_xFe_{2-x}SiO_4$) as a function of the iron content and the temperature. The black diamonds and squares are the laboratory data for $x = 2$ and 1.8, respectively. The open circles are taken from spectra of circumstellar silicates. Clearly, these circumstellar crystalline silicates are very magnesium rich and iron poor as well as very cool ($T < 100$ K). Figure reproduced with permission from F. J. Molster, L. B. F. M. Waters, A. G. G. M. Tielens, C. Koike, and H. Chihara, 2002, *A. & A.*, **382**, p. 241. Laboratory data: J. E. Bowey, *et al.*, 2000, in *ISO Beyond the Peaks*, ed. Salama, ESA-SP-456 p. 339 and V. Mennella, *et al.*, 1998, *Ap. J.*, **496**, p. 1058.

with graphite. This non-uniqueness illustrates the sensitivity of the absorption profile to the detailed characteristics of the grain; e.g., by adjusting the detailed parameters of the grain slightly, different (but related) materials can be made to fit the observations. Of course, this sensitivity poses a problem for all of these fits in view of the observed constancy of the peak position but variable width of the interstellar 2175 Å feature.

The observed strength of the 2175 Å feature corresponds to a required abundance of carbon in the form of graphite of 6×10^{-5} relative to H. This corresponds to a mass fraction of graphite of 7.2×10^{-4} relative to H. For HAC grains, the inferred abundance and mass fractions are a factor of two larger.

Graphite is a (semi)metal and there is a slight overlap between the valence and conduction bands at the edge of the Brillouin zone. Hence, its optical properties are dominated by electronic intraband and interband transitions, even in the infrared. Because of its highly absorbing character, graphite cannot be responsible for the observed interstellar polarization and, indeed, the 2175 Å feature shows no excess polarization. The presence of impurities (e.g., H atoms) can produce localized electronic states and break up the metallic character of graphite. This happens in hydrogenated amorphous carbon (HAC) materials (cf. Section 5.5.3). When the aromatic domain size of the graphitic "islands" becomes less than ∼25 Å, the band gap is some 1 eV and such a material will be transparent in the visible.

5.5.3 Amorphous carbon and HAC

The interstellar 3.4 μm feature (Fig. 5.8) is due to the C–H stretch in aliphatic materials (e.g., CH_2 and CH_3 groups in sp^3 bonded carbon). The weak interstellar 6.8 and 7.2 μm bands are similarly due to the C–H deformation modes of such materials. Laboratory spectra of HAC grains provide a reasonable fit to the observed profiles and relative strength of these features. The observed strength of the 3.4 μm feature requires a carbon abundance of about 1.2×10^{-4} relative to H in this dust component or about 1.4×10^{-3} by mass. There is no evidence for O atoms in their molecular structure (O/H $< 3 \times 10^{-6}$ by number).

The sizes of these hydrocarbon grains are not well constrained. However, if they are also the carriers of the 2175 Å bump (cf. Section 5.5.2), their sizes are quite small (≃50–200 Å). The little information available on the polarization characteristics of the 3.4 μm band suggests that these grains are not polarizing agents in the ISM.

5.5.4 Diamond

Diamond is a homonuclear material and has therefore no IR active fundamental vibrations. The diamond IR spectrum shows weak (and broad) IR activity near the peak of its phonon distribution, 5–8 μm. The presence of impurities can induce IR activity. Particularly relevant are the C–H stretching modes due to H atoms that terminate the sp^3 carbon structure near the surface. This leads to a pattern of emission bands that are very characteristic for the surface structure. The near-IR spectra of two Herbig AeBe stars and one post-asymptotic giant branch object show emission bands at 3.4 and 3.5 μm, which are very well fitted by laboratory measured diamond bands at 1000 K. These bands are unique to these particular objects and demonstrate the presence of diamond grains in, at least, some circumstellar environments. Their absence in the IR emission spectrum of the general ISM may imply that either the surfaces are not hydrogenated in the ISM, possibly because of the low temperature or the bombardment with energetic ions, or that diamonds are a local condensation product of these very distinct environments.

Diamonds also show a strong absorption edge in the FUV. However, unless a major portion of the carbon is locked up in small diamond grains, the presence of interstellar diamond grains is difficult to ascertain from the observed extinction curve.

5.5.5 Silicon

Silicon nanoparticles ($\simeq$15–50 Å), consisting of crystalline silicon cores surrounded by a SiO mantle, have been proposed as the carriers of the ERE. Laboratory studies have shown that such particles luminesce efficiently – up to 50% – and the measured spectrum is in good agreement with the observations. However, this assignment does require a major fraction of the elemental silicon in the form of these nanoparticles, which is difficult to reconcile with the prominence of the 9.7 and 18 μm silicate bands and the limited elemental abundance of silicon.

5.5.6 Carbides

Many C-rich AGB stars show an emission feature near 11.3 μm due to SiC grains. This feature has, however, not been seen in absorption or emission in the diffuse ISM. The limit on the fraction of the elemental C in the form of SiC in the diffuse ISM is $\simeq 6\times10^{-3}$, corresponding to a mass limit of 9×10^{-5} relative to H.

Many low metallicity, post-AGB objects from low mass ($\simeq$1 $M_{\odot}$) progenitors show an emission feature near 21 μm, which has been attributed to TiC grains condensed during the extreme conditions of the superwind that terminates the AGB. This feature has also never been observed in the ISM.

5.5.7 Ices

Interstellar ices have only been observed along lines of sight containing dense molecular cloud material, where they can be a major component of the interstellar dust. Their composition and characteristics are discussed in Section 10.7.5.

5.5.8 Stardust

Using a variety of extraction techniques, genuine individual stardust grains have been recovered from carbonaceous meteorites with an isotopic composition that betrays the nucleosynthetic heritage of their stellar birthsite. These isotopic anomalies often exceed orders of magnitude and, therefore, probably reflect on the isotopic composition of the material from which the grain condensed. For the smaller grains, these isotopic anomalies refer to an average over many grains. However, micron-sized or larger grains can be investigated individually, revealing an incredible detail on the nucleosynthetic products at the level of individual stellar birthsites of dust, which is unattainable by any other means. Table 5.4 summarizes some characteristics of meteoritic stardust, including composition and birthsites. Most of these compounds show evidence for multiple stellar birthsites. Finally, some classes of interplanetary dust particles collected in the stratosphere are thought to derive from comets. Hence, they may actually represent a rather unadulterated record of interstellar dust.

Comparing the characteristics of stardust extracted from meteorites with the characteristics of dust observed in interstellar and circumstellar environments, we notice similarities and differences. First, many of the compounds extracted from meteorites are also observed in circumstellar environments, albeit not always in those indicated by the isotopic composition. Second, the sizes of individual stardust grains often exceed sizes of interstellar grains by an order of magnitude. It should be noted here that, because of various selection effects, stardust studies do favor large grains but there are large abundances of smaller grains present as well and an unbiased view is not available, yet. Nevertheless, micron-sized grains are expected to have very low abundances in the interstellar medium. Third, the abundances of all these compounds are probably severely affected by processing during the star and planet formation process and only the most resilient and refractory phases have survived. Indeed, most of the presolar dust has been destroyed. In particular, in the interstellar medium, most of the silicon is locked up in silicates and most of the carbon in carbonaceous dust; quite in contrast to studies of meteorites. This is particularly obvious for the oxide phases, where much of the meteoritic material has mundane isotopic compositions (at least, compared with stardust), has sizes much in excess of presolar dust, and petrographic structures that reveal extensive exchange with and recondensation from a hot nebular gas. For meteoritic materials, this is not too surprising because the inner Solar System was assembled during a phase when the nebula was very hot and most of the materials vaporized. Largely, only the oxides recondensed (which makes the carbonaceous phases somewhat easier to recognize as presolar). However, this point even seems to be true for interplanetary dust particles (probably) derived from comets. Such grains were thought to be incorporated into cometary bodies far out in the Solar System and hence would have experienced little processing. Recent studies of oxygen isotopic anomalies in silicate grains in IDPs reveal only a small fraction of presolar silicate grains. Many of the silicates in IDPs may thus actually derive from condensation processes in the inner Solar System or, possibly, in other planetary systems formed around the same time as the Solar System and then redistributed in the outer Solar nebula through nebular mixing or outflow

Table 5.4 *Stardust*

Compound	Size range (Å)	Meteoritic abundance (ppm)	Isotopic anomaly	Stellar birthsite	Notes
Diamond	10	1 400	Xe/Kr H-L	SN	a
Graphite	8 000–70 000	14	s-process, C, N, ^{22}Ne	AGB, SN, novae	
Silicon carbide	300–10 000	10	s-process, C, N, Al	AGB	b, c
TiC/ZrC/FeC	50–2 000			AGB, SN type II	d
Si_3N_4	10 000	0.002	Si, C, N, Al	SN type II	c
Al_2O_3	5 000–30 000	0.05	O, Al	RG, AGB	c
$MgAl_2O_4$	1 000–30 000	< 0.05	O, Al	RG, AGB	c
$CaAl_{12}O_{19}$	20 000	0.002	O, Al, Ca	RG, AGB	c
Mg_2SiO_4	3 000	(5 500)	O	RG, AGB	e
Amorphous silicates	3 000	(5 500)	O	RG, AGB	e

(a) Diamond stardust shows rather a mundane C and N isotopic composition; due to mixing in of non-stardust diamond?
(b) Some SiC stardust grains originate from SNe.
(c) Aluminum anomalies actually refer to ^{26}Mg, the radiogenic daughter product of ^{26}Al.
(d) These grains are embedded in graphite stardust.
(e) A few silicate presolar silicate grains have been identified in IDPs.

Table 5.5 *Dust-to-gas mass ratios*

Method	Dust-to-gas ratio	Section
$A_V/N_{(H)}$	1.1×10^{-2}	5.3.5
depletions	9.2×10^{-3}	5.3.6
Kramers–Kronig	7.3×10^{-3}	5.4.1
silicates	9.0×10^{-3}	5.5.1
graphite	7.2×10^{-4}	5.5.2
hydrogenated amorphous carbon	1.4×10^{-3}	5.5.3
silicon carbide	$< 9 \times 10^{-5}$	5.5.6

activity of the protostar. In any case, these caveats should be kept in mind when comparing the properties of stardust with those of dust in the interstellar medium.

5.5.9 Summary

Table 5.5 summarizes various estimates of the dust-to-gas mass ratio. As is evident from the discussion in the relevant sections, these values are not very accurate and a value of 10^{-2} is a reasonable compromise. The estimates for the different dust components are equally uncertain. Judging from these values, silicates and graphite or amorphous carbon are important dust components.

5.6 Further reading

There are excellent textbooks on interstellar dust; [1] is particularly strong from an observational point and also provides a coherent synopsis of the views of interstellar dust in the literature; [2] approaches this subject from an elementary level, including all the relevant physics. Also, several reviews have appeared on the subject of interstellar dust; [3], [4], and [5].

There are several monographs on the interaction of small particles with light, including the classic [6]. I strongly recommend [7], which also includes a discussion of the optical properties of dielectric and metallic materials. For in-depth studies, the reader is referred to the proceedings of light scattering conferences, [8].

The subject of grain temperatures is scattered through the literature. An early but dated reference is [9]. More recent studies include [10] and [11] although neither provides simple estimates of the temperature. The approximations for the Planck-averaged silicate and graphite emissivities were gleaned from [12].

Collisional grain charging, including the handy analytical approximations to the reduced collision rates, is discussed in [13]. The photo-electric study stems from [14].

There is a vast literature on sputtering of materials by high energy ions [15]. This literature has been reviewed from an astrophysical perspective in [16]. [17] is also recommended for its insight into the physics of sputtering and the discussion on the effects of sputtering in various astronomical environments. The physics of grain–grain collisions has been discussed in [16] and [18].

The monograph on interstellar dust [1] should be consulted for an in-depth discussion of observational aspects of interstellar dust. In addition to the review articles cited above, the reader can also consult the proceedings of various conferences on interstellar dust, notably: [19] and [20]. Interstellar depletion studies have recently been reviewed in [21]. The properties of stardust have been reviewed in [22]. Astronomical observations of crystalline silicates have been presented in [23] for young stars and in [24] and [25] for old stars.

The website

http://urania.astro.spbu.ru/DOP/2-MODS/INTROD/

presents extinction properties for small particles of different shapes and optical properties. Reflectance spectra of various ceramics are available at

http://www.aist.go.jp/RIODB/opcc/.

Various laboratory groups measure the optical properties of astrophysically relevant materials. Some of these data can be accessed at

http://www.astro.uni-jena.de/Laboratory/Database/databases.html.

A useful compilation of references to optical properties of astronomically relevant materials is provided by

http://www.astro.spbu.ru/JPDOC/.

The three volumes of the handbook for optical constants of materials, [26], can also be consulted. Specifically the optical properties of amorphous silicates have been studied in [27], [28], and [29]. Crystalline silicates and various oxides have been analyzed in [30], [31], [32], and [33]. Laboratory data of carbonaceous dust have been presented in [34], [35], [36], and [37]. The optical properties of graphite have been reviewed in [12]. Temperature variations in the far-IR wavelength dependence of the extinction by silicates and carbonaceous dust have been measured in [38].

Numerical codes for the calculation of absorption and scattering cross sections of (coated) spheres can be found in an appendix of [7]. A simple Java-based applet calculating Mie extinction efficiencies can be found at

http://www.unternehmen.com/Bernard-Michel/eindex.html.

A discrete dipole approximation scattering code is available at B. Draine's website,

http://www.astro.princeton.edu/~draine/DDSCAT.6.0.html.

Radiative transfer codes for dusty environments can be found at

http://www/pa/uky/~moshe/dusty/,

which is maintained by M. Elitzur. A Java applet for calculating dust spectral energy distributions and fitting observations is available at

http://dustem.astro.umd.edu/,

which is maintained by M. Wolfire. Two-dimensional radiative transfer codes are accessible through

http://www.mpa-garching.mpg.de/PUBLICATIONS/DATA/radtrans/,

which is maintained by C.P. Dullemond and

http://homepage.oma.be/ueta/research/2dust/,

which is maintained by T. Ueta.

References

[1] D. C. B. Whitter, 1992 *Dust in the Galactic Environment* (Bristol: IOP)
[2] E. Krügel, 2002, *The Physics of Interstellar Dust* (Bristol: IOP)
[3] B. D. Savage and J. S. Mathis, 1979, *Ann. Rev. Astron. Astrophys.*, **17**, p. 73
[4] J. S. Mathis, 1990, *Ann. Rev. Astron. Astrophys.*, **28**, p. 37
[5] B. T. Draine, 2003, *Ann. Rev. Astron. Astrophys.*, **41**, p. 241
[6] H. C. van de Hulst, 1957, *Light Scattering by Small Particles* (New York: Wiley)
[7] C. F. Bohren and D. R. Huffman, 1983, *Absorption and Scattering of Light by Small Particles* (New York: Wiley)
[8] M. I. Mischenko, J. W. Hovenier, and L. D. Travis, 2000, ed., *Light Scattering by Non-Spherical Particles: Theory, Measurements, and Applications* (San Diego: Academic Press)
[9] J. M. Greenberg, 1971, *A. & A.*, **12**, p. 240
[10] B. T. Draine and N. Anderson, 1985, *Ap. J.*, **292**, p. 494
[11] F. X. Désert, F. Boulanger, and J. L. Puget, 1990, *A. & A.*, **237**, p. 215
[12] B. T. Draine and H. M. Lee, 1984, *Ap. J.*, **285**, p. 89
[13] B. T. Draine and B. Sutin, 1987, *Ap. J.*, **320**, p. 803
[14] E. L. O. Bakes and A. G. G. M. Tielens, 1994, *Ap. J.*, **427**, p. 822
[15] R. Behrisch, 1981, ed., *Sputtering by Particle Bombardment I* (Berlin: Springer Verlag); 1983, *Sputtering by Particle Bombardment II* (Berlin: Springer Verlag)
[16] A. G. G. M. Tielens, C. F. McKee, C. G. Seab, and D. Hollenbach, 1994, *Ap. J.*, **431**, p. 321
[17] E. M. Bringa and R. E. Johnson, 2003, in *Solid State Astrochemistry*, ed., V. Pirronello, (Dordrecht: Kluwer), p. 357
[18] A. P. Jones, A. G. G. M. Tielens, and D. Hollenbach, 1996, *Ap. J.*, **469**, p. 740
[19] L. J. Allamandola and A. G. G. M. Tielens, 1989, ed. *Interstellar Dust* (Dordrecht: Reidel)

[20] A. N. Witt, G. C. Clayton, and B. T. Draine, ed., 2004, *Astrophysics of Dust* (San Francisco: Astronomical Society of the Pacific)
[21] B. D. Savage and K. R. Sembach, 1996, *Ann. Rev. Astron. Astrophys.*, **34**, p. 279
[22] E. Anders and E. Zinner, 1993, *Meteoritics*, **28**, p. 490
[23] K. Malfait, C. Waelkens, and L. B. F. M. Waters, 1998, *A. & A.*, **332**, p. L25
[24] F. J. Molster, L. B. F. M. Waters, A. G. G. M. Tielens, and M. J. Barlow, 2002, *A. & A.*, **382**, p. 184
[25] F. J. Molster, L. B. F. M. Waters, and A. G. G. M. Tielens, 2002, *A. & A.*, **382**, p. 222
[26] E. D. Palik, 1985, ed., *Handbook of Optical Constants of Solids* (New York: Academic Press); 1991, *Handbook of Optical Constants of Solids II* (New York: Academic Press); 1998, *Handbook of Optical Constants of Solids III* (New York: Academic Press)
[27] J. R. Brucato, L. Colangeli, V. Mennella, P. Palumbo, and E. Bussoletti, 1999, *A. & A.*, **348**, p. 1012
[28] J. Dorschner, *et al.*, 1995, *A. & A.*, **300**, p. 503
[29] T. Henning and H. Mutschke, 1997, *A. & A.*, **327**, p. 743
[30] A. Scott and W. W. Duley, 1996, *Ap. J. S.*, **105**, p. 401
[31] B. Begemann, J. Dorschner, T. Henning, and H. Mutschke, 1997, *Ap. J.*, **746**, p. 199
[32] C. Jaeger, *et al.*, 1998, *A. & A.*, **339**, p. 904
[33] C. Koike, C. Kaito, T. Yamamoto, *et al.*, *Icarus*, **114**, p. 203
[34] C. Koike, *et al.*, 2000, *A. & A.*, **363**, p. 1115
[35] L. Colangeli, V. Mennella, P. Palumbo, A. Rotundi, and E. Bussoletti, 1995, *A. & AO S.*, **113**, p. 561
[36] D. G. Furton, J. W. Laiho, and A. N. Witt, 1999, *Ap. J.*, **526**, p. 752
[37] C. Koike, S. Kimuar, C. Kaito, *et al.*, 1995, *Ap. J.*, **446**, p. 902
[38] M. Schnaiter, H. Mutschke, J. Dorschner, T. Henning, and F. Salama, 1998, *Ap. J.*, **498**, p. 486
[39] V. Mennella, *et al.*, 1998, *Ap. J.*, **496**, p. 1058

6
Interstellar polycyclic aromatic hydrocarbon molecules

6.1 Introduction

While earlier suggestions had appeared over the years, the importance of large molecules in space was first realized on the basis of the observed strong mid-infrared emission in the ISM. The Infrared Astronomical Satellite (IRAS) discovered widespread emission at 12 μm in the diffuse ISM – the so-called IR cirrus – where the expected temperature of dust in radiative equilibrium with the stellar radiation field is expected to be too cool to emit at such short wavelengths (cf. Section 5.2.3). This problem had actually already been recognized in connection with the observed mid-IR emission from PDRs far from the illuminating stars, which is also much brighter than expected for radiatively heated dust grains (Section 9.4). It was then quickly realized that very small dust grains with 20–100 C atoms – actually, large molecules – can be transiently heated to high temperatures, because of their limited heat capacity. Such hot species will cool through emission in their mid-IR vibrational modes. The observed interstellar IR spectrum is very characteristic of aromatic species and hence the carriers are really large polycyclic aromatic hydrocarbon molecules (PAHs).

In this chapter, we will discuss the physics and chemistry of such large molecules. The emphasis will be on their interaction with radiation. However, the presence of large molecules in space will also have profound influence on other aspects of the ISM and these will be examined as well. It should be emphasized that PAHs are molecules and the physics and chemistry of molecules is distinct from that of dust in many ways. In particular, the IR emission by PAHs is really an IR fluorescence process where absorption of a single FUV photon leads to electronic excitation. This electronic energy is transferred to the vibrational manifold of the molecule and is eventually radiated away through the vibrational modes of the species. The concept of temperature during this absorption–emission process is therefore subtly different from that used when analyzing the IR emission of dust grains in Section 5.2.2. Likewise, the ionization and recombination processes of interstellar PAHs differ somewhat from those of grains. In this chapter, we will briefly examine these differences. Finally, in some ways, this discussion is very general for large molecules and nanoparticles, which bridge the molecular-grain gap. This is also briefly touched upon in this chapter.

6.1.1 PAH structure

Polycyclic aromatic hydrocarbon molecules are a family of planar molecules, consisting of carbon atoms arranged in a characteristic honeycombed lattice structure of fused six-membered rings decorated at the edges by H atoms (see Fig. 6.1). With their four valence electrons, carbon atoms in aromatic structures form three covalent σ bonds with neighboring C or H atoms. This leads to a planar structure. The fourth electron is located in a p orbital perpendicular to this plane. The p orbitals on adjacent C atoms overlap to form π bonds and delocalized electron clouds above and below the plane of the molecule. Generally, the more extensive the delocalized electron cloud is, the more stable the molecule. Essentially, for these species to react, the σ bonding has to be broken and the extended aromatic π system has to be disrupted.

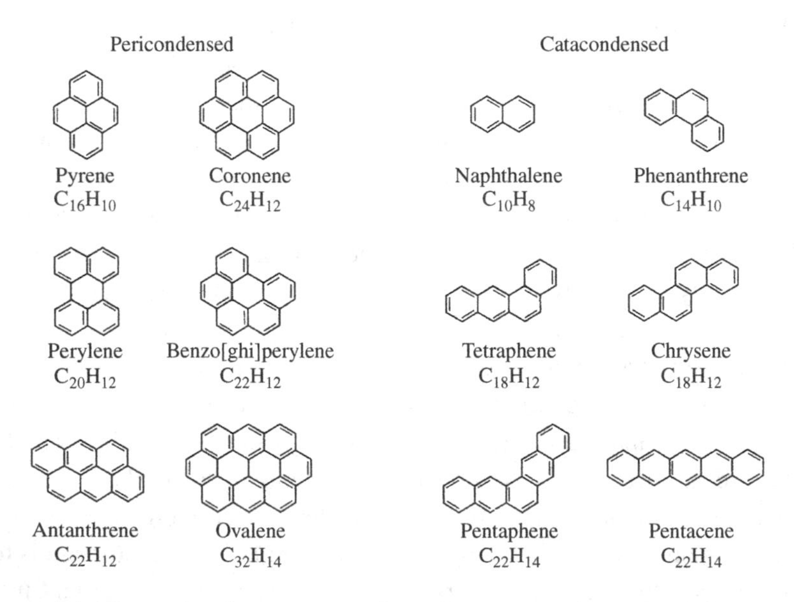

Figure 6.1 The molecular structure of some representative polycyclic aromatic hydrocarbon molecules. Pericondensed PAHs are on the left. Catacondensed PAHs are on the right. Reproduced with permission from F. Salama, E.L.O. Bakes, L.J. Allamandola, and A.G.G.M. Tielens, 1996, *Ap. J.*, **458**, p. 621.

The class of pericondensed PAHs contains C atoms that are members of three separate rings. Within this class, centrally condensed, compact PAHs (e.g., coronene) are sometimes called superaromatics. The class of more open PAH structures – catacondensed PAHs where no C atom belongs to more than two rings – contains various subclasses such as the acenes (linear rows of rings; naphthalene, anthracene, tetracene,...) and the phenes (bent rows of rings; phenanthrene, tetraphene,...). The structure of a PAH is closely linked to its stability. The centrally condensed PAHs are among the most stable PAHs, because this structure generally allows complete electron delocalization and truly aromatic bonding between all adjacent carbon atoms. The general structural formula of centrally condensed PAHs is $C_{6r^2}H_{6r}$ with $3r^2-3r+1$ hexagonal cycles arranged in $r-1$ rings around the central cycle. The first three are coronene ($C_{24}H_{12}$), circumcoronene ($C_{54}H_{18}$), and circumcircumcoronene ($C_{96}H_{24}$). With a C–C bond length of $\simeq$1.4 Å, the surface area of one aromatic cycle is $\simeq$5 Å^2. The total surface area of a compact PAH is then, approximately,

$$\sigma_{\rm PAH} \simeq 3\times10^{-15} r^2 \simeq 5\times10^{-16} N_{\rm c}\,{\rm cm}^2, \tag{6.1}$$

with $N_{\rm c}$ the number of C atoms of the PAH. The radius of this approximate circular PAH is then

$$a \simeq 0.9\times10^{-8} N_{\rm c}^{1/2}\,{\rm cm}. \tag{6.2}$$

So, a "typical" 50 C atom interstellar PAH will have some 20 fused cycles, some 20 H atoms at its periphery, a size of $\sim$6 Å, and a surface area of 200 Å^2. For comparison, recall that the sizes of interstellar dust grains derived from extinction measurements range from 50–3000 Å (Section 5.4), although the small size range is not well probed by these types of studies.

6.2 IR emission by PAH molecules

In order to analyze the observations of the IR emission features and to derive the characteristics of their interstellar carriers (e.g., abundance, sizes), we have to understand the interaction of large molecules with light. Specifically, we have to be able to relate the FUV photon energy absorbed to the "temperature" of the emitter and we have to be able to calculate the IR emission spectrum given the temperature. For dust grains, we have done this in Sections 5.2.1 and 5.2.2. In those sections, we discovered that the internal temperature of a dust grain, which describes the excitation of the phonon modes, is completely decoupled from the kinetic temperature of the gas and is set by a balance between UV photon absorption and IR photon emission. In this chapter, we will determine the vibrational excitation temperature of large molecules. This differs from the procedure we followed for dust grains because the temperature of large molecules is highly time dependent; so, we cannot just balance the time-averaged emission and absorption rates. To phrase it differently, large molecules are very hot immediately after absorption of a single FUV photon and during the emission of their cooling IR photons. After they have cooled down, they will remain cold for a long time until they absorb another FUV photon. Thus, we have to solve the energy balance of an interstellar PAH during the emission process.

In this section, we will first discuss the photophysics of large molecules (Section 6.2.1). Then, we will examine the temperature in a system of limited size, using concepts that are familiar from courses in statistical mechanics: microcanonical and canonical ensembles (Section 6.2.2). Once we have described the temperature of a PAH as a function of its size and absorbed FUV photon energy, we can calculate the IR emission spectrum (Section 6.2.3) and analyze the observations (Sections 6.7.1–6.7.3).

6.2.1 Photophysics of PAHs

The photophysics of large molecules has been discussed in Section 2.1.1. A small, neutral PAH molecule in the singlet electronic ground state (S_0) can absorb UV photons of fairly specific energies, corresponding to discrete electronic transitions, taking the molecule up to the S_1, S_2, or higher state (Fig. 6.2). This electronic excitation can be followed by a variety of de-excitation processes, including ionization and photodissociation. In terrestrial environments, collisional de-excitation is generally dominant. In the collision-free environment of space, the focus is on infrared fluorescence. Internal conversion (IC) from the excited electronic state will leave the molecule highly vibrationally excited in a lower lying electronic state. The molecule can also undergo intersystem crossing (ISC) to an excited vibrational state of an electronic state with a different multiplicity (the triplet system). Either process will eventually be followed by radiative cascade down the vibrational ladder to the ground vibrational state of the electronic state involved. For singlet states, this can be followed by fluorescence. If intersystem crossing has occurred, this will leave the molecule still electronically excited and the final radiative decay step is through phosphorescence; i.e., the emission of a visible photon.

The energy level structure of PAHs will depend on their size (and molecular structure). This is schematically indicated in Fig. 6.3. For a neutral PAH the π-bonding system is completely filled, while the anti-bonding ($\pi^\star$) levels are empty. These two-level systems correspond to the valence and conduction band of a graphite sheet. The total width of these bands is constant but the number of levels increases and their energy separation decreases with the size of the PAH.

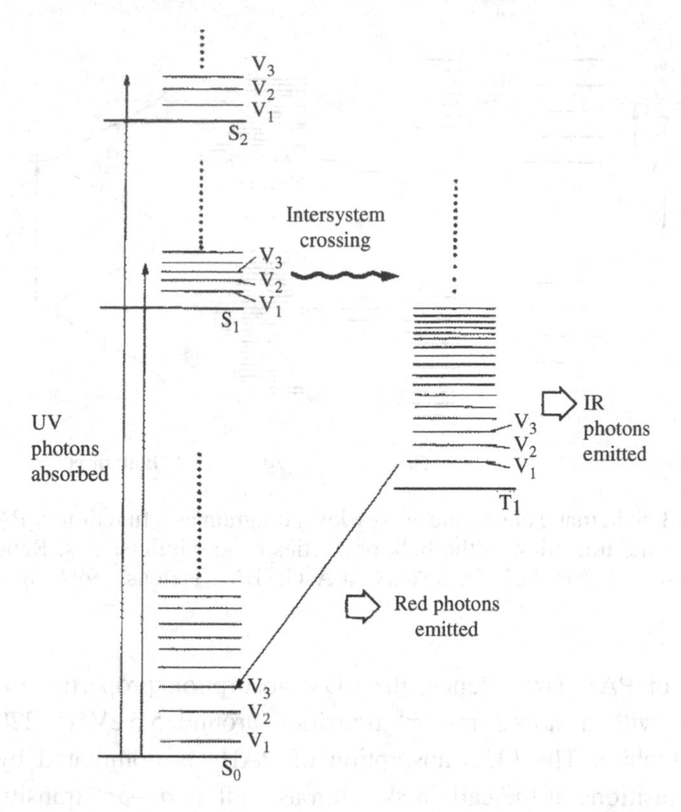

Figure 6.2 Schematic energy level diagram for a neutral PAH, illustrating the various radiative and non-radiative excitation and relaxation channels. Reproduced with permission from L. J. Allamandola, A. G. G. M. Tielens, and J. R. Barker, 1989, *Ap. J. S.*, **71**, p. 733.

This is similar to the energy levels of an electron in a box, which are separated by $(2n+1)\,(h/2a)^2$ with a the size of the box and n the quantum number. At the same time, as the size of the PAH increases, the energy separation between the π and $\pi^\star$ systems decreases until they eventually touch for an infinite graphitic sheet. In graphite, the π and $\pi^\star$ bands actually overlap slightly because of the weak binding between the graphitic sheets in this material. This gives graphite semimetal optical properties whose absorption in the FUV to IR range is dominated by electronic transitions. With an electronic energy level separation in a PAH of $\sim 6/N_c$ eV, only PAHs with more than 10^4 C atoms will have energy levels separated by less than the thermal energy at 10 K. A similar size is required to close the band gap between the π and $\pi^\star$ states. Hence, PAHs are 'semiconductors' and their optical properties in the UV range are dominated by electronic transitions and at IR wavelengths by vibrational transitions. While the band gap closes with PAH size, the peak in the π and $\pi^\star$ electronic energy level distribution is largely

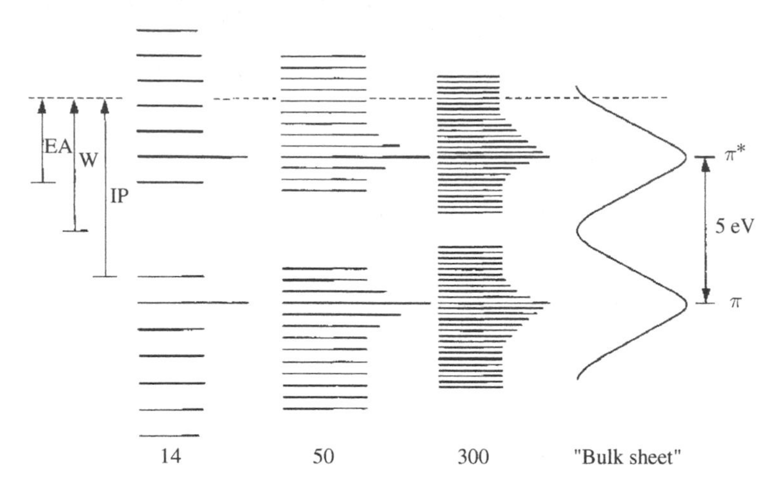

Figure 6.3 Schematic electronic energy level diagram as a function of PAH size, illustrating the transition to the bulk properties of graphitic sheets. Reproduced with permission from E. L. O. Bakes and A. G. G. M. Tielens, 1994, *Ap. J.*, **427**, p. 822.

independent of PAH size. Hence, the FUV absorption properties of PAHs are very similar, with a strong $\pi - \pi^\star$ transition around 5.5 eV ($\simeq$2200 Å), very similar to graphite. The FUV absorption of PAHs is dominated by the onset of $\sigma - \sigma^\star$ transitions of the carbon skeleton as well as $\sigma - \sigma^\star$ transitions associated with C–H bonds and ionizing transitions. Nevertheless, with increasing PAH size, the first transition will shift from the FUV towards the blue and then the red. For small pericondensed PAHs, the wavelength of the first strong absorption is approximately given by

$$\lambda \simeq 1630 + 370 N_c^{1/2}\ \text{Å}. \tag{6.3}$$

A 50 C atom pericondensed PAH will start to absorb around 4300 Å, while a 100 C atom PAH will already absorb at wavelengths longer than 5000 Å. After ionization, a π state will have opened up and electronic transitions involving this state will occur at lower energies (i.e., in the visible). The average absorption cross section will scale approximately with the number of π electrons (i.e., with N_c) and is $\sigma_{\rm FUV} = 7 \times 10^{-18}\ N_c$ cm^2.

Because of the high density of states of a highly vibrationally excited species, the different vibrational states of a molecule are very efficiently coupled and this leads to a rapid redistribution of the energy among all the vibrational modes accessible (internal vibrational redistribution; IVR). In general, modes couple well to some other modes, which then couple to other modes and transfer of energy from one mode to a different specific mode of the molecule may well take several steps. The IVR rate constant depends on the coupling between the various modes and the density of states at the internal energy of the species. For us, all of this is just details and we are only concerned with the point that for large molecules, such as PAHs, the vibrational modes are already well coupled at internal energies as low as 1000 cm^{-1}.

The various time scales involved in the excitation and relaxation processes are very different. Ultraviolet photon absorption in the ISM occurs on a time scale of

$$\tau_{UV} = k_{UV}^{-1} = (4\pi\sigma_{UV}(PAH)\mathcal{N}_{UV})^{-1} \simeq \frac{1.4\times10^9}{N_c G_0}\ \text{s}, \qquad (6.4)$$

with $\sigma_{UV}(PAH)$ the UV absorption cross section of the PAH (approximately equal to 7×10^{-18} cm^2 per carbon atom [N_c]) and $\mathcal{N}_{UV}$ the mean photon intensity of the radiation field – here expressed in terms of the Habing field, G_0. So, a typical interstellar PAH with $N_c = 50$ absorbs a UV photon once a year in the diffuse ISM and every 10 minutes in a PDR such as the Orion Bar. In contrast, non-radiative decay is very rapid; $\sim10^{-12}$ s for internal conversion and $\sim10^{-9}$ s for intersystem crossing. Internal vibrational redistribution occurs also on a very rapid time scale, $\sim10^{-12}$ s. Radiative vibrational relaxation through mid-IR transitions on the other hand is much slower, ~1 s. The radiative relaxation rate of low-lying vibrational states in the far-IR is of course much slower (e.g., $A \sim \nu^3$, cf. Chapter 2). Finally, collisional de-excitation is the slowest process, $\sim10^9/n_0$ s, with n_0 the density of collision partners.

The large difference between the FUV absorption rate and the IR emission rates implies that the temperature of the species will fluctuate widely. It is then clear that, unlike for 1000 Å dust grains, the average radiative equilibrium temperature will not be a good description of the excitation temperature of the species, when it is highly excited. Indeed, a small PAH may reach temperatures in excess of 1000 K immediately after FUV photon absorption but after a few seconds it may have cooled down to 10 K and remain that cold until it absorbs another FUV photon after a day or so (depending on the local FUV flux). While these temperature fluctuations are generally limited to molecular-sized species, they will occur for all species with widely disparate UV absorption and IR emission time scales such as nanoparticles and very small grains.

6.2.2 *The vibrational excitation of PAHs*

We have to evaluate the temperature that describes the vibrational excitation of a PAH as a function of time (e.g., internal energy) during the IR emission process. From a statistical physics point of view, the excitation temperature is connected to the average energy in a mode. Normally, we would average the energy over all species to arrive at the temperature or equivalently average the energy of one particle over a long time period. For PAHs, we have to be careful in this averaging. For a PAH, there are only a few modes available ($\sim$150 for $N_c = 50$ compared with 2×10^{24} for a mole of gas) and the former averaging has to take the finite size into account. Time-dependent averaging, on the other hand, is hampered by the rapid IR cooling. Nevertheless, the redistribution of the energy over the different modes is much faster than the IR emission time scale and the concept of average excitation at the (instantaneous) internal energy is still well defined.

It is important to recall from statistical mechanics the concepts of microcanonical and canonical systems. In a microcanonical ensemble, the energy of a system is fixed. In a canonical system, the energy fluctuates but the average energy – kT – is fixed. When a system is large enough, these fluctuations are relatively small compared with the total energy of the system and the microcanonical and canonical descriptions become essentially equivalent. A PAH that has just absorbed an FUV photon is really in a microcanonical state with a well defined energy. However, it is generally easier to evaluate rates using a canonical description and a temperature. We will therefore have to establish the interrelationship of these two representations of the excitation of the PAH and the limits on the validity of the canonical description.

A microcanonical system with s degrees of freedom can be described by the s coordinates, q_i, and the associated momenta, p_i. Its evolution is governed by its Hamiltonian, $H(\mathbf{p}, \mathbf{q})$. In statistical mechanics all the action happens in the phase space spanned by these coordinates and momenta. Any state of the system can be described by a point in this $2s$-dimensional space. Define the number of states, $W(E)$, with energy between E and $E + \mathrm{d}E$ as

$$W(E) = \frac{1}{h^s} \int \cdots \int \mathrm{d}p_1 \cdots \mathrm{d}p_s \mathrm{d}q_1 \cdots \mathrm{d}q_s, \tag{6.5}$$

where the integral ranges from $H = E$ to $H = E + \mathrm{d}E$. The density of states $\rho(E)$ is then defined as the number of states per unit energy, $\rho(E) = W(E)/\mathrm{d}E$. For the microcanonical ensemble, this density of states is the descriptive tool. For a collection, s, of classical oscillators with frequencies ν_i, the density of states is given by

$$\rho(E) = \frac{E^{s-1}}{(s-1)! \prod_i^s h\nu_i}. \tag{6.6}$$

While this analytical expression for the classical density of states is very handy, it is only accurate at very high energies. In general, finding a simple expression for the density of states is difficult, essentially because of the restriction imposed on the energy of the states that should be counted (i.e., only those with energies in the range $(E, E + \mathrm{d}E)$). Nevertheless, Eq. (6.6) does demonstrate two essential features of the density of states: for a given number of modes, the density of states increases rapidly with internal energy. Likewise, for a given internal energy, the density of states increases rapidly with the number of modes. Defining the geometric average frequency as $\overline{\nu}^s = \prod \nu_i$ and using Stirling's approximation, $\ln[n!] = n \ln[n]$, we can write

$$\rho(E) = \frac{1}{\overline{\nu}} \left(\frac{E}{(s-1)\, h\overline{\nu}} \right)^{s-1}, \tag{6.7}$$

where the density of states is now given in units of per wavenumber.

The difficulty in evaluating the density of states lies in the quantum nature of vibrations and the coupling between them (and with rotational states) through anharmonic interactions. An improvement to this expression for the density of states for classical oscillators is provided by the semi-empirical Whitten–Rabinovitch approximation,

$$\rho(E) = \frac{(E + aE_0)^{s-1}}{(s-1)! \prod_i^s h\nu_i}, \tag{6.8}$$

where E_0 is the total zero-point energy of the molecule, which can be calculated from the vibrational frequencies, and a is a semi-empirical function, which depends in a complicated way on the average and the dispersion of the vibrational frequencies and the energy in units of the zero-point energy. For PAHs, an approximation to the parameter a is provided by

$$a \sim 0.87 + 0.079 \ln[E/E_0]. \tag{6.9}$$

When the frequencies of a molecule are known, the density of states can be evaluated by direct counts, neglecting anharmonic interaction. Initially, such counts were done by grouping frequencies in as few degenerate groups as possible. However, very elegant techniques have been developed to do exact counting. This has been done for a few small PAHs for which the frequencies are known and comparison of the resulting density of states shows good agreement, reflecting the similarity in vibrational frequencies of these systems. Moreover, this density of states can be used to evaluate the temperature–internal-energy relation (see below), which can be compared to that measured for graphite (Fig. 6.4). Because of the overall similarity in binding and vibrational frequencies of PAHs and graphite, the $T - E$ relation should also be very similar, providing another test of the derived density of states. The calculated density of states of PAH molecules can be well approximated by

$$\ln[\rho(E)] = 2.84 \times 10^{-2} s \left(\frac{E}{s} \right)^{0.60} - 6.15, \tag{6.10}$$

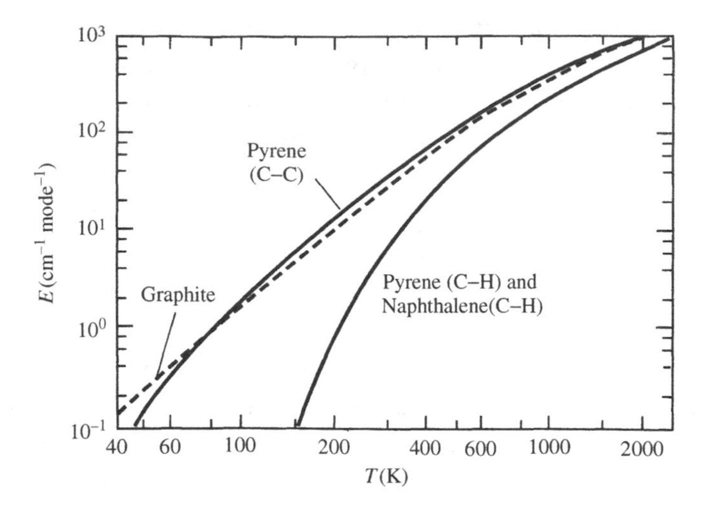

Figure 6.4 The internal energy as a function of temperature of the small PAH, pyrene ($C_{16}H_{10}$), is compared to that of graphite. The contributions due to the C–H modes in pyrene and naphthalene are also shown. Reproduced with permission from W. Schutte, A. G. G. M. Tielens, and L. J. Allamandola, 1993, *Ap. J.*, **415**, p. 397.

with s the number of degrees of freedom ($s = 3\,N_c - 6$), E in units of cm^{-1}, and $\rho(E)$ in units of states per cm^{-1}. This approximation is valid over the energy range 2.5×10^{-2} to $3 \times 10^2\,\mathrm{cm}^{-1}$ per mode ($10^{-5} - 10^{-1}$ eV per C atom). The H atoms at the periphery of the PAH do not contribute much to the internal energy and these are ignored here (see below). We note that there is a similarity between this expression and the classical density of states for s harmonic oscillators (Eq. (6.6)). Because of the similarity in binding for PAHs, the geometric mean frequency can be subsumed in the constant in this expression.

The canonical ensemble describes a system that is in contact with a heat reservoir, which keeps it at a fixed temperature, T (equal to the average energy in units of k). The density of states of this system is then

$$\rho(\mathbf{p},\ \mathbf{q}) = \exp[-H(\mathbf{p},\ \mathbf{q})/kT]. \tag{6.11}$$

The canonical system is described by the partition function at the given temperature; i.e.,

$$Q(T) = \frac{1}{N!h^s} \int \cdots \int \exp[-H(\mathbf{p},\ \mathbf{q})/kT]\mathrm{d}p_1 \cdots \mathrm{d}p_s\, \mathrm{d}q_1 \cdots \mathrm{d}q_s. \tag{6.12}$$

Both the microcanonical and the canonical ensemble provide meaningful descriptions of macroscopic systems. The canonical system with its concepts of average energy and temperature – culminating in the Boltzmann factor – provides more easily used expressions. However, for small systems, the canonical description will break down and care should be used in its application to PAH molecules. We will come back to this point below.

Because energy exchange between the different modes in a PAH is so rapid, the ergodic approximation is justified; i.e., the energy is distributed statistically over all modes. A vibrationally excited interstellar PAH is a system with constant energy and hence – in statistical mechanics terms – forms a microcanonical ensemble. The excitation is described by the density of states; defining $P_i(v)$ as the fraction of species with v excitations of mode i, we have

$$P_i(v) = \frac{N_{v,i}(E)}{N(E)} = \frac{\rho_r(E - vh\nu_i)}{\rho(E)}, \tag{6.13}$$

with $N_{v,i}(E)$ the number of species with v quanta in mode i with frequency ν_i and $N(E)$ the total number of molecules with energy E. The total density of vibrational states, $\rho(E)$, is the number of ways the energy E can be divided over all available states, while the reduced density of states, $\rho_r(E - vh\nu_i)$, represents the number of ways the energy, $E - vh\nu_i$, can be divided over all modes except the mode under consideration.

Consider now the system as a canonical ensemble – e.g., connected to a large heat reservoir with which it exchanges energy – and describe the excitation in terms of the average energy and temperature. The probability, $P(E', T)$, that in a canonical system with mean energy kT an individual mode has an internal energy E' is

$$P(E', T)\, \mathrm{d}E' = \frac{\rho(E')}{\Phi(T)} \exp[-E'/kT]\, \mathrm{d}E', \tag{6.14}$$

with Φ the partition function,

$$\Phi(T) = \int_0^\infty \rho(E') \exp[-E'/kT]\, \mathrm{d}E'. \tag{6.15}$$

The density of states increases exponentially with energy while the Boltzmann factor decreases exponentially with E'. Hence, the probability will show a sharp maximum. To second order, the probability function can be represented by a Gaussian with mean energy kT and a dispersion $\sigma_{E'} \simeq kT/\sqrt{s}$. As the system becomes larger, the distribution function will get broader but, relative to kT, the distribution will actually sharpen. In the limit of an infinitely large system, the distribution approaches a delta function. Thus, in the limit of large s, the canonical approximation will approach the microcanonical distribution with a single energy equal to the mean energy.

We can also approach this discussion from a slightly different point of view. Consider a single mode, i, in a PAH connected to a thermal bath represented by all other modes. Under certain limitations, the system can then be considered a canonical ensemble and the excitation of this mode can be described by a temperature, T, equal to the mean energy in the mode in units of the Boltzmann constant, k. Of course, a truly canonical ensemble requires that all energies are accessible and that is not the case here since the mode can never have more energy than the total energy, E, in the system. In practice, however, the canonical approach leads to reasonable results as long as the average energy in mode i is small compared to the total energy in the system. Or more generally, for a system to be a canonical heat reservoir, its internal temperature should not be affected by the (small) amount of energy exchanged with the system under consideration. Because the energy will fluctuate around the mean energy and because the vibrational excitation is so heavily weighted towards higher energies, the average excitation will be overestimated in the canonical approximation. For a species with 100 C atoms, this results in a factor 2 error in the fractional excitation when the energy in the system is $\simeq 4$ times the energy of the mode. The error decreases to a factor of 1.1 when the energy in the system increases to ~ 10 times the energy of the mode. Thus, the error is largest for the highest frequency modes.

We can pursue this slightly further by introducing the entropy, S, of a species,

$$S = k \ln[\rho(E)]. \tag{6.16}$$

The microcanonical temperature, T_m, can then be defined through the relation

$$\frac{1}{kT_m} = \frac{1}{k}\frac{dS}{dE} = \frac{d\ln[\rho(E)]}{dE}. \tag{6.17}$$

Using the expression for the density of states (Eq. 6.10), this results in

$$T_m \simeq 2000 \left(\frac{E(\text{eV})}{N_c}\right)^{0.4} K, \tag{6.18}$$

with E in electron volts. This equation is approximately valid over the range 35–1000 K (within 10%) but rapidly deteriorates outside of these limits. Because of the (small) skewness of the canonical distribution, there is a small difference between the energy of the system, E, and the average energy, $<E(T_m)>$, which can be evaluated from a Taylor expansion of Eq. (6.14),

$$E \simeq <E(T_m)> - kT_m. \tag{6.19}$$

We can consider the last term as a finite heat bath correction. When the system becomes large, this correction term becomes very small. We can introduce the microcanonical heat capacity, C_m,

$$C_m = \frac{dE}{dT_m} = 2.5\frac{E}{T_m}, \tag{6.20}$$

where the right-hand expression is derived using Eq. (6.18) for the density of states of PAHs. Thus, this energy difference can also be seen as a relationship between the microcanonical and canonical heat capacities, C_m and C_c; viz.

$$C_m \simeq C_c - k. \tag{6.21}$$

Again, this correction factor is small for large systems.

6.2.3 Infrared fluorescence

We have now derived expressions for the excitation of a PAH with a given internal energy through the density of states (Eq. (6.10)) or through the excitation energy (Eq. (6.18)). The emission intensity due to the vibrational transition $v \rightarrow v-1$ in mode i is then given by

$$I(E, i, v) = \frac{N_{v,i}(E) A_{v,i} h\nu_i}{4\pi}, \tag{6.22}$$

with $A_{v,i}$ the Einstein A emission coefficient associated with this transition and $N_{v,i}(E)$ all species with internal energy E with v quanta in mode i along the line of sight. Thus, using Eqs. (6.10) or (6.18) to describe the excitation of a PAH with internal energy E, we can calculate the intensity using, for example, a Monte Carlo technique to follow the energy cascade during the emission process. However, it is advantageous to switch to the canonical description.

For harmonic oscillators, all transitions in mode i occur at the same frequency and the total intensity at this frequency can be obtained by summing over v. Also, for harmonic oscillators, $A_{v,i}$ is equal to $vA_{1,i}$. In the microcanonical description, the excitation is described by Eq. (6.13). In the canonical description, the excitation is described by the Boltzmann equation and this expression becomes

$$I(E, i) = \frac{N(E) h\nu_i A_{1,i}}{4\pi} \sum_{v=1}^{\infty} \frac{v \exp[-vh\nu_i/kT]}{(1-\exp[-h\nu_i/kT])}. \tag{6.23}$$

Define the optical depth, τ_i, as

$$\tau_i = \kappa_i L = B_i N(E), \tag{6.24}$$

with B_i the Einstein coefficient for absorption in mode i, κ_i the absorption coefficient, and L the length through the cloud. Performing the summation in Eq.(6.23) and using the relationship between the Einstein coefficients, the intensity reduces then to the familiar expression for optically thin emission,

$$I(E, i) = \tau_i B(\nu_i, T), \tag{6.25}$$

with $B(T)$ the Planck function (cf. Sections 2.3.2 and 5.2.2). Essential assumptions in this derivation are the canonical approximation, which permits us to use

Boltzmann to describe the excitation, and the harmonic approximation for the oscillator, which permits the summing, which gives rise to the Planck function. Generally, vibrational modes in molecules are anharmonic and, for the detailed profile of an emission band, the full expression (cf. Eq. (6.22)) should be used. Because anharmonicity effects are small, this is often included by adopting a temperature- (and frequency-) dependent absorption coefficient, $\kappa(\nu_i, T)$, which can then be used to calculate the frequency dependent optical depth. The same, actually, holds true for vibrational modes in solid materials.

The total emergent intensity can now be obtained by following the IR cascade. Consider a species that absorbs a photon with energy $h\nu_{\rm UV}$. The temperature, T_1, immediately after photon absorption is given by the integral expression,

$$h\nu_{\rm UV} = \int_{T_0}^{T_1} C_V(T)\, {\rm d}T, \tag{6.26}$$

with C_V the specific heat of the species and T_0 the initial temperature. This species will cool at a rate given by

$$\frac{{\rm d}T}{{\rm d}t} = \frac{1}{C_V}\frac{{\rm d}E}{{\rm d}t} = -\frac{4\pi}{C_V}\sum_i \kappa_i B(\nu_i, T), \tag{6.27}$$

where the sum is over all vibrational modes. For grains with continuous opacity, the summation over the vibrational modes is replaced by an integral over frequency. The temperature decay law will depend on the detailed characteristics of the specific PAH under consideration. However, because of the similarity in the vibrational properties of PAHs, a reasonable approximation for small ($\simeq$50 C atoms) neutral PAHs is given by

$$\frac{{\rm d}T}{{\rm d}t} \simeq -1.1\times 10^{-5}\, T^{2.53}\,{\rm K\ s^{-1}} \quad T > 250\,{\rm K}. \tag{6.28}$$

For two-dimensional materials, the heat capacity is approximately proportional to T^2 and to zeroth order that is also a good approximation for PAHs. The energy decay rate then scales with a high power of the temperature ($\sim T^{4.5}$). At a temperature of 1000 K, this corresponds to a cooling time scale of $\simeq$2 s. Note that, to first order, this is independent of the PAH size because at a given temperature the emission rate as well as the energy content scale similarly with the number of modes. Ionized PAHs are somewhat (factor 3–4) more effective coolers due to the larger intrinsic strength of the C–C modes.

The emission in a given mode, i, is then given by

$$I(i) = \int_{T_0}^{T_1} I(E(T), i)\; G(T)\, {\rm d}T, \tag{6.29}$$

where $G(T)$ is the fraction of the time a cooling species spends at temperature T. Assuming that IR photon emission is a Poisson process, $G(T)$ is given by

$$G(T)\,\mathrm{d}T = \frac{\overline{r}}{\mathrm{d}T/\mathrm{d}t}\exp[-\overline{r}\tau_{\min}(T)]\,\mathrm{d}T, \tag{6.30}$$

where $\tau_{\min}(T)$ is the time it takes to cool down from T_1 to T, which is given by the integral expression,

$$\tau_{\min}(T) = \int_T^{T_1} \frac{1}{\mathrm{d}T/\mathrm{d}t}\,\mathrm{d}T. \tag{6.31}$$

The average time between photon absorptions, $\overline{r}$, is given by

$$\overline{r} = 4\pi\sigma_{\mathrm{PAH}}(\nu_{\mathrm{UV}})\,\mathcal{N}(\nu_{\mathrm{UV}}), \tag{6.32}$$

with $\mathcal{N}(\nu_{\mathrm{UV}})$ the mean photon intensity of the radiation field at frequency, ν_{UV}.

Evaluation of these expressions requires specification of the vibrational modes of a species. These can then be used to calculate the energy decay rate and the heat capacity of the species,

$$C_V = k\sum_{i=1}^{s}\left(\frac{h\nu_i}{kT}\right)^2 \frac{\exp[h\nu_i/kT]}{(\exp[h\nu_i/kT]-1)^2}, \tag{6.33}$$

where the summation is over all s degrees of freedom of the species. When a specific species has not been studied in detail in the laboratory, the heat capacity can be estimated from related species or using group additive expressions (i.e., the species is broken down into groups for which the contribution to the heat capacity is known from studies on related species and the contributions of the different groups are then added). For PAHs, often the heat capacity of graphite is used (cf. Fig. 6.4). For low temperatures, this breaks down because of the lack of low energy vibrational modes in small species, but this is never a serious issue. Also, PAHs have modes associated with H atoms that are not represented in the graphitic carbon skeleton (cf. Fig. 6.4). The effect of the C–H modes is small even for PAHs as small as pyrene and naphthalene and becomes negligible when the PAH size increases and the H/C ratio decreases. The $E-T$ relation evaluated from the density of states (Eqs. (6.10) and (6.18)) is in good agreement with the relation derived from the heat capacity of graphite. In numerically evaluating the IR emission from PAHs, it can be advantageous to fit simple analytical expressions to Eqs. (6.26), (6.27), and (6.31). For the last, the upper boundary can be taken outside of the relevant range and $\tau_{\min}$ evaluated as the difference of appropriate values. A similar approach may help in the evaluation of Eq. (6.26).

For small species, the cooling time scale is so much faster than the UV absorption time scale that T_0 can be set equal to microwave background radiation temperature and calculation of the emission spectrum is rather straightforward. Multiphoton processes become important when $\tau_{UV} < 0.1\tau_{cool}$, where $\tau_{cool} = T/(dT/dt)$. Combining Eqs. (6.4), (6.18), and (6.28), this yields $G_0 > 10^6\,(50/N_c)^{1.61}$ for an absorbed FUV photon energy of 10 eV. In that case, T_0 and T_1 have to be determined in an iterative fashion. One way to do this is by iterating on $G\,(T)$; viz.

$$G_{n+1}(T_0, T) = \frac{1}{\bar{r}} \int_{T_0}^{T} G_n(T_0, T')G(T', T)\, dT', \tag{6.34}$$

where the term in the integral is the probability that, starting at temperature, T_0, the species has a temperature T' after n photon events multiplied by the probability that the absorption of an additional photon takes it to temperature T. Starting with a delta function at the microwave background temperature, this system can then be iterated until convergence is attained.

In reality, an interstellar PAH will be exposed to a spectrum of exciting UV photons. This set of equations should then be averaged over the UV spectrum; i.e., each UV absorption frequency yields a distinct T_1 (Eq. (6.26)). The temperature distribution function, $G(T)$, has to be averaged properly over this range in T_1. Of course, the rate of UV photon absorption (Eq. (6.32)) has to be integrated over the excitation spectrum as well. Finally, the interstellar PAH family may be very diverse and occur in different ionization stages, each with its own intrinsic properties, including distinct vibrational modes. The IR intensity of such a family can then be found by a proper summation; i.e.

$$I(\nu) = \sum_k \sum_j \sum_i n(k) f(j) I_{k,j}(i) \phi_{k,j}(i, \nu), \tag{6.35}$$

where the summation is over the modes i, of ionization stage j, of molecule k. Here, $n(k)$ is the density of the specific PAH molecule k, and $f(j)$ the fractional abundance of ionization stage j. The integrated intensity, $I_{k,j}(i)$, of mode i of molecule k in ionization stage j is given by Eq. (6.29), and $\phi_{k,j}(i, \nu)$ is the intrinsic line profile of this transition.

The evaluation of the line profile requires care. For harmonic oscillators, the intrinsic line profile is given by a Lorentzian. If we assume that the collection of PAHs has peak frequencies that – for a given mode – are randomly distributed around some mean value with a given dispersion, the line profile would be the Voigt profile – well known to stellar spectroscopists – a convolution between a Gaussian and a Lorentzian profile. However, the vibrational modes of PAHs are strongly coupled through anharmonic interactions and the intrinsic line profile is not Lorentzian. Laboratory studies show that, for small PAHs, this anharmonic interaction can be more important than the summation over a collection of PAH species for some bands. Indeed, the observed profiles of the interstellar 6.2 and 11.2 μm bands show a distinctly non-Lorentzian profile, with a sharp blue rise and long red tail, characteristic for anharmonic interactions. The interstellar 7.7 μm band, on the other hand, shows distinct contributions by multiple components. As mentioned earlier in this section, the effects of anharmonicity on the line profile are often included through a temperature- or frequency-dependent absorption coefficient.

6.2.4 Radiative equilibrium versus temperature fluctuations

It is clear from the above discussion that the temperature will fluctuate for small species owing to the infrequent heating events by FUV photons and the rapid IR cooling. Comparing the IR cooling time scale with the IR emission time scale we can estimate the grain size for which fluctuations become important. Adopting

$$\frac{\Delta T}{T} > 0.1, \tag{6.36}$$

for when temperature fluctuations become important, we find

$$\frac{\tau_{\mathrm{UV}}}{T}\left|\frac{\mathrm{d}T}{\mathrm{d}t}\right| > 0.1, \tag{6.37}$$

with τ_{UV} the UV absorption time scale of the PAH (cf. Eq. (6.4)). This results in

$$N_{\mathrm{c}} < 1.6 \times 10^3 T^{1.53} G_0^{-1}. \tag{6.38}$$

For the temperature in this equation, we will adopt the radiative equilibrium temperature of a graphite grain (cf. Section 5.2.2), which results in

$$N_{\mathrm{c}} < 8.3 \times 10^6 \, G_0^{-0.70}. \tag{6.39}$$

Thus, for the diffuse ISM ($G_0 = 1$), we find that temperature fluctuations are important when $N_{\mathrm{c}} < 10^7$ C atoms. For spherical grains with a specific density of $2.2\,\mathrm{g\,cm^{-3}}$, this implies a limiting size of some 275 Å. When the intensity of the UV field increases, there is less time between photon absorptions and smaller grains can still maintain radiative equilibrium; e.g., in the Orion Bar ($G_0 \simeq 10^5$), this limit is 2500 C atoms or $a = 17$ Å. Of course, a 10% variation in temperature has little effect on the emission intensity in the Rayleigh limit but causes major variations on the Wien side of the Planck curve. In any case, all the grains that give rise to the FUV extinction, which have typical sizes <200 Å, will undergo (small) temperature fluctuations.

6.2.5 The temperature distribution of PAHs

The temperature distribution of interstellar PAHs will thus depend on the size of the species. This is illustrated in Fig. 6.5. Classical-sized grains are largely in radiative equilibrium with their environment in the diffuse ISM. As the size

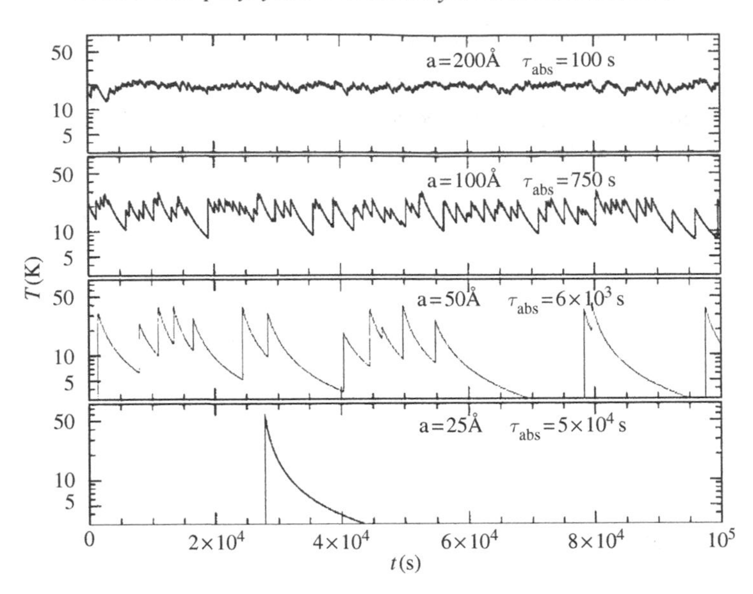

Figure 6.5 The time-dependent behavior of the temperature for various sizes of the species. The time axis corresponds to approximately one day. These calculations pertain to the diffuse ISM. Figure courtesy of B. T. Draine; reprinted, with permission, from the *Ann. Rev. Astron. Astrophys.*, **41**, p. 241 ©2003 by *Ann. Rev.* (www.annualreviews.org).

of the species decreases, temperature fluctuations become more prevalent. Very small grains are very cold except directly after UV photon absorption. For PAH molecules the maximum temperature after FUV photon absorption can reach 2000 K and decays very rapidly (∼2 s).

6.3 PAH charge

As for grains (cf. Section 5.2.3), the charge distribution of interstellar PAHs is set by a balance between photo-ionization and electron and ion collisional charging. The ionization balance can be solved in a manner that is completely analogous to that for dust grains. However, because of their small size, there are a few details that need to be considered.

6.3.1 Ionization potentials

Consider the process

$$\mathrm{PAH}^{Z} \longleftrightarrow \mathrm{PAH}^{Z+1} + e^{-}, \tag{6.40}$$

where the forward and backward arrow corresponds to ionization and electron recombination (attachment), respectively. The ionization potentials of small PAHs

Table 6.1 *Ionization potentials[a] of small PAHs*

PAH	Z	Electron affinity	Ionization potential		
		−1	0	1	2
Pyrene	$C_{16}H_{10}$	0.6	7.4	16.6	
Coronene	$C_{24}H_{12}$	0.6	7.4	13.64	
Ovalene	$C_{32}H_{14}$	0.9	6.9	12.7	
Circumcoronene	$C_{54}H_{18}$	1.3	5.9	8.8	12.9

[a] In units of eV.

are very sensitive to the size and "geometry" of the PAH. Values for some typical PAHs are collected in Table 6.1. In a general way, as for dust, we can write

$$IP(Z) = W + \left(Z + \frac{1}{2}\right)\frac{e^2}{C}, \tag{6.41}$$

with C the capacitance of the PAH and e the electron charge. For the PAHs, the capacitance can be approximated, to first order, by that of a thin circular disk of radius a ($C = 2a/\pi$) – rather than a sphere – and the ionization potential is then

$$IP(Z) \simeq 4.4 + \left(Z + \frac{1}{2}\right)\frac{25.1}{N_c^{1/2}} \text{ eV}. \tag{6.42}$$

These equations predict that for a typical interstellar PAH with $N_c = 50$, the ionization potentials to reach the first three positive ionization stages are 6.2, 9.7, and 13.3 eV. The next one is above 13.6 eV and hence in diffuse clouds, higher ionization stages cannot be reached through photo-ionization. Similarly, the first electron affinity is 2.6 eV. The second electron affinity is −0.9 eV; i.e., this second electron is not bound and this PAH^{2-} readily auto-ionizes back to PAH^{-}. Thus, small PAHs have only a very limited number of ionization states available in the diffuse ISM.

6.3.2 Photo-ionization

The photo-ionization cross section has been studied experimentally only for a handful of PAHs, including pyrene ($C_{16}H_{10}$) and coronene ($C_{24}H_{12}$), and the formalism developed for small grains is in reasonable agreement with these results. The photo-electric ionization rate is then (cf. Eq. (5.63))

$$J_{pe}(Z_d) \simeq 2.5 \times 10^{-13}\,(13.6 - IP(Z_d))^2 N_c f_y(N_c) G_0 \text{ electrons s}^{-1}, \tag{6.43}$$

Table 6.2 *The ionization balance of PAHs in the diffuse ISM*

PAH	Photo-ionization rates[a]			
Z	−1	0	1	2
Pyrene	—	1.7 (−9)	—	—
Coronene	—	2.5 (−9)	—	—
Ovalene	1.4 (−8)	4.0 (−9)	7.1 (−11)	—
Circumcoronene	2.2 (−8)	8.8 (−9)	3.4 (−9)	7.3 (−11)
PAH	**Electron recombination rates[b]**			
Z	0	1	2	3
Pyrene	—	1.2 (−5)	—	—
Coronene	—	1.4 (−5)	—	—
Ovalene	3.4 (−7)	1.6 (−5)	3.2 (−5)	—
Circumcoronene	4.4 (−7)	2.1 (−5)	4.2 (−5)	6.3 (−5)

[a] Calculated from Eq. (6.43) using the ionization potentials in Table 6.1 for $G_0 = 1$ in units of electrons s^{-1}.
[b] Calculated from Eq. (6.44) adopting $\phi_{PAH} = 1/2$ and $T = 100$ K in units of $cm^3\ s^{-1}$.

where the yield enhancement factor, $f_y(a)$, is given by Eq. (5.62). For small PAHs, $f_y \simeq 10$. Photo-electric ionization rates for some typical PAHs are summarized in Table 6.2.

6.3.3 Electron recombination

Electron recombination rates for small PAH cations are very uncertain. Often, PAHs are treated as small spherical grains with a size given by Eq. (6.2). The electron recombination rates are then calculated following the classical formalism (Eq. (5.51) in Section 5.2.3). For a 50 C atom PAH at 100 K, this results in

$$J_e(Z=1) = 4.1 \times 10^{-5} \phi_{PAH} \left(\frac{N_c}{50}\right)^{1/2} \left(\frac{100\,\mathrm{K}}{T}\right)^{1/2} n_e\,\mathrm{s}^{-1}, \tag{6.44}$$

where ϕ_{PAH} is a factor $\lesssim 1$ (see below). Of course, this classical expression is developed for a spherical grain and, for a planar object, a correction factor of 0.8 should be applied. However, the applicability of these classical rates is somewhat in doubt. Measured electron recombination rate coefficients for some small PAH cations, including naphthalene ($C_{10}H_8^+$) and phenanthrene ($C_{14}H_{10}^+$), are $1-1.7 \times 10^{-7}\ cm^3\ s^{-1}$ at 300 K. This is a factor of 8 and 6 smaller than the classical rates would indicate (e.g., $\phi_{PAH} = 0.15$).

6.3.4 Electron attachment

Electron attachment to neutral PAHs can be described in a general sense by the reaction sequence

$$\mathrm{PAH} + \mathrm{e}^- \underset{k_\mathrm{b}}{\overset{k_\mathrm{f}}{\longleftrightarrow}} (\mathrm{PAH}^-)^* \overset{k_\mathrm{r}}{\longrightarrow} \mathrm{PAH}^-, \tag{6.45}$$

where $(\mathrm{PAH}^-)^*$ is an excited PAH anion, k_f is the forward (electron capture) rate coefficient, k_b the backward (auto-ionization) rate coefficient, and k_r the (radiative) stabilization rate coefficient. The rate coefficient for the complete reaction is then given by

$$k_\mathrm{ea} = \left(\frac{k_\mathrm{r}}{k_\mathrm{r} + k_\mathrm{b}}\right) k_\mathrm{f}, \tag{6.46}$$

where the term in brackets can be considered a sticking coefficient. The same actually holds for the radiative recombination of PAH cations. However, in that case, radiative stabilization is very rapid and the sticking coefficient is essentially unity. That is not necessarily the case for electron attachments to neutral PAHs. In equilibrium, the capture and auto-ionization rate coefficients are related through the principle of detailed balance,

$$\frac{k_\mathrm{b}}{k_\mathrm{f}} = \frac{\rho^0}{\rho^-}, \tag{6.47}$$

with ρ^- and ρ^0 the density of states of the anion and that of the neutral plus electron, respectively. Ignoring the small differences in vibrational frequencies, the density of states of the anion can be approximated by that of the neutral (cf. Eq. (6.10)) at an energy corresponding to the electron affinity of the PAH. Then, assuming that the neutral is in its vibrational ground state after loss of the electron, its density of states is that of the free electron,

$$\rho^0 = \rho(\mathrm{e}) = \frac{m^2 v}{\pi^2 \hbar^3}. \tag{6.48}$$

The radiative stabilization rate, k_r, is of the order of $1\ \mathrm{s}^{-1}$ (cf. Section 6.2.3). The electron capture rate is difficult to estimate. At large distances, both electron-induced dipole interaction and image charge interaction contribute. At short distances, other correlation effects, associated with spin and spatial symmetry properties of the molecular and electronic wave functions, may become important. We will assume that the electron capture rate can be approximated by

$$k_\mathrm{f} = 2\pi \left(\frac{\alpha e^2}{m_\mathrm{e}}\right)^{1/2}, \tag{6.49}$$

with α the polarizability. This is of course just the Langevin rate (cf. Section 4.1.3). The limited data on polarizabilities of small, compact PAHs are consistent with

$$\alpha = 1.5 \times 10^{-24} N_c \, \mathrm{cm}^3, \tag{6.50}$$

which is as expected when the polarizability is proportional to the number of π electrons in the system. For very large PAHs, the limit of a thin disk with radius a would be expected where the polarizability scales with a^3 and hence α with $N_c^{3/2}$. However, this limit is not attained.

The overall electron attachment rate, k_{ea}, depends on a variety of parameters. However, the most important one is the electron affinity of the neutral species. The higher the electron affinity, the more states are accessible to divide the energy between and the less likely that auto-ionization will occur before radiative stabilization of the excited anion. As a rule of thumb, the electron sticking coefficient is very small for electron affinities less than 1 eV (cf. Fig. 6.6). The dependence on PAH size – which enters through the degrees of freedom in the density of states as well as into the forward rate – is relatively minor. Finally, we note that, while this discussion is geared towards PAHs, the conclusion is more general. Species with a high electron affinity – even small radicals such as OH – will have large electron attachment rates.

For large PAHs ($N_c > 25$ C atoms), the electron affinity is larger than 1 eV and the electron sticking coefficient is near unity. In that case, the electron capture rate is

$$k_{ea} = 8.5 \times 10^{-7} \, \phi_{PAH} \left(\frac{N_c}{50} \right)^{1/2} \mathrm{cm}^3 \, \mathrm{s}^{-1}. \tag{6.51}$$

Experimentally, the electron attachment rate has been measured for the small PAH anthracene ($C_{14}H_{10}$) to be about 2×10^{-9} cm^3 s^{-1} independent of temperature in

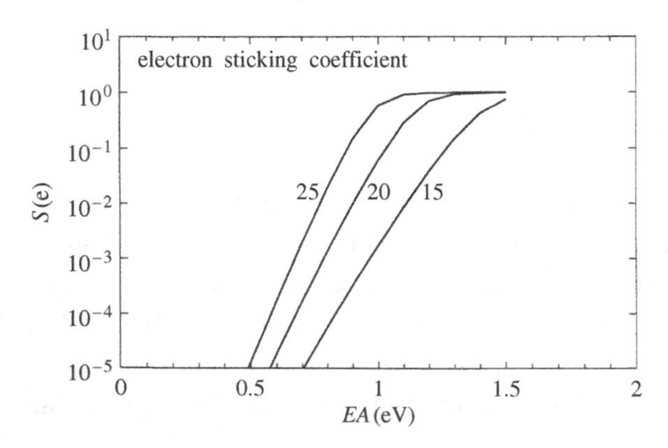

Figure 6.6 The electron sticking coefficient for small PAHs as a function of the electron affinity is plotted for various sized PAHs.

the range 45–300 K. This suggests an electron sticking coefficient of $\simeq 5\times10^{-3}$. This would imply an electron affinity of $\simeq 0.9$ eV, which is in the experimental range of 0.4–1.0 eV.

Finally, some species are known to possess an activation barrier to electron attachment. Fullerenes will be discussed in Section 6.5 but we note here that the fullerenes C_{60} and C_{70} have measured activation barriers of 0.26 and 0.15 eV, respectively. Despite their high electron affinities, the electron attachment rate at low temperatures is very small (e.g., $5\times10^{-7}\exp[-0.26\,\mathrm{eV}/kT]$ for C_{60}). Apparently, for these anions, there is no accessible s state ($L=0$) and association through the p state has a centrifugal barrier of the order of $\hbar^2/m_e R$, with R the radius of C_{60} (5 Å). Whether this also applies to PAHs in the interstellar size range is unknown.

6.3.5 Neutralization reactions

Neutralization of atomic cations with PAH anions is important for the charge balance in the diffuse ISM,

$$\mathrm{X}^+ + \mathrm{PAH}^- \longrightarrow \mathrm{X} + \mathrm{PAH}. \tag{6.52}$$

Approximating the interaction with that for a spherical grain with a radius given by Eq. (6.2), the rate coefficient is given by

$$k_\mathrm{n} = 2.9\times10^{-7}\phi_\mathrm{PAH}\left(\frac{12\,\mathrm{amu}}{m_\mathrm{X}}\right)^{1/2}\left(\frac{100\,\mathrm{K}}{T}\right)^{1/2}\left(\frac{N_\mathrm{c}}{50}\right)^{1/2}\ \mathrm{cm^3\,s^{-1}}, \tag{6.53}$$

where we have again introduced the factor ϕ_PAH – which may have a value in the range 0.1–1 – to remind us that this classical rate is likely to be an overestimate.

6.3.6 Other reactions

A variety of other reactions involving neutral or charged PAHs can be of importance for the charge balance. Of particular importance is the double electron transfer in the reaction with He^+,

$$\mathrm{He}^+ + \mathrm{PAH} \longrightarrow \mathrm{He} + \mathrm{PAH}^{2+} + \mathrm{e}. \tag{6.54}$$

The ionization potential of neutral helium, 24.6 eV, is much larger than the typical ionization potential of ionizing a PAH twice and this reaction is exothermic. The classical rate for this reaction is (Section 5.2.3),

$$k_\mathrm{di} = 1.1\times10^{-8}\phi_\mathrm{PAH}\left(\frac{N_\mathrm{c}}{50}\right)^{1/2}\ \mathrm{cm^3\,s^{-1}}. \tag{6.55}$$

This reaction provides a pathway towards dications in dense clouds, where He^+ ions are formed through the cosmic-ray ionization of neutral helium atoms.

Also of importance is the reaction between cations and neutral PAHs, which we expect will lead to adsorption rather than charge exchange; for example,

$$C^+ + \mathrm{PAH} \longrightarrow (\mathrm{PAH{-}C})^+. \tag{6.56}$$

The rate for this reaction can be derived from the Langevin reaction rate (cf. Section 4.1.3) with the polarizability given by Eq. (6.50),

$$k_{\mathrm{ca}} = 6\times10^{-9}\phi_{\mathrm{PAH}}\left(\frac{N_{\mathrm{c}}}{50}\right)^{1/2}\left(\frac{12\,\mathrm{amu}}{m_{\mathrm{X}}}\right)^{1/2}\ \mathrm{cm^3\,s^{-1}}. \tag{6.57}$$

6.3.7 The ionization balance

As for dust grains, the charge of interstellar PAHs in the diffuse ISM is determined by the balance between photo-ionization and electron recombination. Hence, the ionization balance of PAHs is controlled by the parameter $G_0\,T^{1/2}/n_{\mathrm{e}}$, with G_0 the intensity of the radiation field in units of the Habing field, T the gas temperature, and n_{e} the electron density. It is instructive to consider a PAH, such as pyrene, with only two accessible ionization stages in the ISM, the neutral and the singly charged cation. The neutral fraction, $f(0)$, is then given by (cf. Section 3.3.2)

$$f(0) = [1+\gamma_0]^{-1}, \tag{6.58}$$

with γ_0 the ratio of the ionization rate over the recombination rate, which – adopting the classical expression with $\phi_{\mathrm{PAH}} = 0.5$ – is

$$\gamma_0 = 3.5\times10^{-6}\,N_{\mathrm{c}}^{1/2}\frac{G_0 T^{1/2}}{n_{\mathrm{e}}}. \tag{6.59}$$

Thus, in the diffuse ISM with $G_0 = 1.7$, $T = 80\,\mathrm{K}$, and all the carbon ionized ($n_{\mathrm{e}} = 1.5\times10^{-4} n_0 = 7.5\times10^{-3}\ \mathrm{cm^{-3}}$), $\gamma_0 = 2.9\times10^{-2}$ for a 16 C atom PAH (pyrene) and such a PAH would be mainly neutral. Likewise, for a 50 C atom PAH, we find a neutral fraction of $\simeq$0.95.

This result depends somewhat on the adopted electron recombination rate. If instead of the classical rate, we use the measured rate for phenanthrene, this results in a γ_0 of

$$\gamma_0 = 1.3\times10^{-4}\,N_{\mathrm{c}}^{1/2}\frac{G_0 T^{1/2}}{n_{\mathrm{e}}}, \tag{6.60}$$

where we have assumed a similar dependence of the electron recombination rate on N_c and T. This translates then to $\gamma_0 = 1.9$ and a neutral fraction of 0.49. For a 50 C atom PAH, the estimated neutral fraction is, then, 0.35.

Of course, larger PAHs have larger electron affinities and hence also have "allowed" anionic states and these have to be included in the evaluation of the charge distribution function. Figure 6.7 shows the charge distribution function calculated for circumcoronene ($C_{54}H_{18}$), adopting the rates listed in Table 6.2. Typically, under interstellar conditions, several ionization stages are present. Generally, neutral PAHs are quite abundant but, in addition, an appreciable population of ionic PAHs – either cations or anions – also exists. In interpreting this figure, the uncertainties in the recombination cross sections should be kept in mind. Using the measured rates would shift these curves to the right.

6.3.8 PAHs and the charge balance of the gas

The presence of a large population of molecular anions has direct influence on the charge balance of the interstellar gas because these species provide a non-radiative recombination channel that is much more rapid than the radiative channel open to atomic cations and electrons. This is discussed in Section 8.2.1 for the diffuse ISM and in Section 10.2 for molecular gas.

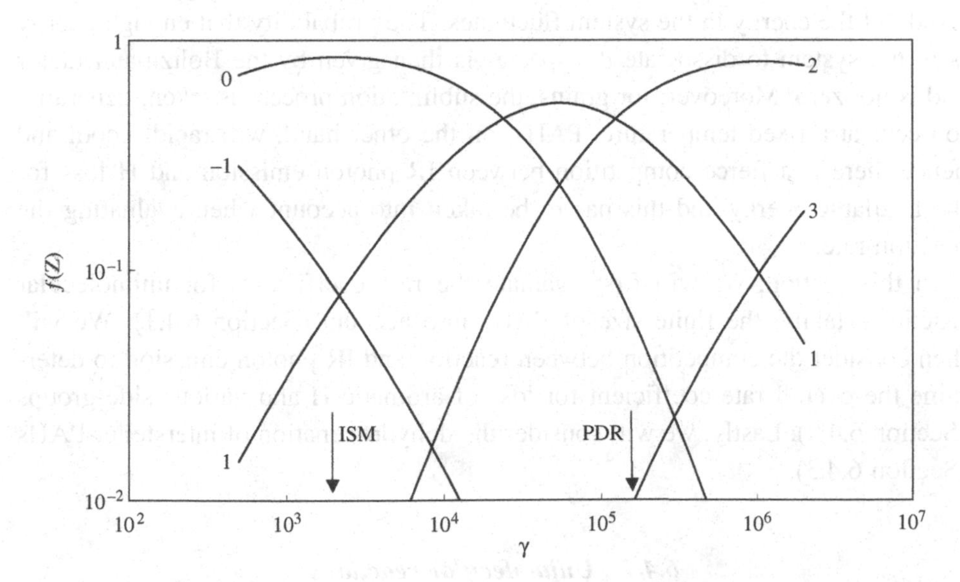

Figure 6.7 The charge distribution function for circumcoronene ($C_{54}H_{18}$) in the interstellar medium. Labels correspond to the charge state. Typical conditions for PDRs (Chapter 9) and diffuse interstellar clouds (Chapter 8) are indicated by arrows. γ is defined as $G_0 T^{1/2}/n_e$.

6.4 Photochemistry of PAHs

Because of the prevalence of FUV photons in the interstellar medium, particularly in regions that are bright in IR emission, chemistry initiated by FUV photon absorption is of key interest for interstellar PAHs. Upon absorption of a UV photon, a highly vibrationally excited PAH molecule can relax through the emission of IR photons. In competition with this process, a sufficiently excited PAH species may instead undergo unimolecular reactions such as the loss of an H atom or, if present, a sidegroup;

$$\mathrm{PAH-R} + h\nu \longrightarrow \mathrm{PAH+R}. \qquad (6.61)$$

For a grain, we could just evaluate the rate of H loss through an Arrhenius expression; i.e., $k(T) = \nu_0 \exp[-E_\mathrm{b}/kT_\mathrm{d}]$, where E_b is the sublimation energy measured for the bulk and T_d is the temperature of the grain. For molecules, however, the small size comes into play. Typical binding energies are $\sim$4 eV, which is a large fraction of the total energy in the system. Hence, the temperature that should be used in this expression has to be evaluated with care (cf. Section 6.2.2). As a telling example, consider a system with a total energy of 4 eV and binding energy of a group of 4.5 eV. Clearly, the (microcanonical) rate is zero: the molecule will not dissociate. However, if we were to calculate an excitation temperature of this species and use the bulk (canonical) approach, we would arrive at a finite rate for dissociation. Recall, in a canonical system the average energy (i.e., kT) is fixed but the energy in the system fluctuates. The probability that enough energy is in the system to dissociate the species is then given by the Boltzmann factor and is not zero. Moreover, for grains, the sublimation process is taken, generally, to occur at a fixed temperature. PAHs, on the other hand, will rapidly cool and hence there is a fierce competition between IR photon emission and H loss for the available energy and this has to be taken into account when evaluating the reaction rate.

In this section, we will first evaluate the rate coefficients for unimolecular reactions, taking the finite size of PAHs into account (Section 6.4.1). We will then consider the competition between reaction and IR photon emission to determine the overall rate coefficient for loss of aromatic H and various side-groups (Section 6.4.2). Lastly, we will consider the dehydrogenation of interstellar PAHs (Section 6.4.3).

6.4.1 Unimolecular reactions

In Section 4.1.5, we have considered molecule formation through radiative association – the inverse of Reaction (6.61) – when two atoms collide and form an activated complex, which then stabilizes through photon emission. The collision energies involved are then much lower and redissociation is likely. Nevertheless, the principle is the same. In initial astrophysical studies of this type of reaction, the reaction rate was evaluated based upon RRK theory, developed in the late 1920s. In this theory, the species is treated as a system of s identical oscillators of frequency ν (cf. Section 4.1.5). Here, a molecule, excited with energy E, will dissociate if one oscillator has an energy in excess of a critical energy, E_0. Defining

$$n \equiv E/h\nu \tag{6.62}$$

and

$$m \equiv E_0/h\nu, \tag{6.63}$$

the probability for dissociation can be found from combinatorial statistics,

$$P = \frac{(n-m+s-1)!n!}{(n-m)!(n+s-1)!}. \tag{6.64}$$

Generally, n and m are not integers. Moreover, a molecule will have a range of oscillator frequencies. In the quantum RRK version of this theory, this expression is rewritten as

$$P = \frac{\Gamma(n-m+s-1)\Gamma(n)}{\Gamma(n-m)\Gamma(n+s-1)}, \tag{6.65}$$

where Γ is the gamma function and ν is now the geometric mean of the vibrational frequencies. If the oscillator frequency is very small (i.e., n and m are both very large) and the number of quanta is much larger than the number of oscillators (i.e., $n - m \gg s$), this probability reduces to

$$P \simeq \left(\frac{n-m}{n}\right)^{s-1}. \tag{6.66}$$

An additional factor ν transforms this probability then to the rate constant,

$$k(E) = \nu \left(\frac{E-E_0}{E}\right)^{s-1}. \tag{6.67}$$

The use of this simple expression has its limitations. Nevertheless, it correctly predicts that the rate rapidly decreases with increasing number of oscillators and increases with the excess energy in the system. Of course, typically, for interstellar PAHs, m and $n - m \simeq 30 < s$ and hence the use of this expression is not really justified.

This early RRK theory has led to the development of RRKM theory where the reaction is described as dissociation through a transition state, or more properly through a saddle point that separates the reactant from the products. The RRKM rate is then given by

$$k_{\mathrm{RRKM}}(E) = \frac{W^{\star}(E - E_0)}{h\rho(E)}, \tag{6.68}$$

where $W^{\star}(E - E_0)$ is the sum of states at the transition state between 0 and $E - E_0$. For a set of classical oscillators (cf. Section 6.4.2), this results in the classical RRKM rate constant,

$$k_{\mathrm{RRKM}}(E) = \left(\frac{E - E_0}{E}\right)^{s-1} \frac{\prod_{i=1}^{s} \nu_i}{\prod_{i=1}^{s-1} \nu_i^{\#}}, \tag{6.69}$$

where the ν_is and the $\nu_i^{\#}$s are the vibrational frequencies of the species and of the transition state, respectively. This ratio of frequencies is itself a frequency and hence we recover Eq. (6.67).

We can also consider the unimolecular dissociation reaction as the reverse of the association reaction of the products forming the parent upon collision. In detailed balance, the rates of these forward and backward reactions are, of course, related. The unimolecular rate is then given by

$$k(E, \epsilon) = A(\epsilon)\frac{\rho_{\mathrm{d}}(E - E_0 - \epsilon)}{\rho_{\mathrm{p}}(E)}, \tag{6.70}$$

where E is the initial excitation energy of the species, E_0 is the binding energy of the fragment, ϵ is the energy carried away by the fragment, and ρ_{p} and ρ_{d} are the densities of states of the parent (PAH-R) and the daughter (PAH), respectively. The factor A is proportional to the cross section for the reverse reaction. If we assume that ϵ is small compared to E, then we can expand the density of states in ϵ and only retain the first term. After integration over ϵ, the total rate becomes

$$k(E) = k_0(T_{\mathrm{d}})\frac{\rho_{\mathrm{d}}(E - E_0)}{\rho_{\mathrm{p}}(E)}, \tag{6.71}$$

where $k_0(T_{\mathrm{d}})$ depends on the interaction potential (in the reverse reaction). As long as there is no activation barrier for the association reaction, the dependence on temperature will be small. This microcanonical daughter temperature, T_{d}, is defined analogously to Eq. (6.17) as

$$\frac{1}{kT_{\mathrm{d}}} = \frac{\mathrm{d}\ln[\rho_{\mathrm{d}}(E)]}{\mathrm{d}E}. \tag{6.72}$$

This expression can be evaluated using the density of states as given by Eq. (6.10) but the equation becomes a bit cumbersome because of the difference in degrees of freedom for the parent and daughter. When the parent and daughter differ by only a few degrees of freedom, we may replace the densities of states by a single function (and include the slowly varying correction factor in the expression for k_0). The unimolecular dissociation rate can then be written in Arrhenius form,

$$k(E) = k_0(T_e)\exp[-E_0/kT_e], \tag{6.73}$$

where T_e is an effective temperature defined as

$$\frac{1}{kT_e} = \frac{\mathrm{d}\ln[\rho_d(E^\star)]}{\mathrm{d}E}, \tag{6.74}$$

with $E^\star$ some energy between E and $E - E_0$. It can be shown that, to first order, this effective temperature is given by

$$T_e \simeq T_m - \frac{E_0}{2C_m}, \tag{6.75}$$

where the second term is the finite heat bath correction. With the expression for the microcanonical heat capacity for PAHs (Eq. (6.20)), this becomes

$$T_e = T_m\left(1 - 0.2\frac{E_0}{E}\right). \tag{6.76}$$

For a given internal energy, the correction factor increases when the (binding) energy involved becomes a larger fraction of the energy in the system. The dependence on size is somewhat hidden: a larger species will require a larger internal energy to attain the same microcanonical temperature (cf. Eq. 6.18) and, hence, the correction factor will be smaller. For PAHs absorbing a given photon energy $h\nu$, this becomes

$$T_e \simeq 2000\left(\frac{h\nu(\mathrm{eV})}{N_c}\right)^{0.4}\left(1 - 0.2\frac{E_0(\mathrm{eV})}{h\nu(\mathrm{eV})}\right) \tag{6.77}$$

and, of course, for a given binding energy and a given photon energy, the correction factor does not change. However, the rate will rapidly decrease with the size of the species. In the limit $E \gg E_0$, and ignoring the difference in number of oscillators between daughter and parent, Expression (6.73) can be shown to be formally equivalent to Eq. (6.70) with $T_e = T_m$.

Finally, the microcanonical RRKM rate can also be obtained from the canonical rate equation using an inverse Laplace transformation. Without derivation, this leads to

$$k_c = A_\infty \frac{\rho(E - E_\infty)}{\rho(E)}, \tag{6.78}$$

with A_∞ and E_∞ the high pressure pre-exponential and activation energy, respectively. The latter is related to the 0 K activation energy E_0 through

$$E_\infty = E_0 + kT + (E^\# - E), \tag{6.79}$$

where the term in brackets refers to the difference in thermal energy between the transition state and the molecule. There is a difference with the RRKM rate in that the density of states refers to the molecule and not the transition state. The attractive part of this is that it allows the calculation of the rate even when the transition state properties are not known.

The frequency ν is often set equal to the vibrational frequency associated with the bond breaking (e.g., $\nu = 10^{14}$ Hz for the C–H mode). It is actually the equivalent of k_0 and hence its value depends on the interaction potential. If we set it equal to

$$k_0 = \frac{kT_e}{h} \exp\left[1 + \frac{\Delta S}{R}\right], \tag{6.80}$$

with ΔS the change in entropy associated with the transition state that depends on T_e, we recover the canonical Arrhenius expression for a gas in thermal equilibrium at temperature T_e. The canonical expression for the entropy change is given by

$$\Delta S = k \ln\left[\frac{\prod \Phi_i^{\#}}{\prod \Phi_i}\right] + \left(\frac{E^{\#} - E}{T}\right), \tag{6.81}$$

where the Φs are the molecular vibrational and rotational partition functions of the transition state and the parent molecule. These can be evaluated from the vibrational frequencies and moments of inertia.

These different evaluations of the unimolecular rate are compared in Fig. 6.8 for an adopted binding energy of 4.45 eV and a change in entropy of 5 cal mole^{-1} K^{-1} (e.g., $k_0 = 7 \times 10^{14}$ s^{-1}). It is clear that RRK and QRKM show a much steeper dependence on energy than the correct microcanonical expression (RRKM). This is a well-known problem with these theories and generally the frequency factor, ν, the critical energy, E_0, and the number of (effective) oscillators, s, are used as free parameters to fit these expressions to the RRKM rate or to experimental data. Good fits can then be obtained. The approximation using the microcanonical temperature provides a good fit to the RRKM rate for large internal energies, as expected from the formal equivalence of these two expressions in the appropriate limits. However, for low internal energies, this rate can be off by an order of magnitude.

In general, given the exponential dependence, small errors in the energies and entropies associated with unimolecular reactions can lead to large uncertainties in the reaction rates. These parameters cannot be estimated accurately enough even with sophisticated ab-initio calculations and, hence, studies generally rely on fitting experimental data to determine these parameters. This is illus-

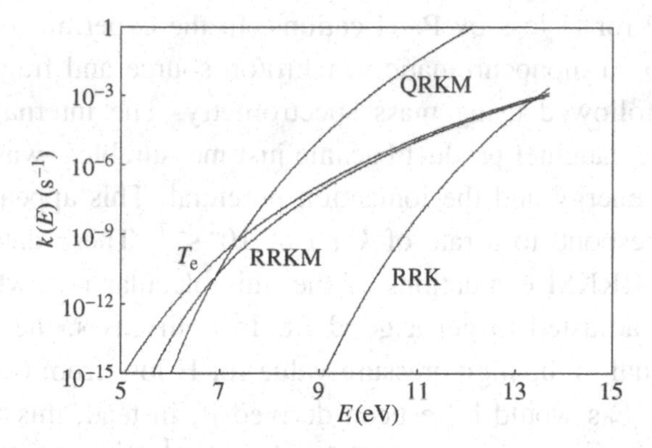

Figure 6.8 The unimolecular dissociation rate as a function of internal energy, E, for coronene $C_{24}H_{12}$. A binding energy of 4.45 eV and a frequency factor of $7 \times 10^{14}\ s^{-1}$ (corresponding to $\Delta S = 5$ cal mole^{-1} K^{-1}) have been assumed. RRK corresponds to Eq. (6.67). QRKM uses the full expression, Eq. (6.65). RRKM refers to Eq. (6.71) while T_e is the Arrhenius approximation involving the microcanonical temperature, Eq. (6.73).

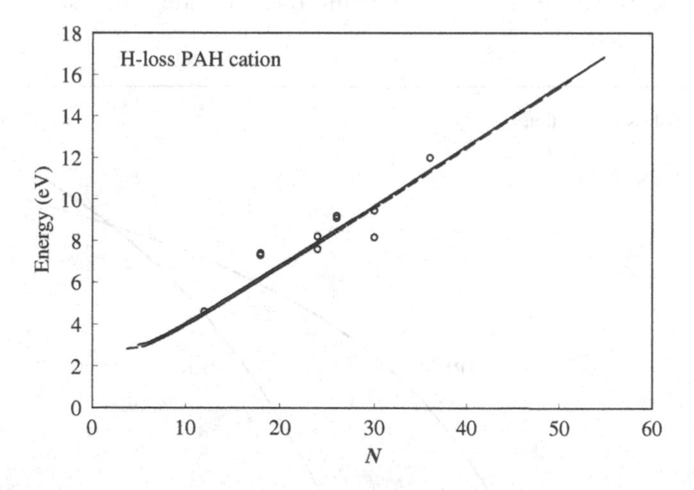

Figure 6.9 The measured internal energy at which the daughter product corresponding to H loss from the PAH cation appears is plotted as a function of the number of atoms in the species. Fits to these data are shown based upon RRKM theory (solid line; Eq. (6.70), ignoring the difference in number of oscillators between daughter and parent, with $E_0 = 3.3$ eV) and RRK theory (dashed line; Eq. (6.67) with $E_0 = 2.8$ eV). In both cases, the pre-exponential factor was set equal to the high pressure value, $A_\infty = 3 \times 10^{16}\ s^{-1}$, measured for H loss from neutral benzene. Data taken from H. W. Jochims, E. Ruhl, H. Baumgartel, S. Tobita, and S. Leach, 1994, *Ap. J.*, **420**, p. 307.

trated in Fig. 6.9 for H loss by PAH cations. In the experiments, neutral PAHs were irradiated by a monochromatic synchroton source and fragment formation of the ion was followed using mass spectrometry. The internal energy of the parent – when the daughter product became just measurable – was then estimated from the photon energy and the ionization potential. This appearance energy is estimated to correspond to a rate of $k(E)$ of $10^4\ \mathrm{s}^{-1}$. These data are compared to the RRK and RRKM evaluations of the unimolecular rate where the critical energy, E_0, was adjusted to get a good fit. In both cases, the pre-exponential factor was set equal to the high pressure value for H loss from (neutral) benzene. Somewhat lower E_0s would have been derived if, instead, this factor had been evaluated using the entropy change of 5 cal mole^{-1} K^{-1} calculated for H loss from cations. Both models provide very good fits to the data, albeit for somewhat different critical energies (3.0 and 2.8 eV). The binding energy of an H atom to various small, neutral and ionized PAHs has been calculated to be approximately 4.5 eV. However, such a high value clearly does not provide a good fit to the data.

While both models can provide good fits to the measured data, when the critical energy is adjusted, the rates have to be extrapolated to interstellar conditions and lead then to very different results. In particular, while the experiments refer to rates of the order of $10^4\ \mathrm{s}^{-1}$, relevant rates in the ISM are in the range

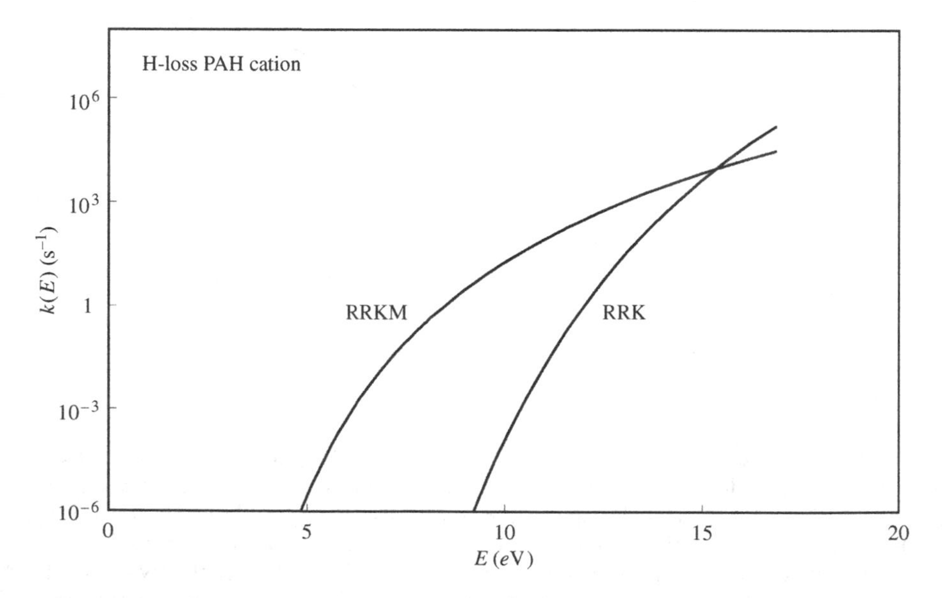

Figure 6.10 The unimolecular dissociation rate for H loss from a 50 C atom PAH cation as a function of internal energy calculated with the RRK and RRKM model. The adopted parameters are taken from the fits shown in Fig. 6.9.

of 10^{-4}–$10^{0}\,\mathrm{s}^{-1}$ (see below) and these different theories show very different energy dependencies (cf. Fig. 6.8). This is illustrated in Fig. 6.10, where the two rates are compared for a 50 C atom PAH. As expected from the fit in Fig. 6.9, the rates are approximately $10^4\,\mathrm{s}^{-1}$ for an internal energy of $\simeq 15\,\mathrm{eV}$. However, the RRK approximation requires a much higher internal energy for a rate of $1\,\mathrm{s}^{-1}$ than the RRKM model.

6.4.2 The astrochemical rate coefficient

In evaluating the photochemistry of interstellar PAHs, the rate for dissociation has to be compared to other decay channels in order to arrive at the probability for dissociation upon photon absorption. Emission of IR photons is generally the dominant channel, and the photodissociation probability after absorption of an FUV photon can be written as

$$p_{\mathrm{d}}(E) = \frac{k(E)}{k(E) + k_{\mathrm{IR}}(E)}, \tag{6.82}$$

where, strictly speaking, the unimolecular dissociation rate should have been evaluated over the cascade. However, because of the steep dependence of this rate on internal energy, it can be evaluated at the internal energy before cooling sets in. The total dissociation rate is

$$k_{\mathrm{t}} = 4\pi \int_{E_{\mathrm{b}}/h}^{E_{\mathrm{max}}/h} \left[\int_{E_{\mathrm{b}}}^{h\nu'} p_{\mathrm{d}}(E)\,\mathrm{d}E \right] \sigma_{\mathrm{UV}}(h\nu')\, \mathcal{N}(h\nu')\,\mathrm{d}\nu'. \tag{6.83}$$

The lower integration limit on these integrals is the critical energy of the reaction. The integration over frequency extends to the maximum energy available (13.6 eV in an HI region).

Relevant parameters for the unimolecular dissociation of PAHs are summarized in Table 6.3. These mainly pertain to sidegroups attached to the periphery of PAHs (Fig. 6.11). The energies listed are the binding energies of the group to the PAH. As exemplified by the discussion above, the actual critical energies may be lower. However, these critical energies are not available for other sidegroups. Calculated RRKM unimolecular dissociation rates for these groups are compared in Fig. 6.12. A number of conclusions can now be drawn. First, in all cases, loss of a sidegroup is only competitive with IR relaxation for small PAHs. Second, these rates depend very strongly on the size of the PAH. Thus, while a 20 C atom PAH may be stable against H loss, a 16 C atom PAH is not. Third, because of their very different binding energies, the rates are very different for different

Table 6.3 *Unimolecular dissociation*

Sidegroup[a]	Bond[b]	E_b (eV)	p_d^c
Hydrogen	PAH – H	4.47	10^{-10}
Methyl	$PAHCH_2 - H$	3.69	10^{-2}
	$PAH - CH_3$	4.0	10^{-4}
Ethyl	$PAHCH_2 - CH_3$	3.1	0.9
Hydroxyl	PAHO – H	3.69	10^{-2}
	PAH – OH	4.5	10^{-10}
Amine	PAHNH – H	3.47	10^{-1}
	$PAH - NH_2$	4.0	10^{-4}
Acetylene	$PAH' - C_2H_2$[d]	8.0	negligible

[a] Structural formulae of these sidegroups are illustrated in Fig. 6.11.
[b] Bond denoted by –.
[c] Estimated photodissociation probability for a 50 carbon atom PAH, assuming $E = 10\,\text{eV}$ and $k_{IR} = 1\,\text{s}^{-1}$ (cf. Eq. 6.82).
[d] Loss of acetylene group from PAH skeleton.

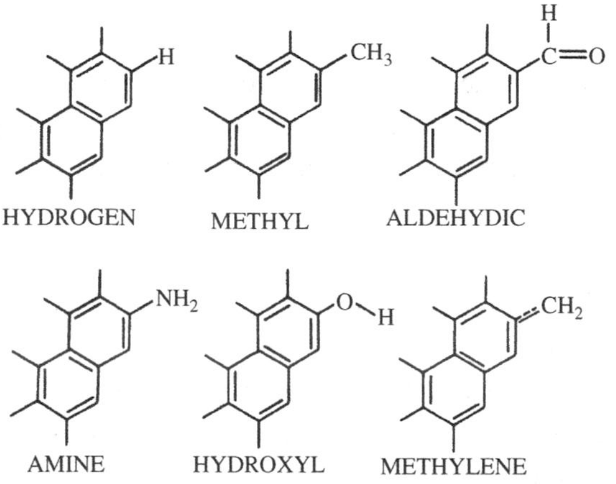

HYDROGEN METHYL ALDEHYDIC

AMINE HYDROXYL METHYLENE

Figure 6.11 Structural formulae of some sidegroups on PAHs. The sidegroups replace an H atom on the periphery of the carbon backbone outlined by the hexagonal network. Figure reproduced with permission from T. R. Geballe, A. G. G. M. Tielens, L. J. Allamandola, A. Moorhouse, and P. W. J. L. Brand, 1989, *Ap. J.*, **341**, p. 278.

sidegroups. Sorting these sidegroups according to their binding energies – but recall the discussion on the critical energy of the H loss channel above – we conclude that, in a "competition," the methyl ($-CH_3$) group of an ethyl ($-CH_2CH_3$) sidegroup is the first to go, followed by H loss from a methyl, hydroxyl, or amine group, the complete methyl or amine group, and finally, the aromatic hydrogen

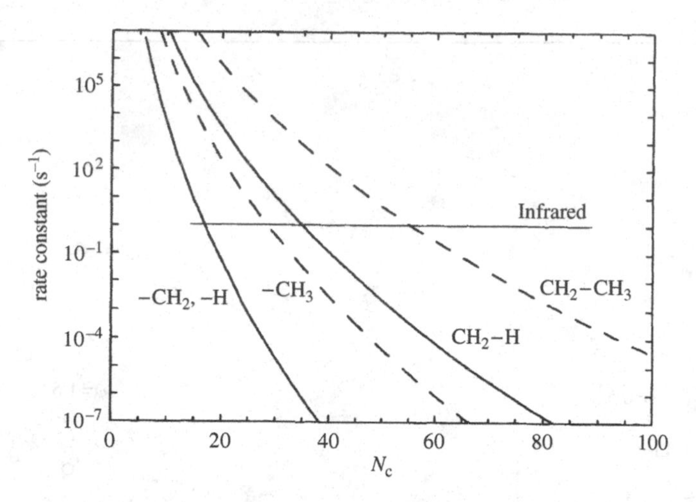

Figure 6.12 Calculated RRKM unimolecular dissociation rates for different side-groups as a function of PAH size for an internal energy of 10 eV. The critical energies have been set equal to the binding energies of the group (Table 6.82). The pre-exponential factor has been set equal to 3×10^{16} s^{-1}. The IR relaxation rate is also indicated. Figure reproduced with permission from C. Joblin, A. G. G. M. Tielens, L. J. Allamandola, and T. R. Geballe, 1996, *Ap. J.*, **458**, p. 610.

and the hydroxyl group. Loss of an acetylene group is calculated to be negligible (which does not fully agree with experiments).

6.4.3 Dehydrogenation of interstellar PAHs

When considering the "surface" coverage of aromatic hydrogen, the reverse association reaction also has to be considered. Here, we will evaluate the H coverage of ovalene ($C_{32}H_{14}$) in the ISM. We can define the dissociation parameter,

$$\psi = \frac{p_d k_{UV}}{k_a n_H}, \tag{6.84}$$

which regulates the dehydrogenation of the PAHs. The association reaction for H has a rate of $k_a = 2 \times 10^{-8}\,(N_c/50)^{1/2}$ cm^3 s^{-1}. The UV photon absorption rate is $k_{UV} \simeq 2.3 \times 10^{-8} G_0$ s^{-1}. This results in

$$\psi = 2.8 \frac{p_d G_0}{n_H}. \tag{6.85}$$

From Fig. 6.12, $k_d(E)$ is $\simeq 10^{-5}$ s^{-1} for ovalene or an H loss probability of $p_d = 10^{-5}$, but note that extrapolating the laboratory experiments shown in Fig. 6.9 to larger PAHs results in a much higher p_d. Here, we will sidestep this issue and consider the H coverage of ovalene for two values of ψ, 0.5 and 1.5 (Fig. 6.13). It is clear that the hydrogenation of interstellar PAHs is a very sensitive function of this dissociation

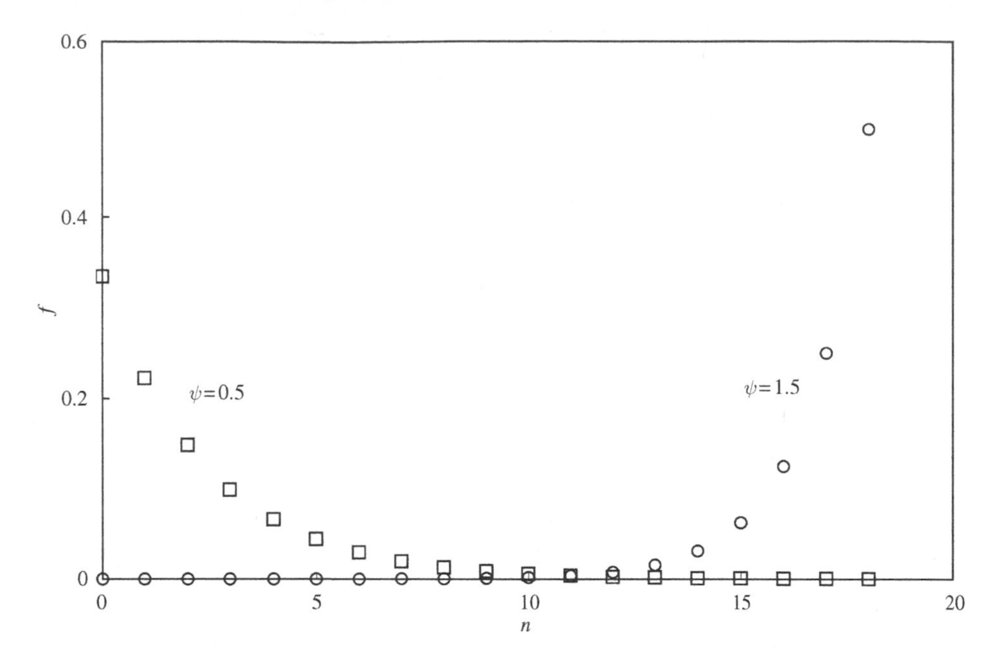

Figure 6.13 The dehydrogenation of ovalene for two different values of the dissociation probability, ψ. The fractional abundance, f, of $C_{23}H_n$ is plotted as a function of n.

parameter and PAHs are either largely fully hydrogenated or completely dehydrogenated. Equivalently, for given physical conditions (G_0/n_H), there is a critical PAH size below which PAHs will be completely dehydrogenated while above this size PAHs will be largely completely hydrogenated. It is likely that highly radicalized, completely dehydrogenated PAHs are unstable and, possibly with some loss of C, rearrange themselves particularly when assisted by UV photons.

In view of the low abundance of atomic O, N, and C, association reactions are of lesser importance for hydroxyl, amine, and methyl groups. In an atomic region, the probability for dissociation can be a factor of $\sim 10^3$ smaller compared with that for aromatic H and yet loss of a sidegroup can be still important. The critical PAH size for loss of such sidegroups is accordingly larger. This may well explain the general absence of hydroxyl and amine groups associated with PAHs in the ISM. In contrast, observations show a strong 3.4 μm feature attributed to methyl groups. As for OH and NH_2 groups, any CH_3 groups initially present on PAHs should be quickly lost photochemically, where we note that their presence in the interstellar spectra conversely implies that they will be excited and can be lost on astronomically relevant time scales. The observed high fraction of methyl groups suggests, then, that the PAH skeleton itself is chemically attacked in the ISM. The break-up of the C skeleton leads to dangling bonds and the preponderance of atomic hydrogen in the ISM will then naturally lead to methyl production. For a bright PDR, such as the Orion Bar, $G_0 = 5 \times 10^4$, and for a typical 50 C atom PAH, $k_{UV} = 0.2\,s^{-1}$ and $p_d = 10^{-10}$, the lifetime of a methyl group is 2×10^3 years; given the approximations and extrapolations, not unreasonably short compared to the lifetime of the region.

6.5 Other large molecules

The concepts developed for interstellar PAH molecules can also be applied directly to other large molecules such as C_{60} and other fullerenes as well as various types of carbon chains. While there is no evidence for these species in the IR spectra of the ISM, electronic transitions of small carbon chains (C_2, C_3) have been detected in absorption towards many diffuse clouds. The abundances of these species are, however, small. Current thinking is that the diffuse interstellar bands are caused by electronic transitions in moderately large molecules and carbon chains and their derivatives are leading candidates for the DIB carriers. Larger carbon chains and their derivatives – up to C_{11} – have been detected inside dense molecular clouds through their pure rotational transitions in the millimeter wavelength region. There is also tantalizing evidence for the presence of C_{60}^+ in the near-IR spectrum of diffuse cloud sight-lines. However, conclusive laboratory confirmation is still lacking. Nanodiamonds have been isolated from meteorites with an isotopic composition of trapped noble gases, which betrays a history predating the formation of the Solar System (cf. Chapter 5). Silicon nanoparticles are leading candidates for the carriers of the extended red emission that is prevalent in the interstellar medium (Chapter 5). The physical properties and the emission characteristics of nanoparticles will differ from those of bulk materials and are in many ways analagous to those of large molecules.

Heats of sublimation and work functions for some of these species are compared in Table 6.4. For PAHs, the limiting values of graphite are given. While the characteristics of carbon clusters will be very dependent on molecular size and structure, the properties of C_{60} are listed as exemplary. This is illustrated in Fig. 6.14 for the most stable isomer within each of five isomer classes, which differ in their geometry and structure. The cohesive binding energies have been calculated using density functional theory and the results depend somewhat on the level of the approximation but some general trends remain. For all structures, the binding energy per C atom increases with increasing size in a very regular way. There is some variation on this for (monocyclic) rings. For the smallest clusters, linear carbon chains are the most stable. At somewhat larger sizes, rings take over. For the largest clusters, graphitic structures and cages are the prevalent structures and the fullerenes eventually dominate (culminating in the high stability of C_{60}).

Table 6.4 *Physical properties*[a]

Species	Heat of sublimation[b] (eV)	Work function (eV)	Debye temperature (K)	Band gap (eV)
PAHs	7.44[a]	4.4	2 570 930[c]	—
Fullerenes	7.4[d]	7.6[d]	—	1.7[d]
Diamond	7.39	5.6	2 200	5.45
Silicon	4.62	4.95	645	1.1

[a] These refer to the bulk material. For PAHs, the values of graphite are given.
[b] Per atom.
[c] Graphite values. The heat capacity of graphite is well represented by that of a two-dimensional material.
[d] Value for C_{60}.

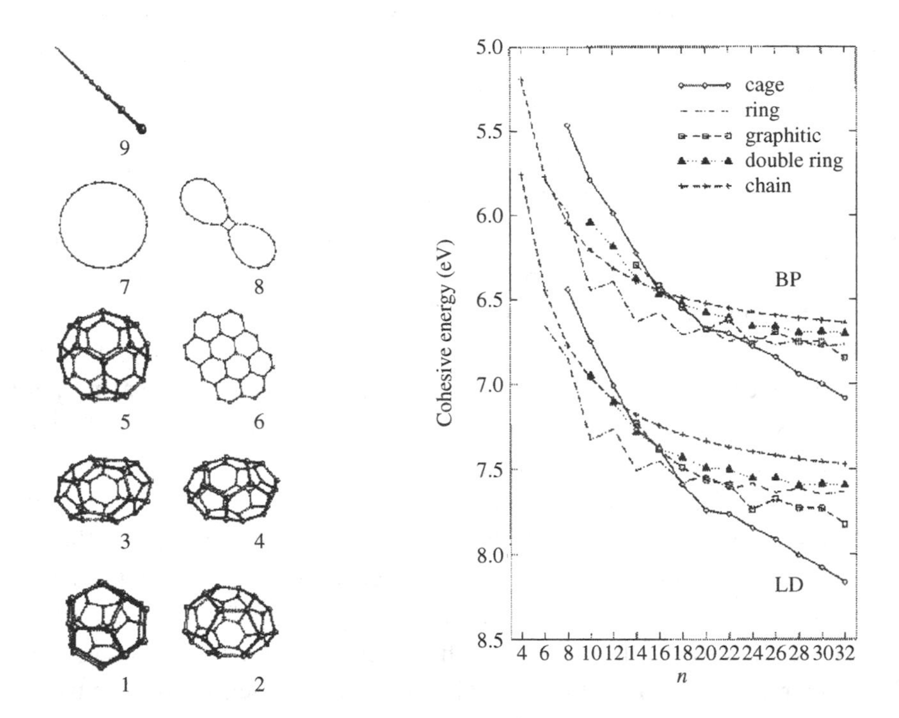

Figure 6.14 The binding energy per atom is plotted as a function of size for different classes of isomers. These are calculated using density functional theory at two different levels of approximation: the local spin density (LD) and a non-local modification involving the gradient of the density (BP). The structures of the various isomers are indicated schematically on the left. They range from cage structures (1 – 5), to graphitic (6), to single (7) and double (8) rings, to chains (9). Reprinted with permission from R. O. Jones, 1999 *J. Chem. Phys.*, **110**, p. 5189. Copyright 1999, American Institute of Physics.

Finally, the energy temperature relationship is of some interest for astrophysical applications. Debye temperatures are listed in Table 6.4. At this temperature, the heat capacity of a material has reached 95% of its classical value ($3NkT$). These Debye temperatures can be used to derive the temperature–internal energy relationship for these species. They are compared in Fig. 6.15. In the evaluation of these nanoparticles, modes associated with surface functional groups terminating the crystal structure have not been taken into account. Given their similarity in binding, PAHs and fullerenes will reach very similar temperatures upon absorption of a UV photon. Diamond nanoparticles will be distinctly hotter while silicon nanoparticles are much cooler.

The band gap provides a measure of the FUV photon absorption of these materials. Silicon as a semiconductor will absorb efficiently in the UV and visible ranges. In contrast, diamonds will not start to absorb until energies above 5.5 eV. However, in the ISM, FUV photons will be prevalent and this may not lead to a large difference in the energy absorption rate. The FUV absorption by PAHs and fullerene species will depend strongly on the size and molecular structure of the species. Typically, the first strong transition of small species will occur

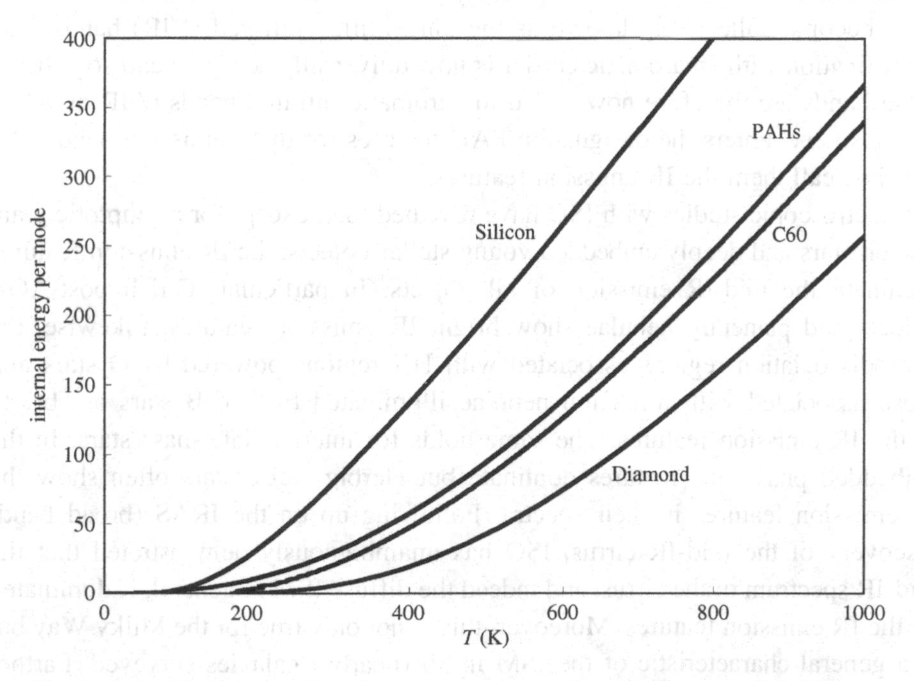

Figure 6.15 The relation between temperature and internal energy for different materials. The energy is given in units of cm^{-1} per mode. For PAHs and C_{60}, these have been derived from the density of states. For diamond and silicon nanoparticles these have been calculated from the heat capacity and the Debye temperatures listed in Table 6.4.

around 5.5 eV and this will shift towards longer wavelengths with increasing size. Cations will possess substantially lower energy electronic transitions.

Switching now to the infrared, these materials can affect the interstellar emission spectra according to the following general guidelines. Diamonds and silicon are homonuclear materials and hence do not possess strong infrared active modes. For crystalline materials, the emission characteristics will be dominated by impurities and crystal defects. In the amorphous state, these materials will show broad IR emission peaks near the peak of the density of (phonon) states. For nanoparticles, surface modes will be predominant. Probably, these are associated with atomic hydrogen, which terminates the binding at the surface in the ISM.

6.6 Infrared observations

Ground-based and airborne observations through the 1970s and 1980s have shown that the IR spectra of bright sources with associated dust and gas are dominated by relatively broad emission features at 3.3, 6.2, 7.7, 8.6, and 11.3 μm, which always appear together. Figure 6.16 shows a sample of such spectra. Because the carriers of these bands remained unidentified for almost a decade, these bands have become collectively known as the unidentified infrared (UIR) bands. The identification with an aromatic carrier is now universally accepted and sometimes these bands are therefore now called the aromatic infrared bands (AIR or AIB). One also encounters the designation PAH features for these emission bands. We will just call them the IR emission features.

Spectroscopic studies with ISO have revealed that, except for asymptotic giant branch stars and deeply embedded young stellar objects, the IR emission features dominate the mid-IR emission of all objects. In particular, C-rich post-AGB objects and planetary nebulae show bright IR emission features. Likewise, the photodissociation regions associated with HII regions powered by O stars and those associated with reflection nebulae illuminated by late B stars are bright in the IR emission features. The same holds for intermediate mass stars. In the embedded phase, ice features dominate but Herbig AeBe stars often show the IR emission features in their spectra. Following up on the IRAS (broad band) discovery of the mid-IR cirrus, ISO has unambiguously demonstrated that the mid-IR spectrum of the cirrus, and indeed the diffuse ISM in general, is dominated by the IR emission features. Moreover, this is not only true for the Milky Way but is a general characteristic of the ISM in all (nearby) galaxies surveyed. Farther away, sensitivity limited ISO to studies of IR bright starburst galaxies, ULIRGs, and AGNs. The presence of bright IR emission features in many of these objects has been taken as a signpost – or even a probe – of massive star formation in these objects.

Besides the well-known IR emission features at 3.3, 6.2, 7.7, 8.6, and 11.3 μm, the observed interstellar spectra show a wealth of weaker features, including bands at 3.4, 3.5, 5.25, 5.65, 6.0, 6.9, 10.5, 11.0, 12.7, 13.5, 14.2, and 16.4 μm. Moreover, many of the well-known features shift in peak position, vary in width, or show substructure. The 6.2 μm band forms a case in point, peaking generally at 6.2 μm but sometimes as red as 6.3 μm. Likewise, even at relatively low resolution, the 7.7 μm band breaks up into two separate features; at 7.6 and 7.8 μm, respectively. At high resolution, this band breaks up into four separate components. It should be emphasized that not all sources show all of these emission components at the same time. Indeed, spectral variation is one important characteristic of the IR emission features, revealing a sensitivity to the local physical conditions.

The IR spectra also show broad plateaus underlying many of the narrow emission bands (cf. Fig. 6.16). In particular, the 3.3 μm band sits on top of a broad pedestal ranging from 3.2 to 3.6 μm. The 6.2 and 7.7 μm bands are perched on a plateau that ranges from 6–9 μm. The 11.3 μm band is located on a broad plateau ranging from 11–14 μm. At longer wavelengths, a broad 15–19 μm plateau is apparent. In view of the observed source-to-source variations, some of these plateaus may represent a blend of many narrower (weak) features. Others may be inherently broader structures. In any case, these plateaus vary independently from the IR emission features and thus represent a different emission component.

6.7 The IR characteristics of PAHs

The use of IR spectroscopy to probe the vibrational modes of materials has been discussed in Chapter 2. The well-known IR emission features at 3.3, 6.2, 7.7, 8.6, and 11.3 μm are characteristic for the stretching and bending vibrations of aromatic hydrocarbon materials (Fig. 6.16). The 3 μm region is characteristic for C–H stretching modes and the 3.3 μm band due to the C–H stretching mode in aromatic species, while the 3.4 μm band may be due to aliphatic groups attached as groups to the PAHs or to PAHs with extra H bonded to some carbons. Pure C–C stretching modes generally fall in the range 6.1–6.5 μm, vibrations involving combinations of C–C in-plane bending modes lie slightly longwards (6.5–8.5 μm), and C–H in-plane wagging modes give rise to bands in the 8.3–8.9 μm range. The 11–15 μm range is characteristic for C–H out-of-plane bending modes. At wavelengths longer than 15 μm, emission is due to in-plane and out-of-plane ring bending modes of the C skeleton and hence these modes are more molecule specific. Table 6.5 summarizes assignments of the observed interstellar IR emission features with specific modes of PAH molecules or clusters of PAH molecules.

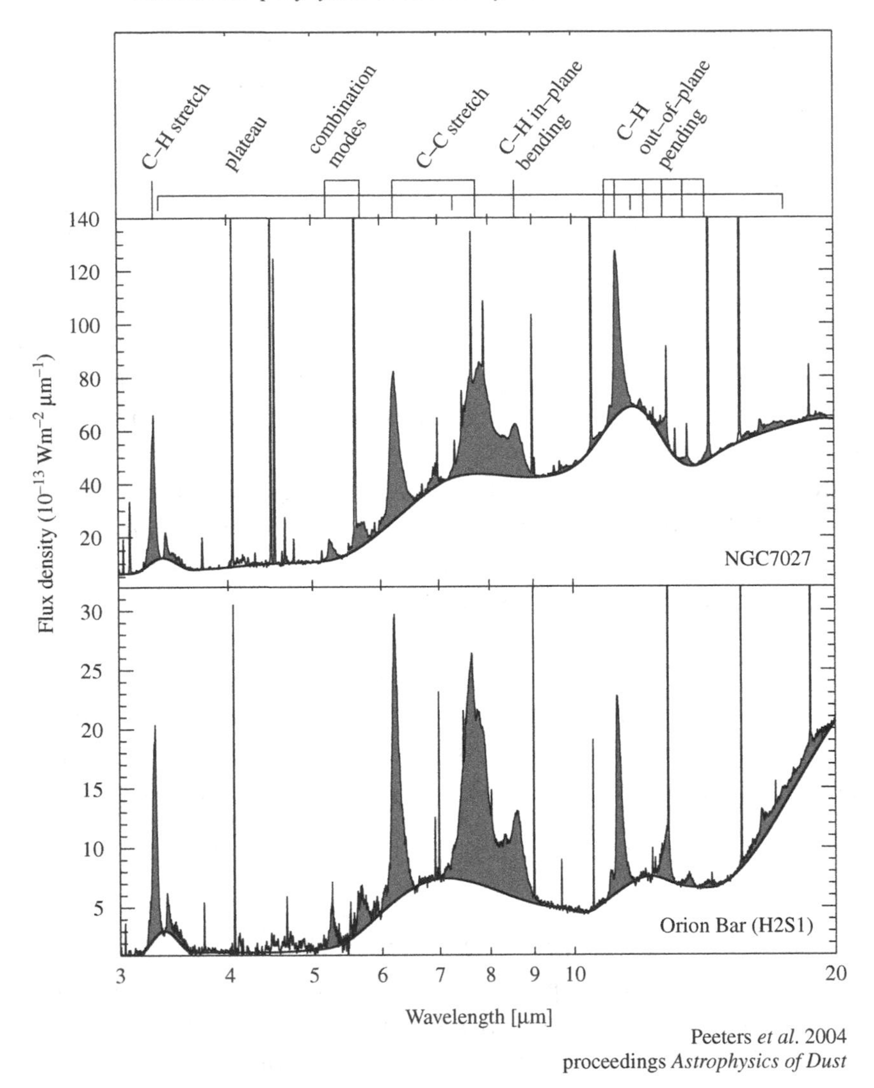

Figure 6.16 The 3–15 μm spectra of the PDRs associated with the Orion Bar and NGC 7027, illustrating the ubiquitous nature as well as the richness of the IR emission features. The IR emission features are shaded. The narrow lines are HI recombination lines, H_2 pure rotational and rotational–vibrational lines, and atomic fine-structure lines. The top panel indicates the PAH vibrational modes associated with each feature. Note the presence of broad plateaus underneath the narrow emission features. Figure adapted from E. Peeters, *et al.*, 2002, *A. & A.*, **390**, p. 1089.

These modes "measure" nearest-neighbor interactions to a large extent. Hence, the infrared signatures of carbonaceous solids with an abundant aromatic component – such as coal, amorphous carbon soots, and charcoals – also bear a general resemblance to the observed interstellar spectra. However, because these are

Table 6.5 *The IR emission features and interstellar PAHs*

Band	Assignment
3.3 μm	aromatic C–H stretching mode
3.4 μm	aliphatic C–H stretching mode in methyl groups
	C–H stretching mode in hydrogenated PAHs
	hot band of the aromatic C–H stretch
5.2 μm	combination mode, C–H bend and C–C stretch
5.65 μm	combination mode, C–H bend and C–C stretch
6.0 μm	C–O stretching mode (?)
6.2 μm	aromatic C–C stretching mode
6.9 μm	aliphatic C–H bending modes
7.6 μm	C–C stretching and C–H in-plane bending modes
7.8 μm	C–C stretching and C–H in-plane bending modes
8.6 μm	C–H in-plane bending modes
11.0 μm	C–H out-of-plane bending modes, solo, cation
11.2 μm	C–H out-of-plane bending modes, solo, neutral
12.7 μm	C–H out-of-plane bending modes, trio, cation (?)
13.6 μm	C–H out-of-plane bending modes, quartet
14.2 μm	C–H out-of-plane bending modes, quartet
16.4 μm	in-plane and out-of-plane C–C–C bending modes in pendant ring (?)
Plateaus	
3.2–3.6 μm	overtone and combination modes, C–C stretch
6–9 μm	blend of many C–C stretch and C–H in-plane bend modes[a]
11–14 μm	blend of C–H out-of-plane bending modes[a]
15–19 μm	in-plane and out-of-plane C–C–C bending modes

[a] In PAH clusters

disordered materials, the features are often much broader than the observed features. Moreover, the feature to continuum ratio for these solids is generally much smaller than observed in (some) sources. Most importantly, grains made of such materials do not reach the temperatures required to emit at mid-IR wavelengths in PDRs far from the illuminating stars or in the diffuse interstellar medium and, when the size is decreased to fulfill the energetic requirements, while at the same time maintaining the aromatic nature characterizing the emission carrier, these materials are essentially PAH molecules.

Large PAHs have many modes near these frequencies. Most of these, particularly for symmetric PAHs, are IR inactive. The precise peak position of IR active modes depends on the size, symmetry, structure, and molecular heterogeneity of the molecule as well as the PAH charge. Hence, guided by laboratory experiments and quantum chemical calculations, the observed spectra can be used to probe the characteristics of the interstellar PAH family. For example, bands in the 11–15 μm region are due to the C–H out-of-plane deformation mode (Fig. 6.17).

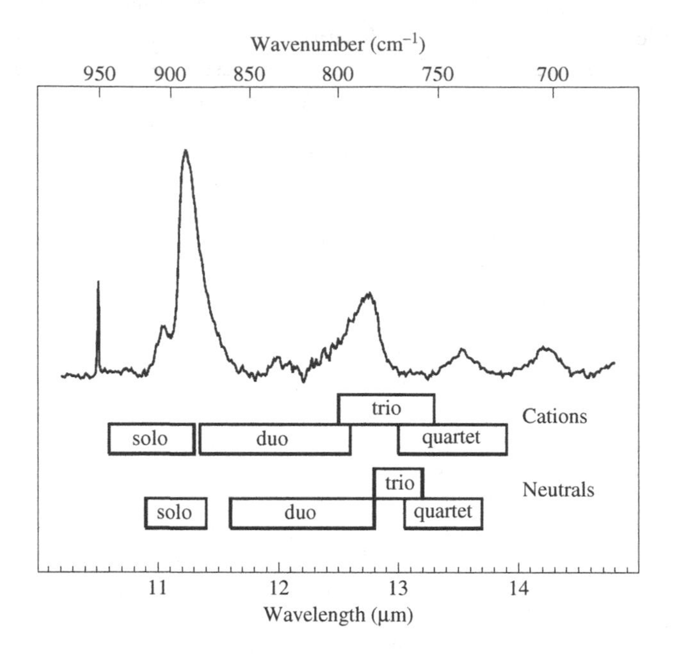

Figure 6.17 A comparison of the average interstellar spectrum (top) with the ranges for the out-of-plane bending modes (bottom). The boxes indicate the wavelength regions associated with the out-of-plane bending vibrations for different types of adjacent hydrogen atoms determined from matrix isolated spectroscopy of neutral and cationic PAHs. Figure reproduced with permission from S. Hony, *et al.*, 2001, *A. & A.*, **370**, p. 1030. Laboratory data from D. Hudgins and L. J. Allamandola, 1999, *Ap. J.*, **516**, p. L41.

Because of coupling between the vibrating H atoms, the exact peak position of these modes depends on the number of directly adjacent H atoms (i.e., bonded to neighboring C atoms on a ring). The well-known 11.3 μm band can be attributed to this mode for isolated Hs in neutral PAHs. The 12.7 μm band, on the other hand, is due to either duos or trios. Thus, the ratio of the 11.3 to the 12.7 μm band is a measure of the abundance of long straight edges (characterized by solo Hs) relative to that of corners (duos or trios). A high 11.2/12.7 μm band ratio indicates the presence of large compact PAHs with long straight edges, while a small ratio suggests a breaking-up or wearing-down of such molecular structures into smaller, irregular structures.

The charge state also has a remarkable effect on the intrinsic IR spectra of PAHs. While peak positions can shift somewhat, the intrinsic strength of modes involving C–C stretching vibrations can increase manifold upon ionization while the C–H stretching and, to a lesser extent, the out-of-plane bending vibrations decrease in strength. This is illustrated in Fig. 6.18, which compares the calculated and measured IR absorption cross sections for a family of PAHs in neutral and cationic states.

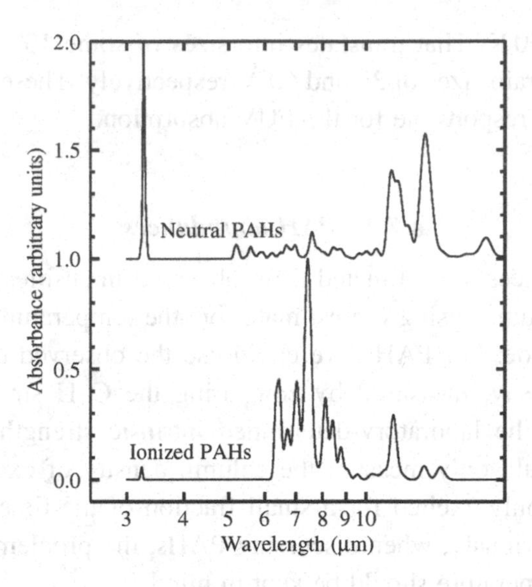

Figure 6.18 The absorption spectrum of a mixture of neutral PAHs (top) compared with the spectrum of the same species in their cationic states. The strength of the C–C modes has increased considerably relative to the C–H modes in the 3 and 11–15 μm region. Figure reproduced with permission from E. Peeters, *et al.,* 2002, *A. & A.*, **390**, p. 1089. Laboratory data from L. J. Allamandola, D. M. Hudgins, and S. A. Sandford, 1999, *Ap. J.*, **511**, p. L115.

6.7.1 The size of interstellar PAHs

As we have seen (Section 6.2.2), the temperature of a small PAH is a sensitive function of its heat capacity and this property can be used to determine the size of the emitting interstellar PAHs. The observed ratio of the 3.3 to the 11.3 μm band – these bands are both due to C–H modes – can be used for this. The observed flux ratio in the Orion Bar ranges from about 0.1 to 0.4, depending on aperture size. Using the latter, derived from the SWS/ISO spectrum, coupled with the intrinsic strength of these modes implies a typical emission temperature of some 625 K. This results in a PAH size of 50–80 C atoms (cf. Eq. 6.18) for photon energies of 6–10 eV. This agrees well with more detailed calculations. Similarly, the plateau emission underneath the 6.2 and 7.7 μm bands coupled with the absence of comparable emission in the C–H stretching mode region, suggests an emission temperature of 400 K and a size for the carrier of this plateau of some 350–600 C atoms. Such large species are probably not single, planar PAH species but rather may represent a three-dimensional cluster of smaller PAHs held together by weak van der Waals bonds or perhaps linked by a few aliphatic chains. The 25 and 60 μm emission bands from the diffuse interstellar medium also show excess emission compared with that expected from "classical" dust grains (cf. Fig. 5.14). These wavelengths correspond to emission temperatures of some 120 and 50 K. That translates into sizes of some 10^4 and 10^5 C atoms, corresponding to grain sizes of 30 and 60 Å, respectively. These very small grains are thus (partially) responsible for the FUV absorption.

6.7.2 PAH abundances

Generally, abundances are estimated from observed intensities of emission lines (or the dust continuum) using (an estimate for) the temperature and the intrinsic strength of the mode. For PAHs, we could use the observed excitation temperature – which is, say, measured by comparing the C–H stretching and bending modes – and the laboratory-determined intrinsic strength of these modes. However, that would only measure the column density of excited PAHs and a PAH is typically only excited for a small fraction of the time, $G_0/3 \times 10^7$ (cf. Section 6.2.1). Obviously, when discussing PAHs, the problems associated with the concept of temperature should be kept in mind.

Instead, the abundance of interstellar PAHs is determined by comparing the observed intensity in the IR emission features with the total dust far-IR continuum emission. PAHs and dust compete for FUV photons and, since the interstellar FUV absorption cross section per H atom has been measured, the PAH abundance relative to H can be determined using an intrinsic PAH FUV absorption cross section, $\sigma_{\rm UV} \simeq 7 \times 10^{-18}$ cm^2 per C atom. Define $f_{\rm IR}$ as the ratio of the flux in the IR emission features to the total far-IR continuum emission; then the fraction of carbon in the form of PAHs, $f_{\rm C}$ is

$$f_{\rm C} = \frac{A_{\rm v}}{N_{\rm H}} \frac{\kappa_{\rm FUV}}{\kappa_{\rm v}} \frac{(1-\omega_{\rm FUV})}{\sigma_{\rm FUV} A_{\rm C}} \frac{f_{\rm IR}}{(1-f_{\rm IR})}, \tag{6.86}$$

where $\kappa_{\rm FUV}/\kappa_{\rm v}$ is the ratio of the FUV to the visual absorption cross section of the dust ($\simeq$3), $\omega_{\rm FUV}$ is the dust albedo in the FUV ($\simeq 0.6$), and $A_{\rm C}$ the elemental abundance of carbon (3.9×10^{-4}),

$$f_{\rm C} = 0.23 \left(\frac{7 \times 10^{-18}\,{\rm cm}^2}{\sigma_{\rm FUV}}\right) \frac{f_{\rm IR}}{(1-f_{\rm IR})}. \tag{6.87}$$

So, with $f_{\rm IR} \simeq 0.13$, typical for bright photodissociation regions such as the Orion Bar and for the diffuse interstellar medium, we find $f_{\rm C} \simeq 3.5 \times 10^{-2}$ or 14 parts per million H atoms. With a typical size of 50 C atoms per PAH, the abundance of PAHs is $\simeq 2.7 \times 10^{-7}$ per H atom. Similarly, for the plateau underneath the IR emission features, with $f_{\rm IR} \simeq 0.08$, we get $\simeq 2 \times 10^{-2}$ or 8 parts per million H atoms and an abundance of its carrier of $\simeq 1.6 \times 10^{-8}$ per H atom. For the 25 ($f_{\rm IR} \simeq 0.07$) and 60 μm ($f_{\rm IR} \simeq 0.15$) cirrus, we arrive at $f_{\rm C} \simeq 0.017$ and $\simeq 0.041$ for their carriers (7 and 16 ppm, respectively).

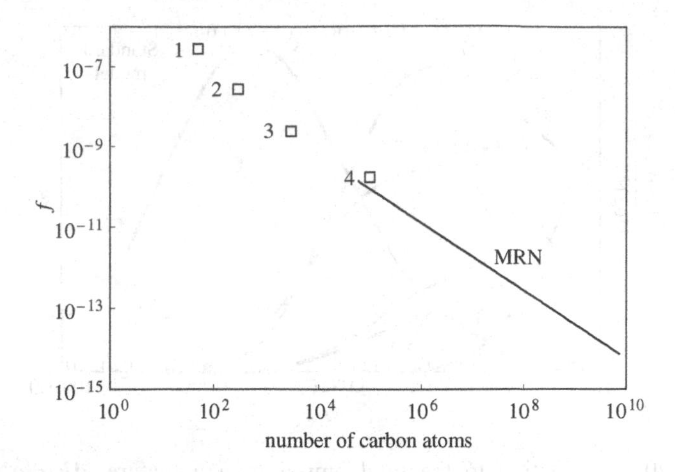

Figure 6.19 The size distribution of PAHs (1), PAH clusters (2), very small grains (3; 25 μm cirrus), and small grains (4; 60 μm cirrus) derived from IR observations: f is the total number of carbon atoms (per H atom) locked up in species or grains containing a given number of carbon atoms. For comparison, the MRN grain size distribution derived from extinction studies is also shown.

6.7.3 The size distribution

Figure 6.19 shows the derived size distribution of the carriers of the IR emission features, and the 25 and 60 μm cirrus. It seems that the PAHs are just the extension of the interstellar grain size distribution into the molecular domain. While such a combined size distribution might be fortuitous, it may also reflect an intrinsic relationship between these molecules and dust grains. On the one hand, PAH molecules may be the building blocks of larger dust grains, either through chemical growth or through coagulation. On the other hand, PAH molecules may be the shattered fragments produced by the collisional cratering process of large dust grains in interstellar shocks (cf. Section 13.4).

Probably, the interstellar size distribution is continuous throughout this size range. Nevertheless, the analysis of the grain size distribution and abundances presented above still remains essentially correct. While the specific characteristics of individual PAHs will depend on size, molecular structure, and charge, for illustrative purposes, Fig. 6.20 shows the contribution to the emission in specific interstellar features calculated as a function of PAH size, adopting generic properties and an MRN size distribution. The curves are plotted in such a way that equal areas under a curve represent equal contributions to that feature. In such a size distribution, the shortest wavelength features are dominated by the smallest PAHs, which reach the highest temperatures in the ISM. This shows that the different features are carried by different sized PAHs. The "emission" temperatures used in

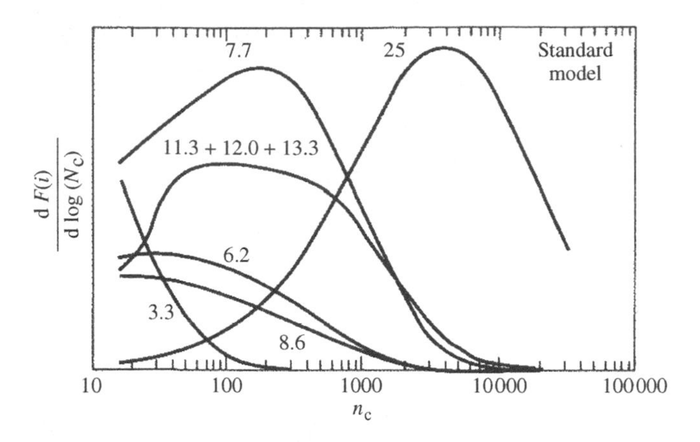

Figure 6.20 Contribution to the total emission in a feature, $dF/d\ln N_c$, as a function of size calculated for an MRN-type size distribution of PAHs with generic properties. The area under each curve is a measure of the total flux of this feature in the spectrum. Also, equal areas under each curve represent equal contributions to the emission in that feature. Figure reproduced with permission from W. A. Schutte, A. G. G. M. Tielens, and L. J. Allamandola, 1993, *Ap. J.*, **415**, p. 397.

the analysis, which lead to Fig. 6.19, represent, therefore, an average over the size distribution and the sizes and abundances derived above should be interpreted as averages as well.

6.7.4 The diffuse interstellar bands

The spectrum of the interstellar medium shows weak absorption bands from the near-UV through the visible into the far-red and near-infrared (Fig. 6.21). The typical width of these bands, 0.5–30 Å, is much larger than the Doppler width and hence these bands have been dubbed the diffuse interstellar bands (DIBs). These DIBs were first discovered some 75 years ago on photographic plates. With the advent of CCD-detector technology, some 200 of these bands are now known with a few strong ones ($W_\lambda/E_{B-V} \simeq 1$ Å magnitude^{-1}), some 50 medium ones ($W_\lambda/E_{B-V} \simeq 0.1$ Å magnitude^{-1}), and (probably) hundreds of very weak ones ($W_\lambda/E_{B-V} \simeq 0.01$ Å magnitude^{-1}). The DIB spectrum shows strong variations in relative strength, as illustrated for the well-known 5780 and 5797 Å DIBs in Fig. 6.21. This seems to be a general characteristic for all strong DIBs, which are well studied, and may also pertain to the weaker ones.

It is now generally accepted that these bands are due to electronic absorptions by molecules rather than to absorption by impurities in dielectric grains. Among the observational evidence is the absence of large variations in the peak position and profile of these bands between sight-lines with very different dust properties

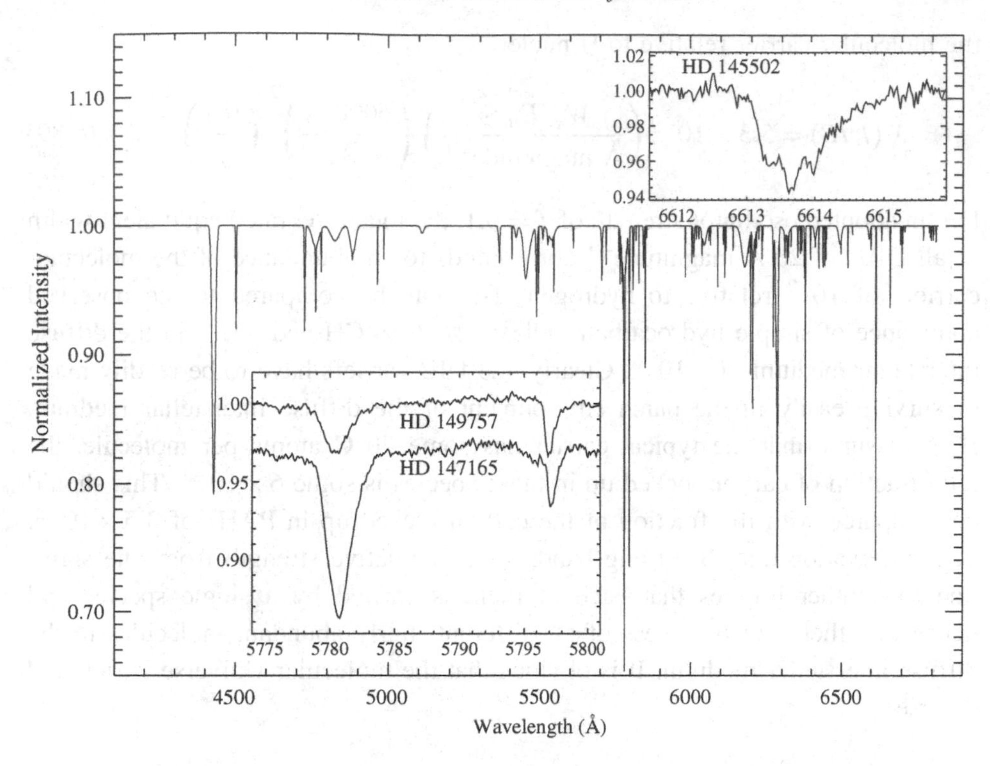

Figure 6.21 The synthesized 4300–6800 Å spectrum of the diffuse interstellar bands as derived from observations of the star BD +63 1964 (P. Ehrenfreund, J. Cami, E. Dartois, and B.H. Foing, 1997, *A. & A.*, **317**, p. L28) illustrates the great variety in relative strength and width of these visual absorption features. The top insert shows the detailed profile of the 6614 Å DIB and its associated substructure observed towards the star HD 145502. The bottom insert illustrates the large variations in the strength of the DIB bands relative to each other for two well-studied DIBs (5780 and 5797 Å) on the basis of observations of the stars HD 149757 and HD 147165. Figure courtesy of J. Cami.

(e.g., grain size). Also, the detailed profiles of some DIBs show substructure that resembles the rotational substructure of molecular transitions (cf. Fig. 6.21). Finally, the enigmatic object, the Red Rectangle, shows *emission* bands at wavelengths that are near to, but not exactly at, the position of some (absorption) DIBs, suggesting that these transitions may be involved in a fluorescence process when the conditions are right.

The column density of absorbers can be found from

$$N_{\rm DIB} = \frac{W_\lambda mc^2}{\lambda^2 f \pi e^2}, \tag{6.88}$$

with W_λ the equivalent width and f the oscillator strength of the transition (cf. Eq. 8.92). Using standard dust parameters, we can write for the abundance of the molecular carrier relative to H nuclei,

$$X(DIB) = 5.3 \times 10^{-9} \left(\frac{W_\lambda / E_{B-V}}{\text{Å magnitude}^{-1}} \right) \left(\frac{6000\ \text{Å}}{\lambda} \right)^2 \left(\frac{0.1}{f} \right). \tag{6.89}$$

For an adopted oscillator strength of $f = 0.1$, the total integrated equivalent width of all DIBs of 20 Å magnitude^{-1} corresponds to an abundance of the molecular carriers of 10^{-7} relative to hydrogen. This can be compared to the observed abundance of simple hydrocarbon radicals, such as CH and CH^+, in the diffuse interstellar medium of $\sim 10^{-8}$. Clearly, the DIB carriers have to be readily made or survive easily in the harsh environment of the diffuse interstellar medium. If we assume that the typical carrier has some 20 C atoms per molecule, the total fraction of carbon locked up in these species is some 6×10^{-3}. This should be compared with the fraction of the carbon locked up in PAHs of 3.5×10^{-2}. The observation that the strong bands vary in relative strength from one sight-line to another implies that each of them is carried by a single species and hence that there are in excess of 50 different, fairly abundant molecules in the diffuse interstellar medium. It is obvious that the molecular Universe is rich and diverse!

6.7.5 Circumstellar diamonds

The infrared spectra of a few sources show, beside the well-known IR emission features due to PAHs, additional bands in the 3 μm region (Fig. 6.22). These bands have been identified with the surface modes associated with hydrogenated diamond surfaces and, for an emission temperature of 1000 ± 50 K, the match between observations and laboratory spectra is very convincing. All known sources exhibiting these features have associated long-lived stable circumstellar disks. For one of these sources, the Herbig Be star, HD 97048, the emission zone of the diamond band is known to be less than 10 AU from the star, placing the diamonds in a putative planet-forming disk. Laboratory studies show that in order for the diamond surfaces to relax to the observed structure, the diamonds need to be quite large, μm-sized. On the other hand, the observed emission temperature coupled with the known size of the emission region and the star's FUV energetics implies nano-sized crystals. This controversy has not yet been resolved. While studies of stardust isolated from meteorites suggests that nanodiamonds are a common constituent of the interstellar grain population (cf. Section 5.5.8), these diamond bands are unique to less than a handful of objects. This may merely imply that in most regions the diamond surfaces are quickly dehydrogenated or relax to a non-hydrogenated surface structure.

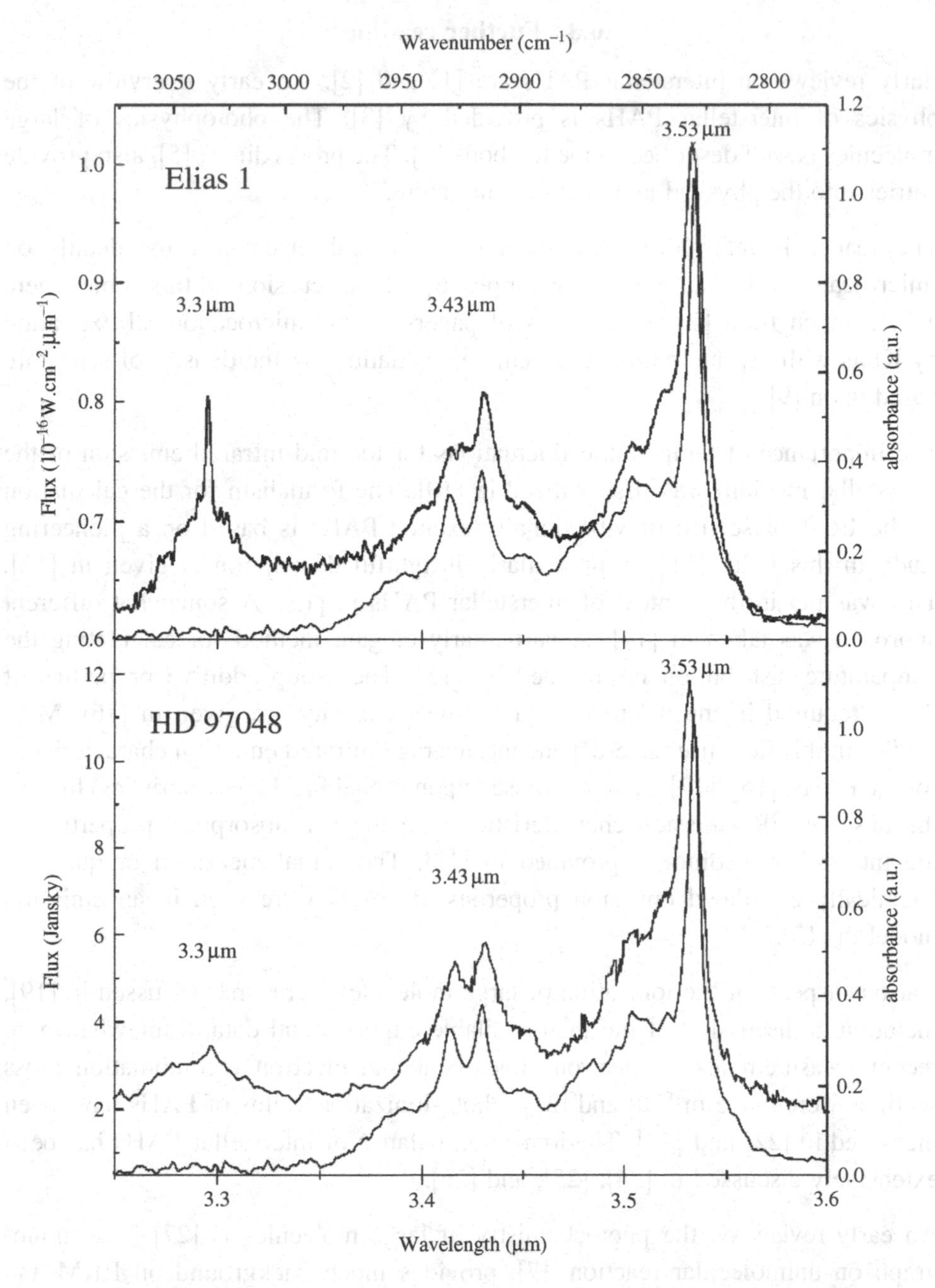

Figure 6.22 The observed near-IR emission spectrum of the Herbig AeBe stars, HD 97048 and Elias 1 (top trace in their panels) reveal the dominant bands due to C–H modes on various H-terminated diamond surfaces (bottom trace in each panel). Figure reproduced with permission from O. Guillois, G. Ledoux, and C. Reynaud, 1999, *Ap. J.*, **521**, p. L133.

6.8 Further reading

Early reviews on interstellar PAHs are [1] and [2]. An early overview of the physics of interstellar PAHs is provided by [3]. The photophysics of large molecules is well described in the textbook [4]. The proceedings, [5], also provide entries into the physical and chemical literature.

The reader is referred to textbooks on statistical mechanics for details on (micro)canonical ensembles, for example [6]. The discussion on this subject here gained much from [7] and a series of papers on the microcanonical excitation by Klots with [8] the main one. Accurate evaluations of the density of states are based upon [9].

The importance of temperature fluctuations for the mid-infrared emission of the interstellar medium was first realized in [10]. The formalism for the calculation of the IR fluorescence of vibrationally excited PAHs is based on a pioneering study in this field, [11]. A particularly insightful description is given in [12]. This was put in the context of interstellar PAHs in [13]. A somewhat different approach was taken in [14]. A particularly elegant method for calculating the temperature distribution is provided by [15]. The group additive properties of PAHs required in the calculation of the heat capacity are given in [16]. Most studies in this field use (size-dependent) average infrared emission characteristics for the PAHs, [15], [17]. A model (based upon global PAH characteristics) linking the observed IR emission characteristics with the UV absorption properties of the interstellar medium is provided in [18]. The actual measured or quantum chemically calculated emission properties of PAHs were used in an emission model in [13].

Various aspects of the ionization of large molecules were first discussed in [19], including a discussion of the then available experimental data. Somewhat more recent measurements of electron attachment and electron recombination cross sections were made in [20] and [21]. Photo-ionization yields of PAHs have been measured in [22] and [23]. The ionization balance of interstellar PAHs has been extensively discussed in [24], [25], and [26].

An early review on the photochemistry of large molecules is [27]. The monograph on unimolecular reaction, [7], provides much background on RKM and RRKM theories. Various aspects involved in the astrochemical problem of loss of hydrogen and various sidegroups are discussed in [28], [29], [30], [25], and [31].

Astronomical and physical aspects of small silicon particles are discussed in [32]. Interstellar nanodiamonds date back to the pioneering stardust-isolation studies from meteorites, [33]. The diamond bands in the spectra of some Herbig AeBe stars were identified in [34]. This is an area of much research – [35], [36], [37]. While the laboratory discovery of fullerenes was initiated by an astronomer's quest for spectral data on large molecules, [38], and the large-scale production of C_{60} resulted from the study of carbonaceous compounds by two astronomers, [39], the discovery of C_{60} in space has remained elusive. The best evidence is the presence of two DIBs at far-red wavelengths, which occur close to the positions of absorption features of C_{60}^+ measured in a Ne matrix, [40]. However, gas phase confirmation is still wanting.

The problem of the identification of the diffuse interstellar bands has been with us for some 80 years. Some of the world's best spectroscopists have devoted a major portion of their careers to this problem with hitherto little avail. A pioneering observational study is [41]. A recent review has been given in [42]. The proceedings of the workshop on the diffuse interstellar bands, [43], provide a good overview of the problem.

A variety of laboratory techniques has been used to study the infrared characteristics of large PAH molecules, [44], [45], [46], [47], culminating in the detection of IR emission from vibrationally excited PAH cations in the gas phase, [48]. Quantum chemical calculations on the IR spectra of PAHs are given in [49], [50], and [51]. A comparative analysis is provided by [47]. A database of matrix isolation spectra of PAHs can be accessed at

http://www.astrochem.org/pahdata/,

which is maintained by D. Hudgins.

A recent overview of the observational characteristics of the infrared emission bands is [52]. Two noteworthy articles on the ISO results and their interpretation, based upon an extensive laboratory database, are [53] and [54].

References

[1] L. J. Allamandola, A. G. G. M. Tielens, and J. R. Barker, 1989, *Ap. J. S.*, **71**, p. 733
[2] J. L. Puget and A. Léger, 1989, *Ann. Rev. Astron. Astrophys.*, **27**, p. 161
[3] A. Omont, 1986, *A. & A.*, **164**, p. 159
[4] N. J. Turro, 1978, *Modern Molecular Photochemistry* (Menlo Park: Benjamin Cummins)
[5] A. Léger, L. d'Hendecourt, and N. Boccara, ed., 1997, *Polycyclic Aromatic Hydrocarbons and Astrophysics* (Dordrecht: Reidel)
[6] L. D. Landau and E. M. Lifshitz, 1988, *Statistical Mechanics* (Oxford: Pergamon Press)
[7] T. Baer and W. L. Hase, 1996, *Unimolecular Reaction Dynamics: Theory and Experiments* (Oxford: Oxford University Press)
[8] C. E. Klots, 1989, *J. Chem. Phys.*, **90**, p. 4470
[9] S. E. Stein and B. S. Rabinovitch, 1973, *J. Chem. Phys.*, **58**, p. 2438

[10] K. Sellgren, 1984, *Ap. J.*, **277**, p. 623
[11] E. M. Purcell, 1979, *Ap. J.*, **206**, p. 685
[12] P. A. Aannestad, 1989, in *Evolution of Interstellar Dust and Related Topics*, ed. A. Bonetti, J. M. Greenberg, and S. Aiello (Amsterdam: Elsevier), p. 121.
[13] E. L. O. Bakes, A. G. G. M. Tielens, and C. W. Bauschlicher, 2001, *Ap. J.*, **556**, p. 501
[14] A. Léger, L. d'Hendecourt, and D. Défourneau, 1989, *A. & A.*, **216**, p. 148
[15] B. T. Draine and A. Li, 2001, *Ap. J.*, **551**, p. 807
[16] S. E. Stein, D. M. Golden, and S. W. Benson, 1977, *J. Phys. Chem.*, **81**, p. 314
[17] W. Schutte, A. G. G. M. Tielens, and L. J. Allamandola, 1993, *Ap. J.*, **415**, p. 397
[18] F. X. Désert, F. Boulanger, and J.-L. Puget, 1990, *A. & A.*, **237**, p. 235
[19] S. Leach, 1987 in *Polycyclic Aromatic Hydrocarbons and Astrophysics*, ed. A. Léger, L. d'Hendecourt, and N. Boccara (Dordrecht: Reidel), p. 89
[20] A. Canosa, D. C. Parent, D. Pasquerault, *et al.*, 1994, *Chem. Phys. Lett.*, **228**, p. 26
[21] H. Abouelaziz, J. C. Gomet, D. Pasquerault, and B. R. Rowe, 1993, *J Chem. Phys.*, **99**, p. 237
[22] L. Verstraete, A. Léger, L. d'Hendecourt, D. Defourneau, and O. Dutuit, 1990, *A. & A.*, **237**, p. 436
[23] F. Salama, E. L. O. Bakes, L. J. Allamandola, and A. G. G. M. Tielens, 1996, *Ap. J.*, **458**, p. 621
[24] E. L. O. Bakes and A. G. G. M. Tielens, 1994, *Ap. J.*, **427**, p. 822
[25] V. Le Page, T. P. Snow, and V. M. Bierbaum, 2003, *Ap. J.*, **584**, p. 316
[26] H. W. Jochims, E. Rühl, H. Baumgärtel, S. Tobita, and S. Leach, 1994, *Ap. J.*, **420**, p. 307
[27] S. Leach, 1996, in *The Diffuse Interstellar Bands*, ed. A. G. G. M. Tielens and T. P. Snow (Dordrecht: Kluwer), p. 281
[28] T. Allain, S. Leach, and E. Sedlmayr, 1996, *A. & A.*, **305**, p. 616
[29] C. Joblin, A. G. G. M. Tielens, L. J. Allamandola, and T. R. Geballe, 1996, *Ap. J.*, **458**, p. 610
[30] H. W. Jochims, H. S. Baumgärtel, and S. Leach, 1996, *A. & A.*, **314**, p. 1003
[31] A. G. G. M. Tielens, *et al.*, 1987, in *Polycyclic Aromatic Hydrocarbons and Astrophysics*, ed. A. Léger, L. d'Hendecourt, and N. Boccara (Dordrecht: Reidel), p. 99
[32] A. N. Witt, K. D. Gordon, and D. G. Furton, 1998, *Ap. J.*, **501**, p. L111
[33] R. S. Lewis, M. Tang, J. F. Wacker, E. Anders, and E. Steel, 1987, *Nature*, **326**, p. 160
[34] O. Guillois, G. Ledoux, and C. Reynaud, 1999, *Ap. J.*, **521**, p. L133
[35] A. P. Jones, L. d'Hendecourt, S.-Y. Sheu, *et al.*, 2004, *A. & A.*, **416**, p. 235
[36] C. van Kerckhoven, A. G. G. M. Tielens, and C.Waelkens, 2002, *A. & A.*, **384**, p. 568
[37] H. G. M. Hill, A. P. Jones, and L. B. d'Hendecourt, 1998, *A. & A.*, **336**, p. L41
[38] H. W. Kroto, J. R. Heath, S. C. O'Brien, R. F. Curl, and R. E. Smalley, 1985, *Nature*, **318**, p. 162
[39] W. Krätschmer, L. D. Lamb, K. Fostiropoulos, and D. R. Hufmann, 1990, *Nature*, **347**, p. 354
[40] B. H. Foing and P. Ehrenfreund, 1997, *A. & A.*, **317**, p. L59
[41] G. H. Herbig, 1975, *Ap. J.*, **199**, p. 702
[42] G. H. Herbig, 1995, *Ann. Rev. Astron. Astrophys.*, **33**, p. 19
[43] A. G. G. M. Tielens and T. P. Snow, 1995, *The Diffuse Interstellar Bands* (Dordrecht: Kluwer)
[44] G. C. Flickinger and T. J. Wdowiak, 1990, *Ap. J.*, **362**, p. L71

[45] L. J. Allamandola, D. M. Hudgins, and S. A. Sandford, 1999, *Ap. J.*, **511**, p. L115
[46] H. Piest, G. von Helden, and G. Meijer, 1999, *Ap. J.*, **520**, p. L75
[47] J. Oomens, A. G. G. M. Tielens, B. G. Sartakov, G. von Helden, and G. Meijer, 2003, *Ap. J.*, **591**, p. 968
[48] H. S. Kim, D. R. Wagner, and R. J. Sakally, 2001, *Phys. Rev. Lett.*, **86**, p. 5691
[49] F. Pauzat, D. Talbi, and Y. Ellinger, 1997, *A. & A.*, **319**, p. 318
[50] S. R. Langhoff, 1996, *J. Phys. Chem. A.*, **100**, p. 2819
[51] C. W. Bauschlicher, 2002, *Ap. J.*, **564**, p. 782
[52] E. Peeters, S. Hony, C. van Kerckhoven, *et al.*, 2002, *A. & A.*, **390**, p. 1089
[53] S. Hony, C. van Kerckhoven, E. Peeters, *et al.*, 2001, *A. & A.*, **370**, p. 1030
[54] E. Peeters, L. J. Allamandola, D. M. Hudgins, S. Hony, and A. G. G. M. Tielens, 2004, in *Astrophysics of Dust*, ed. A. Witt, B. T. Draine, and C. C. Clayton, (San Francisco: PASP), p. 141.

7
HII regions

7.1 Overview

HII regions are created when the extreme ultraviolet radiation from a star ionizes and heats the surrounding gas. Recombination of (thermal) electrons with protons leads to neutralization. The balance between photo-ionization and radiative recombination determines the degree of ionization. The excess energy over the ionization potential is carried away by the photo-electron as kinetic energy. Collisions with other electrons quickly share this energy with the electron bath and lead to a Maxwellian energy distribution. Thermal electrons can excite low-lying levels of trace species and downward radiative transitions cool the nebula. This energy balance sets the temperature of the gas. In this chapter, we will examine these processes and the resulting ionization and temperature structure of the nebula. The observational characteristics of HII regions will also be discussed.

7.2 Ionization balance

7.2.1 Hydrogen

Consider first the ionization of a pure hydrogen nebula. Hydrogen can be ionized by photons more energetic than 13.6 eV (912 Å or 3.3×10^{15} Hz). The ionization balance then reads

$$n_{\rm H} \int_{\nu_{\rm T}}^{\infty} 4\pi \mathcal{N}(\nu) \alpha_{\rm H}(\nu) {\rm d}\nu = n_{\rm e} n_{\rm p} \beta_{\rm A}(T_{\rm e}), \tag{7.1}$$

with $\nu_{\rm T}$ the ionization threshold frequency, $\alpha_{\rm H}(\nu)$ the ionization cross section, and $\beta_{\rm A}(T_{\rm e})$ the recombination rate coefficient to all levels of atomic hydrogen at temperature $T_{\rm e}$. The mean photon intensity, $\mathcal{N}$, is defined as

$$\mathcal{N}(\nu) \equiv \frac{J(\nu)}{h\nu}, \tag{7.2}$$

with $J(\nu)$ the mean intensity of ionizing photons. The neutral hydrogen, $n_{\rm H}$, electron, $n_{\rm e}$, and proton, $n_{\rm p}$, densities are related through the charge conservation and hydrogen abundance equations,

$$n_{\rm e} = n_{\rm p} \tag{7.3}$$

and

$$n_{\rm p} + n_{\rm H} = n, \tag{7.4}$$

with n the total density.

Defining $\epsilon = \sqrt{\nu/\nu_{\rm T} - 1}$, the ionization cross section of hydrogen is given by

$$\alpha_{\rm H}(\nu) = \alpha_0 \left(\frac{\nu_{\rm T}}{\nu}\right)^4 \frac{\exp\left[4\left(1-\tan^{-1}\epsilon\right)/\epsilon\right]}{1-\exp\left[-2\pi/\epsilon\right]}, \tag{7.5}$$

with α_0 the cross section at the threshold ($=6.3 \times 10^{-18}\,{\rm cm}^2$). The ionization cross section is sharply peaked near the threshold and, to a good approximation, decreases as ν^{-3}. Recombination can occur to all excited levels of the neutral hydrogen atom with the simultaneous emission of a photon. Subsequent relaxation of the excited H atom occurs through the emission of further photons. The total recombination rate coefficient is $4.28 \times 10^{-13}\,{\rm cm}^3\,{\rm s}^{-1}$ at a temperature of 10^4 K. The recombination coefficient directly to the ground state is $1.6 \times 10^{-13}\,{\rm cm}^3\,{\rm s}^{-1}$ at an electron temperature of 10^4 K; about 40% of the total recombination coefficient. Because coulomb forces prefer slow encounters, the recombination coefficient will depend inversely on electron temperature (as $T_{\rm e}^{-0.8}$). Radiative recombination is, thus, a slow process ($\sim$100 years at a density of $10^3\,{\rm cm}^{-3}$). Electron–electron collisions occur on a time scale of some 30 s. Hence, photo-electrons can establish a Maxwellian distribution before recombination.

Besides the stellar photons, the radiation field also has contributions from nebular photons that are produced by recombinations. Some of these can re-ionize neutral hydrogen. Hence, in order to solve the ionization balance, we will need to solve the radiative transfer equation for the ionizing photons in the nebula. This turns the local problem of the ionization balance into a global problem. Given the rapid radiative relaxation rates of the HI levels (10^{-4}–10^{-8} s; cf. Chapter 2) compared with the ionization time scale ($\sim 10^6$ s), we can safely assume that all neutral hydrogen is in the ground state under the conditions considered here. When neutral hydrogen is exclusively in the ground state, only recombinations directly to the ground state will produce photons that are able to ionize hydrogen again. Under the assumption that all ionizing photons produced by direct recombinations to the ground level are absorbed on the same spot where they are created – the so-called on-the-spot approximation – the ionization balance becomes (largely) a local problem; i.e.

$$n_{\mathrm{H}} \int_{\nu_{\mathrm{T}}}^{\infty} 4\pi \mathcal{N}_{\star}(\nu) \exp[-\tau(\nu, r)] \alpha_{\mathrm{H}}(\nu) \mathrm{d}\nu = n_{\mathrm{e}} n_{\mathrm{p}} \beta_{\mathrm{B}}(T_{\mathrm{e}}), \tag{7.6}$$

with $4\pi\mathcal{N}_{\star}$ the local photon field due to the star,

$$4\pi \mathcal{N}_{\star}(\nu) = \frac{L_{\star}(\nu)}{4\pi r^2 h\nu}, \tag{7.7}$$

where $L_{\star}(\nu)$ is the stellar luminosity at frequency ν, r is the distance to the star and β_{B} is the recombination coefficient to all levels with $n \geq 2$. The optical depth, $\tau(\nu, r)$, is given by

$$\tau(\nu, r) = \int_0^r n_{\mathrm{H}}(r') \alpha_{\mathrm{H}}(\nu) \mathrm{d}r'. \tag{7.8}$$

Introducing the degree of ionization, x,

$$x = \frac{n_{\mathrm{p}}}{n_{\mathrm{p}} + n_{\mathrm{H}}}, \tag{7.9}$$

the ionization balance can be written as

$$\frac{1-x}{x^2} = \frac{n\beta_{\mathrm{B}}(T_{\mathrm{e}})}{4\pi \overline{\alpha}_{\mathrm{H}} \mathcal{N}}, \tag{7.10}$$

where the average ionization cross section, $\overline{\alpha}_{\mathrm{H}}$, is given by

$$\overline{\alpha}_{\mathrm{H}} \equiv \frac{1}{\mathcal{N}} \int_{\nu_{\mathrm{T}}}^{\infty} \mathcal{N}_{\star}(\nu, r) \alpha_{\mathrm{H}}(\nu) \mathrm{d}\nu, \tag{7.11}$$

with $\mathcal{N}$ the total ionizing photon flux at this position,

$$\mathcal{N} = \int_{\nu_{\mathrm{T}}}^{\infty} \mathcal{N}(\nu) \exp[-\tau(\nu, r)] \mathrm{d}\nu. \tag{7.12}$$

The total number of ionizing photons emitted by the star, $\mathrm{N}_{\mathrm{Lyc}}$, is given by

$$\mathrm{N}_{\mathrm{Lyc}} = \int_{\nu_{\mathrm{T}}}^{\infty} \frac{L_{\star}(\nu)}{h\nu} \mathrm{d}\nu. \tag{7.13}$$

Values for main sequence stars are given in Table 7.1. The average ionization cross section, $\overline{\alpha}_{\mathrm{H}}$, varies slightly as a function of stellar type; from about $2.5 \times 10^{-2} \alpha_0$ to $4.5 \times 10^{-2} \alpha_0$ for black bodies with effective temperatures ranging from 33 000 to 50 000 K.

In an HII region, hydrogen is almost fully ionized. Assuming no attenuation of the stellar photons, we find, for the ionizing photon flux at a distance of 0.5 pc from an O4 star,

$$4\pi \mathcal{N}(r) \simeq \frac{1}{4\pi r^2} \int_{\nu_0}^{\infty} \frac{L_{\star}(\nu)}{h\nu} \mathrm{d}\nu \simeq 2 \times 10^{12} \left(\frac{0.5\mathrm{pc}}{r}\right)^2 \text{ photons cm}^{-2}\,\mathrm{s}^{-1}. \tag{7.14}$$

Table 7.1 *Stellar parameters of O and B stars*[a]

Spectral type	$T_{\rm eff}$(K)	$L(10^5 L_\odot)$	$\mathbb{N}_{\rm Lyc}(10^{49}$ photons s$^{-1})$	$\mathcal{R}_{\rm s}^{b}$ (pc)
O3	51 200	10.8	7.4	1.3
O4	48 700	7.6	5.0	1.2
O5	46 100	5.3	3.4	1.0
O6	43 600	3.7	2.2	0.88
O7	41 000	2.5	1.3	0.75
O8	38 500	1.7	0.74	0.62
O9	35 900	1.2	0.36	0.49
B0	33 300	0.76	0.14	0.36

[a] Stellar parameters for main sequence stars.
[b] Strömgren radius calculated for a density of 10^3 cm^{-3}.

Inserting numerical values for the atomic parameters, $\beta_{\rm B} = 2.6 \times 10^{-13}$ cm^3 s^{-1} and $\overline{\alpha}_{\rm H} = 3 \times 10^{-19}$ cm^2, results in a neutral fraction of

$$1 - x = 4 \times 10^{-4}, \tag{7.15}$$

at a density of 10^3 cm^{-3}. We note that the neutral fraction increases with increasing density. With this high degree of ionization, the average optical depth between the star and this point is very small, $\tau_0 \simeq 0.2$.

Size of the HII region

For an optically thick nebula, the size of the HII region can be found from the global ionization balance, which equates the total recombination rate integrated over the nebula with the total ionizing photon-luminosity, $\mathbb{N}_{\rm Lyc}$. For a constant density nebula, this yields

$$\mathbb{N}_{\rm Lyc} = \frac{4\pi}{3} \mathcal{R}_{\rm s}^3 n^2 \beta_{\rm B} \overline{x}^2, \tag{7.16}$$

with $\overline{x}$ the average degree of ionization in the nebula. Because $x \simeq 1$ throughout the nebula (see below), the radius of the HII region, $\mathcal{R}_{\rm s}$, is

$$\mathcal{R}_{\rm s} = \left(\frac{3\mathbb{N}_{\rm Lyc}}{4\pi n^2 \beta_{\rm B}}\right)^{1/3} \simeq 1.2 \left(\frac{10^3 {\rm cm}^{-3}}{n}\right)^{2/3} \left(\frac{\mathbb{N}_{\rm Lyc}}{5 \times 10^{49}\ {\rm photons\ s}^{-1}}\right)^{1/3} {\rm pc}. \tag{7.17}$$

The radius, $\mathcal{R}_{\rm s}$, is often called the Strömgren radius after Bengt Strömgren, a pioneer in this field. Table 7.1 summarizes relevant properties of O and B stars and their Strömgren radii. Typically, at a density of $10^3\ \mathrm{cm}^{-3}$, the size of an HII region is of the order of a parsec. The mass of ionized gas is then

$$M_{\mathrm{HII}} = \frac{4}{3}\pi n \mathcal{R}_{\mathrm{s}}^3 m_{\mathrm{H}} = \frac{\mathbb{N}_{\mathrm{Lyc}} m_{\mathrm{H}}}{n\beta_{\mathrm{B}}}$$
$$\simeq 155 \left(\frac{\mathbb{N}_{\mathrm{Lyc}}}{5\times 10^{49}\ \mathrm{photons\ s}^{-1}}\right)\left(\frac{10^3\ \mathrm{cm}^{-3}}{n}\right) \mathrm{M}_{\odot}, \tag{7.18}$$

and we find the somewhat counterintuitive result that, for a given ionizing star, the mass of ionized gas increases as the density decreases. Basically, there are fewer recombinations per unit volume at a lower density and hence the neutral fraction is lower. As a result, the same number of ionizing photons can be used to keep a larger number of atoms ionized and the mass of ionized gas increases. The total recombination rate integrated over the nebula is, of course, independent of density and set by the ionizing photon-luminosity of the star.

Degree of ionization

If we make the simple-minded – but incorrect – assumption that the shape of the stellar ionizing radiation field does not change with position in the nebula and that the nebula is almost fully ionized – which is justified – the ionization balance in the on-the-spot approximation can be solved analytically. Introduce the dimensionless variable $z \equiv r/\mathcal{R}_{\mathrm{s}}$ and the optical depth scale $\tau_{\mathrm{s}} \equiv n\overline{\alpha}_{\mathrm{H}}\mathcal{R}_{\mathrm{s}}$, where $\overline{\alpha}_{\mathrm{H}}$ is the ionization cross section averaged over the radiation field. The total optical depth over the HII region is then $\overline{(1-x)}\tau_{\mathrm{s}}$, with $\overline{(1-x)}$ the average neutral fraction. The ionization balance (Eq. (7.6)) can now be written as

$$\frac{1-x}{x^2} = \frac{3z^2 \exp[\tau]}{\tau_{\mathrm{s}}}. \tag{7.19}$$

The optical depth equation (Eq. (7.8)) can be written in slightly different form,

$$\frac{\mathrm{d}\tau}{\mathrm{d}z} = (1-x)\,\tau_{\mathrm{s}}. \tag{7.20}$$

Assuming that $x \simeq 1$, these two equations can be combined to yield for the optical depth

$$\tau = -\ln\left[1 - z^3\right]. \tag{7.21}$$

The neutral fraction is then given by

$$(1-x) = \frac{3z^2}{\tau_{\mathrm{s}}\left(1-z^3\right)}. \tag{7.22}$$

For the optical depth scale, we have

$$\tau_s \simeq 10^3 \left(\frac{n}{10^3\,\mathrm{cm}^{-3}}\right)^{1/3} \left(\frac{\mathbb{N}_{\mathrm{Lyc}}}{5\times 10^{49}\,\mathrm{photons\,s}^{-1}}\right)^{1/3}, \tag{7.23}$$

and hence the neutral fraction is indeed small, $\sim 10^{-3}$ at a density of 10^3 cm^{-3}. The optical depth over most of the HII region is also small. Near the boundary of the HII region (the ionization front), the neutral fraction – and thus also the optical depth – increases rapidly (see below).

Now, absorption of ionizing photons by neutral hydrogen will change the radiation field drastically when $\tau(\nu_T) \gtrsim 1$ owing to the steep frequency dependence of the absorption cross section (cf. Fig. 7.1). Hence, the approximations used in deriving the equations in this subsection break down. The on-the-spot approximation is also not fully justified. Furthermore, the presence of helium atoms can

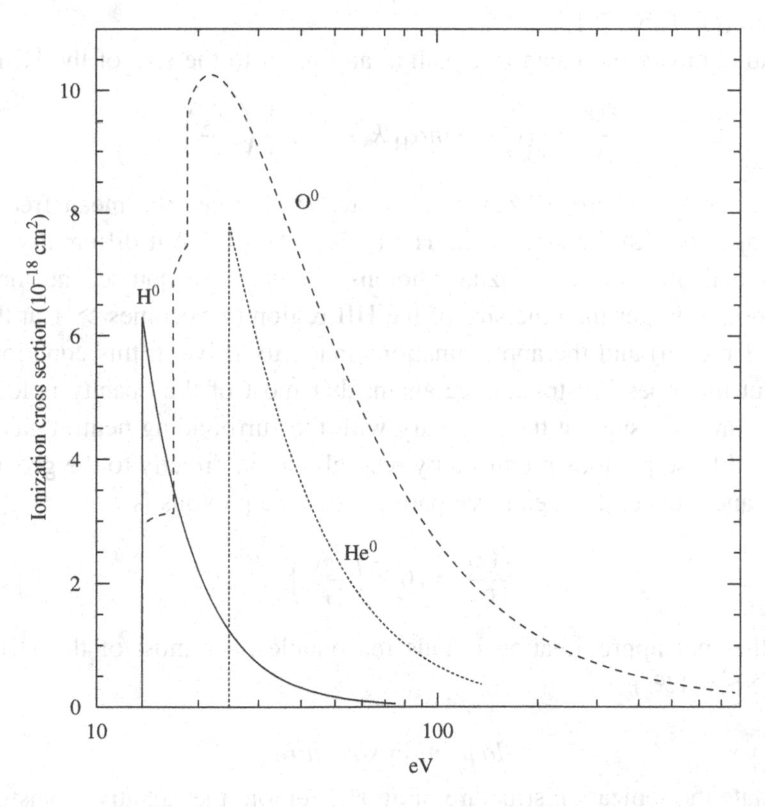

Figure 7.1 The ionization cross sections of hydrogen and helium, showing a maximum at the ionization edge and dropping away approximately as ν^{-3}. The ionization cross section of oxygen shows a more complex behavior due to ionization to excited levels within the same configurations.

influence the ionization balance (cf. Section 7.2.2). Thus, for detailed analysis, the ionization balance has to be solved numerically.

The ionization front

An HII region is terminated by an ionization front where the degree of ionization changes quickly from very close to 1 to very small. The thickness, ℓ, of this transition region is approximately equal to a mean free path for the ionizing photons; i.e.

$$\ell(x) = (n_{\rm H}\overline{\alpha}_{\rm H})^{-1} = ((1-x)n\overline{\alpha}_{\rm H})^{-1}. \tag{7.24}$$

Adopting $x \simeq 0.5$ for the transition region yields

$$\ell = 2 \times 10^{-3} \left(\frac{10^3\,{\rm cm}^{-3}}{n}\right)\ {\rm pc}. \tag{7.25}$$

The thickness of the HII-HI transition is, thus, much smaller than the size of the HII region (cf. Table 7.1).

We can compare the mean free path at any point to the size of the HI region,

$$\frac{\ell(r)}{\mathcal{R}_{\rm s}} = ((1-x)n\overline{\alpha}_{\rm H}\mathcal{R}_{\rm s})^{-1} \simeq \frac{1}{3}\left(\frac{\mathcal{R}_{\rm s}}{r}\right)^2, \tag{7.26}$$

where we have used Eq. (7.22). So, for stellar photons, the mean free path is large compared with the size of the HII region. To phrase it differently: the HII region is optically thin to ionizing photons. Actually, of course, the mean free path cannot be larger than the size of the HII region (τ becomes $\gg 1$ at the edge of the HII region) and the approximations made to arrive at this equation break down. But this does illustrate, once again, that most of the opacity is located in this thin transition shell at the boundary with the surrounding neutral gas.

For the diffuse photons produced by recombination directly to the ground state $\overline{\alpha}_{\rm H} \simeq \alpha_0$ and, hence, the mean free path for diffuse photons is

$$\frac{\ell(r)}{\mathcal{R}_{\rm s}} \simeq 10^{-2}\left(\frac{\mathcal{R}_{\rm s}}{r}\right)^2. \tag{7.27}$$

The on-the-spot approximation is thus reasonable over most of the HII region (i.e., for $r > 0.1\mathcal{R}_{\rm s}$).

Ionization structure

To calculate the ionization structure of an HII region, the radiative transfer equation has to be solved;

$$\frac{{\rm d}I(\nu, r)}{{\rm d}r} = -(1-x)\alpha_{\rm H}(\nu)I(\nu, r) + j(\nu, r), \tag{7.28}$$

with $I(\nu, r)$ the intensity of the radiation field at frequency ν and position r. We can divide the radiation field into a direct stellar part, $I_\star$, and a diffuse nebular part, $I_{\rm neb}$, which is produced by direct recombinations to the ground state (i.e., which can ionize neutral hydrogen). The radiative transfer equation for the stellar radiation can be readily solved to give

$$4\pi I_\star(\nu, r) = \frac{L_\star(\nu)}{4\pi r^2}\exp[-\tau(\nu, r)], \tag{7.29}$$

with $\tau(\nu, r)$ the optical depth,

$$\tau(\nu, r) = n\alpha_{\rm H}(\nu)\int_0^r (1 - x(r'))\mathrm{d}r' = \frac{\alpha_{\rm H}(\nu)}{\alpha_{\rm H}(\nu_T)}\tau(\nu_T, r), \tag{7.30}$$

with $\tau(\nu_{\rm T}, r)$ the optical depth at the threshold. The radiative transfer equation for the diffuse radiation field can be written as

$$\frac{\mathrm{d}I_{\rm neb}}{\mathrm{d}r} = -(1 - x)\alpha_{\rm H}(\nu)I_{\rm neb}(\nu, r) + j(\nu, r). \tag{7.31}$$

At the low temperatures of the nebula, the emission coefficient is due to recombination radiation,

$$j(\nu, r) = \frac{2h\nu^3}{c^2}\left(\frac{h^2}{2\pi mkT}\right)^{3/2}\exp[-h(\nu - \nu_T)/kT_{\rm e}]x^2n^2. \tag{7.32}$$

Because the radiation field depends on the degree of ionization throughout the whole nebula, this set of equations has to be solved iteratively. The total number of photons produced by recombinations is given by $\beta_1 x^2 n^2$, with β_1 the recombination coefficient directly to the first level, which is only $\sim$40% of the total number of ionizations ($\beta_A x^2 n^2$). So, the iteration can be started, in first approximation, with $I_{\rm neb} = 0$.

Figure 7.2 shows the calculated ionization structure of two typical HII regions. As expected, the hotter star produces a much larger HII region. The trends derived from the analytical discussion – complete ionization within the HII region and a sharp transition to neutral gas in the ionization front – are quite apparent in these numerical results.

The ionization parameter

Consider again the ionization balance (Eq. (7.10)). The degree of ionization depends on the ratio $\mathcal{N}/n$ and some atomic parameters. In describing HII regions, it is therefore advantageous to define the ionization parameter as the ratio of mean intensity in the ionizing radiation field to the density. Various specific definitions

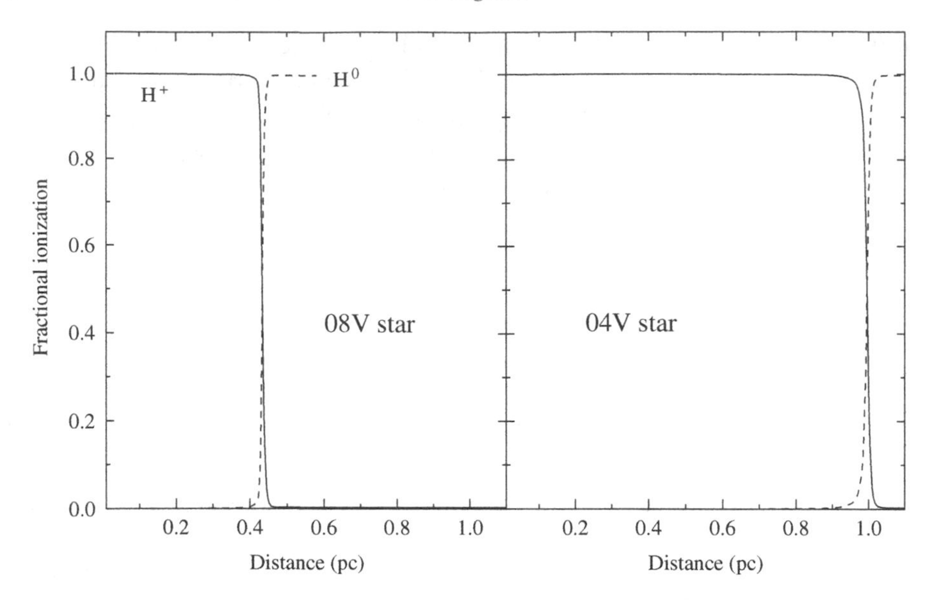

Figure 7.2 The calculated ionization structure of a pure hydrogen nebula ionized by an O4 (right) and an O8 (left) star. The gas is largely ionized throughout the HII region. Near the boundary, the neutral fraction increases sharply.

for this parameter are around in the literature. For plane-parallel models, the ionization parameter is often defined as

$$U = \frac{\mathcal{N}_0}{cn}, \tag{7.33}$$

with $\mathcal{N}_0$ the incident EUV flux; the factor c turns the flux into a radiation energy density. For spherical systems, a slightly different definition is useful,

$$U = \frac{\mathbb{N}_{\rm Lyc}}{4\pi \mathcal{R}_{\rm s}^2 nc}, \tag{7.34}$$

with $\mathbb{N}_{\rm Lyc}$ the stellar ionizing photon luminosity and $\mathcal{R}_{\rm s}$ the Strömgren radius. In this case, the ionization parameter is essentially the ratio of the ionizing photon density at the edge of the HII region (in the absence of absorption) to the hydrogen density. Taking attenuation into account, the actual number of photons available per electron will be much fewer at the edge. The ratio of the number of ionizing photons to the electron density averaged over the volume is, however, given by $< \mathbb{N}_{\rm Lyc}/n_{\rm e} >= 3U$. Whatever the precise definition, U is a measure for the ionization rate over the recombination rate and, thus, controls the H ionization balance. Indeed, as we will see in the next sections, this is a more general conclusion: the abundance ratio of sequential ionization stages of any element will be set by the ionization parameter (and a handful of atomic parameters). Hence, we expect that regions with similar ionization parameters will have similar properties, irrespective of, for example, the detailed geometry. The only parameter we have ignored here is dust, which is, more often, a spoil-sport. For a constant dust abundance per H atom, the fraction of the EUV photons absorbed by dust increases with density (cf. Section 7.2.4). Hence, for dusty HII regions, the effective U for a high density region is less than for a low density region.

7.2.2 Helium

Helium has a typical abundance of some 10% by number in the ISM and its presence can have some influence on the ionization structure of HII regions. The first ionization potential of helium is 24.6 eV. Helium can be doubly ionized when photons with $h\nu > 54.4$ eV are present. However, even the hottest O stars emit little radiation at such high energies and He^{2+} is of lesser importance.[1] Thus, photons with energies between 13.6 and 24.6 eV can ionize only neutral hydrogen while photons with energies above 24.6 eV can ionize both neutral hydrogen and neutral helium. The ionization cross section of neutral helium, $\alpha_{He}(\nu)$, is given by

$$\alpha_{He}(\nu) = \alpha_{He,0} \exp[-(y - 1.807)] \; y \geq 1.807, \tag{7.35}$$

with $y = \nu/\nu_T$, and ν_T the ionization threshold for H; $y = 1.807$ corresponds to the ionization threshold for helium. The helium ionization cross section at threshold is equal to $\alpha_{He,0} = 7.35 \times 10^{-18}\,\text{cm}^2$. The ionization cross section of neutral hydrogen and helium are compared in Fig. 7.1 and it should be noted that, despite the low helium abundance, neutral helium competes well with neutral hydrogen for ionizing photons above 24.6 eV.

Qualitatively, the ionization structure in a nebula with hydrogen and helium depends on the helium abundance as well as the spectrum of the ionizing star. Two extreme cases can be discerned. First, when there are few photons with $h\nu > 24.6$ eV, the neutral fraction of helium will be relatively larger than that of hydrogen. As a result, neutral helium will win the battle and all photons with energies greater than 24.6 eV will ionize helium rather than hydrogen. Furthermore, the He^+ (and H^+) zone will be small and surrounded by a much larger He^0, H^+ zone. Second, when the photon flux with $h\nu > 24.6$ eV is large, then the photons with energies between 13.6 and 24.6 eV play less of a role. In that case, photons with $h\nu > 24.6$ eV will keep both hydrogen and helium ionized.

Thus, the ionization balances of hydrogen and helium are coupled through the ionizing radiation field above 24.6 eV. The situation is further complicated

[1] The nuclei of planetary nebulae and the monsters hiding in the nuclei of AGNs can emit such highly energetic photons and, in those environments, He^{2+} can be very important.

through the diffuse radiation field. First, direct recombinations to the ground state of helium produce photons that can ionize hydrogen as well as helium. Moreover, many of the helium recombinations to higher states produce photons that can ionize hydrogen. How many photons are produced per recombination depends on the fate of the metastable 2^3S and 2^1S He^0 states. At low densities, collisional de-excitation is unimportant and each recombination has a probability of 0.96 to produce a hydrogen ionizing photon. At high densities ($>10^4\,cm^{-3}$), this probability is only 0.66.

An approximate size for the He^+ zone can be found from the global ionization balance, ignoring the presence of hydrogen (cf. Eq. (7.17)),

$$\mathcal{R}_{He} = \left(\frac{3\mathbb{N}_{Lyc}(He)}{4\pi y(1+y)n^2\beta_B(He)} \right)^{1/3}, \tag{7.36}$$

where it is assumed that helium is fully ionized through this zone, y is the helium abundance relative to hydrogen ($y \simeq 0.1$), and $\mathbb{N}_{Lyc}(He)$ is the total number of helium ionizing photons emitted by the star. For an O9 star with $\mathbb{N}_{Lyc}(He) \simeq 1.8 \times 10^{47}$ photons s^{-1}, the He^+-Strömgren radius is 0.36 pc at a density of $10^3\,cm^{-3}$ or only 75% of the H^+-Strömgren radius. For cooler stars, the He^+ zone shrinks very rapidly compared to the H^+ zone. Hotter stars have He^+ zones that are practically coincident with the H^+ zone.

The calculated ionization structure for nebulae containing hydrogen and helium is shown in Fig. 7.3. Compared with the pure hydrogen nebula (cf. Fig. 7.2), the ionized hydrogen zone is slightly smaller, because of the absorption of ionizing photons by helium. The hot O4 star has a small He^{++} zone around it. In this case, the He^+ zone is actually slightly larger than the H^+ zone. Because the ionization cross section is highly peaked towards threshold, the stellar radiation field is least attenuated at the highest frequencies. Since the H cross section for highly energetic photons is smaller than that of He, helium can stay ionized slightly further out than H (cf. Fig. 7.1). As expected, for cooler stars, the He^+ zone shrinks relative to the H^+ zone.

7.2.3 The ionization structure of trace species

We can now generalize the results for the ionization balance of hydrogen and helium to species with many ionization stages. The presence of multiple ionization stages implies, of course, that there will be a set of ionization equations linking sequential ionization stages. Furthermore, the dependence of the ionization cross section on photon energy becomes more complex and a second channel for recombination – dielectronic recombination – opens up. Each of these points will be discussed in turn.

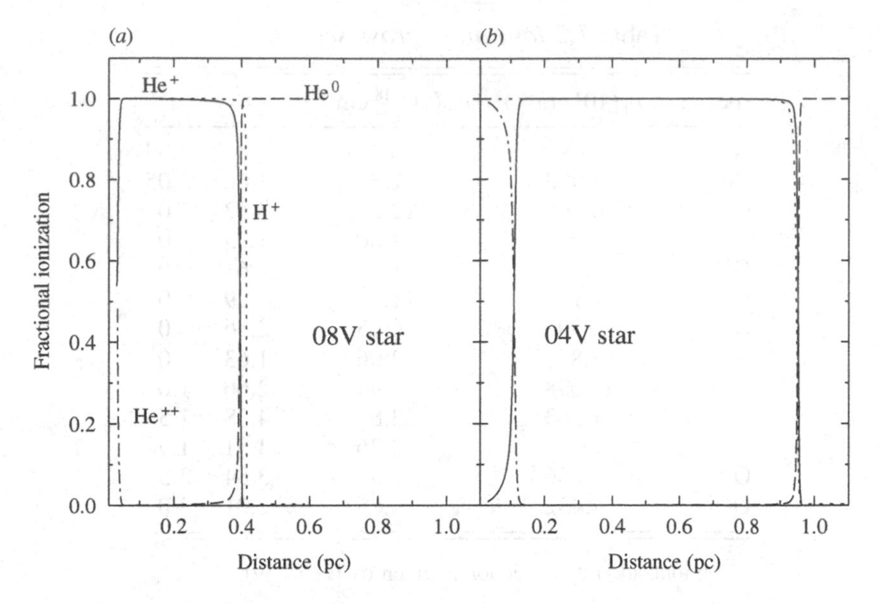

Figure 7.3 The calculated ionization structure of a nebula containing hydrogen and helium. For an O4 star (*a*), the H^+ and He^+ zones are essentially coincident. For an O8 star (*b*) and stars cooler than that, the He^+ zone has shrunk to less than the H^+ zone. Near the star, a small He^{++} zone is evident.

For each species, X_j, the ionization structure can be found by solving a set of ionization balance equations for each charge i,

$$n\left(X_j^{+i}\right)\int_{\nu_j^{+i}}^{\infty} 4\pi\mathcal{N}(\nu, r)\alpha\left(\nu, X_j^i\right)\mathrm{d}\nu = n_{\mathrm{e}}n\left(X_j^{+i+1}\right)\beta\left(T_{\mathrm{e}}, X_j^{+i+1}\right). \qquad (7.37)$$

Because this is a redundant set of equations, for each element, one equation is replaced by a conservation law,

$$\sum_{i=0}^{i_{\max}} n\left(X_j^i\right) = \mathcal{A}_j n, \qquad (7.38)$$

where $i_{\max}$ is the maximum ionization charge considered and $\mathcal{A}_j$ is the gas phase abundance of species X_j. In these calculations, the ionizing radiation field – both the direct stellar and the indirect diffuse radiation field – has to be solved simultaneously with the ionization structure of hydrogen and helium.

Because of resonances, ionization to different levels of the ion, and inner shell ionization, the ionization cross section of heavy species can be complex. The contribution of each threshold, ν_{T}, to the photo-ionization cross section can be approximated by

$$\alpha(\nu) = \alpha_{\mathrm{T}}\left(b\left(\frac{\nu}{\nu_{\mathrm{T}}}\right)^{-a} + (1-b)\left(\frac{\nu}{\nu_{\mathrm{T}}}\right)^{-a-1}\right) \quad \nu > \nu_{\mathrm{T}}, \qquad (7.39)$$

Table 7.2 *Ionization cross section*[a]

species	$\nu_T(10^5\ \mathrm{cm}^{-1})$	$\alpha_T(10^{-18}\ \mathrm{cm}^{-2})$	b	a
H	1.097	6.3	1.34	2.99
He	1.983	7.83	1.66	2.05
C	0.909	12.2	3.32	2.0
C^+	1.97	4.60	1.95	3.0
C^{2+}	3.86	1.6	2.6	3.0
N	1.17	11.4	4.29	2.0
N^+	2.39	6.65	2.86	3.0
N^{2+}	3.83	2.06	1.63	3.0
O	1.098	2.94	2.66	1.0
	1.363	3.85	4.38	1.5
	1.5	2.26	4.31	1.5
O^+	2.836	7.32	3.84	2.5
O^{2+}	4.432	3.65	2.01	3.0

[a] The ionization cross section is given by Eq. (7.39).

where ν_T is the ionization threshold, α_T the cross section at threshold, and a and b are fitting parameters (cf. Table 7.2). The total photo-ionization cross section is then the sum over all thresholds of a species. Figure 7.1 compares the ionization cross section for neutral oxygen to that of neutral hydrogen.

The recombination coefficient includes both radiative recombination and dielectronic recombination. The radiative recombination process is like that for hydrogen and helium where, after capture of the electron, the resulting exiting species stabilizes through the emission of a photon. This is then followed by complete radiative relaxation. To a first approximation, the radiative rate scales approximately with $T^{-0.7}$; slightly steeper than the pure coulomb interaction because of electron correlation effects in these multiple electron systems. In dielectronic recombination, the incoming electron excites a bound electron to an excited electronic state. The long-lived, doubly-excited species then relaxes radiatively. The value of the rate coefficient depends very much on the resonance character of the reaction. For some ions, dielectronic recombination can be very important, while for others, radiative recombination dominates. Some representative values for the radiative and dielectronic recombination coefficients are given in Table 7.3.

Besides photo-ionization, charge exchange reactions of the type

$$X_j^i + Y_k^o \rightleftharpoons X_j^{i-1} + Y_k^{+1}, \tag{7.40}$$

where Y_k is hydrogen or helium, can also be of importance. In view of the small energy difference ($\simeq 220$ K) between the ionization potentials of neutral hydrogen and oxygen, the charge exchange reaction between $O(^3P)$ and H^+ is

Table 7.3 *Recombination coefficients*[a]

ion	$\beta_R(10^{-13}\,cm^3\,s^{-1})$	r [b]	$\beta_D(10^{-13}\,cm^3\,s^{-1})$	I_p [c](eV)
H^+	4.18	0.75	—	13.6
	2.59[d]	0.89	—	
He^+	4.3	0.67	—	24.6
	2.5[d]	0.67	—	
He^{++}	15.2	0.67	—	54.4
C^+	4.66	0.62	1.84	11.3
C^{2+}	24.5	0.65	60.6	24.4
C^{3+}	50.5	0.77	131	47.9
N^+	3.92	0.61	5.22	14.5
N^{2+}	22.8	0.64	20.4	29.6
N^{3+}	54.4	0.68	216	47.4
O^+	3.31	0.68	0.762	13.6
O^{2+}	20.5	0.65	16.6	35.1
O^{3+}	54.3	0.67	114	54.9

[a] Radiative (β_R) and dielectronic (β_D) recombination coefficients for the ion and an electron evaluated at 10^4 K.
[b] Temperature exponent of the radiative recombination rate $((T/10^4)^{-r})$.
[c] Ionization potential at which the ion appears.
[d] Radiative recombination rate to all levels $n \geq 2$.

very important. The forward reaction (ionized oxygen with neutral hydrogen) has a rate coefficient of $k_f \simeq 2 \times 10^{-9}\,cm^3\,s^{-1}$ and the backward rate coefficient is then given by detailed balance as

$$k_b = \frac{8}{9} k_f \exp[-220/T]. \tag{7.41}$$

Thus, when neutral hydrogen is abundant – as in the ionization front – the O/O^+ ratio is very closely tied to the H/H^+ ratio. These charge exchange reactions can be included in the ionization balance through terms of the type

$$k_f n\left(X_j^{+i}\right) n\left(Y_k^{o}\right) - k_b n\left(X_j^{+i+1}\right) n\left(Y_k^{+1}\right). \tag{7.42}$$

Figure 7.4 shows the calculated ionization structure for nitrogen and oxygen. As for hydrogen and helium, the ionization structure of these species has a layered appearance with the highest ionization stage closest to the star and lower ionization stages progressively further away. For a cooler star, the lower ionization stages become relatively more important. The boundaries of these ionization zones depend, of course, on the ionization potentials of the various ionization stages of these species. For an O4 star, O^{++} and N^{++} are the dominant ionization stages of these elements, but these species are not entirely coexistent. Except near the

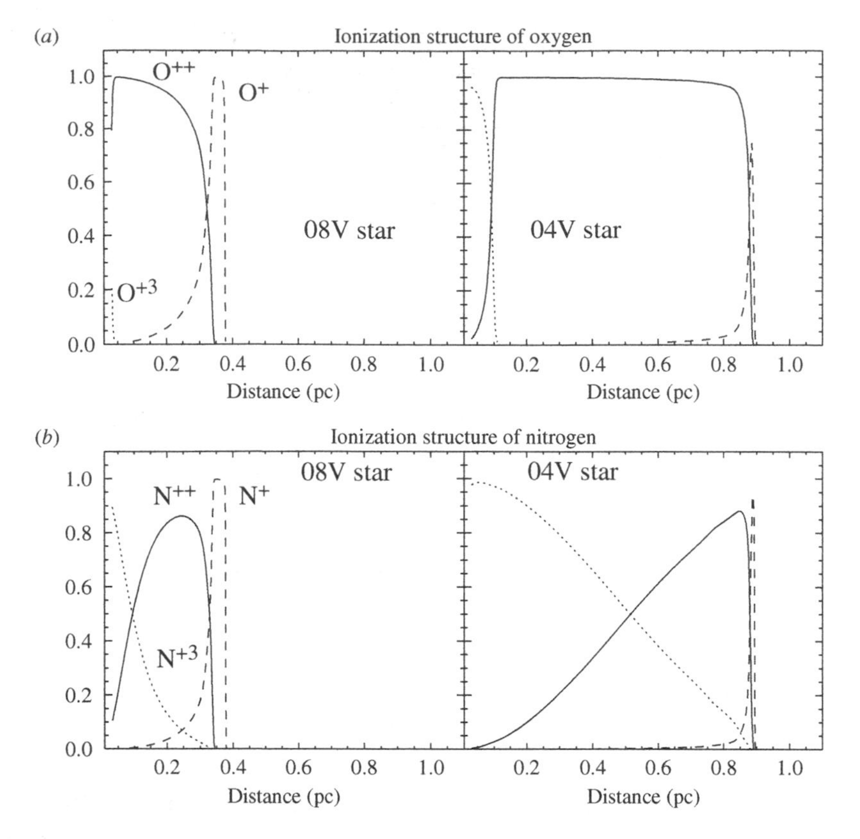

Figure 7.4 The calculated ionization structure of oxygen (*a*) and nitrogen (*b*) for an O4 star (right) and an O8 star (left) at a density of $10^3\,\text{cm}^{-3}$. Note the onion-like shell structure of sequential ionization stages with the highest ionization stages closest to the star. Except near the boundary of the HII region, the transition from one stage of ionization to the next is quite gradual.

edge, extinction of ionizing photons is unimportant and hence the abundance of an ion drops (and rises) gradually, reflecting the r^{-2} dependence of the radiation field. Near the HII/HI boundary, trace species of course also show a sharp edge. Because of the low abundance of trace species, their effects on the ionization structure of hydrogen and helium are small, but not negligible. Indeed, comparing these results with those for a hydrogen and helium nebula (cf. Fig. 7.3), the size of the ionized region has shrunk slightly, by some 10%. The point is that, once ionized, hydrogen does not contribute to the opacity anymore. However, trace species can be multiply ionized and are therefore more effective absorbers on a per atom basis. Moreover, ionization to the highest possible state is never as complete for trace species as it is for H (the neutral fraction of H is typically only 10^{-3}).

7.2.4 Dust and HII regions

Dust and the competition for ionizing photons

The presence of dust inside HII regions can be inferred from scattered light measurements as well as from infrared continuum emission. The abundance of dust in HII regions is, however, not well constrained observationally.

Dust will compete with neutral hydrogen for ionizing photons and, as a result, the presence of dust will shrink the HII region in size. Ignoring this effect for the moment, the absorption optical depth of dust for ionizing photons, $\tau_{\rm d}$, is given by

$$\tau_{\rm d} = \frac{\tau_{\rm d}}{A_{\rm v}} \frac{A_{\rm v}}{N_{\rm H}} n\mathcal{R}_{\rm s}, \tag{7.43}$$

with $A_{\rm v}/N_{\rm H}$ the dust visual extinction cross section per hydrogen atom (see Section 5.3.5). The dust extinction cross section at EUV wavelengths cannot be measured directly. However, reasonable extrapolations from the observed FUV extinction curve can be made. Adopting an EUV-to-visual extinction ratio of 5, an albedo of 0.6, and a dust-to-gas ratio of $N_{\rm H}/A_{\rm v} = 1.9\times 10^{21}/\delta_{\rm d}$ H atom cm^{-2} magnitude^{-1} (where $\delta_{\rm d} = 1$ corresponds to standard dust-to-gas; see Section 5.3.5) yields

$$\tau_{\rm d} \simeq 3\delta_{\rm d} \left(\frac{n}{10^3\ {\rm cm}^{-3}}\right)\left(\frac{\mathcal{R}_{\rm s}}{1\ {\rm pc}}\right). \tag{7.44}$$

Thus, for $\delta_{\rm d} = 1$ and typical conditions, the dust optical depth for EUV photons is ~ 3 over a Strömgren radius. Hence, dust absorption of ionizing photons will be appreciable.

The simple-minded analytical model described in Section 7.2.1 can be adapted by including the dust absorption in the definition of the optical depth (Eq. (7.20)), i.e.

$$\frac{d\tau}{dz} = (1-x)\tau_{\rm s} + \tau_{\rm d}, \tag{7.45}$$

and, under the same assumptions (i.e., $1 - x \ll 1$; e.g., substituting Eq. (7.19) and changing variables to $\tau = \ln u$), we arrive at a linear differential equation,

$$\frac{du}{dz} = -3z^2 - \tau_{\rm d} u, \tag{7.46}$$

with the solution

$$(1-x) = \frac{3z^2}{\tau_{\rm s}} \left[\exp[-\tau_{\rm d} z] + \frac{6}{\tau_{\rm d}^3}\left(\exp[-\tau_{\rm d} z] - 1 + \tau_{\rm d} z - \frac{1}{2}\tau_{\rm d}^2 z^2\right)\right]^{-1}, \tag{7.47}$$

where we have used that for $z = 0$, $\tau = 0$. For small $\tau_{\rm d} z$, the previous result for a pure hydrogen zone is recovered (cf. Eq. 7.22). Qualitatively, the solutions are still the same with a largely ionized HII region bounded by an ionization front.

The number of EUV photons absorbed by the gas between z and $z + \mathrm{d}z$ is

$$(1-x)\,\mathcal{R}_s n \overline{\alpha}_\mathrm{H} \frac{\mathbb{N}_\mathrm{Lyc}}{4\pi z^2} \exp[-\tau] 4\pi z^2 \mathrm{d}z = 3z^2 \mathbb{N}_\mathrm{Lyc} \mathrm{d}z, \tag{7.48}$$

while the number of photons absorbed by dust and gas is

$$((1-x) n \overline{\alpha}_\mathrm{H} \mathcal{R}_s + \tau_\mathrm{d}) \frac{\mathbb{N}_\mathrm{Lyc}}{4\pi z^2} \exp[-\tau] 4\pi z^2 \mathrm{d}z = \mathbb{N}_\mathrm{Lyc} \exp[-\tau] \mathrm{d}\tau. \tag{7.49}$$

The fraction of Lyman continuum photons absorbed by the dust is, thus,

$$f(z) = 1 - \frac{z^3}{1 - \exp[-\tau]}. \tag{7.50}$$

The size of the HII region, z_0, can now be found by realizing that $(1-x)\tau_\mathrm{s}$ becomes large at the boundary. Hence, the expression within the square brackets in Eq. (7.47) will have to go to zero. The resulting transcendental equation can be solved iteratively. For normal dust-to-gas ratios, the effect of dust on the size of an HII region can be substantial (Fig. 7.5). Another way of illustrating this is by considering the fraction of EUV photons, f_d, absorbed by the dust,

$$1 - f_\mathrm{d} = z_0^3 = \left(\frac{\mathcal{R}_\mathrm{d}}{\mathcal{R}_\mathrm{s}}\right)^3, \tag{7.51}$$

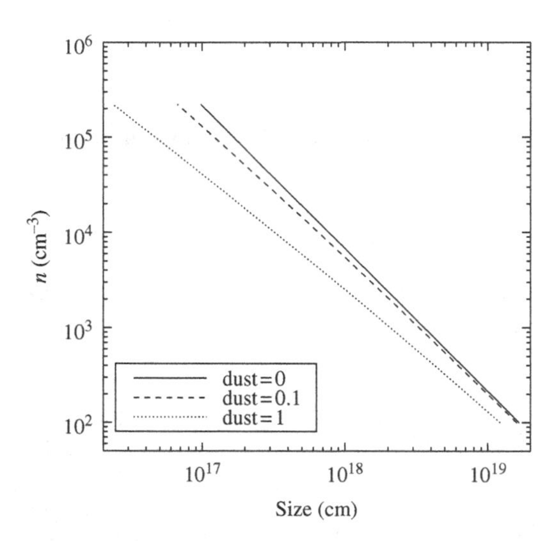

Figure 7.5 The effect of dust on the size of HII regions for a constant density HII region powered by an O4 main sequence star. The curves are labeled by the dust-to-gas ratio, δ_d($\delta_\mathrm{d} = 1$ corresponds to a standard dust-to-gas ratio). The Strömgren sphere solution for a pure hydrogen nebula is also shown. Results for other stars can be scaled by keeping the product $n\mathcal{R}_\mathrm{s}$ constant.

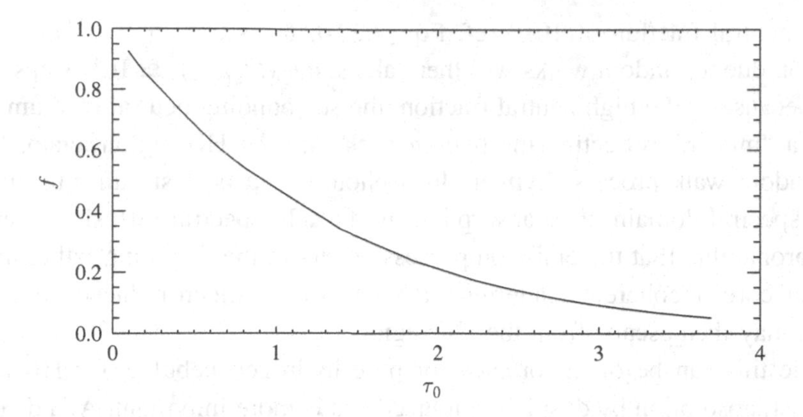

Figure 7.6 The fraction of EUV photons not absorbed by the dust ($1 - f_d = z_0^3$) as a function of the dust optical depth in the HII region, τ_0 ($= \tau_d z_0$).

where $\mathcal{R}_d$ is the size of the HII region when dust is present. This is illustrated in Fig. 7.6. This fraction absorbed by the dust is also apparent from Fig. 7.5 when comparing the appropriate dust-curve with the pure hydrogen curve. The effects of dust become more pronounced when the density increases. This is obvious from Eq. (7.44) when the relationship between the Strömgren radius and the density is taken into account (cf. Eq. (7.17)). We note that, for a density exceeding $10^3\ \mathrm{cm}^{-3}$, more than 90% of the EUV photons go directly into heating the dust rather than ionizing the gas for normal dust-to-gas ratios.

Dust and Lyα radiation

Dust inside the HII region is also important for the absorption of resonantly trapped Lyman alpha radiation. The HI Lyman lines have a very large optical depth within the HII region. In particular, the line-averaged cross section for the Lyman alpha line is $\overline{\sigma}_\alpha \simeq 2 \times 10^{-13}\ \mathrm{cm}^2$ or some 10^6 times the average cross section for ionizing photons in the Lyman continuum. We have already seen that the average optical depth for ionizing photons is ~1 in the ionized gas (cf. Eq. (7.21)). Hence, the optical depth for Lyman alpha photons is 10^6 and a Lyman line photon will be resonantly scattered many times. At each scattering, there is a finite probability that radiative decay will convert an $n \rightarrow 1$ Lyman line photon to an $n' \rightarrow 1$ photon, with $n' < n$, plus a lower series (e.g., Balmer, Pfund, Bracket, ...) photon (cf. Fig. 2.1). Eventually, each Lyman line photon will result in an excitation of the 2^2S or 2^2P level. The former is a metastable level, which can decay through the emission of two photons in the two-photon continuum (the $2^2S \rightarrow 1^2S$ transition).

Lyman alpha photons cannot be further "downgraded" this way. The mean free path of a Lyman alpha photon, ℓ_α, is exceedingly small. At a density of $10^3\ \mathrm{cm}^{-3}$

with a neutral fraction of 10^{-3} (cf. Eq. (7.22)), ℓ_α is only 5×10^{12} cm. Spatial diffusion due to random walks will then take some $(\mathcal{R}_s/\ell_\alpha)^2 \simeq 10^{12}$ steps. Moreover, because of the high neutral fraction, the surrounding neutral medium would act as a "mirror," reflecting the photon back into the HII region again. During this random walk process, Lyman alpha photons can be lost either to diffusion in the spectral domain or to absorption by dust. In spectral diffusion, there is a finite probability that the emission process occurs in the line wing rather than the Doppler core (incoherent scattering). Because of the much reduced opacity, the photon may then escape from the HII region.

While this can be of importance for pure hydrogen nebulae, for HII regions with dust, absorption by dust in the ionized gas is more important. At a density of 10^3 cm^{-3}, the ratio of dust to HI opacity for Lyman alpha photons is approximately $5\times10^{-6}\delta_d$. Hence, a Lyman alpha photon will travel only some 10^{15} cm and will essentially be absorbed on-the-spot by dust. The radiative heating rate per dust grain, Γ_d, is then

$$n_d\Gamma_d(\mathrm{Ly}\alpha) = n^2\beta_B h\nu_\alpha, \tag{7.52}$$

with $h\nu_\alpha$ the Lyman alpha energy (10.2 eV). Note that when the dust abundance drops, this heating rate increases. We can compare this to the heating rate per dust grain by ionizing stellar photons (cf. Section 5.2.2),

$$n_d\Gamma_d\,(\mathrm{EUV}) = \frac{\pi a^2 n_d \overline{Q} N_\star \overline{h\nu}}{4\pi r^2} e^{-\tau}, \tag{7.53}$$

with a the grain size, $\overline{Q}$ the average dust absorption efficiency for ionizing photons, and $\overline{h\nu}$ the average energy of an ionizing photon. The ratio is then

$$\frac{\Gamma_d(\mathrm{Ly}\alpha)}{\Gamma_d(\mathrm{EUV})} \simeq 3\left(\frac{r}{\mathcal{R}_s}\right)^2\left(\frac{h\nu_\alpha}{\overline{h\nu}}\right)\frac{1}{\tau_d} \simeq \left(\frac{r}{\mathcal{R}_s}\right)^2\frac{1}{\tau_d}, \tag{7.54}$$

with τ_d the characteristic optical depth of dust in the HII region (cf. Eq. (7.44)). Thus, despite the fact that Lyman alpha photons contain only some 1/3 of the energy of an ionizing photon, they can still be very important in heating the dust, particularly near the edge of the HII region and when the dust density is low or the dust is heavily depleted.

7.3 Energy balance

The temperature follows from balancing heating and cooling. The cooling and heating of ionized gas has been discussed in Sections 2.4 and 3.2.

7.3.1 A pure hydrogen nebula

For a pure hydrogen nebula, the photo-ionization heating of HII regions is given by

$$n\Gamma_{\rm H}(r) = (1-x)n\int_{\nu_{\rm T}}^{\infty} 4\pi\mathcal{N}(\nu, r)\alpha(\nu)h(\nu-\nu_T)\mathrm{d}\nu. \tag{7.55}$$

Using the ionization balance equation (Eq. 7.1), this yields

$$n\Gamma_{\rm H}(r) = x^2n^2\beta_{\rm A}(T_{\rm e})\frac{3}{2}kT_{\rm H}(r), \tag{7.56}$$

where $3/2kT_{\rm H}$ is the average excess energy of the photo-electron. Focussing first on the stellar component of the radiation field, for a black-body radiation field, $T_{\rm H} \simeq T_\star$ (cf. Section 3.2). Owing to steep spectral dependence of the neutral hydrogen absorption cross section, the stellar radiation field hardens with distance from the star and $3/2kT_{\rm H}$ increases. Because the average energy of an electron is quite low (~0.7 eV) compared to the H ionization potential, the diffuse radiation field is sharply peaked towards the ionization threshold and diffuse photons are generally less effective in heating the gas. In the on-the-spot approximation, the diffuse radiation field can be ignored; each ionization by a diffuse photon yields an electron with the same energy as the recombining electron took away from the gas. The heating rate equation above can then simply be modified by replacing the recombination coefficient to all levels ($\beta_{\rm A}$) by only those with $n \geq 2$ (i.e., $\beta_{\rm B}$).

For a pure hydrogen gas, cooling is due to bound–bound, free–bound, and free–free cooling (cf. Section 2.4). Bound–bound is unimportant compared with the other two. The energy balance then reads

$$n\Gamma_{\rm H} = n^2\Lambda_{\rm fb} + n^2\Lambda_{\rm ff}, \tag{7.57}$$

which can be written as

$$\frac{3}{2}\beta_{\rm B}kT_{\rm H} = \beta_{\rm E}kT_{\rm e} + \beta_{\rm ff}T_{\rm e}^{1/2}, \tag{7.58}$$

where $\beta_{\rm E}$ is the kinetic-energy-averaged recombination coefficient ($\beta_{\rm E} = 1.73 \times 10^{-13}(10^4/T)^{0.9}$; cf. Section 2.4), $\beta_{\rm B}$ is given by $2.6 \times 10^{-13}(10^4/T)^{0.8}$, and $\beta_{\rm ff}$ is the numerical coefficient in the free–free cooling ($\beta_{\rm ff} \simeq 1.4 \times 10^{-27}$ erg cm^3 s K$^{-1/2}$; cf. Eq. (2.50)). This transcendental equation can be solved numerically (Fig. 7.7). Typically, in this case, the temperature of the gas is slightly less than the temperature of the radiation field. Essentially, hydrogen is not a good coolant, because the lowest excited level is ~10 eV (~10^5 K) above ground. As a result, the calculated gas temperature of a pure hydrogen nebula is considerably hotter than the ~10^4 K observed for galactic HII regions. Thus, the ionized gas produced by the first luminous objects in the Universe is expected to be much hotter than present day HII regions. In fact, measurement of the electron

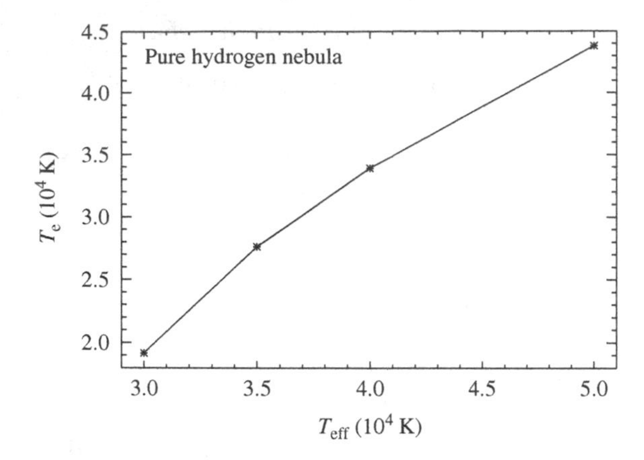

Figure 7.7 The calculated electron temperature, T_e, for a pure H nebula ionized and heated by a star with an effective temperature T_{eff}.

temperature would provide a good measure of the spectral temperature of the ionizing source.

7.3.2 A nebula with trace elements

The presence of trace species will reduce the temperature of the gas considerably owing to the enhanced cooling through bound–bound transitions between low-lying electronic states. For an HII region ionized by an O4 star, $T_H = 3.4 \times 10^4$ K and the heating rate is then

$$n\Gamma \simeq 1.8 \times 10^{-24} n^2 \left(\frac{10^4\,\mathrm{K}}{T}\right)^{0.8} \mathrm{erg\,cm^{-3}\,s^{-1}}. \tag{7.59}$$

We will assume that oxygen is completely doubly ionized and, in the low-density limit, the cooling rate through the 52 and 88 μm [OIII] far-infrared fine-structure lines is then given by (cf. Section 2.4)

$$n^2\Lambda = \sum_i n^2 \mathcal{A}_{\mathrm{O}} \gamma_{\mathrm{lu}} h\nu_{\mathrm{ul}} \simeq 9.8 \times 10^{-25} n^2 \left(\frac{10^4\,\mathrm{K}}{T}\right)^{1/2} \mathrm{erg\,cm^{-3}\,s^{-1}}, \tag{7.60}$$

where the summation is over the two transitions. This is only about half of the heating rate (Eq. (7.59)) and the visual lines (5007 Å, 4959 Å) will have to contribute the remainder. Their cooling is given by (cf. Section 2.4)

$$n^2\Lambda = \sum_i n^2 \mathcal{A}_{\mathrm{O}} \gamma_{\mathrm{lu}} h\nu_{\mathrm{ul}} \simeq 5 \times 10^{-23} n^2 \left(\frac{10^4\,\mathrm{K}}{T}\right)^{1/2} \exp[-29\,000/T]\,\mathrm{erg\,cm^{-3}\,s^{-1}}. \tag{7.61}$$

The electron temperature is then calculated to be approximately $T \simeq 29\,000/\ln[30] \simeq 7200$ K, which should be compared with an electron temperature of $\simeq 35\,000$ K for a pure hydrogen nebula. Two more points should be made here. First, the heating rate is not very sensitive to the stellar parameters; T_H varies only by a factor 2 between an O3 and a B1 star. Furthermore, this only affects the temperature through the ln factor. Second, in the low density limit, cooling by far-infrared fine-structure lines has the same temperature (and density) dependence as the heating rate. These lines are thus not good thermostats. The exponential temperature dependence for the cooling rate by the optical lines, on the other hand, ties the temperature to approximately their energy level separation. To phrase it differently, if the heating rate were a factor of two less or the abundance of O^{2+} a factor of two higher, then low density HII regions would cool down to the energy level separation of the far-infrared fine-structure line energy level separation,

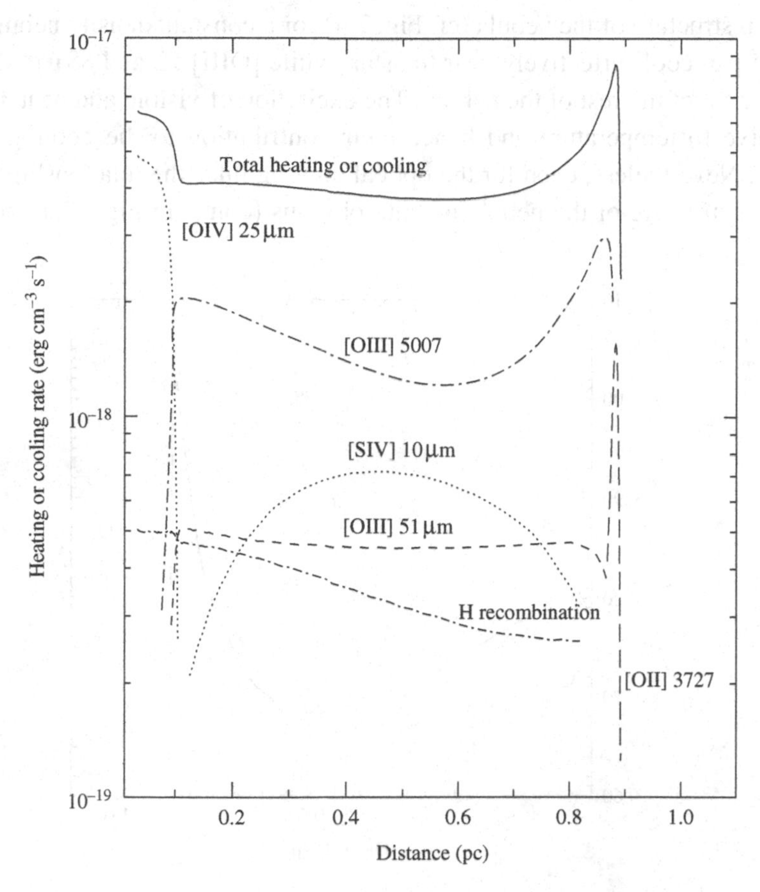

Figure 7.8 Calculated heating and cooling rates for an HII region with a density of 10^3 cm^{-3} powered by an O4 star. The cooling rates by the trace elements show, of course, a shell-like structure.

some 100 K. This can be of importance for high metallicity, low density regions. Of course, in the high-density limit, the far-IR fine-structure lines scale with n (rather than n^2) and hence the optical lines are always more important.

Figure 7.8 shows the dominant heating and cooling rates for an HII region with a density of $10^3\,\mathrm{cm^{-3}}$ powered by an O4 star. The ionization structure of this nebula is illustrated in earlier figures in this chapter. The high heating rate near the star reflects the absorption of highly energetic He^+ ionizing photons (cf. the extent of the He^{++} zone in Fig. 7.3). The rise in the heating rate near the edge of the HII region is due to the hardening of the radiation field because H^0 preferentially absorbs lower energy photons. The presence of trace species allows very effective cooling through low-lying electronic states, including far-infrared fine-structure lines and visible and near-UV lines. Because of their low excitation energies, fine-structure lines are not very sensitive to the temperature of the ionized gas. Hence their contribution to the cooling follows rather directly the ionization structure of the nebula (cf. Fig. 7.4) for a constant density nebula. Thus, [OIV] 25 μm cools effectively near the star, while [OIII] 52 and 88 μm dominate through most of the rest of the nebula. The excitation of visible and near-UV lines is sensitive to temperature and hence their contribution to the cooling is more complex. Nevertheless, even for the optical cooling lines the relationship with the ionization structure of the nebula is quite obvious (compare Figs. 7.4 and 7.8).

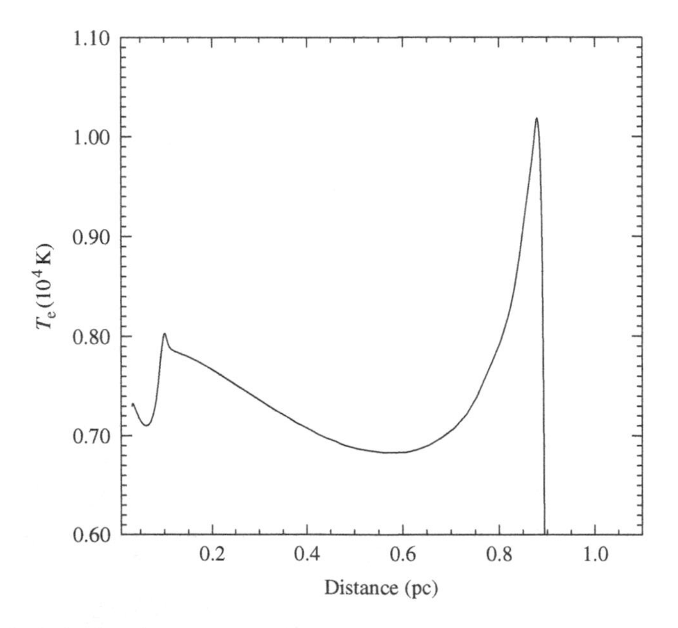

Figure 7.9 Calculated temperature structure for an HII region with a density of $10^3\,\mathrm{cm^{-3}}$ powered by an O4 star. The cooling rates and ionization structure for this nebula are shown earlier in this chapter.

The calculated temperature for this HII region (Fig. 7.9) is about 7000 K over most of the nebula. Near the star, the temperature drops to some 5000 K because of the enhanced cooling by [OIV], which more than compensates for the increased heating due to He^+ ionization. Near the ionization front, the temperature increases to about 10 000 K owing to the hardening of the radiation field.

7.4 Emission characteristics

Figure 7.10 shows the near-UV to visible spectrum of a position in the Orion HII region. The hydrogen recombination lines are quite obvious. In addition, forbidden transitions of trace species with a range of ionization stages are present in this spectrum. Figure 7.11 shows the 2.5–200 μm spectrum of the compact HII region, K3-50, which is dominated by continuum dust emission. Emission features due to PAHs (Chapter 6) as well as absorption features due to silicates (Chapter 5) and ices (Chapter 10) can be recognized. In addition, HI recombination lines as well as atomic fine-structure lines are present. Figure 7.12 shows a composite spectrum of the HII region, W3, in the IR and radio range. The solid line indicates the free–free emission of the ionized gas. The dashed line is indicative of the IR dust continuum.

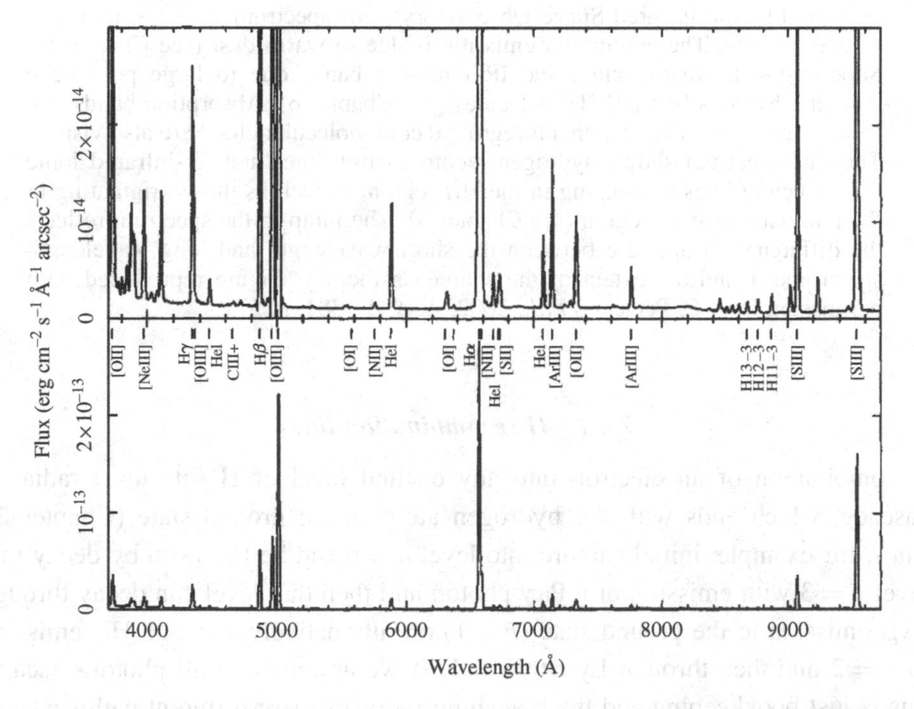

Figure 7.10 The visible–UV spectrum of HII regions. In the upper trace, the spectrum has been scaled up to bring out the underlying continuum. Figure reproduced with permission from J. A. Baldwin, *et al.*, 1991, *Ap. J.*, **374**, p. 580.

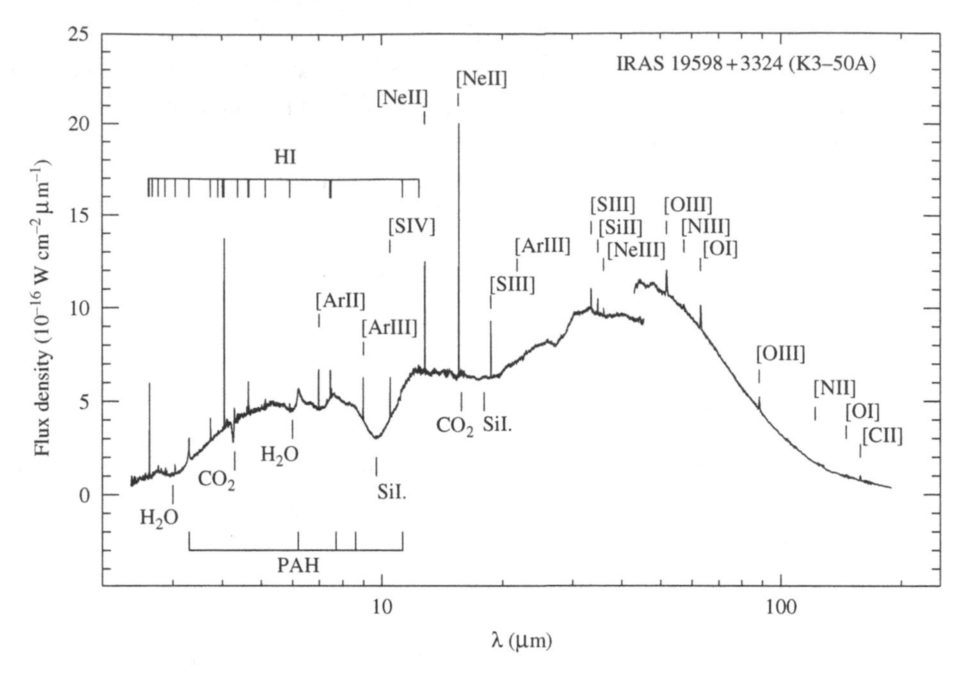

Figure 7.11 The extremely rich infrared spectrum of the HII region, K3-50, measured by the Infrared Space Observatory. This spectrum shows a multitude of components. The continuum emission is due to warm dust (see Chapter 5). Superimposed, we recognize the IR emission bands due to large polycyclic aromatic hydrocarbon (PAH) molecules (see Chapter 6). Absorption bands due to silicates and ices located in a foreground cold molecular cloud are also visible. The line spectrum shows hydrogen recombination lines and far-infrared ionic fine-structure lines originating in the HII region, as well as those originating in the photodissociation region (see Chapter 9). The jump in the spectrum reflects the difference in aperture between the short wavelength and long wavelength spectrometer and the extent of the source on the sky. Figure reproduced with permission from E. Peeters, *et al.*, 2002, *A. & A.*, **381**, p. 571.

7.4.1 H recombination lines

Recombination of an electron into any excited level of H sets up a radiative cascade, which ends with the hydrogen atom in the ground state (Chapter 2). Thus, for example, initial capture into level $n = 6$ can be followed by decay into level $n = 3$ with emission of a Paγ photon and then that level can decay through Lyβ emission to the ground state ($n = 1$) or, alternatively, through Hα emission to $n = 2$ and then through Lyα to $n = 1$. If we assume that all photons escape, this is just bookkeeping and the branching ratios of these different pathways can be evaluated once and for all using the appropriate Einstein A coefficients. This is called case A recombination.

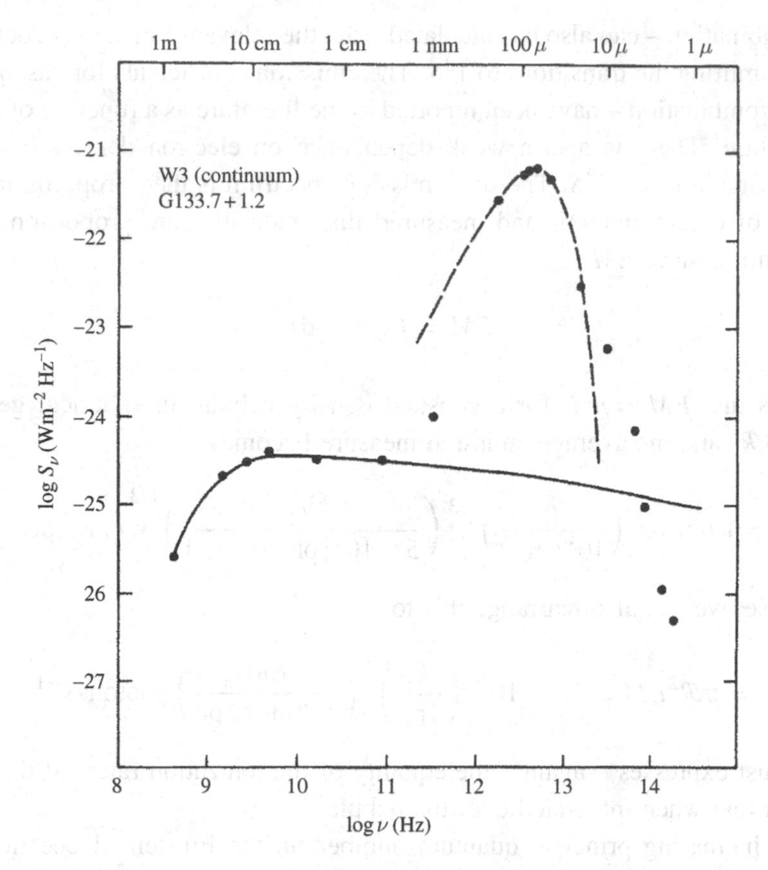

Figure 7.12 The radio–infrared spectrum of the HII region, W3A. The infrared continuum spectrum, peaking around 100 μm (dashed line), is due to dust. The free–free emission at radio wavelengths (solid line) shows the well-known optically thin behavior at long wavelengths (flux $\sim \nu^2$) and the optically thick behavior for shorter wavelengths (flux $\sim \nu^{-0.1}$). The observations drop below the extrapolated free–free (and free–bound) continuum at the shortest wavelength owing to dust absorption. Figure reproduced with permission from G. Wynn Williams, 1974, *I. A. U. S.*, **60**, p. 259.

However, in general, line photons in the Lyman series (i.e., ending in $n = 1$) produced in the cascade do not escape the nebula (cf. Section 7.2.4). Those photons are readily absorbed by neutral H atoms in the nebula, which are essentially all in the ground state. The optical depths of these lines are orders of magnitude larger than the optical depths in the ionizing Lyman continuum (e.g., $\tau(\mathrm{Ly}\alpha) \sim 10^6$ for $n = 10^3\ \mathrm{cm}^{-3}$). After re-absorption, the excited H atom may decay through one of the other branching pathways. Because of the large number of scatterings involved, the upshot will be that any Lyman photon other than Lyman α will be converted into a lower series photon plus a Lyman α photon or a photon in the two-photon continuum ($2^2S \rightarrow 1^2S$). The recombination spectrum in this limit – called case

B recombination – can also be calculated from the relevant Einstein A coefficients by just omitting the transitions to 1^2S. The emission coefficients for case A and B – per recombination – have been reported in the literature as a function of electron temperature. There is also a weak dependence on electron density because of the metastable level, 2^2S. The line emission spectrum is then proportional to the number of recombinations and measured line intensities are proportional to the emission measure, EM,

$$EM = \int_0^L n_e n_p \mathrm{d}\ell, \tag{7.62}$$

which is just $EM = n^2 L$ for a constant density nebula. In spherical geometry, $L = 4/3\mathcal{R}_s$ and the average emission measure becomes

$$EM = 1.6 \times 10^6 \left(\frac{n}{10^3\,\mathrm{cm}^{-3}}\right)^{2/3} \left(\frac{\mathbb{N}_{\mathrm{Lyc}}}{5 \times 10^{49}\,\mathrm{photons\,s}^{-1}}\right)^{1/3} \mathrm{cm}^{-6}\,\mathrm{pc}. \tag{7.63}$$

Of course, we can also rearrange this to

$$\mathbb{N}_{\mathrm{Lyc}} = \pi \mathcal{R}_s^2 EM \simeq 2.4 \times 10^{49} \left(\frac{\mathcal{R}_s}{1\mathrm{pc}}\right)^2 \left(\frac{EM}{10^6\,\mathrm{cm}^{-6}\,\mathrm{pc}}\right) \mathrm{photons\,s}^{-1}, \tag{7.64}$$

which just expresses – again – the equality of the ionization rate and the recombination rate when integrated over the nebula.

With increasing principal quantum number, n, the Einstein A coefficients of the Lyman lines decrease and, hence, the optical depth of the transition decreases. Eventually, for a high enough principal quantum number, n, the recombination cascade will follow case A rather than case B. Of course, the presence of dust can also influence this redistribution of the cascade. For very high densities, the optical depth in the Balmer lines may also become appreciable and the cascade may be further redistributed. A similar principle applies and the cascade can again be calculated in this case (sometimes called case C). For high levels, the Einstein A coefficients decrease (cf. Section 2.1.4) and collisions eventually take over. First, collisions that only change the angular momentum (i.e., L) but keep the principal quantum number (n) the same become important for $n > 15$ for a density of $10^4\,\mathrm{cm}^{-3}$. For higher densities, collisions that change n also become important.

7.4.2 Collisionally excited line radiation

Collisionally excited line radiation of trace species has been discussed extensively in Section 2.4. It is of major importance for the spectral characteristics of HII regions, as is obvious from the cooling discussion (cf. Fig. 7.8).

7.4.3 Radio emission

The ionized plasma in the HII region emits a free–free spectrum. For electrons accelerated in collisions with protons, the emission coefficient is given by

$$\begin{aligned} j(\nu) &= \frac{16e^6}{3m_e^2c^3}\left(\frac{\pi m_e}{6kT_e}\right)^{1/2} g_{ff}\exp\left[-\frac{h\nu}{kT_e}\right] n_e n_p \\ &\simeq 5.4\times10^{-41}\left(\frac{10^4\,\mathrm{K}}{T_e}\right) g_{ff}\, n_e n_p\,\mathrm{erg\,cm^{-3}\,s^{-1}\,sr^{-1}\,Hz^{-1}}, \end{aligned} \tag{7.65}$$

with g_{ff} the free–free gaunt factor, which is a slowly varying function of T_e and ν, given by

$$g_{ff} = 6.9 + 1.90\log\left(\frac{T_e}{10^4\,\mathrm{K}}\right) - 1.26\log\left(\frac{\nu}{\mathrm{GHz}}\right). \tag{7.66}$$

The corresponding free–free absorption coefficient is

$$\kappa(\nu) = \frac{8e^6}{3hm_e^2c\nu^3}\left(\frac{\pi m_e}{6kT_e}\right)^{1/2} g_{ff}\,(1-\exp[-h\nu/kT_e])\, n_e n_p. \tag{7.67}$$

The factor in brackets is the correction for stimulated emission and, at radio frequencies, this is $h\nu/kT_e \ll 1$; i.e., the correction for stimulated emission nearly balances the true absorptions.

In the radio regime, the optical depth is then given by

$$\tau(\nu) = 3.3\times10^{-7}\left(\frac{10^4\,\mathrm{K}}{T_e}\right)^{1.35}\left(\frac{1\,\mathrm{GHz}}{\nu}\right)^{2.1} EM, \tag{7.68}$$

where a power law has been fitted to the temperature and frequency dependence of the gaunt factor, and the emission measure, EM, is given by Eq. (7.62) in units of $\mathrm{cm^{-6}}$ pc. Thus, contrary to most other emission mechanisms, at low frequencies ionized gas nebulae are optically thick in their free–free continuum while at higher frequencies they are optically thin. This just reflects the coulomb interaction, which prefers low collision velocities and, hence, low interaction energies.

For a homogeneous nebula (i.e., constant density and temperature), the spectrum is given by

$$I(\nu) = B(\nu, T_e)(1-\exp[-\tau(\nu)]). \tag{7.69}$$

Define the brightness temperature, T_b, by

$$I(\nu) = \frac{2\nu^2}{c^2}kT_b(\nu). \tag{7.70}$$

Then, for radio waves ($h\nu/kT \ll 1$), the intensity can be simplified to

$$T_b(\nu) = T_e(1-\exp[-\tau(\nu)]) \tag{7.71}$$

and at high optical depth, $T_b \simeq T_e$, while at small optical depth, $T_b \simeq T_e\tau$. Thus, at high frequencies ($\tau \ll 1$), the intensity scales approximately with $\nu^{-0.1}$ and the brightness temperature with $\nu^{-2.1}$, while at low frequencies ($\tau \gg 1$), the intensity scales with ν^2 and the brightness temperature is independent of ν (cf. Fig. 7.12).

7.5 Comparison with observations

There are many ways in which the physical conditions in HII regions can be determined. Each of these has some advantages and some drawbacks. In particular, realistic nebulae will show density and temperature structure and every method averages over these differently.

7.5.1 Electron density

There are several ways to determine the gas density in HII regions. First, the density can be estimated from recombination lines in the optical (i.e., Hα at 6562 Å), infrared (i.e., Brα at 4.05 μm), or radio (i.e., H109α or H137β near 6 cm) or from the free–free continuum in the radio. The strength of these directly yields the emission measure, *EM* (cf. Eq. (7.62)). For a homogeneous nebula, the emission measure can be translated into a density by adopting a scale size,

$$n_{\rm rms} = \left(\left(\frac{EM}{{\rm cm}^{-6}\,{\rm pc}}\right)\left(\frac{\rm pc}{L}\right)\right)^{1/2}. \tag{7.72}$$

The optical and IR lines have to be corrected for dust extinction first. For inhomogeneous nebulae, the emission measure is of course weighted towards higher density regions.

The density can also be estimated by searching for the effects of collisional de-excitation. Consider two levels within the same term (i.e., with almost the same excitation energy) but with different Einstein *A* coefficients or collisional de-excitation rates. The excitation of these levels will then depend differently on the density but not on the temperature. The ratio of the intensities of the lines that originate from these levels are then a good measure of the density in the range between the two critical densities. Line pairs that are often used for this are the [OII] 3729/3726 Å and [SII] 6716/6731 Å lines in the optical and the [OIII] 88/52 μm, [SIII] 33/19 μm, [NeIII] 36/15 μm, and [ArIII] 22/9 μm lines in the infrared. Figure 7.13 illustrates this for the infrared fine-structure lines. For densities much less than the critical density of either line, the intensities of both lines scale with density squared and their ratio is constant ($I_{10}/I_{21} = (\Omega_{01}/\Omega_{02}+1)\,E_{10}/E_{21}$). For high densities, both intensities scale with density and, again, the ratio is independent of the density ($I_{10}/I_{21} = g_1A_{10}E_{10}/g_2A_{21}E_{21}$).

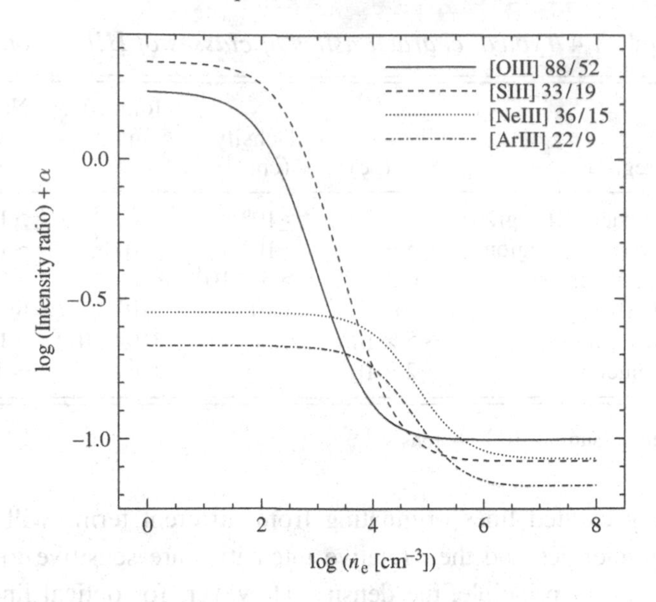

Figure 7.13 The calculated ratio of the intensities of the fine-structure lines of [OIII], [SIII], [NeIII], and [ArIII] in the IR as a function of density. The results for [NeIII] and [ArIII] have been displayed adopting $\alpha = 0.5$.

In between the two critical densities, the ratio is a good measure of the density (cf. Section 2.3). For inhomogeneous nebulae, this method tends to favor the density regime within this range and denser regions will be missed.

HII regions show a wide range in physical conditions. Table 7.4 summarizes typical values for the density and sizes of various classes of HII regions derived from observations. The ionized masses and an estimate of the number of stars involved are also indicated. Part of this diversity in conditions reflects rather directly the interaction of a massive star with its environment. Of importance in this are the high pressures associated with ionized gas – which lead to expansion (Section 12.2) – as well as the powerful winds emanating from such stars – clearing out the environment (Section 12.5). Furthermore, massive stars are often formed in associations and the cumulative effect of these stars is important in shaping the overall interstellar medium of galaxies (see Chapter 8).

7.5.2 Electron temperature

The electron temperature can be determined from the optically thick portion of the radio spectrum, which, in brightness temperature, yields the electron temperature directly. Different density (and, to a lesser extent, temperature) regimes can be discerned with observations at different wavelengths.

Table 7.4 *Typical characteristics of classes of HII regions*

Type of region	Size (pc)	Density (cm^{-3})	Ionized mass ($M_{\odot}$)	Number of ionizing stars[a]
Hypercompact HII region	$\sim 3\times 10^{-3}$	$\geq 10^{6}$	$\sim 10^{-3}$	$\simeq 1$
Ultracompact HII region	$\leq 5\times 10^{-2}$	$\geq 10^{4}$	$\sim 10^{-2}$	~ 1
Compact HII region	≤ 0.5	$\geq 5\times 10^{3}$	~ 1	~ 1
Classical HII regions	~ 10	$\sim 10^{2}$	$\sim 10^{5}$	few
Giant HII regions	$\sim 5\times 10^{1}$	~ 30	$10^{3}-10^{6}$	$\sim 10^{2}$
Starburst nuclei	$\sim 2\times 10^{2}$	~ 10	$10^{6}-10^{8}$	$\sim 10^{3}$

[a] Calculated assuming 10^{48} photons s^{-1}.

Collisionally excited lines originating from different terms will have different excitation energies and their relative intensities are sensitive to the electron temperature and, in principle, the density. However, for optical lines with high critical densities, we can assume the low-density limit (Section 2.3) and these lines become a good measure of the temperature. The [OIII] 4363/(4959+5007) Å lines are often used in this way. The energy levels involved are illustrated in Fig. 2.7. Figure 7.14 gives the ratio of these lines as a function of the temperature.

7.5.3 *The stellar effective temperature*

The energy dependence of the ionizing radiation field can be constrained by measuring the intensity ratio of emission lines from sequential stages of ionization of

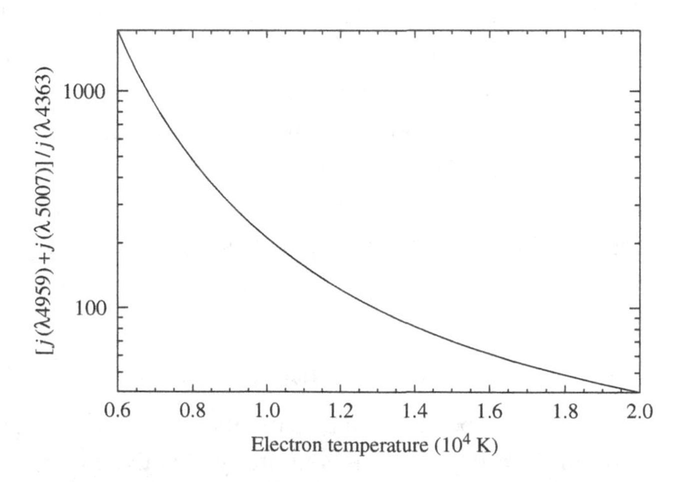

Figure 7.14 The ratio of the intensities of the optically forbidden lines of [OIII] as a function of electron temperature.

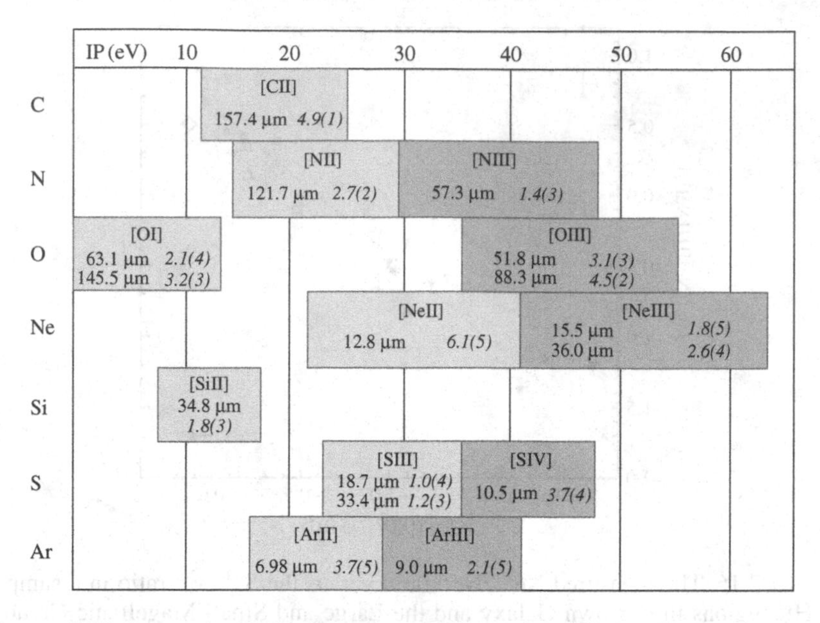

Figure 7.15 Infrared atomic-fine-structure lines for different ionization stages of some elements. Along the horizontal axis, the relevant range of the ionization potential is indicated. The critical density for each transition, indicated in italics, is given in units of cm^{-3} and is expressed as $a(b) = a \times 10^b$. Underneath each line, wavelength and critical density of each transition is indicated. Figure reproduced with permission from N. L. Martín-Hernández, *et al.*, 2002, *A. & A.*, **381**, p. 606.

a single element. Figure 7.15 illustrates various ion pairs with infrared lines that can be used this way to probe different portions of the stellar ionizing flux. Of course, the intensity of a line pair can also depend on density and this can hamper the analysis. It is then better to concentrate on line pairs for which it can be assumed that they are in either the high- or the low-density limit (Section 2.3). Figure 7.16 plots the observed Ne^{2+}/Ne^{+} ratio as a function of the observed S^{3+}/S^{2+} ratio. The neon ratio is sensitive to the ionizing flux above 40 eV relative to that below 40 eV. The sulfur ratio measures the SED above and below 35 eV. The good correlation illustrates that the ionizing radiation field dominates the extent of the ionization structure. The dependence on galactocentric distance evident in the observations (Fig. 7.16) may reflect a dependence of the stellar ionizing spectrum on elemental abundances, which are known to vary across the Galaxy.

For optical and far-red studies, one often relies on the double ratio,

$$\eta = \frac{[\mathrm{OII}]3727/[\mathrm{OIII}]4959, 5007}{[\mathrm{SII}]6717, 6730/[\mathrm{SIII}]9068, 9532}, \tag{7.73}$$

which is sensitive to the slope of the stellar ionizing photon flux between the ionization potentials of S^{+} and O^{+}, but not the total ionizing photon flux (or more precisely, not to the ionizing photon flux relative to the density).

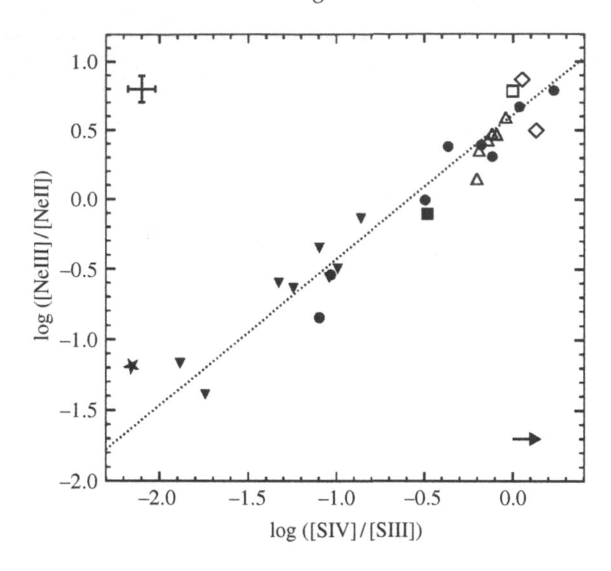

Figure 7.16 The measured Ne^{2+}/Ne^{+} ratio versus the S^{3+}/S^{2+} ratio in a sample of HII regions in our own Galaxy and the Large and Small Magellanic Clouds. Note, besides the good correlation between these two ionization ratios, the dependence on the elemental abundance. Regions with lower elemental abundances (e.g., the SMC, LMC, and the outer Galaxy) have preferentially higher ionization ratios. Figure adapted with permission from N. L. Martín-Hernández, *et al.*, 2002, *A. & A.*, **389**, p. 286.

The total number of ionizing photons can be determined from the intensity of recombination lines or the free–free continuum. Integrated over the nebula, these are a direct measure of the total number of recombinations and, hence, the total ionizing flux. These can be compared to model atmosphere calculations, which yield the number of ionizing photons as a function of effective temperature (cf. Table 7.1). With a measurement of the flux from the star at a given wavelength, and assuming a stellar model atmosphere (i.e., the stellar flux at that wavelength as a function of the effective temperature of the star), the effective temperature can be determined. This method was first employed by Zanstra, who adopted a black body for the stellar radiation field, and the effective temperature measured this way is often called the Zanstra temperature. For a homogeneous nebula, dust extinction corrections will be small if the line and stellar continuum are measured at nearly the same wavelength. However, many nebulae show a patchy extinction distribution and infrared measurements may then be preferred. Of course, this method will fail if a large fraction of the ionizing photons is absorbed by dust.

Stars are far from black bodies in their ionizing continua due to the effects of line blanketing. Moreover, their spectra may change considerably during stellar evolution, are dependent on the strength of their winds, and, furthermore, depend on metallicity. This area of research is still very much in flux. Figure 7.17

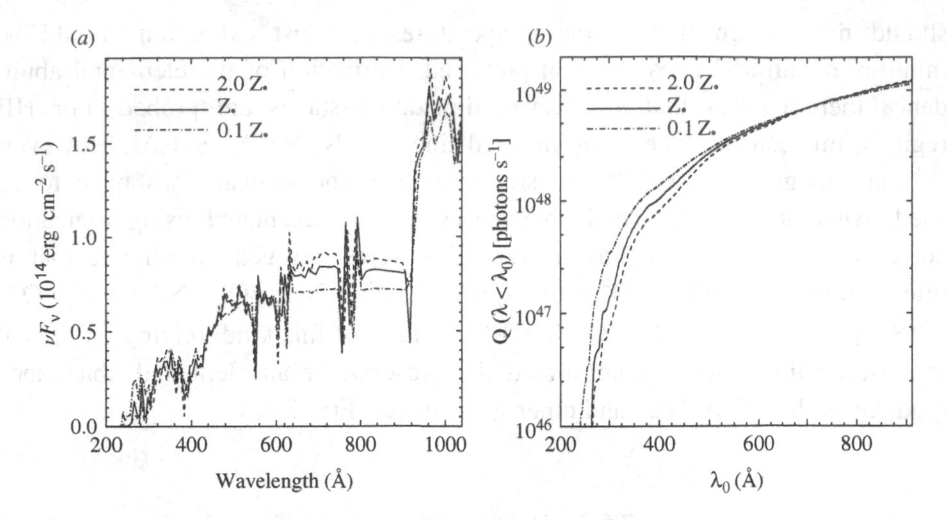

Figure 7.17 (*a*) Non-LTE unified photosphere/wind (CMFGEN) model for an O6 dwarf ($T_{\rm eff} = 41\,010$, $\log g = 4.014$, $L = 2.54 \times 10^5$ $L_\odot$) for three different metallicities. All spectra have a strong metal-line forest contributing to the absorption, in particular below 700 Å. To bring out the overall effect of these line forests the spectra have been re-binned. The prominent features around 550 and 600 and between 700 and 800 Å are blends of wind lines from different ionization stages of carbon, nitrogen, oxygen, and silicon. (*b*) The cumulative photon luminosity for wavelengths smaller than λ_0 as a function of λ_0. Figure adapted from R. Mokiem, *et al.*, 2004, *A. & A.*, **419**, p. 319.

illustrates some of these points based upon stellar atmosphere models, which include the effects of stellar winds. The effects of the line forest are somewhat masked owing to the re-binning, which brings out the overall effect on the flux in the various spectral regions better. The lines, which are still obvious, originate in the wind of this star. We see that as the metallicity increases, the shorter wavelength flux decreases because of the line blanketing and this flux gets redistributed to longer wavelengths but still remains in the ionizing continuum. The effect on the total H-ionizing photon luminosity is small. But the ionizing photon luminosity at the shortest wavelengths is very sensitive. It should be emphasized that the only way to probe these details of the stellar spectral energy distributions is through observations of the strength of lines of atoms that are ionized mainly by photons in this wavelength region; e.g., Ne^{2+}/Ne^{+} and S^{3+}/S^{2+}.

7.5.4 Elemental abundances

Ionic abundances can be determined from the ratio of ionic lines to a hydrogen recombination line or the radio continuum. To limit collisional de-excitation effects, the lines should preferably be in the low-density limit. Also, the line should not be sensitive to the temperature and dust extinction should be minimized. Infrared lines are then preferred. Derivation of the elemental abundance then requires that all relevant ionization stages are probed. For HII regions, this can be done using infrared lines for N, Ne, S, and Ar. However, O^+ has no ground state IR fine-structure lines and optical lines have to be used. Alternatively, elemental abundances can be calculated using ionization correction factors for the unseen ionization stages. A general scheme that is often adopted is $O/H = (O^+ + O^{2+})/H^+$, $Ne/O = Ne^{2+}/O^+$, $N/O = N^+/O^+$, $S/(S^+ + S^{2+}) = 1/\left(1 - (1 - O^+/O)^3\right)^{1/3}$. Studies of this kind, relying on optical or infrared lines, have demonstrated the presence of an elemental abundance gradient in the Milky Way and other galaxies (cf. Fig. 7.18).

7.5.5 Dust in HII regions

Most HII regions show a substantial IR excess. This IR emission is due to warm dust grains heated by EUV and FUV stellar radiation and trapped Lyman alpha photons and other nebular emission. For compact HII regions, much of the dust continuum, particularly at the longer wavelengths, originates from outside the ionized volume. It seems that in this stage, some of the dust has been cleared

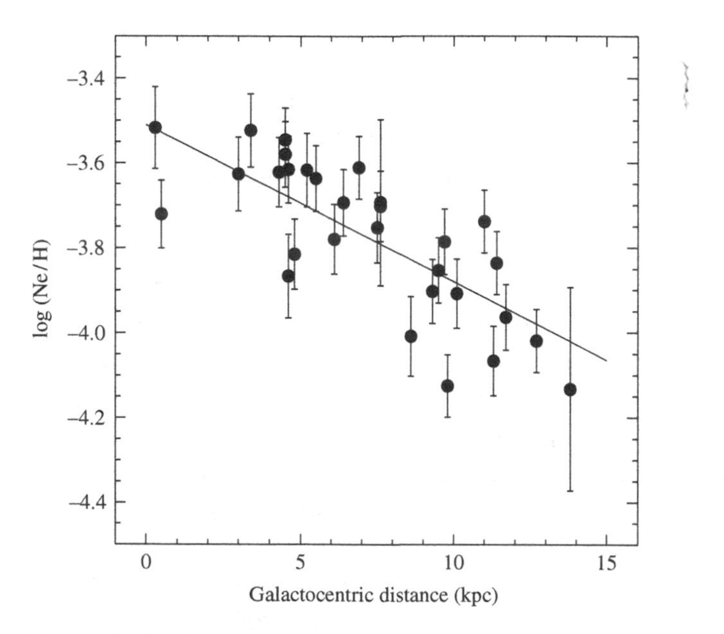

Figure 7.18 The neon abundance gradient in the Milky Way determined from the 12.8 μm [NeII] and 15.5 μm [NeIII] fine-structure lines of HII regions. The Solar Ne abundance (log(Ne/H) is −3.9. Figure reproduced with permission from N. L. Martín-Hernández, *et al.*, 2002, *A. & A.*, **381**, p. 606.

out of the ionized gas volume, either through radiation pressure or through actual destruction by the ionized gas. While not well determined, a value of $\delta_{\rm d} \simeq 0.1$ seems to be in line with observations. The peak of the emission shifts to shorter wavelengths for ultracompact HII regions. During this phase, much of the stellar radiation, EUV and FUV, may be absorbed by dust in the ionized volume.

7.6 Further reading

An excellent, in-depth discussion of the physics of ionized gas is provided by the textbook [1]. In many ways, this chapter is adapted from this book. This book also provides entries into the literature.

Stellar parameters of O and B stars have been taken from [2]. Stellar model atmospheres are provided by [3], [4], and [5].

The classic paper on the ionization structure of HII regions is [6]. Important contributions were also made in [7]. An early compilation of relevant results for the physical processes in ionized gas is [8]. The discussion on the effects of dust on the ionization structure of HII regions was taken from [9]. Detailed grids of models for HII regions are provided by [10] and [11].

All the calculations on the structure of HII regions were made using the ionization code, Cloudy, by N. L. Martín-Hernández. This code is maintained by G. Ferland and is available at

http://www.nublado.org/.

A discussion of various aspects of this code is provided in [12]. An IDL environment for this code is available at

http://isc.astro.cornell.edu/~spoon/mice.html/,

which is maintained by H. Spoon. A compilation of atomic data – including photo-ionization cross section, radiative and dielectronic recombination coefficients, collisional ionization rate coefficients, energy levels, collision strength, and excitation rate coefficients – is available at

http://www.pa.uky.edu/~verner/atom.html,

which is maintained by D. Verner. A compilation of databases for atomic physics is also to be found at

http://plasma-gate.weizmann.ac.il/DBfAPP.html,

which is maintained by Y. Ralchenko. Of course, there is also the NIST site,

http://www.nist.gov/srd/atomic.htm.

The results of the opacity project – an international collaboration to calculate the extensive atomic data required to estimate stellar envelope opacities and to compute Rosseland-mean opacities and other related quantities – can be accessed through their website,

http://vizier.u-strasbg.fr/OP.html#OPP.

A convenient tabulation of case B HI and HeI recombination line intensities is provided by [13].

The galactic elemental abundance gradients and their implications for galactic chemical evolution are discussed in the monograph [14].

References

[1] D. Osterbrock, 1989, *Astrophysics of Gaseous Nebulae and Active Galactic Nuclei* (Mill Valley: University Science Books)
[2] W. D. Vacca, C. D. Garmany, and J. M. Shull, 1996, *Ap. J.*, **460**, p. 914
[3] D. J. Hillier and D. L. Miller, 1998, *Ap. J.*, **496**, p. 407
[4] A. W. A. Pauldrach, T. L. Hoffmann, and M. Lennon, 2001, *A. & A.*, **375**, p. 161
[5] D. Schaerer and A. de Koter, 1997, *A. & A.*, **322**, p. 598
[6] B. Strömgren, 1939, *Ap. J.*, **89**, p. 529
[7] D. G. Hummer and M. J. Seaton, 1963, *M. N. R. A. S.*, **125**, p. 437; 1964, *M. N. R. A. S.*, **127**, p. 217
[8] D. H. Menzel, 1962, *Selected Papers on Physical Processes in Ionized Nebulae* (New York: Dover)
[9] V. Petrosian, J. Silk, and G. B. Field, 1972, *Ap. J.*, **177**, p. L69
[10] R. H. Rubin, 1985, *Ap. J. S.*, **57**, p. 349; 1988, *Ap. J. S.*, **68**, p. 519
[11] G. Stasinska, 1982, *A. & A. S.*, **48**, p. 299; 1990, *A. & A. S.*, **83**, p. 501
[12] G. J. Ferland, 2003, *Ann. Rev. Astron. Astrophys.*, **41**, p. 517
[13] D. G. Hummer and O. J. Storey, 1987, *M. N. R. A. S.*, **224**, p. 801
[14] B. E. J. Pagel, 1997, *Nucleosynthesis and Chemical Evolution of Galaxies* (Cambridge: Cambridge University Press)

8
The phases of the ISM

8.1 Introduction

The ISM contains a number of phases characterized by different temperatures, densities, and ionization fractions (cf. Section 1.2 and Table 1.1). The origin and interrelationship of these phases in the ISM, and their energy and ionization sources, are among the most fundamental subjects of investigation in the field. Quite generally, a new stable phase reflects the onset of a new cooling mechanism or the decline of a heating source. Hence, cold HI clouds and the warm intercloud medium result from the increased importance of [CII] cooling at higher densities and Lyα and [OI] 6300 Å cooling at higher temperatures, respectively (cf. Section 2.6.1). The hot phase reflects the recent input of supernova energy (cf. Section 12.3). Cold molecular clouds result from the increased cooling due to rotational transitions in molecules (cf. Section 2.6.2). However, the latter fall somewhat outside this classification scheme since their existence also reflects the importance of self-gravity. Understanding the structure of the ISM thus requires an understanding of its sources of heating and cooling.

Heating and cooling processes have been discussed in Chapters 3 and 2, respectively. Here, we apply these to the neutral diffuse interstellar medium. We will first examine the ionization and energy balance of the cold and warm phases of the interstellar medium (Section 8.2). Then, we will focus on the physical principles that allow several phases to coexist in thermal and pressure equilibrium in the interstellar medium (Section 8.3). This is the basis of the two-phase model for the interstellar medium that explains the cold neutral medium and the warm neutral medium. We then make an excursion, allowing ionizing photons to enter the mix, to create the warm ionized medium (Section 8.4). The two-phase model does not account for mechanical energy input from stars (e.g., supernova explosions), which drives the interstellar medium away from equilibrium, dominating the vertical structure of the ISM and introducing a new phase consisting of tenuous hot gas, the hot intercloud medium (Section 8.5). This constant stirring up of the interstellar medium by exploding supernovae forms the basis for the three-phase model for the interstellar medium where the cold neutral medium, the warm intercloud media, and the hot intercloud medium coexist in steady state. Correlated supernova explosions in an OB association can lead to the formation of superbubbles, which may blow out of the plane setting up a circulation with the lower halo or even venting all of their hot gas into high altitudes forming a hot corona around the Galaxy. These non-steady-state aspects are also covered in Section 8.5. A short summary is provided in Section 8.6. The chemistry of the diffuse interstellar medium (e.g., of diffuse interstellar clouds) is discussed in Section 8.7. This chapter closes with a discussion of the observations of the various phases of the diffuse interstellar medium.

8.2 Physical processes in atomic gas

As discussed in Chapter 7, The EUV photons from hot stars can keep hydrogen largely ionized in HII regions with a very sharp transition to neutral gas. Conversely, there are essentially no EUV photons outside HII regions and photo-ionization and heating are due to stellar FUV photons with $h\nu < 13.6\,\text{eV}$. In this chapter, we are concerned with the gas that is located outside these HII regions and that is consequently largely neutral. Some electrons are present because of the ionization of trace species with ionization potentials below 13.6 eV (cf. Table 8.1). These include C, S, and Si; because of abundance considerations, C is the most important one. Some ionization of H can result from cosmic rays or X-rays. The gas is again heated through photo-electrons but now from trace species – these atoms as well as molecules made from these atoms – and this photo-electric heating is much less efficient than in HII regions (Sections 3.2 and 3.3). Some heating also results from cosmic-ray and X-ray interaction with the gas (Sections 3.6 and 3.7). This gas can cool through fine-structure transitions in trace species and through the HI Lyman α and the [OI] 6300 Å lines (cf. Section 2.5).

8.2.1 The ionization balance

We will focus first on the degree of ionization in diffuse clouds. Elements with an ionization potential less than 13.6 eV can still be ionized by non-hydrogen-ionizing photons. We can write for the ionization rate of species i,

$$n_i\, k_{\text{ion},i} = 4\pi n_i \int_{\nu_i}^{\nu_\text{H}} \mathcal{N}_\text{ISRF}(\nu)\alpha_i(\nu)\text{d}\nu, \qquad (8.1)$$

where $\mathcal{N}_\text{ISRF}$ is the incident interstellar radiation field and the integral runs from the ionization limit of species i to the H-ionization limit. Because in this case

Table 8.1 *Ionization of trace species*[a]

Element	IP (eV)	Ionization rate[a] (10^{-10} s^{-1})		Recombination rate[b] (10^{-12} cm^3 s^{-1})	
		a	b	α	β
C	11.3	2.1	2.6	9.5	−0.6
S	10.4	3.9	2.5	7.8	−0.63
Si	8.2	12.0	1.6	9.5	−0.6
Fe	7.9	1.0	2.3	7.6	−0.65
Mg	7.6	0.4	1.4	7.2	−0.86
Ca	6.1	3.1	1.9	6.9	−0.9
Ca^+	11.9	0.02	2.9	27.0	−0.8
Na	5.1	0.06	2.0	5.8	−0.69
K	4.3	0.4	1.8	11.0	−0.8

[a] Photo-ionization rates are given as $k_{ion} = a\exp[-bA_v]$, with a the photo-ionization rate in the average interstellar radiation field, and the exponential factor reflects dust extinction into a cloud at a depth corresponding to a visual extinction of A_v.

[b] Recombination rates are given as $k_{rec} = \alpha(T/100)^{\beta}$.

dust extinction into a cloud can be important, the ionization rate coefficient is generally written as (cf. Section 4.1)

$$k_{ion,i} = G_0 a_i \exp[-b_i A_v], \tag{8.2}$$

where the a_i and b_i coefficients represent the ionization rate in the unshielded average interstellar radiation field and the depth dependence of the radiation field due to dust attenuation, respectively. In this equation, G_0 is the intensity of the radiation field in Habing units of the average interstellar radiation field (Section 1.3.1). Values for a_i and b_i are given in Table 8.1. In atomic H regions, this ionization is balanced by radiative recombination (Table 8.1). Because of abundance considerations, neutral carbon dominates the ionization balance. As for H in HII (cf. Section 7.2), we can write for the degree of C ionization, x, in HI regions

$$\frac{1-x}{x^2} = \frac{\mathcal{A}_C n k_{rec}(T)}{k_{ion}} \simeq 3.2\times10^{-4}\left(\frac{1}{G_0}\right)\left(\frac{n}{50\,\mathrm{cm}^{-3}}\right)\left(\frac{100\,\mathrm{K}}{T}\right)^{0.6}, \tag{8.3}$$

where we have ignored other sources of ionization and assumed that all electrons originate from C ionization. This equation is of course completely analogous to the H-ionization equation in HII regions (cf. Eq. (7.9)). Thus, carbon will be largely ionized in HI clouds with a very small neutral carbon fraction. Because the fraction of heavy elements in solids is comparable to that in the gas, absorption of ionizing radiation by dust is important; much more so than for HII regions. Nevertheless, the ionized-to-neutral transition is still very sharp. Similar considerations apply to other trace species. Hence, there will be sequential zones in a diffuse cloud where the role of C^+ as the dominant cation is taken over, first by S and then by Si, Fe, Mg, ...

The ionization balance of atomic gas can be affected by the presence of large molecular anions because such species provide a non-radiative recombination channel that is much faster than the radiative recombination process of atomic cations with electrons (Section 6.3.5). The ionization "cycle" is now as follows: ionization of neutral carbon produces a carbon cation and an electron. Some of these electrons associatively attach to PAHs. PAH anion formation is balanced by photo-ionization. Recombination of the carbon cations with PAH anions then closes the loop. A very similar ionization–recombination cycle is described in Section 10.5 for molecular clouds. There are now two coupled equations (for carbon and for PAHs) that describe the degree of ionization of atomic gas. Combining these equations yields

$$\frac{1-x}{x^2} = \frac{\mathcal{A}_{\rm C}\mathcal{A}_{\rm PAH}n^2 k_{\rm ea}k_{\rm n}}{k_{\rm ion}({\rm C})J_{\rm pe}(-1)}, \tag{8.4}$$

where $\mathcal{A}_{\rm PAH}$ is the PAH abundance by number, $k_{\rm ea}$ the electron attachment rate coefficient for PAHs (Section 6.3.4), $k_{\rm n}$ the C^+-PAH^- neutralization rate coefficient (Section 6.3.5), and $J_{\rm pe}$ the PAH anion photo-ionization rate (Section 6.3.2). Inserting numerical values (with $\phi_{\rm PAH} = 0.5$), and adopting a PAH abundance of 4×10^{-7}, then results in

$$\frac{1-x}{x^2} \simeq 2.2 \times 10^{-3} \left(\frac{1}{G_0}\right)^2 \left(\frac{n}{50\,{\rm cm}^{-3}}\right)^2 \left(\frac{100\,{\rm K}}{T}\right)^{1/2}. \tag{8.5}$$

Thus, because of the presence of large molecules, the neutral fraction will typically be a factor of 7 larger than calculated above (cf. Eq. (8.3)).

For low column densities ($\lesssim 10^{19}$ cm^{-2}), penetrating soft X-rays can also contribute to the degree of ionization (of H; e.g., Section 3.7). In that case, ignoring other ionization processes, we can write for the ionization balance of hydrogen

$$n_{\rm H}\xi_{\rm XR}(N_{\rm w}) = n_{\rm e}n_{\rm p}\beta_{\rm B}(T), \tag{8.6}$$

where $\xi_{\rm XR}(N_{\rm w})$ is the X-ray ionization rate at a hydrogen column of $N_{\rm w}$ (cf. Section 3.7) and $\beta_{\rm B}(T)$ is the H^+ radiative recombination rate coefficient to all levels with $n \geq 2$ (cf. Chapter 7). We can write this as

$$\frac{x^2}{1-x} = \frac{\xi_{\rm XR}(N_{\rm w})}{n\beta(T)}, \tag{8.7}$$

which, realizing that the degree of ionization will be small ($x \ll 1$), can be written as

$$x \simeq 5.1 \times 10^{-4} \left(\frac{\xi_{\rm XR}}{10^{-16}\ {\rm s^{-1}\ (H\ atom)^{-1}}} \right)^{1/2} \left(\frac{50\ {\rm cm^{-3}}}{n} \right)^{1/2} \left(\frac{T}{100\,{\rm K}} \right)^{0.35}, \tag{8.8}$$

where an X-ray ionization rate of 10^{-16} s^{-1} (H atom)$^{-1}$ is appropriate for an absorbing foreground column of 10^{19} H atoms cm^{-2}. Thus, for small columns, X-ray ionization of H can appreciably increase the degree of ionization over that from the FUV ionization of carbon.

Soft X-rays do not penetrate deeply into clouds and the flux of hard X-rays is small in the general diffuse ISM. Cosmic-ray particles, on the other hand, can contribute to the ionization throughout a diffuse interstellar cloud. Like the X-ray ionization, ignoring other sources of ionization, we can write for the degree of ionization by cosmic rays

$$\begin{aligned} x &= \left(\frac{\zeta_{\rm CR}}{n\beta_{\rm B}(T)} \right)^{1/2} \\ &\simeq 2 \times 10^{-3} \left(\frac{\zeta_{\rm CR}}{2 \times 10^{-16}\ {\rm s^{-1}\ (H\ atom)^{-1}}} \right)^{1/2} \\ &\quad \times \left(\frac{50\ {\rm cm^{-3}}}{n} \right)^{1/2} \left(\frac{T}{100\,{\rm K}} \right)^{0.35}, \end{aligned} \tag{8.9}$$

with $\zeta_{\rm CR}$ the primary cosmic-ray ionization rate. Thus, with this high ionization rate (cf. Section 8.8), cosmic-ray ionization dominates the charge balance. Of course, in the evaluation of the ionization balance all relevant processes need to be included simultaneously. As for the C-ionization balance, the presence of large molecular anions can affect the ionization balance for H. We leave that as an exercise for the reader.

In the warm neutral medium, ionization is dominated by cosmic-ray ionization. The calculated degree of ionization, $x \simeq 0.1$, is a factor of 50 higher than for diffuse clouds ($n \simeq 0.5$ cm^{-3}, $T \simeq 8000$ K).

8.2.2 The energy balance

Cooling of atomic gas has been discussed in detail in Chapter 2. The cooling curve is summarized in Section 2.5, Fig. 2.10. The [CII] 158 μm fine-structure line is the dominant coolant at low temperature in diffuse clouds, while the Lyman alpha line dominates in the warm neutral medium.

The characteristics of various possible energy sources – FUV photons, cosmic rays, turbulence, magnetic fields, and the cosmic background radiation – for the interstellar gas are summarized in Chapter 1, Table 1.2, and in Section 3.11. The energy densities – stellar photons, cosmic rays, turbulence – are typically of the same order but their contribution to the energy balance of the neutral ISM depends on their coupling to the gas. Of these energy sources, the interstellar FUV field is the most important and the coupling process is the ionization of trace elements and of PAHs and dust grains.

As emphasized in Section 3.11.1, ionization of neutral carbon is not an important energy source for diffuse clouds. Following the discussion on HII regions, we can put that on a quantitative footing by evaluating the temperature in a CII zone by solving the energy balance due to C-ionization heating (Section 3.2) and [CII] 158 μm cooling (Section 2.5). Neglecting all other heating or cooling processes, and assuming H atom excitation of the C^+ fine-structure levels, the temperature follows from

$$\exp\left[\frac{-92\,\mathrm{K}}{T}\right] \simeq 10\frac{1-x}{x}\left(\frac{G_0}{n}\right) \simeq 6.4\times10^{-5}\left(\frac{100\,\mathrm{K}}{T}\right)^{0.6}, \qquad (8.10)$$

where the right-hand side follows from the C-ionization equation, for which we have adopted Eq. (8.3) with $x \sim 1$. The temperature is then calculated to be about 10 K, independent of the density, FUV field, and the C abundance (as long as C dominates the energy and ionization balances). This is much less than generally observed in the CNM and other heating processes have to be important.

In the early Field–Goldsmith–Habing model for the structure of the ISM, cosmic-ray ionization was assumed to be the dominant heating agent. We can again solve the energy balance (still assuming that ionization is due to photo-ionization of C). In that case,

$$\exp\left[\frac{-92\,\mathrm{K}}{T}\right] = 2\times10^{-2}\left(\frac{50\,\mathrm{cm}^{-3}}{n}\right)\left(\frac{1.4\times10^{-4}}{\mathcal{A}_\mathrm{C}}\right)\left(\frac{\zeta_\mathrm{CR}}{2\times10^{-16}\,\mathrm{s}^{-1}}\right)\mathrm{K}, \qquad (8.11)$$

which is about 25 K for typical conditions. Turning this argument around, to achieve a temperature of some 80 K, typical of diffuse interstellar clouds, the primary cosmic-ray ionization rate, ζ_{CR}, has to be a factor of 15 larger. X-rays do not penetrate deeply into diffuse clouds and, hence, are unimportant as a heating source.

By default, heating of diffuse interstellar clouds has, therefore, to be dominated by the photo-electric effect on large molecules and small dust grains. Let's evaluate the gas temperature assuming maximum efficiency ($\simeq$5%, cf. Section 3.3.3) for the photo-electric effect. In that case, balancing photo-electric heating with [CII] cooling, we arrive at

$$\exp\left[\frac{-92\,\mathrm{K}}{T}\right] = 0.3\left(\frac{50\,\mathrm{cm}^{-3}}{n}\right)\left(\frac{1.4\times10^{-4}}{\mathcal{A}_\mathrm{C}}\right)\mathrm{K}. \qquad (8.12)$$

Thus, when the photo-electric effect operates near the (calculated) maximum efficiency, a gas temperature of some 100 K can be expected for diffuse clouds.

We have focussed here on the energy balance of diffuse clouds. We can also examine the energy balance in the warm neutral medium. In that case, the cooling is dominated by Ly α emission. We will again assume that the photo-electric effect dominates the heating. However, the efficiency will be reduced, because of the charging of the PAHs or dust grains (e.g., $\epsilon \simeq 10^{-2}$). The temperature is then given by

$$\exp[-118,400\,\mathrm{K}/T] \simeq 4.5\times 10^{-7}\left(\frac{0.1}{x}\right)\left(\frac{0.3\,\mathrm{cm}^{-3}}{n}\right)\left(\frac{\epsilon}{10^{-2}}\right)\mathrm{K}, \quad (8.13)$$

which results in a temperature of 8000 K. It should be mentioned here that the photo-electric effect on PAHs and small dust grains is somewhat mitigated by the cooling associated with the recombination process; that is, recombination of positively charged PAHs and dust grains with electrons removes about 1 eV from the gas thermal energy at 10^4 K. Nevertheless, the photo-electric effect still dominates the heating. Note also that, because of the exponential temperature dependence of the cooling rate, the temperature is not very sensitive to the precise value of the heating rate; that is, with a slight adjustment of the temperature, the cooling can balance a wide range of heating rates.

8.3 The CNM and WNM phases of the ISM

In the previous section, we have examined the energy balance of the gas in the neutral medium. The results show that we can understand the observed temperature characteristics of the interstellar medium in terms of a balance between the relevant heating and cooling processes. However, the relationship between the energy balance and the phases of the ISM is deeper. *Two* phases – a cold, dense cloud phase and a warm, neutral intercloud phase – can exist side-by-side in the ISM in, and indeed because of, thermal and pressure equilibrium. In this section, we will examine this deeper relationship.

8.3.1 Two-phase model

The presence of more than one phase in the ISM directly reflects the cooling curve of the interstellar gas (Fig. 2.10). Summarizing the discussion in the previous section, at low temperatures, cooling is dominated by the excitation of the fine-structure level of C^+. For heating rates less than $\simeq 3\times 10^{-27}$ erg s^{-1} (H atom)$^{-1}$ a balance with C^+ cooling can be obtained by adjusting the temperature (cf. Section 2.5). For higher heating rates, however, the heating has to be balanced by additional cooling processes – visible electronic transitions of O and HI Lyman alpha – and the temperature will then have to rise to some 5000–10 000 K for this to be effective (cf. Section 2.6.1). Thus, in essence, [CII] and Lyman α cooling act as thermostats that stabilize the temperatures at about 100 K and about 8000 K, respectively. Of course, if both phases are present, pressure equilibrium requires that the high temperature phase has a lower density. Conversely, if we take the maximum efficiency of the photo-electric effect on PAHs and small dust grains – $\Gamma_{\rm pe} \simeq 5 \times 10^{-26}$ erg s^{-1} (H atom)$^{-1}$ – C^+ cooling can balance the heating for densities in excess of some 20 cm^{-3}. For lower densities, other cooling processes have to contribute or the heating efficiency should be less because of charging effects. Both of these effects correspond to lower densities. These simple considerations go a long way to help us understand the physical characteristics of the phases summarized in Table 1.1.

We can also cast this in a more rigorous form. Consider the net cooling, $\mathcal{L}$, defined as

$$\mathcal{L} \equiv n^2 \Lambda - n\Gamma. \tag{8.14}$$

In general, $\mathcal{L}$ will depend on the physical conditions (e.g., n and T), as do Λ and Γ. In the n–T plane, $\mathcal{L} = 0$ defines the thermal equilibrium curve. For this curve, we can write

$$n\,\Gamma = n^2\,\Lambda. \tag{8.15}$$

In the low density limit, Λ is independent of the density and we can rewrite this as

$$\frac{\Gamma}{nkT} = \frac{\Lambda}{kT}, \tag{8.16}$$

where the right-hand side depends only on the temperature. Thus, the locus of thermal equilibrium can be found directly from the cooling curve (cf. Fig. 8.1). For a point above this curve, heating exceeds cooling while for a point below, cooling exceeds heating. For simplicity, we will concentrate on cosmic-ray heating, as did the first studies in this field. Heating is then independent of density and constant heating rates at constant pressure correspond to a single value of Γ/P and are horizontal lines in this figure. The intersections of this horizontal line and the cooling curve are the equilibrium points. For a range of pressures, there are four equilibrium points. Two of these points are stable points and two are unstable. For the latter two, small perturbations will move the gas away from these points; i.e., the cooling increases when the temperature decreases and vice versa. In a more general sense, the stability criterion for isobaric perturbations becomes

$$\left(\frac{\partial \mathcal{L}}{\partial T}\right)_P = \left(\frac{\partial \mathcal{L}}{\partial T}\right)_\rho - \left(\frac{\rho_0}{T_0}\right)\left(\frac{\partial \mathcal{L}}{\partial \rho}\right)_{\rm T} < 0. \tag{8.17}$$

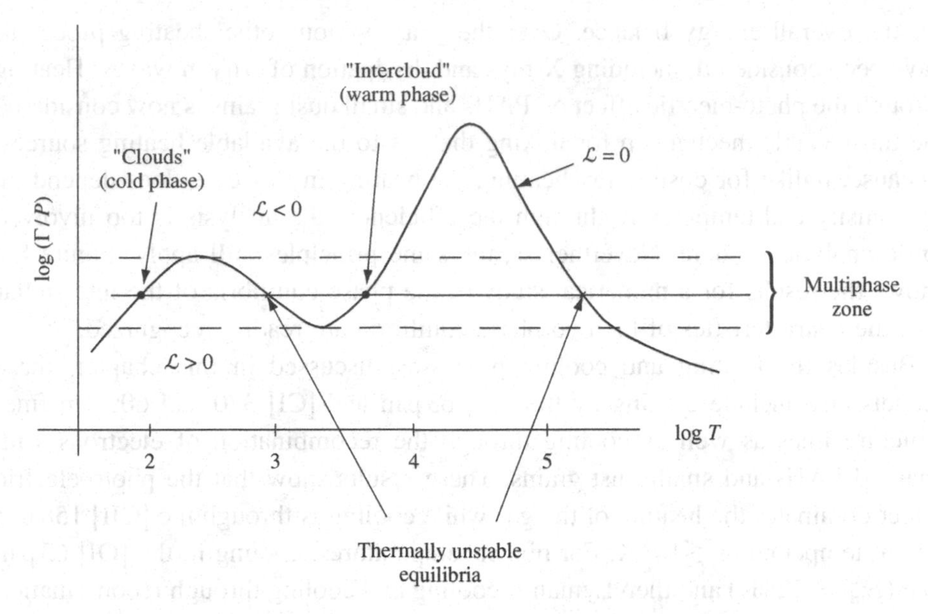

Figure 8.1 A schematic representation of the cooling curve, plotted as Λ/kT. (In thermal equilibrium [$\mathcal{L}=0$, where $\mathcal{L}$ is the net cooling], $\Lambda/kT=\Gamma/P$.) Above this curve, heating exceeds cooling and vice versa below. At constant pressure and constant heating rate (per H atom), the equilibrium points are given by the intersection of a horizontal line with this equilibrium curve. There is then a region in which more than one solution exists: the multiphase zone indicated by dashed lines. Within this zone, two of the equilibrium points are thermally stable and two are thermally unstable. Reproduced with permission from J. M. Shull, 1987, in *Interstellar Processes*, ed. D. Hollenbach and H. Thronson (Dordrecht: Reidel), p. 225.

The two stable points can be identified with a tenuous warm intercloud phase and a dense cool cloud phase. When the pressure drops below a minimum pressure, only the intercloud phase remains, while at high pressures, only the cloud phase is left. Thus, we realize now that the presence of two phases in pressure and thermal equilibrium is a natural consequence of the cooling curve of the gas, which itself reflects the large jump in energy scale from fine-structure levels to the next set of electronic levels in abundant species and nothing in between to cool atomic gas.

8.3.2 Detailed models

The early two-phase models for the ISM, dating back to the late 1960s, adopted a very high cosmic-ray ionization and heating rate ($\zeta_{CR}\sim10^{-15}\ \mathrm{s}^{-1}$). Molecular observations have shown that the actual cosmic ray-ionization rate is an order of magnitude lower in the ISM (cf. Section 8.8) and cosmic rays are unimportant for the overall energy balance. Over the years various other heating processes have been considered, including X-rays and dissipation of Alfvén waves. Heating through the photo-electric effect on PAHs and small dust grains is now considered the most viable mechanism for linking the gas to the available heating sources. Because, unlike for cosmic-ray heating, the heating in this case does depend on the density and temperature through the efficiency, the analysis is too involved to do analytically here. Nevertheless, the same principles still apply. Figure 8.2 shows the results for a numerical study of the phase equilibria of the interstellar gas; the characteristics of the two-phase equilibria are readily recognized.

Besides the heating and cooling processes discussed in this chapter, these models also include cooling by the [OI] 63 μm and [CI] 370 and 609 μm fine-structure lines as well as cooling through the recombination of electrons with charged PAHs and small dust grains. These results show that the photo-electric effect dominates the heating of the gas while cooling is through the [CII] 158 μm line at temperatures $\lesssim 10^3$ K. For higher temperatures, cooling in the [OI] 63 μm line ($E_{ul} = 228$ K) and then Lyman α cooling and cooling through recombination of electrons with charged PAHs or grains takes over. We see that the photo-electric heating per H atom increases with density because of the decreased importance of charging. At the higher densities, ionization of the gas is due to photo-ionization of neutral carbon. At lower densities, X-ray ionization dominates and the degree of ionization depends on the attenuating column.

Stable regimes are now characterized by $d(\log P)/d(\log n) > 0$. So a warm intercloud medium and a cold cloud medium are in equilibrium in the pressure range of $1000 < P/k < 3600$ K cm^{-3}. A typical pressure in the two phase medium would be about 3000 K cm^{-3}, which is perhaps slightly less than observed in the ISM (cf. Tables 1.1 and 1.2). For this pressure these models predict a CNM with a density of 60 cm^{-3} and a temperature of 50 K. The WNM has a density of 0.4 cm^{-3} and a temperature of 7500 K. These values are close to the observations.

8.3.3 Time scales

Anticipating the discussion on time-dependent effects, we will examine the time scales for equilibrium to establish itself. The thermal time scale for these phases is given by

$$\tau_{th} = \frac{nkT}{n^2 \Lambda}, \tag{8.18}$$

which can be written as

$$\tau_{th} \simeq 1.4 \times 10^6 \left[\frac{P}{3000\,\text{cm}^{-3}\,\text{K}}\right] \left[\frac{1\,\text{cm}^{-3}}{n}\right] \left[\frac{10^{-26}\,\text{erg (H atom)}^{-1}\,\text{s}^{-1}}{n\,\Lambda}\right] \text{years.} \tag{8.19}$$

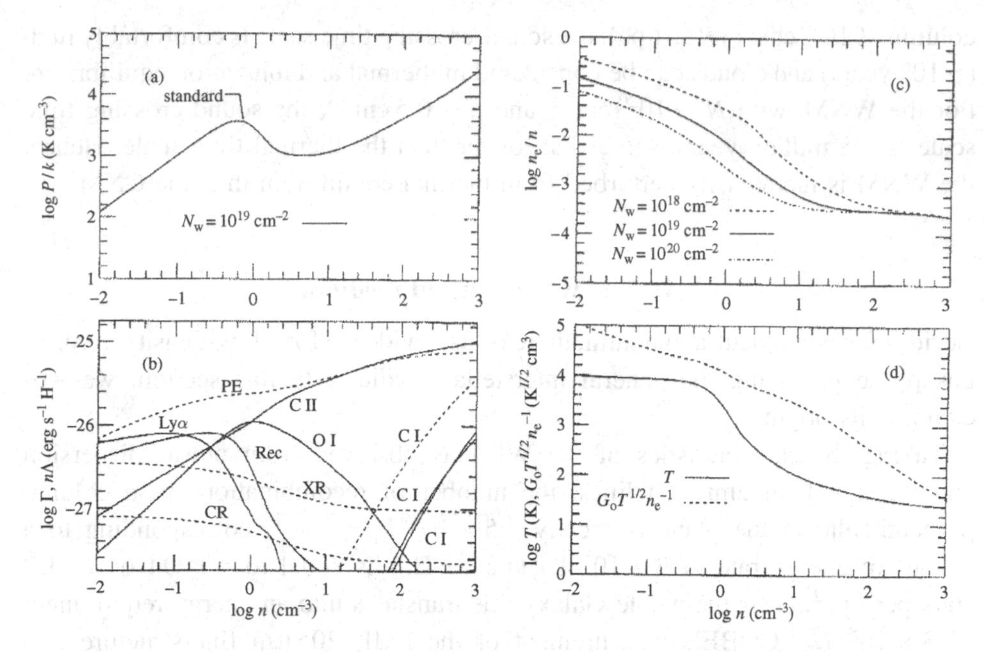

Figure 8.2 Detailed two-phase models for the ISM. (a) The pressure–density relation for interstellar gas in thermal equilibrium. (b) The individual heating (dashed) and cooling processes (solid) at different densities in thermal equilibrium (i.e., for the pressure–density relation of (a)). (c) The degree of ionization for these conditions at three X-ray attenuation depths. (d) The temperature (solid) and ionization parameter $\gamma = G_0 T^{1/2}/n_e$ (dashed) for these models. Reproduced with permission from M. Wolfire, D. Hollenbach, C. F. Mckee, A. G. G. M. Tielens, and E. L. O. Bakes, 1995, *Ap. J.*, **443**, p. 152.

For the CNM, $n = 60$ cm^{-3} and $n\Lambda$ is observed to be $\simeq 6\times10^{-26}$ erg (H atom)$^{-1}$ s^{-1} (Section 8.9.7), resulting in a thermal time scale of about 4000 years. In the WNM ($n = 0.4$ cm^{-3} and $n\Lambda \simeq 2\times10^{-26}$ erg (H atom)$^{-1}$ s^{-1}), this becomes about 2 million years. The radiative recombination time scale is given by

$$\tau_{\rm rec} = \frac{n_+}{n_e n_+ k_{\rm rec}}, \tag{8.20}$$

which becomes

$$\tau_{\rm rec} \simeq 2\times10^2 \frac{T^{0.6}}{n_e} \text{ years.} \tag{8.21}$$

For the CNM with $x \sim 10^{-3}$, the recombination time scale is 6×10^4 years, while in the WNM ($x \sim 0.1$), this is about 10 times longer. Quite generally, it takes longer to establish ionization equilibrium than thermal equilibrium. One time scale to compare these time scales with is the sound-crossing time scale; $\tau_s = R/C_s$, where the sound velocity is $C_s \simeq 1.4\times10^4 T^{1/2}$ cm s^{-1}. For the CNM with a typical HI column of 10^{20} cm^{-2} ($R \simeq 1$ pc), the sound-crossing time scale is comfortably high ($\sim 10^6$ years) and clouds can be considered in thermal and ionization equilibrium. For the WNM, with $N \sim 10^{20}$ cm^{-2} and $n \sim 0.5$ cm^{-3}, the sound-crossing time scale is $\sim$5 million years; somewhat longer than the thermal time scale. Hence, the WNM is more easily perturbed from thermal equilibrium than the CNM.

8.4 The warm ionized medium

Besides the warm neutral medium, there is also evidence for a low-density, ionized gas phase pervading the general interstellar medium. In this section, we will examine its origin.

Taking the characteristics of the WIM as observed from pulsar dispersion and through faint emission lines, the number of recombinations in a column perpendicular to the plane is locally $\simeq 5 \times 10^{49}$ kpc^{-2} s^{-1}, corresponding to a minimum energy rate of 3×10^5 L$_\odot$ kpc^{-2}. This is equal to $\sim$1 O4 or 15 B0 stars per kpc^2. Over the whole Galaxy this translates into an energy requirement of 3×10^8 L$_\odot$. COBE's measurement of the [NII] 205 μm fine-structure line implies a total ionizing photon flux of 3.5×10^{53} photons s^{-1}; about a factor of 5 larger than this extrapolation for the local WIM. This energy requirement is well in excess of the total kinetic energy injected into the ISM by massive stars (cf. Table 1.3). Thus, only photo-ionization by OB stars is a viable source for the WIM and the physics of the WIM is akin to that of HII regions (cf. Chapter 7).

The spectrum of the large-scale diffuse HII is characterized by strong emission from low stages of ionization (i.e., [NII] 6584 Å, [SII] 6717 Å) and weak emission from high stages of ionization (i.e., [OIII] 5007 Å) relative to the HI recombination lines. The [OI] 6300 Å emission is also weak compared with Hα. These observed characteristics are quite different from those of identifiable HII regions around bright O stars. Specifically, for bright HII regions such as Orion, the ionization parameter, U, is $>3 \times 10^{-2}$. In contrast, the observed spectrum of the WIM indicates an ionization parameter of $U \sim 10^{-4}$. The [OI]/Hα ratio is a measure of the degree of ionization of the WIM because of the rapid charge exchange reactions between O^0 and H$^+$. Unfortunately, it is also sensitive to the temperature of the ionized gas, which is not well constrained observationally. For the nominal temperature of 8000 K, the observed ratio of 0.3 corresponds to a neutral fraction of $\simeq$0.1. For the minimum temperature allowed by the observed linewidth (6000 K), this ratio still implies a largely ionized gas ($1 - x \simeq 0.3$).

At a density of 0.1 cm^{-3}, the derived ionization parameter of $U = 10^{-4}$ corresponds to an O7 star at a distance of 500 pc. Transport of ionizing photons over such a large distance poses a severe problem for photo-ionization models for the WIM: the more so since about 15% of all the stellar ionizing photons are involved.

This transport problem is also highlighted by the large-scale height of the WIM, $\simeq$1 kpc compared to that of OB stars. In that respect, consider also the 1 kpc-large Hα filament sticking out of the galactic plane in Fig. 1.5. Finally, detailed studies of specific regions of the WIM, taking into account all the observed OB stars and hot white dwarfs, arrive at the same conclusion: if the WIM is photo-ionized, the ISM has to be transparent to ionizing photons on large scales ($\sim$kpc). At the average density of the ISM, 1 cm^{-3}, the Strömgren radius of an O star is $\simeq$50 pc (cf. Section 7.2), an order of magnitude smaller than the distance required. From cloud statistics, on average five HI clouds with column densities of 3×10^{19} cm^{-2} – optically thick to ionizing radiation – are expected per 500 pc. The conclusion has, therefore, to be that (when discussing the escape of ionizing radiation) HI clouds are not statistically distributed in the Milky Way but that, instead, large portions of the ISM have been cleared of HI. Moreover, these must have very low densities or they must not be in photo-ionization equilibrium, or both. Such holes in the HI seem to be present near OB associations where the concerted action of stellar winds and supernova explosions of all the massive stars create a superbubble that can break out of the disk and vent into the halo. Such "HII chimneys" above OB associations, filled with highly (collisionally) ionized gas, may form natural pathways for ionizing photons to escape to large-scale heights.

8.5 The hot intercloud medium

The interstellar medium also contains a phase of very hot but tenuous gas ($T \simeq 10^6$ K, $n \simeq 10^{-3}$ cm^{-3}). This gas cannot be in thermal equilibrium with the energy sources we have been examining. Rather, this phase owes its existence to the presence of strong shock waves, driven by supernova explosions, that constantly "plow" the ISM. Because of the low density, radiative cooling is low and this mechanical heating of the ISM leads to a coronal phase coexisting with the other phases.

8.5.1 Mechanical energy

In the previous sections, we have discussed the ISM in all quietness. The only energy input considered was radiative energy. However, the ISM is actually a very violent place, where massive stars are constantly stirring up the medium through the action of their winds and supernova explosions (Chapter 12). The total mechanical energy injected by massive stars into the ISM is only a small fraction of the radiative energy budget (cf. Table 1.3) and can be neglected for the energy and ionization budgets of the cold cloud and warm intercloud phases. However, it is very important for hydrostatic equilibrium and the overall structure of the ISM. The interstellar medium reacts to this mechanical energy by becoming very turbulent, the gas disk puffs up, and part of the volume becomes filled with hot gas. The latter may, even in the plane, become a dominant, separate phase of the ISM, the hot ionized (intercloud) medium (HIM) in which the WNM, WIM, and CNM are embedded. The filling factor of this phase in the plane is, however, controversial. Here, we will consider the effect of the mechanical energy input on the ISM.

8.5.2 The vertical structure of the ISM

For simplicity consider a planar, one-dimensional, isothermal ISM, with a total pressure P, including cosmic rays, magnetic field, and thermal. The equation for hydrostatic equilibrium is then given by

$$\frac{1}{\rho(z)}\frac{\mathrm{d}P(z)}{\mathrm{d}z} = z\frac{\mathrm{d}g(z)}{\mathrm{d}z}, \tag{8.22}$$

where $g(z)$ is the gravitational acceleration in the z direction. For an "isothermal" layer, the pressure scales linearly with the density ($P = C_\mathrm{s}^2\rho$, with C_s the sound speed) and the density is then given by a Gaussian,

$$\rho(z) = \rho(0)\exp\left[-\left(\frac{z}{H}\right)^2\right], \tag{8.23}$$

with a scale height

$$H = \left(\frac{2C_\mathrm{s}^2}{-\dot{g}(z)}\right)^{1/2}, \tag{8.24}$$

where $\dot{g}(z)$ is the derivative of the gravitational acceleration,

$$\dot{g}(z) = \frac{\mathrm{d}g(z)}{\mathrm{d}z} \simeq -2.2\times 10^{-11}\,\mathrm{cm\,s^{-2}\,pc^{-1}}. \tag{8.25}$$

The numerical value applies to the solar neighborhood. The scale height then becomes

$$H \simeq 170\left(\frac{C_\mathrm{s}}{10\,\mathrm{km\,s^{-1}}}\right)\,\mathrm{pc}. \tag{8.26}$$

In a cold-phase ISM, C_s is only $\simeq$1 km s^{-1} and all the clouds would collect in a thin ($H \simeq 20$ pc) layer. A warm-intercloud medium would have a scale height of $\simeq$200 pc. In a two-phase ISM, cold clouds would condense out of the warm intercloud phase even at higher latitudes as long as the thermal pressure were above the minimum pressure required for clouds to exist. But while clouds can exist at higher altitudes, they would rain out into a thin disk in the mid-plane. This structure is quite different from the observed structure of the interstellar medium (cf. Table 1.1). One way to avoid such a thin disk of clouds is by letting the clouds have a non-thermal velocity distribution; i.e., by including a turbulent pressure. The observed, turbulent, cloud velocity dispersion of $\simeq 7$ km s^{-1} yields a scale height quite comparable to the observed one. The increased scale height of clouds can also reflect the importance of magnetic fields. After all, the magnetic pressure is quite high as well. The magnetic field might, however, be prone to instabilities curving it. Turbulently cold clouds would then again slide down the field lines to a thin cold disk. In the real ISM, the problem of the vertical structure is more complicated than has been considered in this section but the same conclusion remains: the vertical distribution of clouds is governed by turbulence. The same must hold true for those components in the WNM/WIM phases that show a larger scale height then Eq. (8.24) would indicate.

8.5.3 Three-phase model

The turbulent motions of the clouds result from the mechanical energy injection by massive stars. In a typical supernova explosion, a few solar masses are ejected with some 10^{51} erg, corresponding to velocities of 10 000 km s^{-1}. These ejecta will sweep up the surrounding medium, while slowly decelerating. During the evolution of the supernova remnant, the energy of the explosion will be partly radiated away by the shocked, swept-up, ambient ISM, partly converted into thermal energy of very hot but tenuous gas filling the supernova cavity, and partly converted into kinetic motion of the shell of swept-up ISM. The details of the evolution of supernova remnants will be discussed in Sections 12.3 and 12.4. For the global structure of the ISM, there are two points to be made. First, supernova explosions will pump energy into turbulent motions of the ISM. Second, supernovae will create regions of very hot but tenuous gas that cool only very slowly. As a result, when supernovae are abundant, the disks of galaxies will be puffed up by the turbulent motions while a significant fraction of the ISM will be filled by hot tenuous gas. In such a supernova-dominated ISM, there will be three phases coexisting in the plane of the Galaxy: the cloud (CNM) and intercloud (WNM and WIM) phases discussed in the previous sections, and this hot tenuous gas (the hot intercloud medium, HIM). Since cold clouds are confined to the plane, there are two phases in the low halo (the HIM and the WNM and WIM). At high latitudes, only the HIM remains. In this section, we will examine the effects of supernovae on the structure of the ISM in some detail.[1]

[1] Actually, fast winds from massive stars also pump large amounts of mechanical energy into the ISM (Section 12.5), creating large bubbles filled with hot tenuous gas. Because this occurs in stellar evolutionary phases, which on a galactic time scale are only an an "eye-blink" before the SN explosion, we may consider their effects as priming the surrounding medium and consider the overall effects of massive stars.

Supernova ejecta shock the surrounding material to temperatures of the order of

$$T \simeq 3 \times 10^5 \left(\frac{v_s^2}{100\,\mathrm{km\,s^{-1}}} \right)^2 \mathrm{K}, \tag{8.27}$$

where v_s is the shock velocity (cf. Section 11.2.1). In a two-phase medium, the supernova shock will propagate faster into the lower-density medium ($v_{wnm}/v_{cnm} = \sqrt{\rho_{cnm}/\rho_{wnm}}$; Section 12.4); e.g., a supernova will predominantly expand in the low-density intercloud phase (which also has the highest filling factor). For shock velocities in excess of ~250 km s^{-1}, the shocked gas is so hot ($T \gtrsim 10^6$ K) that radiative cooling is very slow ($\tau_{th} \propto T/n\Lambda \gtrsim 10^6$ years). Such supernova remnants cool mainly through adiabatic expansion; work is done against displacing the surrounding ISM (Section 12.3.2). Thus, for a high supernova rate per unit volume, k_{sn}, a long supernova remnant lifetime, τ_{snr}, and a large final remnant volume, V_{snr}, the expanding supernova remnants would start to overlap and form a connected tunnel network filled with hot gas. We can now define the porosity parameter, Q,

$$Q = k_{sn} V_{snr} \tau_{snr}. \tag{8.28}$$

For low SN rates, Q is just the fraction of the total volume of the ISM filled by hot SNRs. When k_{sn} increases, the effects of individual supernovae are moderated by SNRs running into each other – in which case, the "new" SN energy may mainly go into reheating the "old" one. This rejuvenation process will lead to a tunnel network of hot gas, which may fill a large portion of the ISM. In this case, the mechanical energy of supernovae will go into maintaining this phase of hot gas against the effects of radiative cooling and, particularly, energy conduction into embedded colder structures (cf. Section 12.4).

The time-evolution of supernova remnants (e.g., V_{snr}, τ_{snr}) is discussed in detail in Sections 12.3 and 12.4. Here, we will anticipate the final phase in the supernova remnant evolution, the radiative expansion phase (Section 12.3.3), which is of importance for the ISM structure analysis. The time, τ_{sn}, and the volume, V_{snr}, at which an SNR will merge with the ISM are given by

$$\tau_{sn} = 1.7 \times 10^6 \left(\frac{E_{sn}}{10^{51}\,\mathrm{erg}} \right)^{11/45} n_0^{-11/45}\ \mathrm{years} \tag{8.29}$$

and

$$V_{snr} = 1.7 \times 10^{-3} \left(\frac{E_{sn}}{10^{51}\,\mathrm{erg}} \right)^{11/15} n_0^{-11/15}\ \mathrm{kpc}^3, \tag{8.30}$$

with n_0 the initial density of the surrounding medium and E_{sn} the ejection energy of a supernova. This corresponds to a final size of 87 pc for an ambient (WNM) density of $0.5\,\text{cm}^{-3}$. Adopting an SN rate of

$$k_{sn} = 6.7 \times 10^{-5}\, N_{sn}\, \text{years}^{-1}\, \text{kpc}^{-3}, \tag{8.31}$$

with N_{sn} the number of SN per 100 years in the whole Galaxy (with total volume 150 kpc^3) and assuming an SN energy of $E_{sn} = 10^{51}$ erg, then the porosity parameter is given by

$$Q \simeq 0.12 N_{sn} \left(\frac{E_{sn}}{10^{51}\,\text{erg}}\right)^{44/45} n_0^{-44/45}. \tag{8.32}$$

So, for a two-phase medium dominated by an intercloud medium with density $0.5\,\text{cm}^{-3}$ and a supernova rate of 1 per century – allowing for some SN clustering, see below – we have $Q = 0.37$. Thus, if supernovae are distributed randomly, 37% of them will explode in a pre-existing supernova remnant. The probability that a supernova will explode within ~90 pc of an existing SNR and then expand to overlap is $1 - \exp[-8Q] \simeq 0.95$. Clearly, this hot phase may be very important.

For a uniform medium with constant Q, the time-averaged fraction of the medium filled with hot gas is given by

$$f_{\text{HIM}} = \frac{Q}{1+Q}. \tag{8.33}$$

So, for a Q of 0.37, calculated above for a two-phase medium, the HIM filling fraction is 0.27. However, this is very sensitive to the density of the medium in which the supernovae are expanding; e.g., the value of the HIM filling factor itself. For example, for a density of the medium of $0.1\,\text{cm}^{-3}$, $Q \simeq 1.8$ and $f_{\text{HIM}} \simeq 0.6$ and supernova remnants will very much overlap. Clearly, this is a runaway process: supernova activity clears out cavities, lowering the density of the medium in which supernovae expand and allowing larger remnant volumes. For large Q, supernovae will sweep up the warm neutral and ionized phases of the ISM into thin but dense walls surrounding a tenuous hot phase, the HIM. Perhaps assisted by the thermal instability, these walls will lead to the CNM phase. Thus, in the SN dominated regime, the ISM will appear very filamentary.

The exact value of Q is quite uncertain. Besides the dependence on the ambient density discussed above, Q also depends sensitively on the details of the expansion of the supernova remnant. The analytical results described above are only approximate and various effects have not been included. This includes radiative energy loss of the hot gas, the detailed rearrangement of the internal structure of the SNR during the transition between the various expansion phases (cf. Section 12.3), and, most importantly, the mitigating effects of a magnetic field. In particular, a high magnetic field pressure in the ISM will stall the expansion of a supernova remnant at a much smaller radius than calculated above. Hence, the derived value for the porosity parameter and the volume filling factor are then much reduced. More detailed evaluations of Q result in values ranging from 0.25 to 0.75 and so the hot intercloud medium is expected to be an important component, even in the disk of the Galaxy.

Getting back to the pressure support of the interstellar medium, the turbulent pressure due to supernova kinetic energy injection follows from the momentum left over at the end of the radiative expansion phase when a thin shell of mass M_{snr} is coasting at the sound speed of the ambient medium, C_0 (cf. Section 12.3.3). This momentum is conserved until the SNR collides with another SNR. This will happen when the shell has expanded to a radius R_f, after a time R_f/C_0. This size and time scale are related by

$$\frac{4\pi}{3} R_f^3 = \frac{C_0}{R_f k_{sn}}. \tag{8.34}$$

The momentum flux density, $\rho v^2 = \rho_0 V_{snr} C_0 / 4\pi r^2$, averaged over volume and time, is then

$$< \rho v^2 > = \left\langle \frac{\rho_0 V_{snr} C_0}{4\pi r^2} \right\rangle = \rho_0 V_{snr}\, C_0\, R_f k_{sn}. \tag{8.35}$$

Using Eq. (8.34), we find for the turbulent pressure in terms of the total pressure

$$\frac{P_t}{P_0} = \frac{<\rho v^2>}{3\rho_0 C_0^2} = \frac{Q^{3/4}}{3\eta^{3/4}} \simeq 0.9 Q^{3/4}, \tag{8.36}$$

where we have used $C_0 = \eta R_{snr}/\tau$ (with η the self-similarity parameter; Section 12.3) and $Q = V_{snr}\, k_{sn}\, \tau$ and $\eta \simeq 0.3$ during this phase. Thus, when SN stir up the ISM, Q becomes large and the turbulent pressure dominates the total pressure of the ISM. Conversely, when the turbulent pressure is large, the scale height becomes large and the disk puffs up.

In the popular McKee–Ostriker model for a supernova-remnant-dominated ISM, there will then be three phases (cf. Fig. 8.3): the hot and tenuous gas in the SNR (HIM), the cold clouds (CNM), and the warm intercloud phases (WNM and WIM). The latter are envisioned as the surfaces of the clouds in this model, heated and ionized by electron conduction from the HIM in addition to the FUV photons from hot stars. The energy conducted into the (inter)cloud phase leads to a mass flux out (Section 12.4). This evaporation of the clouds is the dominant mass input into the HIM. The supernova remnants, on the other hand, sweep up the surrounding warm intercloud material and compress it and, helped by the thermal instability, form new clouds. Thus the mechanical action of stars on their environment fluffs up the cloud distribution and leads to the formation of a hot

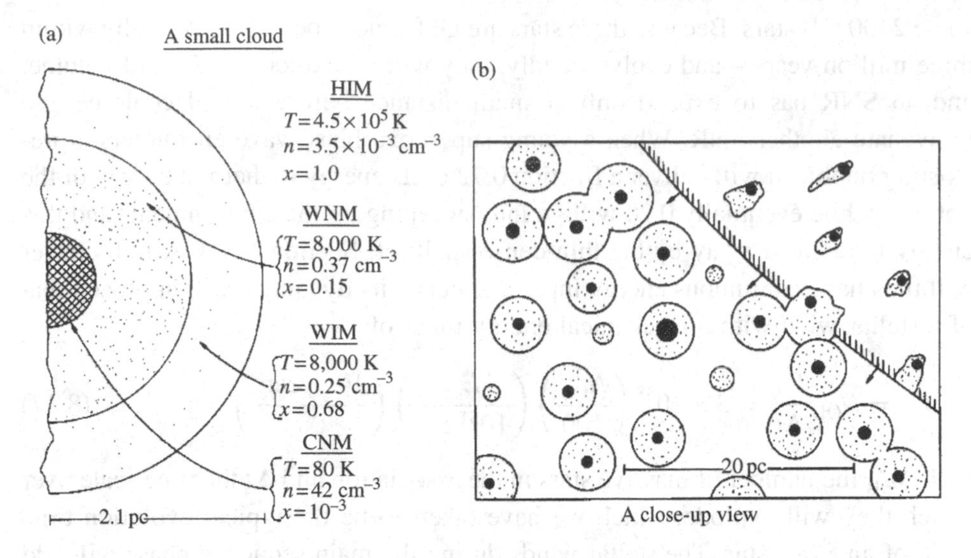

Figure 8.3 A schematic diagram of the three-phase model for the ISM stirred up by a supernova. (*a*) A blow-up of a cloud (CNM) surrounded by the warm intercloud (WNM and WIM) and embedded in the hot intercloud medium. Typical densities, temperatures, and degrees of ionization for these phases calculated in this model are also indicated. (*b*) The effect of an expanding SNR on the cloud population in the ISM (e.g., crunching and evaporating the warm intercloud media surfaces; cf. Section 12.4). Figure reproduced with permission from C. F. McKee and J. P. Ostriker, 1997, *Ap. J.*, **218**, p. 148.

and tenuous phase, the HIM. The importance of the HIM (i.e., its filling factor) depends on the porosity factor, Q. The value of this factor for the ISM in the Milky Way is highly controversial, in particular if the interstellar magnetic field pressure is large. This will limit the expansion of the SNR and, hence, can reduce Q and F_{HIM} substantially.

8.5.4 The role of the halo

In the evaluation of the HIM filling factor in terms of the porosity parameter, a random distribution of the supernovae was assumed. Actually, many SN occur in OB associations. The concerted effort of the many supernovae exploding in OB associations (and the winds of their OB star progenitors; Section 12.5) may lead to the formation of superbubbles in the ISM. A typical OB association has 10–20 stars with $M > 10\,\mathrm{M}_{\odot}$. Some very rich galactic associations – such as NGC 3603 and Cyg OB2 – have up to 100 massive stars, which will eventually explode as supernovae. Giant HII regions and starburst nuclei contain $\sim 10^2 - 10^3$ early type stars, albeit in a somewhat larger-sized region. The NGC 2070 cluster, powering the 30 Dor region in the LMC, is a prime example of a giant HII region containing some 2400 OB stars. Because these stars are all formed coevally – typically within three million years – and evolve rapidly, they will all explode in a small volume, and an SNR has to expand only a small distance before it will coalesce and rejuvenate another SNR. When a young supernova blast wave encounters a pre-existing bubble, it will deliver a fraction 0.72 of its energy to thermal energy of the hot gas, while eventually 0.28 will go into sweeping up the environment (and this energy is radiated away by the thin dense shell; cf. Section 12.3). A rich cluster will thus have continuous energy input and hence its dynamics will approach that of a stellar wind with a mechanical energy input of

$$L_{\mathrm{w}} = N_{\mathrm{OB}} \frac{E_{\mathrm{sn}}}{\Delta t} = 3 \times 10^{38} \left(\frac{N_{\mathrm{OB}}}{100} \right) \left(\frac{E_{\mathrm{sn}}}{10^{51}\,\mathrm{erg}} \right) \left(\frac{10^{7}\,\mathrm{years}}{\Delta t} \right) \mathrm{erg\,s^{-1}}, \tag{8.37}$$

with N_{OB} the number of massive stars in the association and Δt the time scale over which they will explode, which we have taken to be the typical evolution time scale of an $8\,\mathrm{M_{\odot}}$ star. The stellar winds during the main sequence phase will add an additional 50% to this mechanical luminosity. The dynamics of wind-blown bubbles are discussed in Section 12.5. For typical parameters, we have, for the size of the superbubble,

$$R_{\mathrm{sb}} = 400 \left(\frac{N_{\mathrm{OB}}}{100} \right)^{1/5} \left(\frac{E_{\mathrm{sn}}}{10^{51}\,\mathrm{erg}} \right)^{1/5} \left(\frac{\Delta t}{10^{7}\,\mathrm{years}} \right)^{2/5} \mathrm{pc}, \tag{8.38}$$

(cf. Section 12.5.1). This is indeed much larger than the sizes of individual supernova remnants and of OB associations and a superbubble will evolve. Initially, this superbubble will expand spherically symmetric but, when its size exceeds some three times the scale height of the ISM, it will accelerate in the z direction, down the pressure gradient, and break out of the disk. The scale height of the cloud layer is only ~100 pc and a typical OB association will break out of this layer fairly readily. However, the intercloud (WNM and WIM) layer is much thicker, ~200 pc, and only the richest OB associations will create "chimneys," which vent their hot gas into the halo. Such chimneys have been observed in galactic regions of massive star formation such as the W4 region. The circumnuclear starburst region in the Galaxy, M82, provides another example of venting into the halo. The Sun is located in the local bubble of low density, hot gas, which seems to be part of a chimney system venting into the halo. In addition, there are the nearby (radio) loops I, II, and III, which are also superbubble structures produced by massive star formation in the Scorpius–Centaurus association. Numerical calculations show that this process can be very effective, particularly when the explosion occurs somewhat offset from the mid-plane. Upon breakout, the dense expanding shell will become Rayleigh–Taylor unstable and some 5–10% of the mass of the shell will be injected as cloud-projectiles onto ballistic trajectories at velocities of 50–100 km s^{-1}. With a measured gravitational acceleration at a galactic altitude of 200 pc of 6×10^{-9} cm s^{-2}, the residence time of these clouds in the halo is some 80 million years and they may reach an altitude of 1–2 kpc above the plane. This residence time is some 4–10 times the lifetime of the cluster (10–20 million years). The total mass swept up is some $100\rho_0 H^3$ and, hence, some 10^6 $M_\odot$ is injected as cloudlets into the halo by a single blow-out. Thus, in order to keep the disk-halo circulation of 5 $M_\odot$ $year^{-1}$ running, about five large OB clusters have to be formed per million years. With a cluster lifetime of $\sim$20 Myears, about 100 clusters should be present, corresponding to a surface density of 0.15 kpc^{-2}, which is typical for an Sb galaxy. These chimneys will also vent hot gas into the halo. This gas will reach a height that is commensurate with its temperature and the local gravitational potential. Eventually, this gas will cool and condense into (inter)clouds, which will rain down on the disk again (cf. Fig. 8.4). These chimneys will also form natural pathways for ionizing photons to flow from OB associations to the lower halo, ionizing the WIM. Moreover, this flow will mix heavy elements, produced by the supernovae, over a similar scale length.

Thus, correlated supernovae will set up a galactic fountain, which dominates the mechanical energy and mass flow into the halo, and will not contribute to the formation of an all-pervasive HIM in the plane. This will lead to high temperature coronal gas at high latitudes, lead to intermediate or high velocity clouds at high latitudes, and in general provide a way of distributing freshly synthesized elements over a large portion of the galactic disk. An ISM dominated by supernovae in OB association will therefore look quite different from one dominated by isolated SN and the halo will play a role akin to the HIM phase, regulating the (turbulent) pressure, the vertical structure of the ISM, and the mass balance of the phases.

8.6 Summary: the violent ISM

Recapitulating the discussions in the previous sections, the interstellar medium will contain several phases. A cold, dense neutral cloud phase and a lower density, partly ionized intercloud phase can coexist in thermal and pressure equilibrium. These two phases reflect the intrinsic characteristics of the heating and the cooling of the gas and are thus closely tied to the microphysics of the medium. In addition, there is a pervasive, hot, tenuous, intercloud phase, which is created by the mechanical energy pumped into the medium by massive stars and hence reflects macroscopic processes in the medium, coupled with the long cooling time scales of such gas. These supernovae will also create a very turbulent medium with a puffed-up disk and large pressure excursions. So, the action of supernovae will drive the medium away from thermal and pressure equilibrium. Considering the

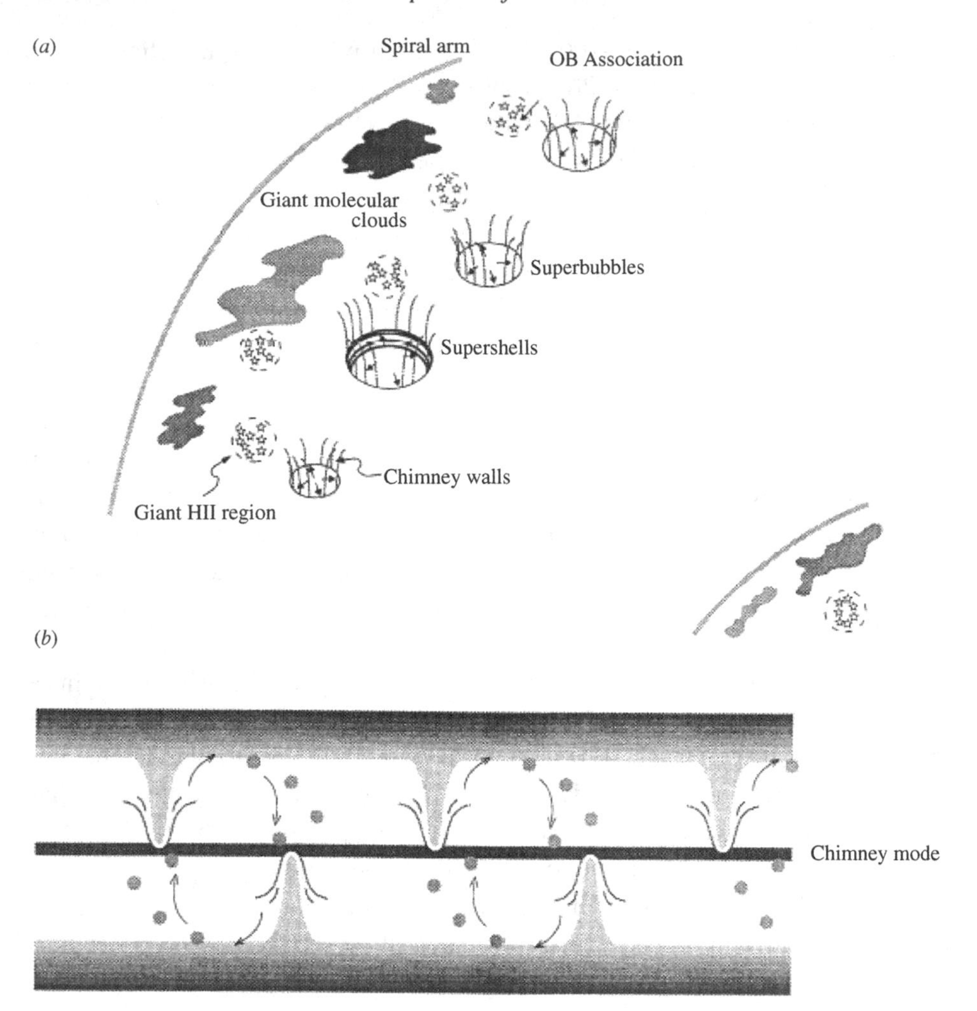

Figure 8.4 A schematic of the formation of superbubbles (*a*) along a spiral arm in the disk and the superbubble evolution (*b*) in the halo. Figure reproduced with permission from C. A. Norman and S. Ikeuchi, 1991, *Ap. J.*, **375**, 479.

time scales involved, the cold medium can be considered in thermal and pressure equilibrium but, for the warm intercloud media, this is only marginally true and memory effects due to the mechanical action of supernovae may play a major role. Indeed, HI observations show that a substantial fraction of the WNM is in the thermally unstable regime. Likewise, the ionization fraction seems to be larger than steady-state models would predict. The hot phase is of course very much a "memory" phase. Now, while clouds are in equilibrium, the large-scale structure of the CNM will reflect the effects of supernovae. Expanding SNR will sweep up the ambient medium into walls and very filamentary structures, leading to a very frothy appearance of the medium. The large pressure fluctuations associated with supernovae will also convert intercloud media to cloud media and back. The concerted efforts of supernovae in an association leads to the formation of superbubbles, which may blow out into the halo, setting up a galactic fountain. The shells of gas swept up in the plane will cool rapidly, compress, form molecules, and fragment into giant molecular clouds on a time scale of some 30 million years. A new generation of massive stars is then born, which turn supernova in another 10 million years. These supernovae then expand in the fossil remnant of the earlier generation. The interstellar medium is thus a place where the microscopic processes of the gas meet the macroscopic actions of the stars and this drives the evolution of the Galaxy.

Now, while the important phases of the ISM have long been identified, their physical properties are still not always well known. Their interrelationship and interaction in a dynamical universe is also not well understood. The feedback from massive stars on the surrounding ISM has been particularly resistant to global characterization. The porosity of the interstellar medium, which plays such a major role in global models for the ISM, is difficult to measure observationally and hard to calculate from first principles because it depends so much on the fine details. In particular, magnetic fields may have a stifling effect on supernova remnant expansion and hence the structure of the ISM. Thus, many of the key questions are still open: "What is the evolution of the interstellar medium over cosmic time as gas is recycled from one generation of stars to the next?," "How does the star formation rate depend on the environment?," and "What is the role of the Halo?"

8.7 Chemistry of diffuse clouds

A number of simple species have been observed in diffuse clouds. Initially, such studies were limited to the visible and near-UV wavelength ranges accessible from the ground and, hence, focussed on transitions of species with low-lying electronic states (e.g., radicals and ions). The first interstellar molecules, CH, CH^+, and CN, were discovered this way in the late 1930s and early 1940s of the previous century. With the advent of space-based FUV absorption spectroscopy, using Copernicus, IUE, HST, and FUSE, observations of species with filled electronic shells also became possible, leading to the detection of the most abundant molecular species, H_2, as well as a slew of others (HD, CO, OH, CH_2, H_2O, C_2). The simple carbon chains, C_2 and C_3, have also been observed in the visible and far-red wavelength ranges. Over the last decade, the emission from molecules in diffuse clouds has been studied through their rotational transitions at submillimeter wavelengths (e.g., CO, HCO^+). Finally, rotational–vibrational transitions of the important H_3^+ ion have been discovered in the near-IR ($\sim 3.7\,\mu$m). Besides these simple species, the presence of large polycyclic aromatic hydrocarbon molecules in the diffuse ISM is well attested for by the strong IR emission bands that dominate the near- and mid-IR emission spectrum. Their chemistry has been discussed in Chapter 6. Here, we will focus on the small molecules that are traditionally considered in models for the chemistry of the diffuse ISM.

The chemistry of diffuse clouds is initiated by cosmic-ray ionization of hydrogen and therefore shares characteristics with the chemistry of dark clouds (cf. Chapter 10). However, the ambient FUV radiation field opens up chemical channels that are not available in dark clouds. The chemistry of diffuse clouds also resembles the chemistry of photodissociation regions (cf. Chapter 9), except that the FUV radiation field is weaker and the typical densities and temperatures are lower.

8.7.1 Molecular hydrogen

Molecular hydrogen takes a special place in interstellar chemistry. Molecular hydrogen is a key intermediary in the formation of most gas phase species. Also, formation and destruction of H_2 differs in some aspects from that of many other species. In particular, photodissociation of H_2 is initiated through bound–bound transitions and, because of its high abundance, self-shielding against dissociating FUV radiation is very important. Furthermore, gas phase routes towards H_2 are notoriously slow and it is generally assumed to be formed on grain surfaces. Finally, because of its abundance, H_2 can dominate the spectral characteristics of molecular clouds in the FUV and near-IR. For these reasons, we will discuss H_2 in some depth.

H_2 photodissociation

Photodissociation of H_2 proceeds through FUV absorption in the Lyman and Werner transitions in the 912–1100 Å range, followed by fluorescence to the vibrational continuum of the ground electronic state about 10–15% of the time (see Chapter 3.4, Fig. 3.5). The H_2 photodissociation rate then follows from a summation over all lines. When the H_2 column density exceeds $10^{14}\,\text{cm}^{-2}$, the FUV absorption lines become optically thick and self-shielding becomes important. The photodissociation rate then depends on the H_2 abundance and level population distribution as a function of depth in the cloud.

It is instructive to evaluate the HI / H_2 transition using a simple self-shielding approximation. Consider a plane-parallel, constant density slab, illuminated from one side. The photodissociation rate of H_2 is given as

$$k_{\text{UV}}(H_2)n\,(H_2) = \beta_{\text{ss}}(N(H_2))\exp[-\tau_{\text{d}}]k_{\text{UV}}(0)n(H_2), \tag{8.39}$$

where $n(\mathrm{H_2})$ is the $\mathrm{H_2}$ density, τ_d is the dust optical depth at 1000 Å, $N(\mathrm{H_2})$ is the $\mathrm{H_2}$ column density into the cloud, and $k_\mathrm{UV}(0)$ is the unshielded photodissociation rate ($\simeq 4\times10^{-11}G_0\ \mathrm{s^{-1}}$). The self-shielding factor can be approximated by

$$\beta_\mathrm{ss} = \left(\frac{N(\mathrm{H_2})}{N_0}\right)^{-0.75}, \tag{8.40}$$

in the column density range $N_0 = 10^{14} \lesssim N(\mathrm{H_2}) \lesssim 10^{21}\ \mathrm{cm^{-2}}$. This dependence of β_ss on the column is somewhat steeper than the dependence expected for heavily saturated lines on the square-root portion of the curve of growth, because of the effects of line overlap. In the steady state, the $\mathrm{H_2}$ formation (see Section 8.7.1) and photodissociation rate are balanced,

$$k_\mathrm{UV}(\mathrm{H_2})n(\mathrm{H_2}) = k_\mathrm{d}(\mathrm{H_2})nn_\mathrm{H}. \tag{8.41}$$

Neglecting dust opacity ($N \lesssim 10^{21}\ \mathrm{cm^{-2}}$) and using the relation between total density and the density of H and $\mathrm{H_2}$, this yields

$$n(\mathrm{H_2}) = n\left(\frac{k_\mathrm{UV}N_0^{3/4}}{k_\mathrm{d}nN(\mathrm{H_2})^{3/4}}+2\right)^{-1}. \tag{8.42}$$

Realizing that $\mathrm{d}N(\mathrm{H_2})/\mathrm{d}z = n(\mathrm{H_2})$, we find

$$\frac{4k_\mathrm{UV}(0)N_0^{3/4}}{k_\mathrm{d}n}\left(N\left(N_{\mathrm{H_2}}\right)\right)^{1/4}+2N\left(\mathrm{H_{H_2}}\right) = N, \tag{8.43}$$

where N is the column of hydrogen nuclei ($N = nz$). Ignoring the second term on the left-hand side, we find

$$N(\mathrm{H_2}) = \left(\frac{k_\mathrm{d}n}{4k_\mathrm{UV}(0)}\right)^4\left(\frac{N}{N_0}\right)^4 N_0, \tag{8.44}$$

and, for the abundance of $\mathrm{H_2}$,

$$X(\mathrm{H_2}) = \frac{\mathrm{d}N(\mathrm{H_2})}{\mathrm{d}N} = \frac{4N(\mathrm{H_2})}{N} = 4\left(\frac{k_\mathrm{d}n}{4k_\mathrm{UV}(0)}\right)^4\left(\frac{N}{N_0}\right)^3. \tag{8.45}$$

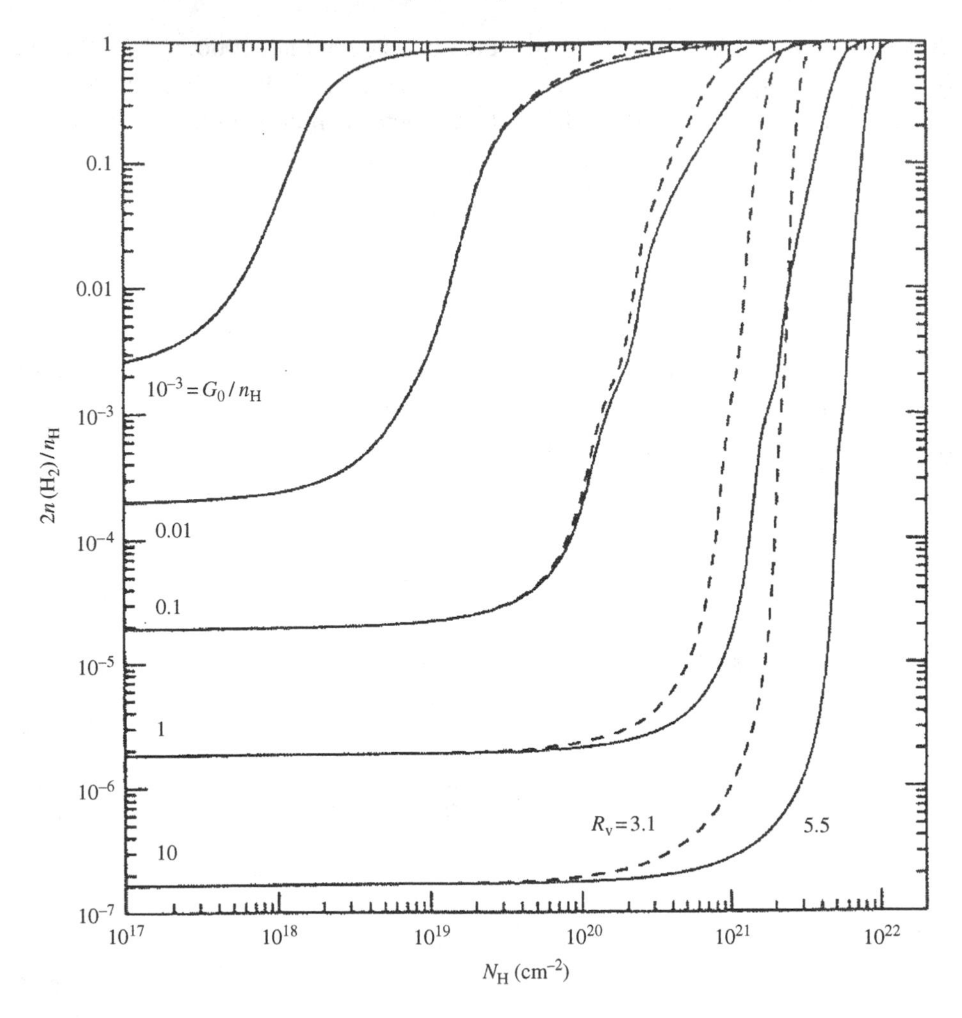

Figure 8.5 Calculated H_2 abundances as a function of total column density. The total gas density and temperature have been set equal to 200 cm^{-3} and 200 K, respectively. The different curves are labeled by the ratio of incident FUV field and the density. The dashed ($\tau_d(1000\,\text{Å}) = 2 \times 10^{-21} N$) and solid ($\tau_d(1000\,\text{Å}) = 6 \times 10^{-22} N$) lines distinguish two different FUV dust extinction cross sections used. Figure reprinted with permission from B. T. Draine and F. Bertoldi, 1996, *Ap. J.*, **468**, p. 269.

Detailed evaluation of the H_2 abundance as a function of column density into clouds illuminated by different FUV fields are shown in Fig. 8.5. Because

$$\frac{k_d(H_2)n}{4k_{UV}(0)} = 1.9 \times 10^{-7} \frac{n}{G_0}, \tag{8.46}$$

the structure of the HI / H_2 transition zone will be regulated by G_0/n; the equivalent of the ionization parameter in studies of ionized gas. While this derivation has limitations built in ($N_0 = 10^{14}\,\mathrm{cm^{-2}} \lesssim N_{H_2} \lesssim 10^{21}\,\mathrm{cm^{-2}}$ and $N \lesssim 10^{21}\,\mathrm{cm^{-2}}$), it illustrates the sharp transition in the molecular hydrogen abundance with depth in the cloud (Fig. 8.5) due to self-shielding (e.g., n ($N(\mathrm{H_2}) \propto N^4$ and $X(\mathrm{H_2}) \propto N^3$).

Equating the HI / H_2 interface or dissociation front with $X(\mathrm{H_2}) = 1/4$, we obtain the hydrogen nucleus column density, N_{DF}, into the cloud where the gas is half molecular and half atomic,

$$N_{\mathrm{DF}} \simeq 3.7 \times 10^{22} \left(\frac{G_0}{n}\right)^{4/3} \mathrm{cm}^2. \tag{8.47}$$

Note that this equation adopts $k_{\mathrm{d}}(\mathrm{H_2}) = 3 \times 10^{-17}\,\mathrm{cm^3\,s^{-1}}$.
Dust opacity becomes important when $N_{\mathrm{DF}} \gtrsim 5 \times 10^{20}\,\mathrm{cm^{-2}}$, or when

$$G_0/n \gtrsim 4 \times 10^{-2}\,\mathrm{cm^3}. \tag{8.48}$$

Self-shielding alone dominates H_2 dissociation and hence the location of the HI / H_2 transition when $G_0/n \lesssim 4 \times 10^{-2}\,\mathrm{cm^3}$ (Fig. 8.5). This includes diffuse clouds and molecular clouds exposed to the interstellar radiation field and dense clumps in PDRs (cf. Chapter 9) with higher FUV fluxes. PDRs associated with bright FUV sources typically have $G_0/n \sim 1\,\mathrm{cm^3}$ and the location of the HI / H_2 transition is then dominated by dust absorption and typically occurs at $A_{\mathrm{v}} \simeq 2$ ($N \simeq 4 \times 10^{21}\,\mathrm{cm^{-2}}$). At that point, dust has reduced the H_2 photodissociation rate sufficiently that an appreciable column of H_2 can build up, H_2 self-shielding takes over, and the HI / H_2 transition will still be very sharp.

H_2 formation

The H_2 formation rate can be derived semi-empirically from H_2 / H column density ratios observed in diffuse clouds (cf. Eq. (8.44) and Fig. 8.6), which gives

$$R_{\mathrm{d}}(\mathrm{H_2}) = 1 - 3 \times 10^{-17} n n(\mathrm{H})\,\mathrm{cm^{-3}\,s^{-1}}. \tag{8.49}$$

These observations refer to gas with temperatures in the range 50-100 K and dust with a temperature of about 15 K. Probably, the H_2 formation rate depends to some extent on these temperatures.

There is no good way to produce molecular hydrogen through gas phase reactions. Radiative association of two H atoms,

$$\mathrm{H} + \mathrm{H} \longrightarrow \mathrm{H_2} + h\nu, \tag{8.50}$$

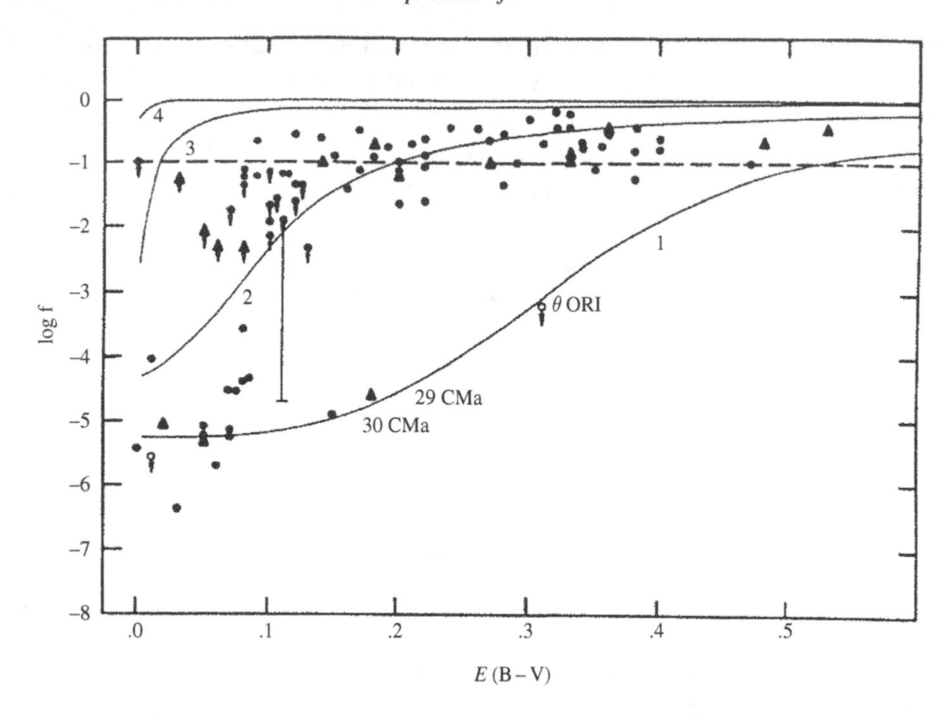

Figure 8.6 The fractional abundance of H_2 derived from UV absorption lines as a function of dust extinction (e.g., total gas column). Triangles indicate Wolf–Rayet, Oe, Be, and peculiar stars. Open circles are upper limits. Arrows on solid symbols indicate stars for which the abundance determination is uncertain. The solid curves are for different, homogeneous models with densities increasing from 10 to 10^4 cm^{-3}, T is 100 K (except for model 4 (20 K), and $k_d(H_2) = 1\times$ (1,2) or 3×10^{-17} cm^{-3} s^{-1} (3,4). Figure reprinted with permission from B. D. Savage, J. F. Drake, W. Budich, and R. C. Bohlin, 1977, *Ap. J.*, **216**, p. 291.

is a slow process ($\lesssim 10^{-23}$ cm^3 s^{-1}) because radiative transitions in a homonuclear molecule are strongly forbidden. Molecular hydrogen can also be formed through the H^- ion,

$$H + H^- \longrightarrow H_2 + e, \tag{8.51}$$

which occurs at a rate of $k = 1.3\times10^{-9}$ cm^3 s^{-1}. The hydrogen anion is formed through the radiative association reaction

$$H + e \longrightarrow H^- + h\nu \qquad k = 10^{-18}\,T\,\mathrm{cm^3 s^{-1}}, \tag{8.52}$$

and destroyed through photodissociation,

$$H^- + h\nu \longrightarrow H + e \qquad k = 2.4\times10^{-7}\ \ \mathrm{s^{-1}}, \tag{8.53}$$

where the latter rate is for the average ambient interstellar radiation field. The H_2 formation rate through this route is given by

$$R_{H^-}(H_2) = 1.3\times10^{-9} n(H^-) n(H)\,\mathrm{cm^{-3}\,s^{-1}}, \tag{8.54}$$

which we can write as

$$R_{H^-}(H_2) \simeq 7.5\times10^{-23}\left[\frac{T}{100\,\mathrm{K}}\right]\left[\frac{x}{1.4\times10^{-4}}\right] n n(H)\,\mathrm{cm^{-3}\,s^{-1}}. \tag{8.55}$$

Comparing the rates of these two gas phase routes for the formation of H_2 with that required to explain the observed H_2 abundances, it is clear that there has to be an efficient alternative.[2]

Because of the difficulty of forming H_2 in the gas phase, it is now generally accepted that the formation of H_2 in the ISM proceeds on the surfaces of interstellar dust grains. The H_2 formation rate per unit volume, $R_d(H_2)$, can then be expressed as

$$R_d(H_2) = \frac{1}{2} S(T, T_d)\,\eta\, n_d \sigma_d\, n(H) v_H, \tag{8.56}$$

where $S(T, T_d)$ is the sticking probability of an H atom with temperature T colliding with a grain of temperature T_d, η is the probability that an adsorbed H atom will migrate over the grain surface, find another H atom, and form H_2 before evaporating from the grain surface, $n_d\sigma_d$ is the total grain surface area per unit volume, n_H is the H atom density, and $v_H = 1.5\times10^4\,T^{1/2}$ cm s^{-1}, is the thermal speed of the H atoms. Typically, $n_d\sigma_d \simeq 10^{-21}\,n$ cm^{-2} and Eq. (8.56) then becomes

$$R_d(H_2) \simeq 5\times10^{-17}\left(\frac{T}{100\,\mathrm{K}}\right)^{1/2} S(T, T_d)\,\eta(T_d)\, n\, n_H\,\mathrm{cm^{-3}\,s^{-1}}. \tag{8.57}$$

For sufficiently low temperatures, S and $\eta \sim 1$ (see below) and H_2 formation on interstellar grain surfaces can explain quantitatively the observed H_2 abundances. Because S and η tend to decrease with increasing temperature as atoms bounce from the grain surfaces or evaporate before reacting, this rate will decrease with T and T_d and the details of this dependence will depend on the interaction of atomic H with astrophysically relevant surfaces.

[2] The H^- channel may have played an important role in H_2 formation in the early Universe prior to heavy element synthesis in the first generation of stars.

The formation of H_2 has been the subject of several experimental studies. Based on those studies, a kinetic model for H_2 formation under astrophysically relevant conditions has been developed, which is very instructive. For the diffuse interstellar medium, bare silicate and graphitic grains are thought to be the dominant grain surfaces.[3] Experiments have shown the existence of physisorbed and chemisorbed sites for atomic H on such surfaces. H atoms will accrete into physisorbed sites, migrate around, and either react to form H_2 with another atomic H, transfer into a chemisorbed site, or evaporate. The relative probabilities for these different results depend on the barriers involved. The well depths for these sites are very different ($\sim 500\,\mathrm{K}$ versus $\sim 10\,000\,\mathrm{K}$) and only H in physisorbed sites will be mobile at low T. Hence, at low temperatures, by far the most efficient way to form H_2 is when a physisorbed atom reacts with a chemisorbed atom. Other routes to form H_2 as well as the evaporation of chemisorbed atoms are negligible. Realizing that the surface coverage of physisorbed H is always small compared to 1 (physisorbed H quickly migrates to a chemisorbed well), the steady-state equations can be simplified to

$$r_{\mathrm{H}}(1-\theta_{\mathrm{p}}(\mathrm{H}_2)) - k_{\mathrm{pc}}\theta_{\mathrm{p}}(\mathrm{H}) - k_{\mathrm{ev}}(\mathrm{H})\theta_{\mathrm{p}}(\mathrm{H}) = 0, \tag{8.58}$$

$$k_{\mathrm{pc}}\theta_{\mathrm{p}}(\mathrm{H})(1-2\theta_{\mathrm{c}}(\mathrm{H})) = 0, \tag{8.59}$$

$$\mu k_{\mathrm{pc}}\theta_{\mathrm{p}}(\mathrm{H})\theta_{\mathrm{c}}(\mathrm{H}) - k_{\mathrm{ev}}(\mathrm{H}_2)\theta_{\mathrm{p}}(\mathrm{H}_2) = 0, \tag{8.60}$$

where r_{H} is the H atom accretion rate per site ($= n(\mathrm{H})\pi a^2\, v(\mathrm{H})/N_{\mathrm{s}}$, with N_{s} the number of sites per unit surface area ($2\times 10^{15}\,\mathrm{cm}^{-2}$), $\theta_{\mathrm{p,c}}$ are the surface coverages in the physisorbed and chemisorbed sites, k_{ev} is the evaporation rate ($= \nu_0 \exp[-E_{\mathrm{b}}/kT]$). The rate at which H transfers from physisorbed sites to chemisorbed sites, k_{pc}, depends on the migration barrier involved and whether tunneling or thermal hopping dominates. This results in

$$\theta_{\mathrm{c}}(\mathrm{H}) = \frac{1}{2}, \tag{8.61}$$

$$\theta_{\mathrm{p}}(\mathrm{H}) = \left(\frac{\mu k_{\mathrm{pc}}}{2k_{\mathrm{ev}}(\mathrm{H}_2)} + \frac{k_{\mathrm{pc}}}{r_{\mathrm{H}}} + \frac{k_{\mathrm{ev}}(\mathrm{H})}{r_{\mathrm{H}}}\right)^{-1}, \tag{8.62}$$

$$\theta_{\mathrm{p}}(\mathrm{H}_2) = \frac{\mu k_{\mathrm{pc}}}{2k_{\mathrm{ev}}(\mathrm{H}_2)}\theta_{\mathrm{p}}(\mathrm{H}). \tag{8.63}$$

[3] In molecular clouds, such surfaces may be covered by simple ice molecules (e.g., H_2O) and this will affect the mode of H_2 formation (e.g., H abstraction from species such as N_2H_2 by atomic H rather than direct atomic H recombination). But then hydrogen will be largely molecular in such an environment anyway.

The H_2 formation efficiency is, then,

$$\eta = \left(\frac{\mu r_H}{2k_{ev}(H_2)} + 1 + \frac{k_{ev}(H)}{k_{pc}} \right)^{-1}. \quad (8.64)$$

The three terms in this expression dominate in different temperature regimes. The first term of this expression – important at low T – represents H_2 formation when newly formed molecules do not evaporate. The surface then becomes quickly saturated with molecules, and incoming H atoms cannot populate the physisorbed sites. Further H adsorption may be possible in physisorbed sites associated with this "sea" of H_2 covering the grain surface. At present, the experiments provide no information on this and this is not included in this model. Nevertheless, it is probable that H recombination stays efficient even at the lowest temperatures. The second term of this expression dominates at somewhat higher grain temperatures. The newly formed molecules can evaporate and the physisorbed atoms recombine. This range of temperature (around 5–20 K) is extremely efficient and all the incoming H atoms leave the grain as H_2. The third term represents the temperature range where evaporation of physisorbed H atoms competes with hopping of physisorbed H into chemisorbed sites and the H_2 formation then declines with increasing temperature.

At very high temperatures, when physisorbed H quickly evaporates, H_2 forms through "collisions" by chemisorbed H. A similar system of equations can be set up and solved for the H_2 recombination efficiency. This results in a correction factor,

$$\eta_c = \left(1 + \frac{k_{ev,c}^2(H) k_{ev}(H)}{2 r_H k_{pc} k_{cc}} \right)^{-1}, \quad (8.65)$$

which shuts down the recombination efficiency at high temperatures. Here, $k_{ev,c}$ is the evaporation rate of H from chemisorbed sites and k_{cc} is the rate at which H transfers between chemisorbed sites.

The resulting H_2 formation efficiency thus depends very much on the interaction of H atoms with the surface and their mobility at different temperatures. Much progress has been made in recent years in this area through both experiments and quantum chemical calculations. The results of an analysis of the experimental data based upon kinetic models are summarized in Fig. 8.7. The global characteristics of the H_2 recombination efficiency are well established and at low temperature the interaction of physisorbed H with these surfaces is well understood. Hence, H_2 formation operates at unit efficiency for T less than 20 K (silicates) and 25 K (carbonaceous). However, the details of the interaction of chemisorbed H with these surfaces are less clear. In particular, the transfer between physisorbed and chemisorbed sites at higher temperatures, which is

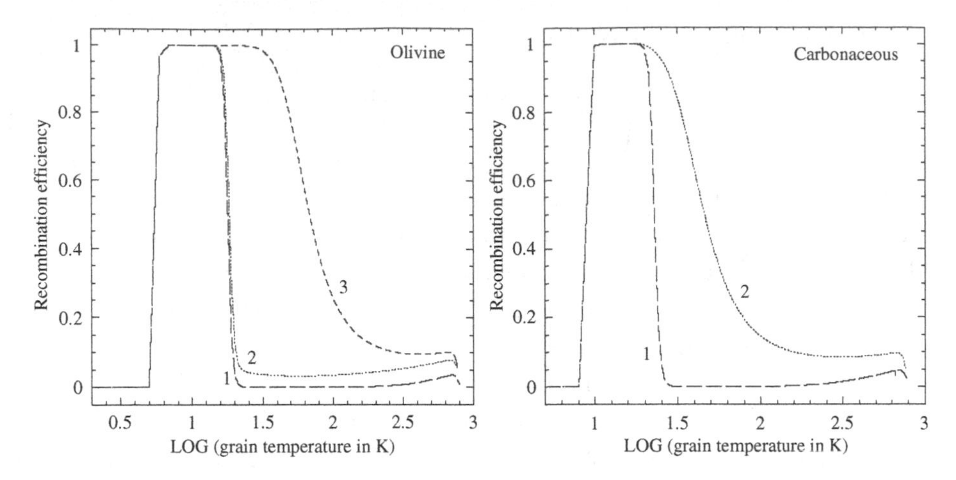

Figure 8.7 Recombination of atomic H on silicate and carbonaceous surfaces as a function of temperature. The three regimes described in the text are readily apparent. While the presence of chemisorbed sites is well established on these surfaces, the details of the transfer between these sites and between physisorbed and chemisorbed sites are still somewhat open. The results for three (olivine) and two (carbonaceous) interaction potentials consistent with present experiments are indicated in the figure.

crucial for H_2 formation at elevated temperatures, is still open. From an astronomical point of view, observations of warm PDRs (e.g., the Orion Bar) have shown that H_2 formation is efficient even at dust temperatures up to 75 K. The observations thus seem to prefer curves 2 (carbonaceous) or 3 (silicate) in Fig. 8.7. If we adopt this point of view, then we can use the analytical expressions for the H_2 recombination efficiency (Eqs. (8.64) and (8.65)) with the following approximations,

$$\frac{k_{\rm ev}(\rm H)}{k_{\rm pc}} = \frac{1}{4}\left(1+\sqrt{\frac{E_{\rm c}-E_{\rm s}}{E_{\rm p}-E_{\rm s}}}\right)^2 \exp[-E_{\rm s}/kT] \tag{8.66}$$

and

$$\eta_{\rm c} = \left(1+\frac{\nu_{\rm z}}{2r_{\rm H}}\left(1+\sqrt{\frac{E_{\rm c}-E_{\rm s}}{E_{\rm p}-E_{\rm s}}}\right)^2 \exp[-1.5E_{\rm c}/kT]\right)^{-1}, \tag{8.67}$$

with E_c, E_p, and E_s the chemisorbed, physisorbed, and saddle-point energy, respectively, and ν_z the frequency factor in the chemisorbed site (Eq. (4.30)). The parameters in these fits are provided in Table 8.2. The frequency factors can be found from the harmonic potential discussion (Section 4.2.2; $\nu_c = 2.3 \times 10^{13}$ s^{-1}, $\nu_p = 3.7 \times 10^{12}$ s^{-1} [carbonaceous] and $\nu_p = 3.2 \times 10^{12}$ s^{-1} [silicate]). Note that

Table 8.2 H_2 *formation on interstellar grain surfaces*[a]

Surface	E_c [b](K)	E_p [b](K)	E_s(K)	$(E_c - E_s)$(K)	$(E_p - E_s)$(K)	a (Å)	μ
Carbon	29 460	712	177	29 283	555	3	0.4
Silicates	29 460	551	208	29 252	327	4	0.3

[a] Parameters describing the fits 2 (carbon) and 3 (silicate) in Fig. 8.7. For details on these results, the original reference (S. Cazaux and A. G. G. M. Tielens, 2004, *Ap. J.*, **604**, p. 222) should be consulted.

[b] Includes the zero-point energy.

these analytical expressions do not provide a good fit to the other results shown in this figure because tunneling is more important.

8.7.2 Oxygen chemistry

The high abundance of atomic H in diffuse clouds makes H^+ – produced through cosmic-ray ionization – an important starting point of the O chemistry. Charge exchange of neutral O with H^+ is rapid because of the near-resonance character (Section 4.1.4). Reactions of O^+ with H_2 then form OH^+ (Fig. 8.8). Further reactions with H_2 lead to H_3O^+, which dissociatively recombines to OH and H_2O. Reactions with C^+ then lead to CO^+ and HCO^+ and, eventually, CO. Of course, molecule formation is strongly counteracted by FUV photodestruction. Deeper in the cloud, where H_2 is an important species, cosmic-ray ionization of H_2 leads eventually to the formation of H_3^+, which readily reacts with O to form OH^+. By then, we are entering dark cloud chemistry (cf. Section 10.4.3).

8.7.3 Carbon chemistry

In diffuse clouds, C^+ is the main reservoir of gas phase carbon. The reaction of C^+ with H_2 to form CH^+ – analogous to the start of the O-bearing chemistry – is strongly inhibited by an activation barrier ($\Delta E \simeq 4600$ K) at low temperatures. There may be some non-thermal H_2 pumped by FUV photons, which could promote this reaction (Section 4.1.9). However, this reaction is not as important as in bright PDRs. Instead, diffuse cloud C chemistry is generally thought to be initiated by the radiative association reaction of C^+ with H_2 forming CH_2^+ ($k \simeq 10^{-15}\ \mathrm{cm^3\,s^{-1}}$). This then reacts with H_2 to form CH_3^+. CH_3^+ reacts with H_2 only through a very slow radiative association reaction. Instead, CH_3^+ dissociatively recombines with electrons to form the small hydrocarbon radicals CH and CH_2. Further reactions of C^+ with these radicals then lead to a build-up of acetylenic species, which can react further to form complex hydrocarbons. However, because

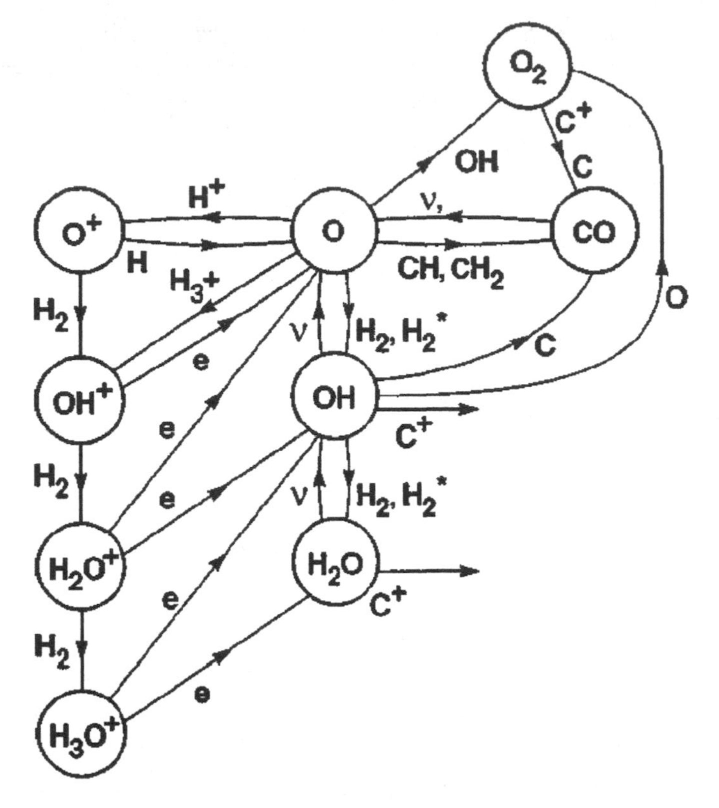

Figure 8.8 Schematic reaction network involved in O chemistry in diffuse clouds and photodissociation regions. In diffuse clouds, oxygen chemistry is initiated by the charge exchange reaction between O and H^+. In dark clouds, the reaction with H_3^+ takes over, while in photodissociation regions, the reactions of O with H_2 and H_2^* can be important.

of the strong FUV radiation field, the abundance of these species is very limited. Deeper in, where the FUV radiation field has been reduced by dust attenuation, carbon is rapidly locked up in the very stable molecule, CO. Complex hydrocarbon formation in dark clouds is then limited by the rate at which carbon can be broken out of CO.

8.7.4 Nitrogen chemistry

Nitrogen is in its neutral atomic form in diffuse clouds and this is very unreactive. The reaction of N with H_3^+ to NH^+ is endothermic. Hence, gas phase N chemistry in diffuse clouds may have to be initiated by cosmic-ray ionization of N. Sequential reactions of N^+ with H_2 interrupted by electron recombination then leads to the formation of NH, NH_2, and NH_3. NH_3 formation may actually be inhibited at

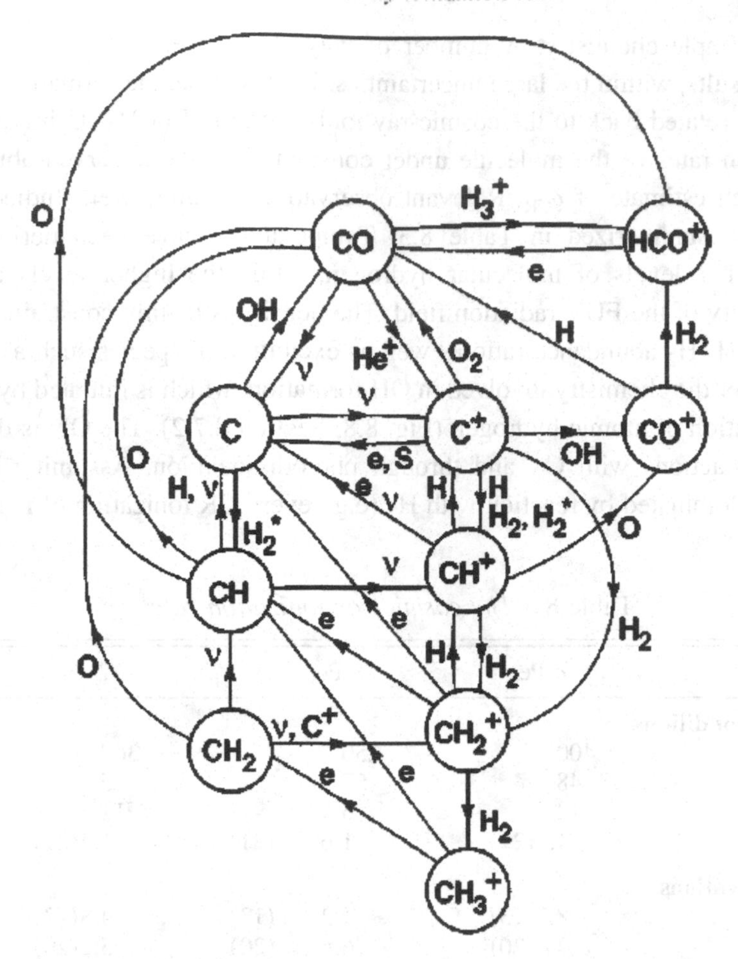

Figure 8.9 Schematic reaction network involved in C chemistry in diffuse clouds and photodissociation regions. In diffuse clouds, carbon chemistry is initiated by the radiative association reaction of C^+ with H_2. In photodissociation regions, the reactions of C^+ with H_2 and H_2^* can be important.

low temperatures because the reaction of NH_3^+ with H_2 has a 1000 K barrier and electrons are very abundant in diffuse clouds (cf. Section 10.4.2). When H_2 is abundant, this chemistry can be equally well initiated by the reaction of cosmic-ray produced H_2^+ with N to form NH^+ at a rate of $1.9 \times 10^{-9}\ \mathrm{cm^3\ s^{-1}}$.

8.8 The cosmic-ray ionization rate

Cosmic-ray ionization is the cornerstone of gas-phase chemistry routes. Conversely, molecular observations can be used to determine the cosmic-ray ionization rate. Diffuse molecular clouds are often used for this purpose because of their simple chemistry. A number of ways have been devised, resulting in similar results, within the large uncertainties. In each case, the formation rate of a species is related back to the cosmic-ray ionization of H or H_2. Using estimated destruction rates of the molecule under consideration, the observed abundances result in an estimate of ζ_{CR}. Relevant observations of three well-studied diffuse clouds are summarized in Table 8.3. Temperatures have been derived from the lowest J levels of molecular hydrogen, while the higher levels constrain the intensity of the FUV radiation field. The density is mainly constrained by the observed H/H_2 abundance ratio as well as excitation of species such as C_2.

Consider the chemistry involved in OH formation, which is initiated by cosmic-ray ionization of atomic hydrogen (Fig. 8.8, Section 8.7.2). The OH is destroyed through reactions with C^+ and through photodissociation. Assuming that loss of O^+ is dominated by reaction with H_2 (e.g., every CR ionization of H results in

Table 8.3 *The cosmic-ray ionization rate*[a]

	o Per	ζ Per	ζ Oph	units
Physical conditions				
n	400	250	300	cm^{-3}
T	48	57	75	K
G_0	12	6	11	Habing
N_0	1.6(21)	1.6 (21)	1.4(21)	cm^{-2}
OH observations				
N (OH)	8.2(13)	4.2 (13)	4.8(13)	cm^{-2}
N (H)	7.4(20)	6.5 (20)	5.2(20)	cm^{-2}
ζ_{CR}	7.0(−16)	2.0 (−16)	5.(−16)	s^{-1}
HD observations				
N (HD)	5.9(15)	3.8 (15)	2.1(14)	cm^{-2}
N (O)	7.4(17)	4.8 (17)	7.1(17)	cm^{-2}
ζ_{CR}	6.0(−16)	4.0 (−16)	1.(−16)	s^{-1}
NH observations				
N (N)		1.6 (17)	1.1(17)	cm^{-2}
N (NH)		9.0 (11)	9.(11)	cm^{-2}
ζ_{CR}		4.0 (−15)	1.(−14)	s^{-1}
H_3^+ observations				
N (H_3^+)		8 (13)		cm^{-2}
N (H_2)		5 (20)		cm^{-2}
N (C^+)		1.8 (17)		cm^{-2}
ζ_{CR}		2 (−16)		s^{-1}

[a] A summary of the molecular observations of three well-studied diffuse clouds. The cosmic-ray ionization rates derived from different molecular studies are listed separately (for details see text).

the formation of OH) and that photodissociation dominates OH destruction, we can estimate the primary cosmic-ray ionization rate, ζ_{CR}, per H nucleus

$$\zeta_{CR} = 0.67 k_{UV}(\text{OH}) \frac{N(\text{OH})}{N(\text{H})}, \tag{8.68}$$

where the numerical factor takes secondary ionizations into account. The unshielded photodissociation rate of OH is $7.6 \times 10^{-10} G_0\ \text{s}^{-1}$. The total FUV dust optical depth of these clouds is $\simeq 2$ and, hence, dust shielding would reduce this rate by less than a factor of ~ 3.

The chemical scheme for HD also starts with charge exchange with H^+. The resulting D^+ reacts quickly with H_2 and HD is destroyed by FUV photons. The H^+ abundance is regulated through fast charge exchange reactions with O/O^+. In the steady state, we have for the cosmic-ray ionization rate

$$\begin{aligned} \zeta(\text{HD}) &= \frac{n(\text{HD})\, n(\text{O})\, k_{\text{O}}\, k_{UV}(\text{HD})}{n(\text{H})\, n(\text{O})\, k_{\text{D}}} \\ &= 3.3 \times 10^{-10} \exp[-191/T]\, G_0 \frac{N(\text{HD})}{N(\text{H})}\ \text{s}^{-1}, \end{aligned} \tag{8.69}$$

where we have adopted an unshielded HD photodissociation rate of $2.6 \times 10^{-11}\ \text{s}^{-1}$, and O and D abundances of 3×10^{-4} and 1.5×10^{-5}. For the ratio of the rates of O and D with H^+ we have used $k_{\text{O}}/k_{\text{D}} = 0.64 \exp[-191/T]$. In this scheme, the HD chemistry is thus also derived from H^+ and the derived ionization rates are quite similar to those derived from OH. It should be noted that for columns in excess of $10^{15}\ \text{cm}^{-2}$, HD self-shielding becomes important and for O Per and ζ Per this will decrease the derived rates.

The cosmic-ray ionization rate can also be estimated from the observed abundance of NH. Setting up a steady-state balance where every cosmic-ray ionization of N – at a rate $4.5\,\zeta_{CR}$ – leads eventually to NH, which is predominantly destroyed through FUV photons (Section 8.7.4), yields

$$\zeta_{CR} = \frac{k_{UV}}{4.5} \frac{N(\text{NH})}{N(\text{N})}, \tag{8.70}$$

where we have assumed that any NH_2 produced is dissociated first to NH and then to N. Also, half of the dissociative recombination of molecular cation intermediaries lead directly to N and there is an about equal contribution to NH^+ formation through the H_2^+ route. We have canceled these two factors against each other. Observations for ζ Per and ζ Oph are again summarized in Table 8.3. The unshielded UV dissociation rate is $5 \times 10^{-10}\ \text{s}^{-1}$. With this gas-phase scheme, adopting the unshielded rate, the observations of NH require a very high cosmic-ray ionization rate. There is some uncertainty due to FUV extinction but that amounts to less than a factor of 3. This is a problem inherent to gas-phase chemistry schemes, because nitrogen chemistry is difficult to initiate and NH is easy to dissociate.

Recently, H_3^+ has been observed in the diffuse ISM through its near-IR vibrational transitions. In diffuse clouds, the chemistry of H_3^+ is particularly simple. Cosmic-ray ionization of H_2 leads to H_2^+, which reacts with H_2, to form H_3^+. H_3^+ is lost through dissociative electron recombination as well as through neutralization with large PAH anions. Just taking electron recombination into account leads, in the steady state, to

$$\zeta_{CR} = \frac{k_e n(e) n(H_3^+)}{2.3 n(H_2)}, \tag{8.71}$$

where the numerical factor takes secondary ionization into account. Assuming that the electrons come from C^+ and that the cloud is uniform (e.g., H_2 is coexistent) and adopting an electron recombination coefficient of $10^{-6}\,T^{-0.45}$, we have

$$\zeta_{CR} = 7\times 10^{-8}\frac{N(C^+)N(H_3^+)}{N(H_2)}\frac{1}{L} = 7\times 10^{-16}\left(\frac{pc}{L}\right). \tag{8.72}$$

where L is the pathlength through the cloud and the right-hand side pertains to ζ Per (Table 8.3). Various tracers indicate that this cloud is quite dense, 200–400 cm^{-3}, for a diffuse cloud. Adopting a density of 250 cm^{-3} and a total hydrogen column of 1.6×10^{21} cm^{-2}, L is 2 pc and $\zeta_{CR} = 3\times 10^{-16}$.

The values for the cosmic-ray ionization rate are somewhat uncertain because of dust attenuation – which would decrease the CR ionization rate – and because there are other loss channels for H^+ (e.g., recombination) and for OH (e.g., reaction with C^+), both of which tend to increase the estimate for the cosmic-ray ionization rate. In particular, the key to the analysis of the OH observations is the assumption that charge exchange with O is a dominant loss channel for H^+ and radiative electron recombination is unimportant. So, a large fraction of the ionizations of atomic H flow to OH. However, if large PAHs are present, PAH anions will dominate the charge balance of atomic ions (Section 8.2.1) and the H^+ abundance is reduced. In that case, the abundance ratio of H^+ to H_3^+ is simply $0.7\,n(H)/n(H_2)$, which is of the order of unity. Hence, OH can be as rapidly formed through reactions of H_3^+ with O as through the H^+ channel. If we ignore for the moment the H^+ channel, we have

$$\frac{N(OH)}{N(H_3^+)} = \frac{n(C^+)k_{C^+}}{n(O)k_{H_3^+}} \simeq 0.3. \tag{8.73}$$

For ζ Per, the observed value is slightly higher, perhaps revealing the input from the H^+ channel. Hence, the OH observations towards this star may refer to the molecular gas. Of course, we should also take recombination of H_3^+ with

PAH anions into account, but, in this case, the effect is much smaller because dissociative recombination is already a fast process.

The only discordant ionization rate in Table 8.3 is the one derived from NH. Interstellar nitrogen chemistry, in general, is poorly understood and, probably, this is merely one expression of this. Possibly, the high abundance of NH reflects the presence of a different formation route for this radical. Formation on grain surfaces has been suggested as one alternative route. Scaling the formation rate on grain surfaces from the H_2 formation rate (Section 8.7.1), we have

$$N(\mathrm{NH}) = 3 \times 10^{-8} \left(\frac{n}{G_0}\right) N(\mathrm{N}) \qquad \mathrm{cm}^{-2}. \tag{8.74}$$

Hence, the densities would have to be substantially higher (1000 and 3000 cm^{-3}) than estimated (Table 8.3) for this to be important. An alternative would be to have atomic N react with vibrationally excited H_2. The rapid FUV pumping and the slow radiative decay can lead to a small abundance of vibrationally excited H_2, $\simeq 5 \times 10^{-9} - 2 \times 10^{-8}$ in $v = 1$ according to detailed models for ζ Per and ζ Oph. Assuming a rate of $10^{-9}\,\mathrm{cm}^3\,\mathrm{s}^{-1}$ for the reaction of N with vibrationally excited H_2 (which is a stretch) and a density of $300\,\mathrm{cm}^{-3}$ results in a predicted column density ratio of $N(\mathrm{NH})/N(\mathrm{N}) \simeq 10^{-5}$, which is quite comparable to the observed ratio.

In summary, the actual cosmic-ray ionization rate in diffuse clouds is probably $\sim 2 \times 10^{-16}\,\mathrm{s}^{-1}$ and the main uncertainties stem from the dust attenuation of the dissociating radiation field and the internal structure (density, temperature, and abundance gradients) of the cloud.

The cosmic-ray ionization rate can also be estimated from observations of H_3^+ in dense clouds. The reaction sequence is then

$$\mathrm{H_2} \xrightarrow{\mathrm{CR}} \mathrm{H_2^+} \xrightarrow{\mathrm{H_2}} \mathrm{H_3^+} \xrightarrow{\mathrm{CO,O}} \text{products}. \tag{8.75}$$

The primary cosmic-ray ionization rate is then

$$\zeta_{\mathrm{CR}} = 0.43 X(\mathrm{H_3^+}) \left(X(\mathrm{CO})\, k_{\mathrm{CO}} + X(\mathrm{O})\, k_{\mathrm{O}} \right) n(\mathrm{H_2}). \tag{8.76}$$

The reaction rates for H_3^+ are $k_{\mathrm{CO}} = 1.7 \times 10^{-9}\ \mathrm{cm}^3\,\mathrm{s}^{-1}$ and $k_{\mathrm{O}} = 8 \times 10^{-10}\,\mathrm{cm}^3\,\mathrm{s}^{-1}$. CO and O are the main reservoirs of carbon and oxygen in molecular clouds with abundances, assuming no freeze-out, of 1.5×10^{-4}. In dark clouds, H_3^+ has been observed towards massive protostars through its rotational–vibrational transitions in the near-IR with abundances in the range of $1\text{–}2 \times 10^{-9}$. Converting these to cosmic-ray ionization rates requires an estimate of the density structure around these protostars. Adopting a density of $10^6\,\mathrm{cm}^{-3}$ implies $\zeta_{\mathrm{CR}} = 1 - 2 \times 10^{-16}\,\mathrm{s}^{-1}$, in agreement with the diffuse cloud observations. Detailed models, taking available molecular observations into account, result in a somewhat lower primary cosmic-ray ionization rate inside dense clouds, $\zeta_{CR} = 3\times10^{-17}\,\mathrm{s}^{-1}$. All of this depends very much on the assumed density structure and is thus more uncertain than for the diffuse cloud studies.

8.9 Observations

8.9.1 HI observations

The 21 cm line connects the two hyper-fine-structure levels of the ground state of atomic hydrogen – with antiparallel and parallel electron and nuclear spin. Because of its low frequency and its forbidden character, the spontaneous radiative transition probability is very small ($2.9\times10^{-15}\,\mathrm{s}^{-1}$), which results in a very low critical density ($2\times10^{-5}\,\mathrm{cm}^{-3}$). As a result, the excitation of these levels is well described by the Boltzmann distribution. Furthermore, given the small energy separation ($\simeq 0.07$ K), the excitation simplifies to the ratio of the statistical weights ($n_u/n_l = g_u/g_l = 3$). The line-averaged optical depth is then (cf. Section 2.3.2)

$$\begin{aligned}\tau_{HI} &= \frac{A_{ul}c^3}{8\pi\nu_{ul}^3}\frac{n_u L}{b}\left(\frac{n_l g_u}{n_u g_l} - 1\right)\\ &\simeq 5.5\times10^{-21} N_H \left(\frac{\mathrm{km\,s^{-1}}}{b}\right)\left(\frac{100\,\mathrm{K}}{T}\right),\end{aligned} \tag{8.77}$$

with N_H the total column of hydrogen and b the Doppler broadening parameter. Thus, for a diffuse cloud, the 21 cm line becomes optically thick for HI columns exceeding $\sim10^{20}\,\mathrm{cm}^{-2}$. For the intercloud medium, that is a factor 100 larger and hence never important.

We are in the Rayleigh regime and, in the optically thin limit, we can write (cf. Section 2.3.2)

$$N_H \simeq 3\times10^{18}\left(\frac{T_b\Delta v}{\mathrm{K\,km\,s^{-1}}}\right)\mathrm{cm}^{-2}, \tag{8.78}$$

where we have used that the FWHM of the line, $\Delta v = 1.67\,b$. Thus, in the low optical depth limit, the column density derived from the integrated intensity, $T_b\Delta v$, is independent of the kinetic gas temperature. The mass of a cloud (or, more general, an HI-structure) follows immediately from the observed brightness distribution on the sky.

Assuming a size for the emitted region – say, adopting the projected size on the sky – the density of the gas can be estimated. Of course, this density is only a lower limit if the gas is very filamentally distributed. A limit on the temperature can be derived from the observed linewidth, but turbulent motions may make this meaningless. In the marginally optically thick limit, $\tau\sim3$, the brightness temperature is equal to the kinetic temperature of the gas.[4] The optical depth can be estimated whenever a suitable continuum background source is present,

$$\tau(\mathrm{HI}) = \ln\left[\frac{T_{\mathrm{c}}}{T_{\mathrm{line}}}\right], \qquad (8.79)$$

with T_{c} the continuum temperature of the source and T_{line} the observed (absorption) line temperature. Then, assuming that the HI properties are smoothly distributed over the sky, an average of the HI emission over spatially adjacent positions can be used to arrive at the kinetic temperature of the gas (Fig. 8.10). This method emphasizes, of course, low temperature regions (e.g., clouds) the most (cf. Eq. (8.77)).

HI studies have provided much insight into the structure of the interstellar medium. In particular, such studies have revealed the existence of two physically distinct phases: the warm and cold media (WNM and CNM). About 60% of the HI is in the WNM filling about 60% of the volume in the plane (and more at higher latitudes) with a typical column density of $10^{20}\,\mathrm{cm}^{-2}$.

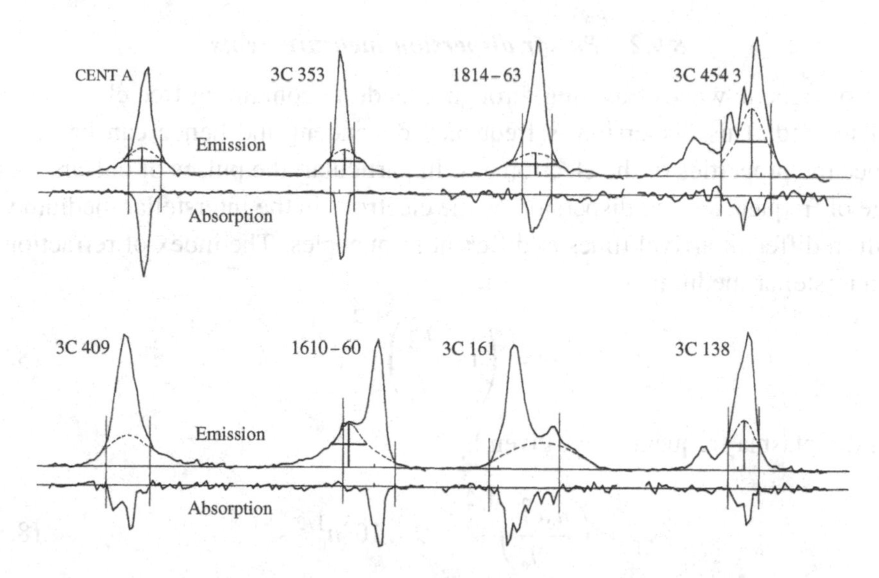

Figure 8.10 HI emission and absorption lines towards high galactic latitudes. The absorption is due to cold diffuse clouds. The broad components and wings also present in the emission profiles originate in much warmer (WNM) HI gas, which contributes little to the absorption. Figure reproduced with permission from V. Radhakrishnan, J. D. Murray, P. Lockhart, and R. P. J Whittle, 1972, *Ap. J. S.*, **24**, p. 15.

[4] For a very optically thick line, the line will be on the square-root portion of the curve of growth but that never really applies for HI (cf. Section 2.3.2).

A significant fraction ($\sim$50 %) of the WNM occurs in the thermally unstable temperature-regime (500–5000 K), illustrating the importance of time evolution. The CNM has a typical column density of $5 \times 10^{19}\,\mathrm{cm}^{-2}$ and a column density weighted average temperature of 70 K. However, some gas is as cold as 15 K. These clouds are highly turbulent and structurally they resemble sheets rather than spherical clouds. The distribution of HI is complex with a CNM layer with a scale height of 100 pc. The WNM layer consists of two layers: a Gaussian layer with scale height of 250 pc and an exponential layer with scale height of some 500 pc. In addition to these components, there is also intermediate and high velocity gas (e.g., gas that does not partake in galactic rotation), generally at high latitudes, suggesting large-scale disturbances of the interstellar medium. Indeed, some 10% of the sky is covered by high velocity ($>90\,\mathrm{km\,s^{-1}}$) HI gas with column densities exceeding $10^{18}\,\mathrm{cm}^{-2}$. The intermediate velocity ($90 > v_{\mathrm{lsr}} > 30\,\mathrm{km\,s^{-1}}$) gas probably represents gas partaking in a galactic fountain. At least some of the high velocity gas, on the other hand, is extragalactic in origin; notably the Magellanic Stream and the low-metallicity gas in the high velocity cloud complex (C).

8.9.2 *Pulsar dispersion measurements*

Electromagnetic waves traveling through a medium containing free electrons will be dispersed. This dispersion is frequency dependent and hence can be used to deduce the properties of the electron gas. In particular, the pulses of pulsars span a range of frequencies and dispersion by the electrons in the interstellar medium will result in different arrival times at different frequencies. The index of refraction of the interstellar medium is

$$m = \left(1 - \frac{\nu_{\mathrm{p}}^2}{\nu^2}\right)^{1/2}, \tag{8.80}$$

with the plasma frequency, ν_{p}, given by

$$\nu_{\mathrm{p}} = \left(\frac{n_{\mathrm{e}} e^2}{\pi m_{\mathrm{e}}}\right)^{1/2} \simeq 9 \times 10^3 n_{\mathrm{e}}^{1/2}\,\mathrm{s}^{-1}, \tag{8.81}$$

which is of the order of 1 kHz in the ISM; e.g., much less than the frequency of the radio waves studied (GHz range). A wave packet will travel at the group velocity. In a medium with refractive index m, the group velocity will differ from the phase velocity (c) by a factor

$$\frac{v_{\mathrm{group}}}{c} = \left(1 - \frac{\nu_{\mathrm{p}}^2}{\nu^2}\right)^{1/2}. \tag{8.82}$$

The arrival time of a pulse is then given by

$$t = \int_0^d v_{\text{group}}^{-1} \mathrm{d}\ell, \tag{8.83}$$

where d is the distance to the pulsar. At low frequencies, we can expand Eq. (8.82) and the time delay, $\Delta t = t - d/c$, is then

$$\begin{aligned} \Delta t &= \frac{e^2}{2\pi m_e c} \frac{D}{\nu^2} \\ &\simeq 4.1 \times 10^{-3} \left(\frac{D}{\text{cm}^{-3}\,\text{pc}}\right) \left(\frac{\text{GHz}}{\nu}\right)^2 \text{s}, \end{aligned} \tag{8.84}$$

where the dispersion measure, D, is the integral of the electron density over the line of sight,

$$D = \int n_e \, \mathrm{d}\ell. \tag{8.85}$$

Actually, what we can measure is the change in the time delay with frequency (i.e., the derivative of Eq. (8.84)),

$$\begin{aligned} \frac{\mathrm{d}\Delta t}{\mathrm{d}\nu} &= \frac{e^2}{\pi m_e c} \frac{D}{\nu^3} \\ &\simeq 8.3 \times 10^{-3} \left(\frac{D}{\text{cm}^{-3}\,\text{pc}}\right) \left(\frac{\text{GHz}}{\nu}\right)^3 \text{s}\,\text{GHz}^{-1}. \end{aligned} \tag{8.86}$$

Dispersion measures are one of the principal ways to determine the mean electron density in the ISM. Over a thousand pulsars have been observed, resulting in an average electron density of $\simeq 0.03\,\text{cm}^{-3}$ in the ISM and a scale height for the ionized gas component of almost 1 kpc. This average electron density corresponds to a typical dispersion measure of $30\,\text{cm}^{-3}\,\text{pc}$ for a distance of 1 kpc and a delay of 12 s at 100 MHz.

In addition, studies of pulsars also allow a measurement of the interstellar magnetic field. The presence of a magnetic field will make the plasma anisotropic, with an index of refraction given by

$$m_{\text{R,L}} = \left(1 - \frac{\nu_p^2}{\nu\,(\nu \pm \nu_B)}\right)^{1/2}, \tag{8.87}$$

for right-handed and left-handed polarization. The cyclotron frequency is given by $\nu_B = eB_{\parallel}/2\pi m_e c$. This results in a Faraday rotation angle given by

$$\Delta\theta_F = \frac{\pi}{c\nu^2} \int_0^d \nu_p^2 \omega_B \mathrm{d}\ell. \tag{8.88}$$

We can then define the rotation measure as

$$R = 0.81 \int_0^d n_e B(\mu\text{G}) \cos\theta \, \mathrm{d}\ell \, \text{rad m}^{-2}. \tag{8.89}$$

The units here are traditional – density in cm^{-3}, magnetic field strength in μG, and the pathlength in pc. The rotation is then given by $R/\lambda^2(\text{m})$. The magnetic field then follows from $\overline{B}_{\parallel} = 1.2R/D\,\mu\text{G}$. Alternatively, we can also take the derivative of Eq. (8.88) and compare that with the derivative of the time delay,

$$\frac{\mathrm{d}\Delta\theta_F/\mathrm{d}\nu}{\mathrm{d}\Delta t/\mathrm{d}\nu} = 20\,B_{\parallel}. \tag{8.90}$$

Pulsar studies have shown that there is an ordered magnetic field with a strength of 1–2 μG and a random magnetic field of some 5–6 μG on a scale length of some 50 pc.

8.9.3 Hα emission

Recombinations of electrons and protons in the warm ionized medium give rise to weak Hα emission at 6563 Å (cf. Fig. 1.5),

$$\begin{aligned} I(\text{H}\alpha) &= \frac{\alpha_{32} h\nu_\alpha}{4\pi} \int n_e n_p \mathrm{d}\ell \\ &\simeq 8.8 \times 10^{-8} EM \qquad \text{erg cm}^{-2}\,\text{s}^{-1}\,\text{sr}^{-1}, \end{aligned} \tag{8.91}$$

with α_{32}($= 1.2 \times 10^{-13}\,\text{cm}^3\,\text{s}^{-1}$ at $T = 10^4$ K) the "effective" recombination coefficient (actually, α_{32}/α_A is the number of Hα photons produced per recombination with α_A the total recombination coefficient to all levels) and EM the emission measure ($\int n_e^2 \mathrm{d}\ell$; cf. Section 7.4.1). Often, the Hα intensity is expressed in units of rayleighs, which is the intensity equivalent to 10^6 recombinations ($2.4 \times 10^{-7}\,\text{erg cm}^{-2}\,\text{s}^{-1}\,\text{sr}^{-1}$). This H$\alpha$ emission from the diffuse ISM has been observed using wide-field Fabry–Perot interferometers. In addition, the [SII] 6716 Å and [NII] 6584 Å lines have been observed. In combination with the pulsar dispersion measures – which are proportional to $\int n_e \mathrm{d}\ell$ – these studies reveal that 90% of the ionized gas of the Galaxy is in low-density fully ionized regions, which fill about 0.25 of the volume of the ISM, and has a scale height of about 1 kpc.

8.9.4 Atomic and molecular absorption lines

Interstellar gas will produce visible and UV absorption lines in the spectra of stars. Often, the lines are not fully resolved and the strength can be measured in terms of an equivalent width, W_λ,

$$W_\lambda = \int \frac{I_c(\nu) - I_{obs}(\nu)}{I_c(\nu)} \mathrm{d}\lambda = \int (1 - \exp[-\tau(\lambda)])\mathrm{d}\lambda, \tag{8.92}$$

with $I_c(\nu)$ the continuum intensity and $I_{obs}(\nu)$ the observed intensity. For small columns ($\tau \ll 1$), the equivalent width is given by

$$\frac{W_\lambda}{\lambda} = \frac{\pi e^2}{m_e c^2} N_l \lambda f_{lu}. \tag{8.93}$$

Hence, in this limit, the equivalent width of a line increases linearly with the column density (the linear part of the curve of growth). From an observed line, we can then derive the column density,

$$N_l = 1.1 \times 10^{17} \left(\frac{\text{Å}}{\lambda}\right)^2 \left(\frac{1}{f_{lu}}\right) \left(\frac{W_\lambda}{\text{mÅ}}\right) \text{cm}^{-2}, \tag{8.94}$$

with f_{lu} the oscillator strength. For large columns the line saturates and the equivalent width does not increase much more with increasing column width,

$$W_\lambda \simeq \frac{2b\lambda}{c} (\ln[\tau])^{1/2}, \tag{8.95}$$

where τ is given by

$$\begin{aligned} \tau &= \frac{\pi e^2}{m_e c} f_{lu} \frac{\lambda}{\pi^{1/2} b} N_l \\ &\simeq 7.5 \left(\frac{\text{kms}^{-1}}{b}\right) \left(\frac{N_l}{10^{12}\,\text{cm}^{-2}}\right) \left(\frac{5000\,\text{Å}}{\lambda}\right) f_{lu}. \end{aligned} \tag{8.96}$$

For very high columns, absorption in the line damping wings becomes important. In contrast with stellar atmospheres, pressure broadening is unimportant and the damping wing is then dominated by the natural linewidth. On this square-root portion of the curve of growth, the relation between equivalent width and column density of absorbers is given by

$$\begin{aligned} N_l &= \frac{m_e c^3}{e^2 \lambda^4} \frac{W_\lambda^2}{f_{lu} \gamma} \\ &= 1.8 \times 10^{18} \left(\frac{W_\lambda}{\text{mÅ}}\right)^2 \left(\frac{5000\,\text{Å}}{\lambda}\right)^4 \frac{1}{\gamma f_{lu}}, \end{aligned} \tag{8.97}$$

where γ is the radiation damping constant. Generally, this is only important for the HI Lyα line.

By necessity, optical and FUV studies are limited to bright, relatively nearby stars. These confirm the presence of cold neutral clouds and warm ionized and neutral intercloud media in the Galaxy. In addition, these studies provide information on the abundances of atoms, ions, and molecular species in the ISM. This has revealed large depletion factors for most heavy elements – it is likely that the missing elements are locked up in dust (cf. Section 5.3.6) – as well as systematic differences between the amount of depletion in the CNM and WNM. It is also possible to probe the fine-structure level excitation of various species through FUV absorption lines originating from these levels; notably C^+ and C^0. The 1336 Å line originating in the $^2P_{3/2}$ level measures the column density of excited C^+ and the radiative cooling emission of the neutral ISM (cf. Section 8.9.7). The corresponding ground state absorption line is highly saturated but there is a weak absorption line at 2325 Å, which measures the gas-phase C abundance. The excitation of the neutral carbon levels has been used to derive the pressure of the interstellar gas (Section 8.9.6).

8.9.5 Hot gas

A hot gas will emit in the X-ray regime through free–free emission, radiative recombination, and the two-photon process in the continuum and through collisionally excited lines and the radiative recombination cascade. As for the ionized gas in HII regions, the emission spectrum is characterized by the temperature of the plasma and the emission measure, $\int n^2 d\ell$ (and the elemental abundances), but, in addition, foreground gas can severely attenuate the X-rays, particularly at low energies. The observed X-ray emission of the Galaxy consists of three components: (1) A local component associated with the local bubble with a temperature of 1.5×10^6 K and an emission measure of $5 \times 10^{-3}\,\mathrm{cm^{-6}\,pc}$; (2) a partly absorbed component associated with halo gas at a temperature of 2×10^6 K, an emission measure of $2.5 \times 10^{-3}\,\mathrm{cm^{-6}\,pc}$, and a foreground H column of $3.6 \times 10^{20}\,\mathrm{cm^{-2}}$; (3) an extragalactic component having a power law spectrum due to AGNs, which dominates at the higher energies. The location of the gas is, of course, difficult to ascertain. However, clouds at high latitudes shadow the halo component and hence the scale height has to be substantial (> 3 kpc). Adopting a 3 kpc scale height, the plasma density has to be $\sim 10^{-3}\,\mathrm{cm^{-3}}$.

Absorption lines of highly ionized atoms are a telltale sign of hot gas. In particular, OVI, NV, and CIV lines have been used to trace the distribution of gas with temperatures in the range $1\text{–}5 \times 10^5$ K in the disk and halo of the Galaxy. The ionization potentials involved in the creation and destruction of these ions lie in the range $\sim 50 - 150$ eV and hence these ions reach a peak abundance in a collisional plasma in this temperature range (cf. Table 8.4). Also, these ions still have absorption originating from the ground state, which fall in the FUV range (rather than the X-rays) and hence their abundance can be tracked by FUV satellites such as Copernicus, the International Ultraviolet Explorer, the Hubble Space Telescope, the Hopkins Ultraviolet Telescope, and the Far Ultraviolet Spectroscopic Explorer. Particularly, the latter has been instrumental in "mapping"

Table 8.4 *Highly ionized atoms*

ion	$IP(i-1 \rightarrow i)^a$(eV)	$IP(i \rightarrow i+1)^b$(eV)	T^c_{peak}(K)
OVI	113.9	138.1	3×10^5
NV	77.5	97.9	2×10^5
CIV	47.9	64.5	1×10^5
SiIV	33.5	45.1	8×10^4

[a] Ionization potential to create ion *i*.
[b] Ionization potential to destroy ion *i*.
[c] Temperature at which abundance of ion peaks in collisional ionization equilibrium.

the distribution of the hot gas in the Galaxy through these ions. The high ionization potential of OVI, coupled with the strong He^+ absorption edge in the spectra of even the hottest star, ensures that OVI is an indicator of collisionally ionized gas rather than photo-ionized gas. In contrast, in view of its ionization potential, CIV can also be created in HII regions. Of course, OVI can be generated in a gas irradiated by soft X-rays where it would trace a high ionization parameter (cf. Section 7.2.1; e.g., a low density or a high radiation field).

Absorption line studies of about 100 extragalactic objects have revealed a widespread but highly irregularly distributed OVI layer. The average column density of OVI perpendicular to the plane is 1.6×10^{14} cm^{-2}. The distribution can be represented by a plane-parallel layer of hot gas (3×10^5 K) with a mid-plane OVI density of 1.7×10^{-8} cm^{-3} and a scale height $\sim$2 kpc, coupled with some excess OVI absorbing gas possibly associated with local structures such as the local bubble, nearby superbubbles, or Galactic fountain activity in the Perseus or Sagittarius arms.

Gas with a temperature of $\sim 3\times10^5$ cools very rapidly. For Solar abundances, the cooling time scale is $\tau_{cool} = 3kT/n\Omega \simeq 9000/n$ years or some 10 Myears at a density of 10^{-3} cm^{-3}. Hence, the OVI absorption lines are associated with gas that either is in non-equilibrium – e.g., cooling down in a bubble or a cooling flow – or in the interfaces between hot (10^6 K) and cool (10^4 K) gas where thermal conduction or turbulent mixing produces intermediate temperature gas. The total column density of cooling gas is then $N_{cool} = nv_{cool}\tau_{cool}$, where v_{cool} is some characteristic velocity of the gas. Thus, the total column density is independent of the density. The column density of OVI is, then,

$$N(\mathrm{OVI}) \simeq 3\times10^{18} \mathcal{A}_{\mathrm{O}} X(\mathrm{OVI}) \left(\frac{v_{cool}}{10^2\,\mathrm{km\,s^{-1}}}\right) \mathrm{cm}^{-2}, \tag{8.98}$$

where $\mathcal{A}_{\mathrm{O}}$ is the oxygen abundance and X(OVI) the fraction of the oxygen in the form of OVI. The cooling velocity will depend on the characteristics of the physical situation under consideration. Hot postshock gas produced by shocks in the range 200–500 km s^{-1} will flow at speeds 50–125 km s^{-1} (cf. Section 11.2.1). For cooling flows and mixing layers, the sound speed, C_s, is most relevant, 75 km s^{-1} at $T = 3 \times 10^5$ K. In collisional ionization equilibrium, X(OVI) reaches a peak value of $\simeq 0.2$ around a temperature of 3×10^5 K. Thus, for an oxygen abundance of 3×10^{-4}, we arrive at a column density $N(\text{OVI}) \simeq 2 \times 10^{14}$ cm^{-2} for cooling halo gas. If the halo OVI is dominated by a fountain flow, a mass flux of about $\simeq 1.5\,M_\odot$ year^{-1} to each side of the Galaxy is implied.

For a conducting interface, the column density of evaporating gas is somewhat smaller. In this case, the heat flux conducted in is largely balanced by cooling associated with cloud evaporation, rather than radiative cooling. This is discussed in Section 12.4.2. The column density of gas is then

$$N(\text{OVI}) = \frac{\dot{M}_c}{4\pi R_c^2} \Delta t \mathcal{A}_O X(\text{OVI}). \tag{8.99}$$

Adopting for the mass flux from a cloud, $\dot{M}_c$, the classical expression of the heat conduction (Section 12.4.2), and $\Delta t = R_c/C_s$ and assuming the same parameters as above, we arrive at a column density of OVI in a conducting interface of 3×10^{13} cm^{-2} when the temperature of the hot gas in the bubble is 2×10^6 K. Detailed calculations yield somewhat smaller column densities, $\sim 10^{13}$ cm^{-2}, because the heat flux is saturated (cf. Section 12.4.2). In the plane, there are, on average, about six OVI components per kpc with an average column density of 10^{13} cm^{-2}. Thus, a large number of filamentary clouds embedded in a hot intercloud medium may be able to explain these observations.

This intermediate temperature gas will also show absorption lines from NV and CIV (Table 8.4). The interpretation of these lines is somewhat more ambiguous because, unlike OVI, these two ions may also result from photo-ionization. The observed column density ratios of CIV/OVI and NV/OVI in the halo are in the ranges 0.5–0.8 and 0.15–0.25, respectively. None of the proposed models – radiative cooling of collisionally ionized plasma, conducting interfaces, or turbulent mixing layers – can explain both of these ratios. Besides a mixture of different components, particularly near the plane, these ratios may also be influenced by ionizing EUV or soft X-ray photons.

8.9.6 Interstellar pressures

The pressure of the interstellar medium can be determined from observations of the excitation of the fine-structure levels of neutral carbon ($^3P_0$, $^3P_1$, and $^3P_2$). The level populations can be determined from FUV absorption lines originating from these levels (cf. Section 8.9.4). Future space-based instrumentation may also be able to measure the very weak submillimeter emission lines connecting these levels in the diffuse ISM directly. The C^0 level populations are set by a balance between collisional excitation and radiative decay (e.g., the density in interstellar clouds is below the critical density; cf. Section 2.3). The collisional rates depend on the density and the temperature. Hence, the FUV absorption line data can be used to probe the physical conditions of the gas. From a large set of data, a typical interstellar pressure of $\simeq 2000\,\mathrm{K\,cm^{-3}}$ is derived. However, there are intrinsic problems with this method, since the physical conditions derived from the observed excitation of the $^3P_1$ and $^3P_2$ levels do not agree. Hence, this method is somewhat in doubt. This method could also be applied to other fine-structure levels. However, generally, the levels involved are at much higher energies and hence are much more sensitive to the actual gas temperature. The C^+ level populations act, of course, as thermostats and so adjust themselves until heating balances cooling (cf. Section 8.9.7).

8.9.7 The emission spectra of the CNM and WNM

As emphasized throughout this chapter, the CNM will be dominantly cooled by the [CII] 158 μm line. The calculated cooling rate of this line in the CNM is 6×10^{-26} erg (H-atom)$^{-1}$ s^{-1}. This [CII] cooling line has been observed directly at high latitudes (where there is little confusion) using sounding rockets and the Infrared Space Observatory (ISO). Comparison with HI column densities then results in typical [CII] cooling rates of 3×10^{-26} erg (H atom)$^{-1}$ s^{-1}, which is in good agreement with the models. In the plane (or out of the plane), the CII cooling can also be measured indirectly through the strength of the FUV absorption line originating from the upper level $^2P_{3/2}$ of this IR transition (the CII $\lambda 1336$ line) against bright background sources for which the H column density has been measured from Lyman α or been estimated from observed S^+ column densities, which are presumably undepleted in the diffuse ISM. In the plane, the observed cooling rate is then 10^{-25} erg (H atom)$^{-1}$ s^{-1}, while for halo sight-lines the rate is about 3×10^{-26} erg (H atom)$^{-1}$ s^{-1}, again in good agreement with the models. Of course, such studies suffer somewhat from selection effects since, in the plane, these sight-lines tend to be associated with FUV-bright stars, which can give rise to localized higher heating rates. The higher latitude sight-lines, on the other hand, may probe more WNM than CNM material. Using these calculated cooling rates, we can estimate the total CII luminosity of the Milky Way to be (cf. Table 1.1) about 8×10^7 and $2 \times 10^7\,L_\odot$ for the CNM and WNM, respectively. Together, this is actually somewhat larger than the total [CII] 158 μm luminosity of the Galaxy observed by COBE (cf. Table 1.3). Of course, in this estimate we adopted the results for the high pressures appropriate for the galactic plane and some of the gas may be at lower pressures at higher altitudes. In summary, the dominant cooling line of the HI phases of the ISM has been identified observationally and theoretical models of the photo-electric effect on PAHs and small dust grains are able to provide the required heating.

For the CNM, the next cooling line is the [OI] 63 μm line but its intensity will be about two orders of magnitude less than the [CII] line. Because of the higher temperature, the [OI] cooling line is about as important a coolant for the WNM as the [CII] line. This line has not been searched for yet from the WNM, although its luminosity on a Galaxy-wide base may be quite substantial (of course, PDRs can be very important contributors to this line as well; cf. Chapter 9). Finally, the Lyα line is also an important coolant for the WNM. This line will, however, be resonantly scattered until it is absorbed by the dust (Section 7.2.4).

8.10 Further reading

Ionization and recombination rates have been taken from [1] and [2].

The classic study of pressure and thermal equilibrium of the diffuse interstellar medium is [3]. Over the years, various studies have improved on the physics of the heating, cooling, ionization, and recombination processes involved. A recent study is [4]. The ionization of the warm ionized medium is described in [5]. The influence of supernovae on the global structure of the interstellar medium was realized in [6]. This led to the seminal paper, [7], on the three-phase model for the interstellar medium. The importance of concerted supernovae in OB associations, blow-outs from the Galactic plane, and the circulation between the disk and the lower halo (Galactic fountain) is described in [8], [9], [10], [11], and [12]. An assessment of the status of the three-phase model can be found in [13]. Theories of the hot phase have been reviewed in [14].

There is a wealth of literature on molecular observations of diffuse clouds, dating back some 75 years. Some relevant articles are [15], [16], [17], [18], [19], and [20]. The photodissociation of molecular hydrogen in the interstellar radiation field has been studied in [21], [22], [23], and, most recently, [24] and [25]. The formation of molecular hydrogen on interstellar grain surfaces has a long history and an important reference is [26]. Elegant experimental studies on H recombination on astrophysically relevant surfaces have been published in [27]. This has led to the theoretical discussion on H_2 formation described here, [28]. Detailed models for the chemistry of diffuse clouds are [29], [30], [31], and [32].

Observations of the CNM, WNM, and WIM have been reviewed in [33]. A more recent study of the characteristics of neutral hydrogen in the interstellar medium is [34]. The HI galactic plane survey using the DRAO interferometer at Penticton can be assessed through http://www.ras.ucalgary.ca/CGPS/.

The results of the WHAM survey of the WIM are described in [35] and can be found at

http://www.astro.wisc.edu/wham/index.html.

The early Copernicus results on absorption lines from the hot gas are reviewed in [36]. Recent FUSE results are described in [37] and can be found at

http://fuse.pha.jhu.edu/.

An early Copernicus study on the pressure of the ISM as measured by CI fine-structure lines is [38] and a more recent one based on HST observations is [39]. Observations of the X-ray background are discussed in [40]. X-ray observations using Chandra can be found at

http://chandra.harvard.edu/.

The clever approach to the cooling of the interstellar gas through the fine-structure level excitation of C^+ is described in [41] and has more recently been applied in [42]. Direct observations of the [CII] 158 μm cooling line of the Galaxy were made using sounding rockets, balloons, and COBE, and can be found in [43], [44], and [45]. The OVI column density discussion derives from [46].

References

[1] W. G. A. Roberge, S. Jones, S. Lepp, and A. Dalgarno, 1991, *Ap. J. S.*, **243**, p. 817
[2] D. A. Verner and G. J. Ferland, 1996, *Ap. J. S.*, **103**, p. 467
[3] G. B. Field, D. W. Goldsmith, and H. J. Habing, 1969, *Ap. J.*, **155**, p. L149
[4] M. G. Wolfire, C. F. McKee, D. J. Hollenbach, and A. G. G. M. Tielens, 2003, *Ap. J.*, **587**, p. 278
[5] J. S. Mathis, 1986, *Ap. J.*, **301**, p. 423
[6] D. Cox and B. Smith, 1974, *Ap. J.*, **189**, p. L105
[7] C. F. McKee and J. P. Ostriker, 1977, *Ap. J.*, **218**, p. 148
[8] P. R. Shapiro and G. B. Field, 1976, *Ap. J.*, **205**, p. 762
[9] J. N. Bregman, 1980, *Ap. J.*, **236**, p. 577
[10] M. Mac Low, R. McCray, and M. L. Norman, 1989, *Ap. J.*, **337**, p. 141
[11] C. Heiles, 1990, *Ap. J.*, **354**, p. 483
[12] C. A. Norman and S. Ikeuchi, 1989, *Ap. J.*, **345**, p. 372
[13] C. F. McKee, 1990, in *The Evolution of the Interstellar Medium* (San Francisco: ASP), p. 3
[14] L. Spitzer, Jr, 1990, *Ann. Rev. Astron. Astrophys.*, **28**, p. 71
[15] A. C. Danks, S. R. Federman, and D. L. Lambert, 1984, *A. & A.*, **130**, p. 62
[16] B. E. Turner, 1994, *Ap. J.*, **437**, p. 658
[17] B. E. Turner, R. Terzieva, and E. Herbst, 1999, *Ap. J.*, **518**, p. 699
[18] S. R. Federman, J. A. Cardelli, Y. Sheffer, D. L. Lambert, and D. C. Morton, 1994, *Ap. J.*, **432**, p. L139
[19] H. Liszt and R. Lucas, 2001, *A. & A.*, **370**, p. 576; 2000, *A. & A.*, **358**, p. 1069
[20] B. J. McCall, *et al.*, 2003, *Nature*, **422**, p. 500; 2002, *Ap. J.*, **567**, p. 391

[21] J. Black and A. Dalgarno, 1976, *Ap. J.*, **203**, p. 132
[22] J. Black and E. F. van Dishoeck, 1987, *Ap. J.*, **322**, p. 412
[23] H. Abgrall, J. Le Bourlot, G. Pinau des Forets, *et al.*, 1992, *A. & A.*, **253**, p. 525
[24] M. K. Browning, J. Tumlinson, and J. M. Shull, 2003, *Ap. J.*, **582**, p. 810
[25] B. T. Draine and F. Bertoldi, 1996, *Ap. J.*, **468**, p. 269
[26] D. Hollenbach and E. E. Salpeter, 1970, *Ap. J.*, **163**, p. 155
[27] V. Pirronello, C. Liu, J. Roser, and G. Vidali, 1999, *A. & A.*, **344**, p. 681; V. Pirronello, O. Biham, C. Liu, L. Shen, and G. Vidali, 1997, *Ap. J.*, **483**, p. L131
[28] S. Cazaux and A. G. G. M. Tielens, 2004, *Ap. J.*, **504**, p. 222
[29] J. Black and A. Dalgarno, 1976, *Ap. J. S.*, **34**, p. 405
[30] E. F. van Dishoeck and J. Black, 1988, *Ap. J.*, **334**, p. 711
[31] Y. P. Viala, E. Roueff, and H. Abgrall, 1988, *A. & A.*, **190**, p. 215
[32] B. E. Turner, E. Herbst, and R. Terzieva, 2000, *Ap. J. S.*, **126**, p. 427
[33] S. Kulkarni and C. Heiles, 1987, in *Interstellar Processes*, ed. D. Hollenbach and H. Thronson (Dordrecht: Reidel), p. 87
[34] C. Heiles and T. H. Troland, 2003, *Ap. J.*, **586**, p. 1067
[35] L. M. Haffner, R. J. Reynolds, S. L. Tufte, *et al.*, 2003, *Ap. J. S.*, **149**, p. 405
[36] R. McCray and T. P. Snow, 1979, *Ann. Rev. Astron. Astrophys.*, **17**, p. 213
[37] B. D. Savage, K. R. Sembach, and B. P. Wakker, *et al.*, 2003, *Ap. J. S.*, **146**, p. 125
[38] E. B. Jenkins, M. Jura, and M. Loewenstein, 1983, *Ap. J.*, **270**, p. 88
[39] E. B. Jenkins and T. M. Tripp, 2001, *Ap. J.*, **137**, p. 297
[40] G. P. Garmire, J. A. Nousek, K. M. V. Apparao, *et al.*, 1992, *Ap. J.*, **399**, p. 694
[41] S. R. Pottasch, P. R. Wesselius, and R. J. van Duinen, 1979, *A. & A.*, **74**, p. L15
[42] C. Gry, J. Lequeux, and F. Boulanger, 1992, *A. & A.*, **266**, p. 457
[43] J. J. Bock, V. V. Hristov, M. Kawada, *et al.*, 1993, *Ap. J.*, **410**, p. L115
[44] T. Nakagawa, Y. Y. Yui, Y. Doi, *et al.*, 1998, *Ap. J. S.*, **115**, p. 259
[45] C. L. Bennett, *et al.*, 1994, *Ap. J.*, **434**, p. 587
[46] T. M. Heckman, C. A. Norman, D. K. Strickland, and K. R. Sembach, 2002, *Ap. J.*, **577**, p. 691

9

Photodissociation regions

9.1 Introduction

Photodissociation regions (PDRs) are regions where FUV photons dominate the energy balance or chemistry of the gas. In this chapter, we will examine the physical characteristics of dense and luminous PDRs near bright O and B stars, starting with the ionization balance (Section 9.2) and the energy balance (Section 9.3) of the gas in PDRs. We follow that with a discussion of the dust temperature in PDRs. The chemistry of PDRs (Section 9.5) is very similar to that of diffuse clouds (cf. Section 8.7), except for possible time-dependent effects. We have, then, all the ingredients to understand the structure of PDRs (Section 9.6). The remainder of this chapter focusses on the analysis and interpretation of observations of PDRs. We will start this off with back-of-the-envelope estimates of the incident FUV field, density, temperature, and mass based on a few key observations (Section 9.7). More thorough analysis tools are discussed in Section 9.8. These different techniques for estimating the physical conditions in PDRs are then compared, based on a case study of the Orion Bar (Section 9.9). Section 9.10 contrasts the physical conditions derived for various well-known PDRs. Finally, we examine the H_2 IR fluorescence spectrum of PDRs in Section 9.11.

Figure 9.1 illustrates the structure of a PDR. FUV photons penetrate a molecular cloud, ionizing, dissociating, and heating the gas. Any H-ionizing photons are absorbed in a thin ($N_H \simeq 10^{19}$ cm^{-2} or $\Delta A_v \simeq 10^{-2}$) transition zone (cf. Section 7.2.1) in which the ionization structure changes from almost fully ionized ($1 - x \simeq 10^{-4}$) to almost fully neutral ($x \simeq 10^{-4}$). FUV photons with energies less than 13.6 eV will dissociate molecular hydrogen (Section 8.7.1) and ionize carbon (Section 8.2.1), forming an HI/CII region. Once the flux of H_2-dissociating photons is substantially attenuated (in dense PDRs that is generally by dust at $A_v \simeq 2$ magnitudes), the composition is dominated by H_2. Somewhat deeper in the cloud ($A_v \simeq 4$ magnitudes), the carbon-ionizing flux has dropped sufficiently and ionized carbon recombines and forms carbon monoxide

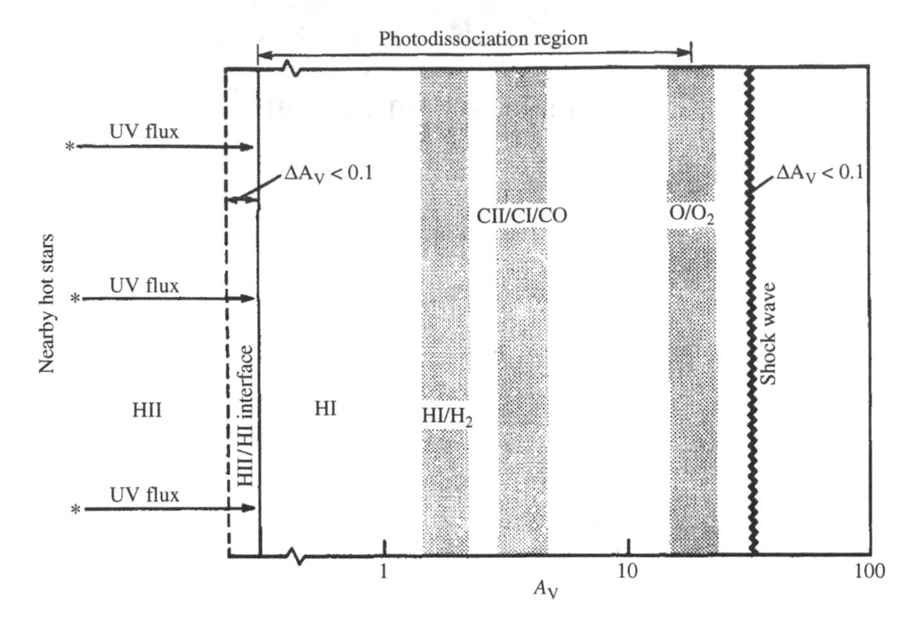

Figure 9.1 A schematic view of a photodissociation region. The PDR is illuminated from the left by a strong FUV field. The PDR extends from the H^+/H transition region through the H/H_2 and $C^+/C/CO$ transitions until the O/O_2 boundary. It thus includes the predominantly neutral atomic surface layer as well as large columns of molecular gas.

(Fig. 9.1). The transition from atomic to molecular oxygen occurs even deeper in the cloud.

Initially, PDRs were associated with regions of high FUV flux and high density. However, now, PDRs have come to encompass all regions where the physics and chemistry is dominated by penetrating FUV photons and, thus, PDRs include diffuse and translucent interstellar clouds; i.e., the general diffuse ISM. Most of the mass in molecular clouds is in regions with $A_V < 8$ magnitudes where, for example, FUV photons still influence the chemistry of oxygen and the ionization balance through ionization of trace species. Hence, in terms of physical and chemical processes, most of the ISM is, in essence, in PDRs. We have examined the physics and chemistry of the diffuse interstellar medium already in Chapter 8. Compared with the diffuse ISM, PDRs offer the advantage that FUV photons obviously dominate the heating, ionization, and chemistry of the region. Hence, PDRs provide ideal astronomical "laboratories," which can be used to study the interaction of FUV photons and neutral matter in detail. In turn, this knowledge can then be applied to, for example, the general diffuse ISM.

The structure of a PDR is governed by the density, n, and the intensity of the incident FUV radiation field. Expressed in terms of the average interstellar radiation field – corresponding to a unidirectional radiation field of 1.6×10^{-3} erg cm^{-2} s^{-1} – we have for a star at a distance d,

$$G_0 = 625 \frac{L_\star \chi}{4\pi d^2}, \tag{9.1}$$

where χ is the fraction of the luminosity above 6 eV. For a PDR surrounding an HII region, we can calculate the incident flux adopting the Strömgren sphere solution for the density–size relation (cf. Section 7.2.1),

$$G_0 \simeq 1.4 \times 10^4 \left(\frac{5 \times 10^{49} \text{ photons s}^{-1}}{\mathbb{N}_{\text{Lyc}}} \right)^{2/3} \times \left(\frac{n_e}{10^3 \text{ cm}^{-3}} \right)^{4/3} \left(\frac{L_\star}{7.6 \times 10^5 \text{ L}_\odot} \right) \left(\frac{\chi}{1} \right). \tag{9.2}$$

While the adopted stellar values are applicable for an O4 star, G_0 is very insensitive to these values – varying by less than 20% for O stars – because a decrease in luminosity is compensated by a decrease in ionizing photon flux and hence size. If we now assume approximate pressure equilibrium between the HII region and the PDR,

$$2\, n_e\, kT_e \simeq n_0 k\, T_{\text{PDR}}, \tag{9.3}$$

and realize that the electron temperature of the ionized gas is $T_e \simeq 8000$ K while the PDR is only $T_{\text{PDR}} \simeq 400$ K, then we find

$$G_0 \simeq 10^2 \left(\frac{n_0}{10^3 \text{ cm}^{-3}} \right)^{4/3}. \tag{9.4}$$

There is thus a natural relationship between the intensity of the incident FUV field and the density of a PDR. The ionization parameter, γ, plays an important role in PDRs. In terms of these estimates, we have, typically

$$\gamma = G_0 T^{0.5}/n_e \simeq 1.5 \times 10^4 \left(\frac{n}{10^3 \text{ cm}^{-3}} \right)^{1/3} \text{cm}^3\, \text{K}^{1/2}, \tag{9.5}$$

which is an order of magnitude higher than for the diffuse ISM ($\gamma \simeq 1500$ cm^3 K$^{1/2}$).

9.2 Ionization balance

As for the diffuse ISM (cf. Section 8.2.1), the ionization balance is dominated by FUV ionization of trace species. Near the surface, carbon ionization is the most important. Define the ionized fraction as $x = n_{C^+}/n_C$. When carbon dominates the ionization balance, the neutral fraction is given by

$$\frac{1-x}{x^2} = 3.3 \times 10^{-6} \left(\frac{\mathcal{A}_C}{1.4 \times 10^{-4}}\right) \left(\frac{n}{10^4\,\mathrm{cm}^{-3}}\right) \times \left(\frac{300\,\mathrm{K}}{T}\right)^{0.6} \left(\frac{10^4}{G_0}\right) \exp[2.6 A_v]. \tag{9.6}$$

The exponential factor takes care of the dust attenuation. Setting $x = 0.5$, the size of the C^+ zone is approximately $A_v \simeq 5$ magnitudes for these typical values of n, G_0, and T. Absorption of ionizing photons by neutral carbon as well as by H_2 can be of some importance as well and the actual size is somewhat less ($A_v \simeq 4$). The transition from ionized to neutral carbon – set by the exponential – is quite sharp. As for diffuse clouds, this C^+ zone transits into successive zones of ionized trace species with lower ionization potentials and lower abundances: a S^+ zone, a Si^+/ Fe^+/ Mg^+ zone, and a Na^+ zone. Because Si will participate in ion–molecule chemistry and molecular ions recombine much faster with electrons than atoms, the Si^+ zone may be somewhat limited in size.

Deeper in, cosmic-ray ionization of H_2 will take over as the dominant ionization source. Equating the cosmic-ray ionization rate with the photo-ionization rate, this will occur at a depth of about

$$A_v = \frac{1}{b} \ln \left[\frac{a G_0 \mathcal{A}_i}{\zeta_{CR}}\right], \tag{9.7}$$

where the coefficients in the FUV ionization rate, $k_{ion} = a G_0 \exp[-b A_v]$, are given in Table 8.1. Taking Mg as our example, the depth at which this occurs is $A_v \simeq 6$ magnitudes for $G_0 = 10^4$. Beyond that point, the ionization balance is that of dark clouds (cf. Section 10.5).

9.3 Energy balance

9.3.1 Heating and cooling processes

The importance of the photo-electric effect in coupling the gas to stellar FUV photons has already been extensively discussed in Section 3.3. Compared with the diffuse ISM, the higher ratio of the FUV flux to the electron density, G_0/n_e, implies that PAHs and grains will be more highly charged and hence the photo-electric effect is somewhat less efficient (cf. Fig. 3.4). The other important heating source in the atomic zone of PDRs is photopumping of H_2 followed by collisional de-excitation of the resulting vibrationally excited species (cf. Section 3.4). Deeper in a PDR, IR pumping of the [OI] fine-structure line followed by collisional de-excitation can also be of importance. Photo-ionization of carbon and other trace species is generally of lesser importance (cf. Section 8.2.2).

Cooling of PDRs occurs predominantly through fine-structure lines of abundant atoms and ions (cf. Section 2.5). This was discussed in Section 8.2.2 for the [CII] 158 μm line in the diffuse ISM. Because of the general higher densities, cooling through the [OI] 63 μm fine-structure line is often very important in PDRs. Figure 9.2 compares the cooling in the [OI] 63 μm line with that in the [CII] 158 μm line as a function of density for various temperatures. This shows the behavior characteristic for two levels with very different critical densities. We have discussed this in general terms in Section 2.3 and have applied this in the study of HII regions in Section 7.5.1. For very high densities ($n \gg n_{\rm cr}([{\rm OI}]) = 5\times10^5\,{\rm cm}^{-3}$), both species are in local thermodynamic equilibrium and the ratio of their cooling rates is independent of density. In that limit, oxygen cooling dominates. In the low density limit ($n \ll n_{\rm cr}([{\rm CII}])$), both cooling rates scale quadratically with density – each upward collision results in the emission of a cooling photon – and again the ratio depends only on temperature. In this low density limit – appropriate for the diffuse ISM – [CII] emission dominates the cooling. It is only in the intermediate regime ($n_{\rm cr}([{\rm CII}]) < n < n_{\rm cr}([{\rm OI}])$) that the ratio of these cooling lines depends on density. These two lines are also sensitive to somewhat different gas temperatures, because of the difference in excitation energy of the upper levels, but by 250 K this difference has disappeared.

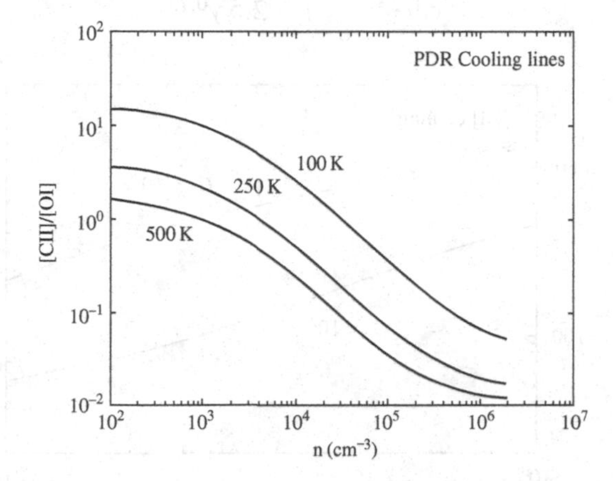

Figure 9.2 The ratio of the cooling in the [OI] 63 μm line to the cooling in the [CII] 158 μm line as a function of the density for various temperatures. These ratios have been calculated in the two-level approximation using the parameters in Table 2.7 and an O/C gas-phase abundance ratio of 2.3.

9.3.2 The gas temperature

As an illustrative example, consider photo-electric heating (given by Eqs. (3.16) and (3.17)). For the cooling, we will only take the [OI] 63 μm line into consideration (i.e., $T \gtrsim 100$ K and $n > 10^4$ cm^{-3}), but we will consider this in the low density limit ($n < n_{cr} \simeq 9 \times 10^5 (100/T)^{0.67}$ cm^{-3}, cf. Table 2.7); viz.

$$n^2 \Lambda_{63\mu m} = 2.5 \times 10^{-29} \, n^2 \, T^{0.67} \exp[-228/T] \, \mathrm{erg\,cm^{-3}\,s^{-1}}. \tag{9.8}$$

Using only the low temperature ($T < 10^4$ K) expression for the photo-electric effect, and assuming that the electrons come from (complete) C ionization, the energy balance can then be rewritten to the following transcendental equation

$$\frac{\exp[-y]}{y^{1.17}} = \frac{3.4\gamma}{1 + 2.5\gamma^{0.73}}, \tag{9.9}$$

with

$$y = \frac{228\,\mathrm{K}}{T} \tag{9.10}$$

and γ the ionization parameter (defined here as $G_0 T^{1/2}/n$ with $n_e = 1.4 \times 10^{-4}\, n$), which measures the ionization rate over the recombination rate and, hence, the efficiency of the photo-electric effect (cf. Section 3.3). Figure 9.3 shows the temperature as a function of density for a few values of the incident FUV field. Typically, bright, dense PDRs are fairly warm (up to 1000 K).

For lower density PDRs, [CII] cooling dominates. In the low-density limit ($n < n_{cr} \simeq 4 \times 10^3$ cm^{-3}), a similar equation can be derived,

$$\frac{\exp[-y']}{y'^{1/2}} = \frac{1.7\gamma}{1 + 2.5\gamma^{0.73}}, \tag{9.11}$$

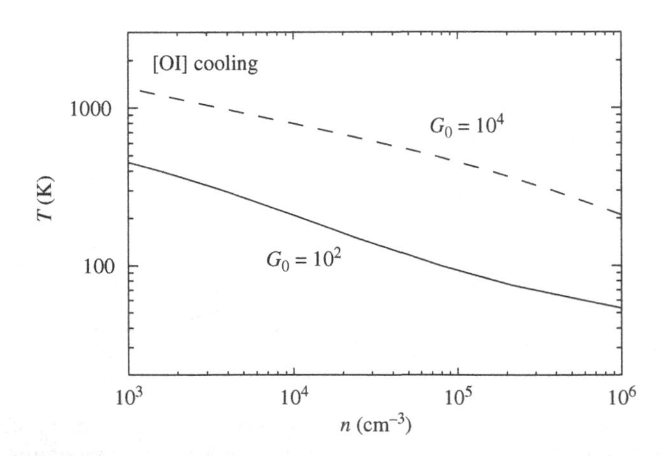

Figure 9.3 The gas temperature in a (high density) PDR heated by the photo-electric effect and cooled through the [OI] 63 μm line.

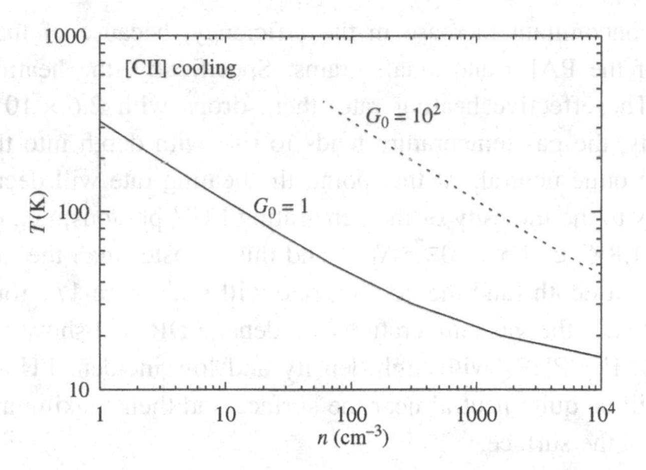

Figure 9.4 The gas temperature in a (low density) PDR heated by the photo-electric effect and cooled through the [CII] 158 μm line.

with y' given by

$$y' = \frac{92\,\mathrm{K}}{T}. \tag{9.12}$$

This relationship is illustrated in Fig. 9.4. For low values of γ, photo-electric heating reaches the maximum efficiency and this expression simplifies to

$$T = \frac{92}{\ln[6\times 10^{-2}\,n/G_0]}\,\mathrm{K}. \tag{9.13}$$

As noted in Section 8.2.2, for the diffuse ISM ($G_0 = 1$), [CII] cooling can balance photo-electric heating for densities in excess of about 20 cm^{-3} and this is concomitantly higher when G_0 is larger. Typically, $G_0/n \sim 0.1$–1 in PDRs (Eq. (9.4) and PDRs are quite warm, ~300–1000 K.

It is of some interest to consider the depth dependence of the gas temperature. For dense PDRs cooled by the [OI] 63 μm line, the cooling rate drops generally faster than the heating rate. The optical depth in the [OI] line is given by (cf. Section 2.2.2)

$$\tau_{\mathrm{OI}} = \frac{A_{\mathrm{ul}}\,c^3}{8\pi\,\nu_{\mathrm{ul}}^3}\,\frac{n_{\mathrm{u}}}{b/\Delta z}\left[\frac{n_{\mathrm{l}}\,g_{\mathrm{u}}}{n_{\mathrm{u}}\,g_{\mathrm{l}}} - 1\right] \simeq 1.05\times 10^{-21}\,N_{\mathrm{H}}\,f_x(n, T), \tag{9.14}$$

where the factor f_x contains all the excitation dependencies. For $T = 300$ K and $n = 10^5$ cm^{-3}, f_x is about 0.5 and is not very sensitive to the exact values of density and temperature. So, the optical depth in the [OI] line is proportional to $5.4\times 10^{-22}\,N_{\mathrm{H}}$. In contrast, the photo-electric heating rate drops more slowly with depth into the PDR. Basically, the decrease in heating photons is partly offset by a concomitant increase in the efficiency, because of the decrease in the charge of the PAHs and small grains. Specifically, the heating rate drops with $G_0^{0.27}$. The effective heating rate, then, drops with $2.6 \times 10^{-22} N_H$. As a result, initially, the gas temperature tends to rise with depth into the PDR until the grains are quite neutral. At that point, the heating rate will decrease directly proportionally to the intensity of the penetrating FUV photons; i.e., exponentially with $\tau_{FUV} = 1.8\, A_v \simeq 9.5 \times 10^{-22} N_H$ – and this is faster than the increase in the [OI] line optical depth (and the cooling rate will scale with $1/\tau$ for an optically thick line). Hence, the gas temperature in a dense PDR will show a maximum at about $A_v \simeq 1$. For PDRs with high density and low incident FUV field, PAHs and grains will be quite neutral near the surface and their maximum temperature occurs close to the surface.

9.4 Dust temperature

The temperature of dust in the interstellar medium has been discussed in Section 5.2.2. The penetrating stellar photons will create a surface layer of warm dust in a PDR, which will radiate copiously at infrared wavelengths. The radial temperature distribution of the dust is now governed by the absorption of the stellar and these infrared photons. For a dust grain, the energy balance is given by

$$\begin{aligned} 4\pi a^2 \int Q_{abs}(\nu)\pi B(\nu, T_d)\, d\nu = \pi a^2 \int Q_{abs}(\nu)\, F_\star(\nu)\, \exp[-\tau(\nu)]\, d\nu \\ + 4\pi a^2 \int Q_{abs}(\nu)\pi J_d(\nu)\, d\nu \\ + 4\pi a^2 \int Q_{abs}(\nu)\pi B(\nu, T = 2.78K)\, d\nu, \end{aligned} \tag{9.15}$$

where $F_\star$ is the incident stellar flux, τ is the optical depth from the cloud surface, J_d is the mean intensity of the re-emitted dust radiation, and the last term represents the microwave background radiation. We will assume a dust absorption law with $\beta = 1$ for $\nu < \nu_0$ and an absorption efficiency of 1 elsewhere (cf. Section 5.2.1); for ν_0 we will adopt 3×10^{15}, corresponding to 1000 Å. We will further assume that the dust emission is characterized by a single slab with a constant temperature, T_0, and an "effective" emission optical depth,

$$\tau_d(\nu) = \tau_{100\mu m} \left(\frac{\nu}{3 \times 10^{12}\ \mathrm{Hz}}\right), \tag{9.16}$$

where $\tau_{100\mu m}$ is the "effective" optical depth at 100 μm. The optical depth, τ_d, will be small, because it is set by the FUV optical depth scale. The dust infrared radiation field is, then,

$$J_{\rm d}(\nu) = (0.42 - \ln[\tau_{\rm d}(\nu)])\tau_{\rm d}(\nu)\, B(\nu, T_0), \tag{9.17}$$

where the term in brackets results from the angle average over this warm slab. The "effective" dust optical depth, $\tau_{\rm d}$, and the temperature, T_0, together characterize the infrared radiation field. The logarithmic factor is approximately constant over the dust emission spectrum and we can replace it by the value at the black-body peak ($\ln[3.45 \times 10^{-2}\tau_{100\mu\rm m}\, T_0]$). The dust temperature is then given by

$$\begin{aligned} T_{\rm d}^5 &= 8.9 \times 10^{-11}\nu_0\, G_0 \exp[-1.8A_{\rm v}] + 2.78^5 \\ &\quad + 3.4 \times 10^{-2}(0.42 - \ln[3.5 \times 10^{-2}\tau_{100\mu\rm m}\, T_0])\tau_{100\mu\rm m}\, T_0^6\ \rm K. \end{aligned} \tag{9.18}$$

The temperature of the slab is found by evaluating the dust temperature caused by the FUV radiation field only (Eq. (9.15) and Section 5.2.2),

$$T_0 = 12.2\, G_0^{1/5}\ \rm K. \tag{9.19}$$

The "effective" optical depth at 100 μm follows from the requirement that the IR emission from the slab equals the incident FUV flux,

$$1.2 \times 10^{-4}\, G_0 = \int \tau_{\rm d}(\nu)\, B(\nu, T_0)\, {\rm d}\nu, \tag{9.20}$$

which yields

$$\tau_{100\mu\rm m} = 2.7 \times 10^2 \frac{G_0}{T_0^5} = 10^{-3}. \tag{9.21}$$

Figure 9.5 shows the dust temperature as a function of the depth in the PDR for three different intensities of the incident FUV field. For bright PDRs, the dust temperature at the surface can reach some 80 K but rapidly drops, because of the attenuation of the FUV radiation. The temperature reached in the deep interior of the PDR will be higher when G_0 is larger because of the brighter IR radiation field by the surface slab, which is the main heating source at depth. These dust temperatures are much less than the gas temperature despite the fact that much more of the photon energy goes into heating the dust. Basically, the dust has many more low-lying (phonon) modes available for efficient cooling. In contrast, the gas can only cool through relatively high-lying fine-structure levels ($\Delta E \simeq 100$–300 K).

The infrared intensity from the PDR is now given by

$$\begin{aligned} I(\nu) &= \int B(\nu, T_{\rm d}(\tau))\, {\rm d}\tau \\ &= 6.8 \times 10^{-16}\nu \int B(\nu, T_{\rm d}(A_{\rm v}))\, {\rm d}A_{\rm v}. \end{aligned} \tag{9.22}$$

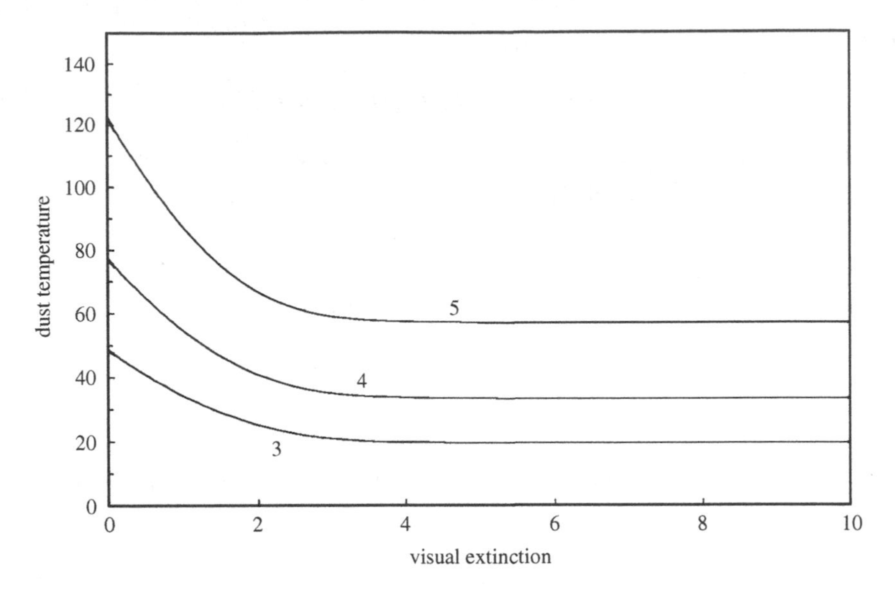

Figure 9.5 The dust temperature as a function of visual extinction in a PDR. The three curves are for a G_0 of 10^3 (3), 10^4 (4), and 10^5 (5).

Of course, this intensity only describes the PDR emission by dust grains in radiative equilibrium with the radiation field. Hence, the near- and mid-IR emission due to transiently heated small dust grains and large PAH molecules (Chapter 6) is not accounted for.

9.5 Chemistry

The chemistry of PDRs is very similar to that of diffuse clouds and edges of molecular clouds (cf. Sections 8.7 and 10.4.3). Penetrating FUV photons keep hydrogen and oxygen atomic and carbon ionized. As the molecular hydrogen abundance increases, the chemistry takes off. One difference with diffuse clouds is that various reactions with molecular hydrogen, which have small activation energies (e.g., C^+, O, OH), can proceed at an appreciable rate in the warm gas of dense PDRs. Likewise, vibrationally excited H_2 resulting from FUV pumping can also rapidly react with these radicals. Another difference is that time-dependent effects can be very important near the surface where the high intensity of the FUV field results in rapid photodissociation.

9.5.1 Molecular hydrogen dissociation front

The HI/H_2 and its location as a function of depth has been discussed in detail in Section 8.7.1. In PDRs, where $G_0/n \sim 1$, the location of the H_2 dissociation front corresponds to $A_v \simeq 2$ or a column of $N_{DF} = 4 \times 10^{21}$ cm^{-2}. At that depth, H_2 self-shielding takes over and the transition from H to H_2 is quite rapid. This assumes a steady state (Section 8.7.1) and HII regions can show rapid dynamical evolution (Section 12.2). Consider first the case where a star suddenly "turns on" and illuminates a molecular cloud. This will send a photodissociation front through the gas. The time-dependent evolution of the H_2 abundance is given by (Section 8.7.1)

$$\frac{dn(H_2)}{dt} = k_d(H_2) n n_H - k_{UV}(0)\, \beta_{ss}(H_2) \exp[-\tau_d]\, n(H_2), \tag{9.23}$$

with k_d and k_{UV} the H_2 formation rate on grains and the unshielded UV dissociation rate, β_{ss} the H_2 self-shielding factor, and τ_d the dust optical depth. The steady state will be reached near N_{DF}, on a time scale

$$\tau = (2n\, k_d)^{-1} \simeq 5 \times 10^4 \left(\frac{10^5 \text{ cm}^{-3}}{n} \right) \text{ years.} \tag{9.24}$$

This is evaluated deep in the slab. Near the surface, where the intensity of dissociating photons is much higher, this time scale will be much shorter. This is illustrated in Fig. 9.6. The steep photodissociation front initially accelerates ($v_{pdf} \propto t$) into the slab because, as H_2 near the surface is dissociated, the self-shielding rapidly decreases. Given this time scale, in general, this sudden turn-on has little influence on PDRs.

In addition, there can be time-dependent effects associated with the dynamics of the region. In particular, an HII region will expand with time because of the

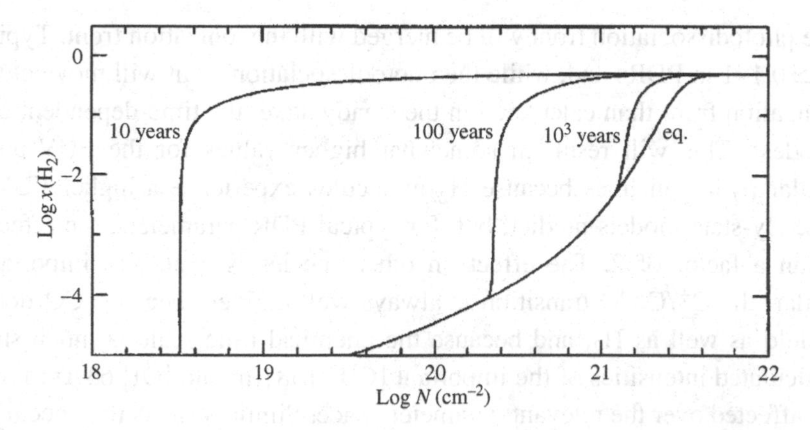

Figure 9.6 The abundance of H_2 is plotted as a function of total column density in a PDR for different times after a star is suddenly "switched on." The PDR parameters are $n = 10^5$ cm^{-3} and $G_0 = 10^5$. The curve labeled eq. is the steady-state distribution. Figure reproduced with permission from D. Hollenbach and A. Natta, 1995, *Ap. J.*, **455**, p. 133.

over-pressure of the ionized gas. This is discussed in detail in Section 12.2. As a result, in the frame of the ionization front, neutral gas moves slowly through the PDR, is ionized at the ionization front, and then rapidly flows away into the HII region (with respect to the molecular cloud, the ionization front is moving into the surroundings at some fraction of the speed of sound; Section 12.2). The velocity at which the neutral gas flows through the PDR is given by

$$v_{\rm pdr} = \frac{C_{\rm I}^2}{2C_{\rm II}} \simeq 0.5\,{\rm km\,s^{-1}}, \tag{9.25}$$

where we have used $C_{\rm I} \simeq 3$ km s^{-1} for the sound speed in the neutral gas, and $C_{\rm II} \simeq 10$ km s^{-1} for the sound speed in the ionized gas. We can then evaluate the second term in Eq. (9.23), adopting the expression for the self-shielding factor given in Section 8.7.1 and neglecting dust attenuation. The H_2 dissociation time scale is then given by

$$\tau_{\rm dis} = 8.5 \times 10^2 \left(\frac{10^5}{G_0}\right) \left(\frac{N({\rm H_2})}{5 \times 10^{20}\,{\rm cm^{-2}}}\right)^{3/4} \ {\rm years}. \tag{9.26}$$

Comparing the chemical time scale with the dynamical time scale, $\tau_{\rm dyn} = N/nv_{\rm pdr}$, we see that advection of the material through the PDR is faster than dissociation when

$$\frac{G_0}{n} < 0.15 \left(\frac{v_{\rm pdr}}{0.5\,{\rm km\,s^{-1}}}\right) {\rm cm}^3, \tag{9.27}$$

and the photodissociation front will be merged with the ionization front. Typically, $G_0/n \simeq 0.1$–1 in PDRs and, while the photodissociation front will move closer to the ionization front than calculated in the steady state, the time-dependent effects are modest. This will result in somewhat higher values for the FUV-pumped molecular hydrogen lines because H_2 molecules experience a higher FUV flux than steady-state models predict, but, for typical PDR parameters, the effects are less than a factor of 2. The effect on other species is even less important. In particular, the C^+/C/CO transition is always well defined, because CO does not self-shield as well as H_2, and because the chemical time scale is much shorter. The calculated intensities of the important [CII] 158 μm and [OI] 63 μm lines are hardly affected over the relevant parameter space. Similar effects may occur, from turbulence. In that case, the appropriate dynamical time scale is $\tau_{\rm dyn} = \ell/v(\ell)$ with $v(\ell)$ the turbulent velocity of an eddy with size scale ℓ (cf. Section 3.8). Adopting a turbulent velocity field of 0.5 km s^{-1} on a length scale of 0.1 pc, we find that turbulence will affect the chemistry when $G_0/n < 0.15\,{\rm cm}^3$.

9.5.2 Chemical zones

A number of zones can be distinguished in PDRs, each characterized by the chemical reactions that control its molecular composition.

The warm surface layers of PDRs consist largely of atomic H, C^+, and O and the chemistry of note involves various radicals. Atomic O reacts with traces of molecular hydrogen present to form the radical, OH. This reaction has an activation barrier but may be driven by the warm temperature of the gas or by the vibrational excitation of H_2 from FUV pumping. The OH is mainly photodissociated but some reacts with C^+ to form CO^+, which charge-transfers with H to form CO. By reacting with H_2, C^+ and a small fraction of neutral C present form CH^+ and CH, which flow back to C^+ and C through reaction with H. Atomic nitrogen similarly forms NH, which is channeled by C^+ to (eventually) HCN. As the depth increases, the concentration of radicals initially increases and the dominant reaction channels may change somewhat – i.e., burning of hydrocarbons can become an important channel for CO formation. Eventually, because of the drop in temperature and in the abundance of vibrationally excited H_2, the abundance of radicals decreases again. Thus, in this surface layer ($N_H \sim 10^{21}$ cm^{-2}), the chemistry is dominated by radicals and their reactions, and the abundance of stable molecules is very much limited by photodissociation.

The C^+/C/CO transition occurs deeper in the PDR when the dissociating radiation field has been sufficiently attenuated (see above). This transition is also influenced by the presence of neutral sulfur, which charge exchanges with C^+. The degree of ionization is still relatively high in this zone because penetrating FUV photons can still ionize metals such as silicon. For PDRs characterized by high densities and incident FUV fields, the high temperature can lead to appreciable amounts of CO ($X(\mathrm{CO}) \sim 10^{-6}$) in the surface layers.

Deep in the cloud, ion–molecule chemistry – initiated by cosmic-ray ionization of H_2 – takes over. The chemical composition will then revert to that of dark clouds (cf. Section 10.5). Oxygen, for example, will be channeled to OH and H_2O through the reaction of H_3^+ with O. At this point, nitrogen is mainly in the form of N_2. Cosmic-ray produced He^+ can break N out of N_2, which is then available for formation of HCN and NH_3.

9.6 PDR structure

Generally, PDR models consider a plane-parallel, semi-infinite slab illuminated from one side by an intense FUV field. The penetrating FUV photons create an atomic surface layer (Fig. 9.7). At a depth corresponding to $A_v \simeq 2$, the transition from atomic H to molecular H_2 occurs. Because of rapid photodestruction, the abundance of vibrationally excited molecular hydrogen, H_2^*, does not peak

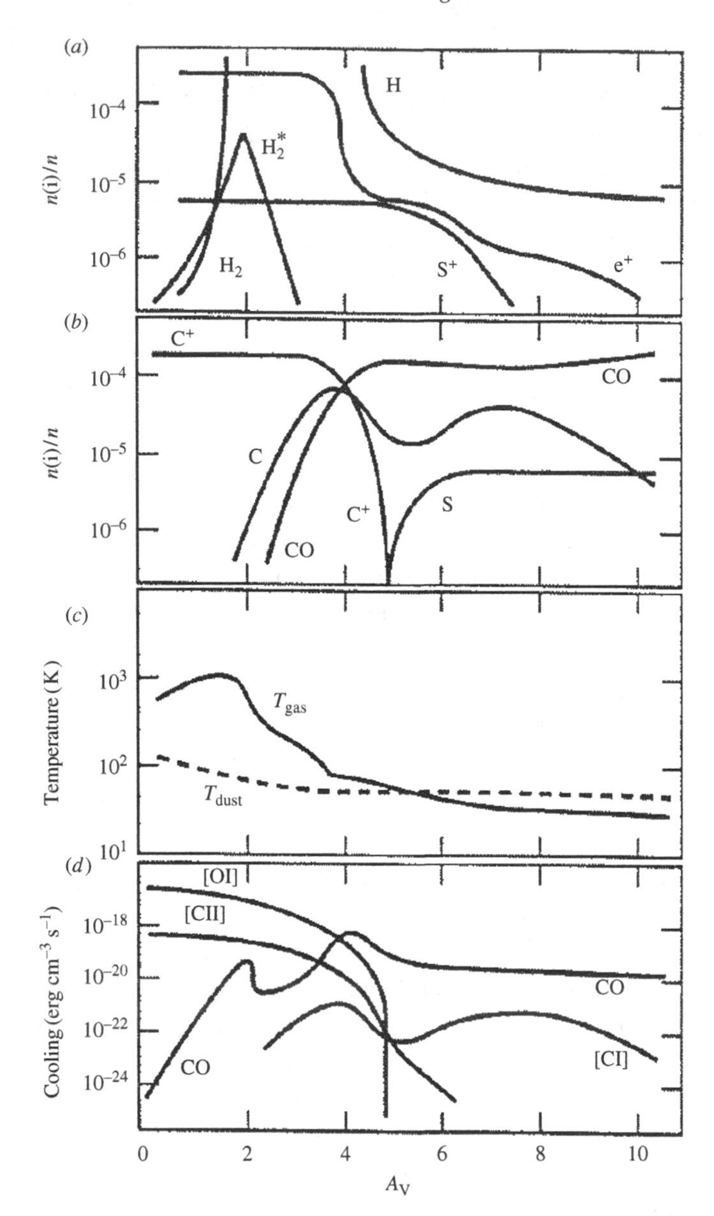

Figure 9.7 The physical and chemical structure calculated for the PDR in Orion ($n = 2.3 \times 10^5\ \mathrm{cm}^{-3}$, $G_0 = 10^5$) as a function of depth in units of visual extinction. The illuminating source is to the left. (*a*) and (*b*) Abundances of various species. (*c*) The gas and dust temperatures. (*d*) Calculated cooling rates for the [OI] 63 μm, [CII] 158 μm, and [CI] 610 μm lines and the total CO rotational cooling.

until $A_v \sim 2$. Because of dust attenuation, the carbon balance shifts from C^+ to C and CO at $A_v \simeq 4$. The second peak in the neutral carbon abundance results from charge exchange between C^+ and S. Except for the O locked up in CO, essentially all the oxygen is in atomic form until very deep in the cloud; $A_v \sim 8$. Because of their low ionization potential, trace species such as S can remain ionized through a substantial portion of the PDR.

Besides the chemical composition, the FUV photons also control the energy balance of the gas through the photo-electric effect (cf. Section 9.3). Typically, about 0.1–1% of the FUV energy is converted into gas heating this way. The rest is emitted as far-IR dust continuum radiation. The gas in the surface layer is then much warmer ($\simeq$500 K) than the dust (30–75 K; cf. Fig. 9.7). Somewhat deeper into the PDR ($A_v > 4$), penetrating red and near-IR photons keep the dust warm and gas–grain collisions maintain the gas at slightly below the dust temperature. In PDRs, the gas cools through the far-IR fine-structure lines of mainly [OI] 63 μm and [CII] 158 μm at the surface and the rotational lines of CO deeper into the PDR (Fig. 9.7).

9.7 Comparison with observations

PDRs are bright in the far-IR dust continuum, the PAH emission features, the far-IR fine-structure lines of [OI] and [CII], and the rotational lines of CO. Besides these dominant cooling processes, PDRs are also the source of (fluorescent or collisionally excited) rotational–vibrational transitions of H_2 in the near-IR, the atomic fine-structure lines of [CI] 609 μm, 370 μm, and [SiII] 35 μm, and the recombination lines of CI in the radio (e.g., C91α) and the far-red (e.g., 9850 Å and 8727 Å). PDRs also stand out in rotational transitions of trace molecules like CO^+, CN, and C_2, originating from the radical zone. We will first consider the analysis of observations using simple analytical approximations and compare and then contrast these with results obtained from full PDR models.

9.7.1 Incident FUV field

The intensity of the incident FUV field can be estimated rather directly from the observed far-IR continuum emission. As long as the PDR is optically thick in FUV, all the incident radiation is absorbed by the dust and reradiated into the far IR. Assuming, for simplicity, a single temperature, the total far-IR luminosity is then

$$L_{\mathrm{FIR}} = n_{\mathrm{d}} \pi a^2 \, 4\pi \, V \int Q(\nu) \, B(\nu, T_{\mathrm{d}}) \, \mathrm{d}\nu, \qquad (9.28)$$

with $n_{\rm d}$, a, and $T_{\rm d}$ the density, size, and temperature of the dust, $Q(\nu)$ the dust absorption efficiency at frequency ν, and V the volume of the source. We will adopt $Q(\nu) = Q_0(\nu/\nu_0)$, which seems reasonable for such regions. Assuming now that $G_0 = 1.3 \times 10^3 \, L_{\rm FIR}/S$, where S is the surface area of the cloud facing the star and the numerical factor converts the flux into Habing units taking the unidirectional nature of the incident field into account, this can be rewritten as

$$G_0 = 8.3 \times 10^3 \frac{V}{\ell S} \tau(\nu_0) \nu_0^{-1} \int \nu B(\nu, T_{\rm d}) \, {\rm d}\nu, \tag{9.29}$$

with $\tau(\nu_0)$ the dust optical depth at the reference frequency ν_0, and ℓ the pathlength along the line of sight. Far-infrared observations at multiple wavelengths can be used to derive the optical depth and the integral and hence G_0. The greatest uncertainty in this determination is the assumed geometry of the cloud. For an edge-on disk, the geometry factor is equal to d/ℓ, with d the thickness of the disk. For a sphere, on the other hand, the geometry factor is equal to unity.

If the relative orientation of the cloud and illuminating star are known, G_0 can also be determined from the observations of the FUV flux. Often, this is done by adopting a spectral type for the star and using the projected distance. By definition, G_0 refers only to photons in the 6–13.6 eV energy range. For B-type stars, a correction factor is required – typically 0.5 – for energy emitted at longer wavelengths, which does heat the dust but not the gas. O stars emit about half of their energy at wavelengths shorter than the Lyman limit. However, those photons are absorbed in the HII region and downgraded, often to the FUV range, and this correction factor is generally ignored.

The observed intensity of H_2 rotational–vibrational lines can also be a useful tool for determining G_0. When the IR spectrum is due to "pure" fluorescence ($n \lesssim 10^4 \, {\rm cm}^{-3}$), these intensities can be calculated directly from the transition probabilities of the levels involved, taking the radiative transfer (in the pumping FUV lines) into account. For $G_0/n < 10^{-2} \, {\rm cm}^3$, H_2 self-shielding dominates the opacity in the 912–1100 Å range (see Section 8.7.1) and the line intensity scales directly with the intensity of the incident FUV radiation field (Fig. 9.8). For $G_0/n \gtrsim 0.1 \, {\rm cm}^3$, dust opacity is more important and the line intensities become largely independent of G_0.

9.7.2 Density

For HII regions, the density can often be easily determined from pairs of fine-structure lines such as [OIII] 52/88 μm. Unfortunately, for PDRs there are no suitable pairs from one species. The two [OI] lines have very comparable critical densities, which makes them less useful for this type of analysis. Generally, the [OI] 63 μm/[CII] 158 μm ratio is used instead, but this requires an assumption

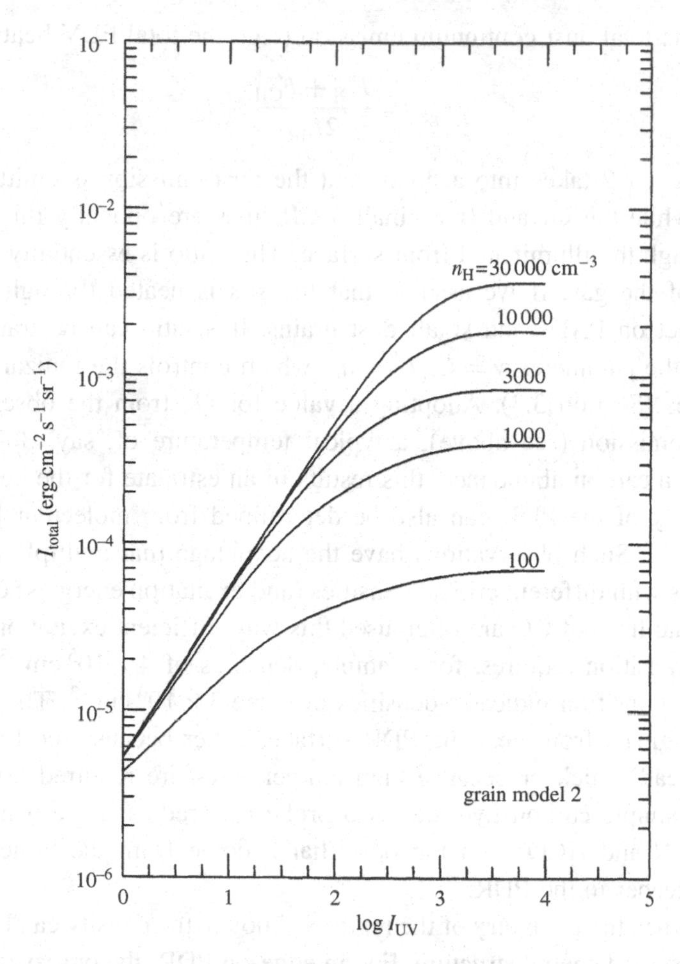

Figure 9.8 The intensity of the H_2 1–0 S(1) line as a function of the intensity of the incident FUV light (here labeled as $I_{\rm uv}$ rather than G_0). The different curves are labeled by density. Figure reproduced with permission from J. H. Black and E. F. van Dishoeck, 1987, *Ap. J.*, **322**, p. 412.

of the abundance ratio. Fortunately, both OI and CII are the dominant ionization stages of these elements and hence observed interstellar abundances can be assumed. For optically thin emission, Fig. 9.2 shows that this line ratio is a good diagnostic for the density in the range 10^3–10^6 cm^{-3}. However, the 63 μm line has typically an optical depth, τ, of a 1–3 while the 158 μm line is marginally optically thin ($\tau \lesssim 1$) in the emitting gas. This limits the usefulness of this line ratio and often, despite its weakness, the [OI] 146 μm line is substituted for the 63 μm line.

An estimate of the density can also be obtained from the heating efficiency, which can be derived from observations by comparing the fluxes in the gas cooling lines with the total dust continuum emission (e.g., the total FUV heating flux),

$$\epsilon_{\text{gas}} = \frac{F_{\text{OI}} + F_{\text{CII}}}{2F_{\text{IR}}}, \tag{9.30}$$

where the factor 2 takes into account that the dust emission is emitted into 4π steradians while the OI and (marginally) CII lines are optically thick and only escape through the illuminated front surface. This ratio is essentially the heating efficiency of the gas. If we assume that the gas is heated through the photoelectric effect on PAHs and small dust grains, this ratio can be translated into a value for the parameter $\gamma = G_0 T^{1/2}/n_e$, which controls the ionization state of these species (Section 3.3). Adopting a value for G_0 from the observed far-IR continuum emission (see above), a typical temperature of, say, 300 K (or see below), and a carbon abundance, this results in an estimate for the density.

The density of the PDR can also be determined from molecular line studies (Chapter 10.7). Such observations have the advantage that multiple lines of the same species with different critical densities (and excitation energies) can be used. The rotational lines of CO are often used this way. Efficient excitation of the CO $J = 7-6$ transition requires, for example, densities of 4×10^5 cm^{-3}, while the $J = 14-13$ transition indicates densities of some 3×10^6 cm^{-3}. These CO lines typically originate from near the PDR surface, either because the lines become rapidly optically thick or because high temperatures are required. Species such as CN and simple carbon hydrides also probe (the radical zone near) the PDR surface. HCN and HCO^+, on the other hand, come from the molecular zone, somewhat deeper in the PDR.

Finally, when the geometry of the region is known, the density can be estimated from the observed spatial structure. For an edge-on PDR, the observed separation between two atomic or molecular tracers can be translated into an absolute scale size using the distance. Because the scale size of the PDR is set by FUV penetration in a dusty medium, this physical scale provides an estimate for the density using standard dust-to-gas ratios (i.e., $N_H/A_v = 1.9 \times 10^{21}$ H atom cm^{-2} magnitude^{-1}) and the average extinction curve (e.g., $\kappa_d(\text{FUV})/\kappa_d(V) = 1.8$).

9.7.3 *Temperature*

The temperature of the emitted gas can be estimated from the observed intensity of optically thick lines (Section 2.3.2). Recalling the discussion on a two-level system, the intensity of an optically thick line is given by the Planck function multiplied by the width of the line (cf. Eq. 2.48),

$$I = B(T)\frac{\nu b}{c}, \tag{9.31}$$

with b the Doppler broadening parameter, which can be estimated from optically thin lines. Theoretical studies suggest that the [OI] 63 μm line is optically thick in PDRs and this line is often used for temperature studies. However, because of its high critical density, this line may actually measure the excitation temperature, which is less than the kinetic temperature of the gas (cf. Section 2.3.1). The [CII] 158 μm line has a low critical density, however, this line is generally not really optically thick.

The temperature of the gas can also be estimated from the observed intensities of H_2 lines. Because these are quadrupole lines with small Einstein A coefficients, optical depth effects are unimportant and the observed intensities can be directly converted into column densities of the upper state (cf. Eq. 2.49). In thermodynamic equilibrium, these are given by

$$N_u = N_{H_2} \frac{g_u \exp[-E_u/kT]}{\Phi(T)}, \tag{9.32}$$

with N_{H_2} the total H_2 column density and $\Phi(T)$ the partition function. The main temperature dependence is in the exponential term and a log-linear plot of N_u/g_u versus E_u therefore directly provides the temperature through the slope and the total H_2 column density through the intercept. Of course, this measures the excitation temperature of the levels. The low-lying pure rotational levels of H_2 have very small critical densities and, hence, directly measure the kinetic gas temperature. However, the populations of the higher pure rotational levels and the vibrational levels are strongly influenced by radiative pumping through the Lyman–Werner bands (cf. Section 3.4). The rotational level populations are much less influenced that way and, within each vibrational level, the rotational level population is still largely characterized by the kinetic gas temperature.

9.7.4 Mass

The mass of warm gas in the PDR can best be estimated from the observed [CII] 158 μm line intensity. This line is marginally optically thin ($\tau \simeq 1$) and all the ions are "seen." It also has a relatively low critical density and the LTE assumption is reasonable. Moreover, C^+ is the dominant ionization stage; so, the conversion to hydrogen does not require an ionization correction factor. In these limits, the total gas mass in the PDR is given by

$$M_H = \frac{4\pi d^2 m_H}{\mathcal{A}_c A_{ul} E_{lu}} \left(\left(\frac{g_l}{g_u} \right) \exp[E_{lu}/kT] + 1 \right) F_{CII}, \tag{9.33}$$

with $\mathcal{A}_c$ the gas phase C abundance, g_l, g_u, A_{ul}, and E_{lu} the statistical weights, Einstein A, and energy difference of the levels involved, F_{CII} the observed [CII]

158 μm flux, and d the distance of the source. Taking the high density, high temperature limit and assuming a C^+ abundance of 1.4×10^{-4}, this simplifies to

$$M_H = 3.4 \left(\frac{F_{CII}}{10^{-10}\,\mathrm{erg\,cm^{-2}\,s^{-1}}} \right) \left(\frac{d}{\mathrm{kpc}} \right)^2 \mathrm{M}_\odot. \qquad (9.34)$$

The CII line measures only the atomic surface layer of the PDR. Low-lying CO rotational transitions can be used to probe the deeper molecular layers. Of course, the lowest levels will generally have an important contribution from the general molecular cloud. To circumvent this problem, for dense PDRs, the ^{13}C isotope of the CO $J = 6-5$ or $J = 7-6$ lines can be used, which originate in the warm gas near the surface layer. The low-lying pure rotational lines of molecular hydrogen can be used to derive the mass of warm molecular gas. Because of FUV pumping, the vibrational levels, on the other hand, are only very temporarily excited and do not measure the PDR mass accurately. Other species, atomic or molecular, require uncertain correction factors for their abundances.

9.8 PDR diagnostic model diagrams

Over the years, a number of analysis tools have been devised, which allow direct comparison of the observations with extensive model calculations in the form of diagnostic diagrams. In this way, the physical conditions in the emitting gas can be derived in a relatively painless manner. Of course, use of such diagrams entails "buying" the whole enchilada; i.e., all the assumptions involved in the models. Nevertheless, use of these diagrams – coupled with some common sense and an understanding of the physics and chemistry of PDRs – can be very illuminating. Here, we will discuss three particularly useful diagrams.

9.8.1 The [CII] and [OI] lines

The [CII] and [OI] lines are bright and hence easily measured. The diagnostic diagram involving these lines is based on the difference in critical densities and excitation energies, which makes their line ratio an effective probe of density and temperatures in the typical range of PDR conditions (100–500 K and 10^3–10^5 cm^{-3}; cf. Fig. 9.2). Furthermore, they are the dominant cooling lines and hence measure the total gas heating. Figure 9.9 plots the [CII] 158 μm to [OI] 63 μm ratio versus the heating efficiency, $(I_{OI} + I_{CII})/2I_{FIR}$. The results illustrate that, as the density increases or G_0 decreases, the heating efficiency increases. The [CII]/[OI] ratio increases as the density decreases or the FUV field decreases. Both of these ratios can be determined directly from observations and hence the density and incident FUV field can be estimated.

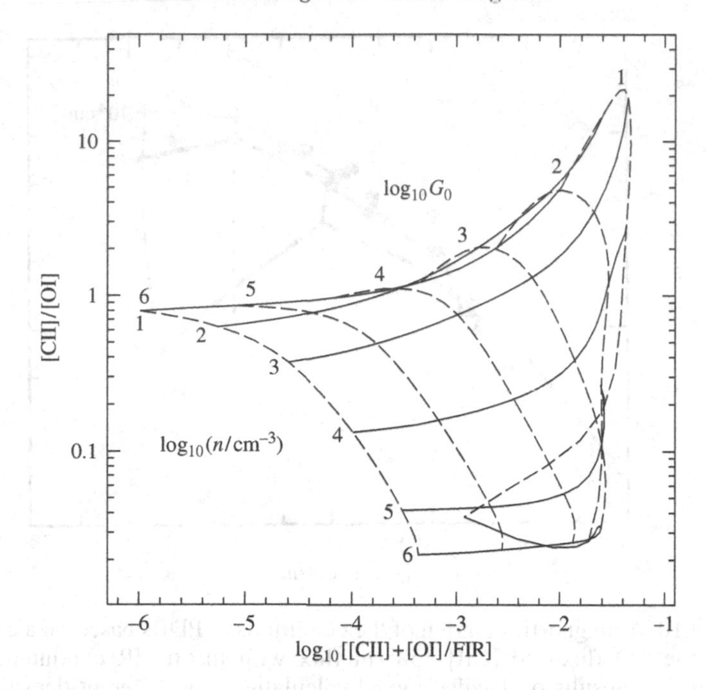

Figure 9.9 A diagnostic diagram for PDRs based on the observed intensity ratio of the [CII] 158 μm and [OI] 63 μm lines and the overall cooling efficiency. The lines present the results of detailed model calculations for different densities and incident FUV fields. Figure kindly provided by M. J. Kaufman; derived from the models described in M. J. Kaufman, M. G. Wolfire, D. Hollenbach, and M.L. Luhman, 1999, *Ap. J.*, **527**, p. 795.

9.8.2 The [CII]–CO relation

Because penetration of FUV photons in the PDR is controlled by dust absorption, the C^+ column density is nearly constant over much of the parameter space relevant for PDRs. Then, when the density of the PDR is above the critical density ($n > 10^3$ cm^{-3}) and $G_0 > 10^2$ (i.e., $T > 100$ K), the C^+ intensity is not very sensitive to the physical conditions any more. (At that point, [OI] takes over as the dominant cooling line.) The CO $J = 1-0$ line always becomes optically thick very close to the C^+/C/CO transition region. Again, the depth in the PDR where this transition occurs depends on the dust attenuation and, because the attenuated FUV field in this transition region is largely independent of the model parameters, the gas temperature does not vary much. As a result, the CO intensity is very insensitive to n and G_0. As a coincidence, the [CII] and CO $J = 1-0$ lines have very similar critical densities and hence, even at low densities, the two lines scale similarly.

In model–observation comparisons, it is advantageous to take the ratios of these lines to the far-IR continuum intensity to eliminate any beam dilution effects for

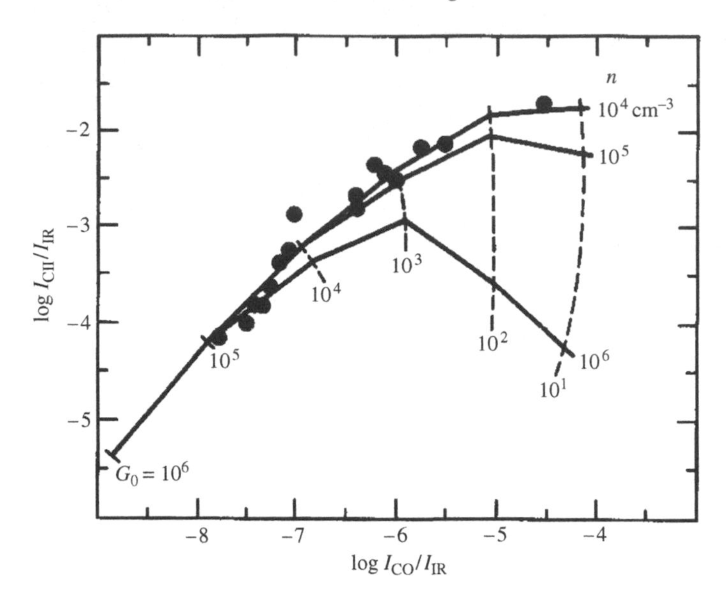

Figure 9.10 A diagnostic diagram of the conditions in PDRs based on a comparison of the CO flux and [CII] 158 μm flux with the far-IR continuum flux. Lines give the results of detailed model calculations for different densities and incident FUV fields. Dots are observations from a variety of galactic PDRs. Figure reproduced with permission from M. G. Wolfire, D. Hollenbach, and A. G. G. M. Tielens, 1989, *Ap. J.*, **344**, p. 770.

unresolved sources. Figure 9.10 compares observations to the models. Clearly, this relationship is a good measure of the intensity of the incident FUV field and not very sensitive to the density. Both ratios increase inversely with G_0. Essentially, of course, this method for the determination of G_0 boils down to the direct determination of G_0 from the far-IR continuum whereby the [CII] (or CO) observations are used to "measure" the size of the illuminated surface of the PDR.

The observations of PDRs show nice agreement with these theoretical curves. Even sources such as B35, a dense Bok globule illuminated by the nearby λ-Ori star cluster, agree well. However, observations of low metallicity systems form an exception. They tend to fall to the left of the theoretical relationship in the region of low CO $J = 1 - 0$ emission for their observed [CII] intensity. Clearly, in this case, the CO emitting surface is much smaller than the [CII] emitting surface. This reflects the lower metallicity. For a constant gas column, the C-ionizing photons can penetrate much deeper into the cloud, in terms of physical size. For this effect to be important, this must mean that the emitting clouds are relatively modest in size with $A_v < 4$ magnitudes, or

$$\ell < 2.6 \left(\frac{10^3\,\text{cm}^{-3}}{n} \right) \left(\frac{Z_\odot}{Z} \right) \text{pc}, \tag{9.35}$$

with $Z/Z_\odot$ the metallicity relative to that of the Sun. For comparison, the dense ($\simeq 10^5\,\mathrm{cm}^{-3}$) PDR in Orion illuminated by θ^1C Ori is ~1 pc in size. A PDR in the LMC ($Z = Z_\odot/2$) and particularly the SMC ($Z = Z_\odot/6$) may have a much smaller molecular zone than the C^+ zone.

9.8.3 [CII] and [OI] line intensities

The absolute intensities of the atomic fine-structure lines can also be used to derive the physical conditions. Generally, such analysis is limited to the bright [OI] and [CII] lines, although other lines can be included. In Fig. 9.11, observations of the [CII] 158 μm and [OI] 63 μm lines in a sample of PDRs are plotted as a function of the intensity of the incident FUV field determined from the far-IR continuum intensity. Examining these observations, we conclude that the [CII] line dominates the cooling for $G_0 \lesssim 10^3$. The [OI] line dominates at higher G_0. Typically, the total cooling line intensity is in the range 10^{-2}–10^{-3} of the total far-IR continuum intensity. Clearly, the gas is fairly well coupled to the incident FUV photon energy, the PDRs are the source of much of the IR luminosity, [CII] and [OI] dominate the cooling, and the observed efficiency is in good agreement with theoretical models of the photo-electric effect working on PAHs and small dust grains (cf. Section 3.3).

Also shown in Fig. 9.11 are PDR model calculations for different densities. In the models, the [CII] intensity scales with the incident FUV flux (or far-IR intensity) at low G_0, where [CII] is the dominant cooling line. The [CII] intensity levels off at high G_0 when the temperature exceeds ~100 K. At that point, [OI] 63 μm becomes a better coolant and its intensity is proportional to G_0 (cf. the discussion in Section 9.3.1). The models are in good agreement with the observations for densities in the range 10^3–$10^5\,\mathrm{cm}^{-3}$. We note that, as expected, the PDRs in reflection nebulae such as NGC 7023, and bright rim clouds, such as S140, are characterized by somewhat lower densities ($3 \times 10^3\,\mathrm{cm}^{-3}$) and incident FUV fields than those surrounding HII regions, such as M42 in Orion ($10^5\,\mathrm{cm}^{-3}$).

9.9 The Orion Bar

The edge-on geometry of the Orion Bar PDR lends itself particularly well to detailed studies of the spatial structure of PDRs. In this region, the H/H_2 transition, as outlined by the peak emission from vibrationally excited H_2, occurs at about 10″ into the cloud from the ionization front (i.e., the location of the Bar in [OI] 6300 Å and [SII] 6731 Å). The CO 1–0 and [CI] 609 μm lines peak rather abruptly at a distance of 20″ from the ionization front. These observations agree very well with predictions of the global characteristics of PDR models. In such models,

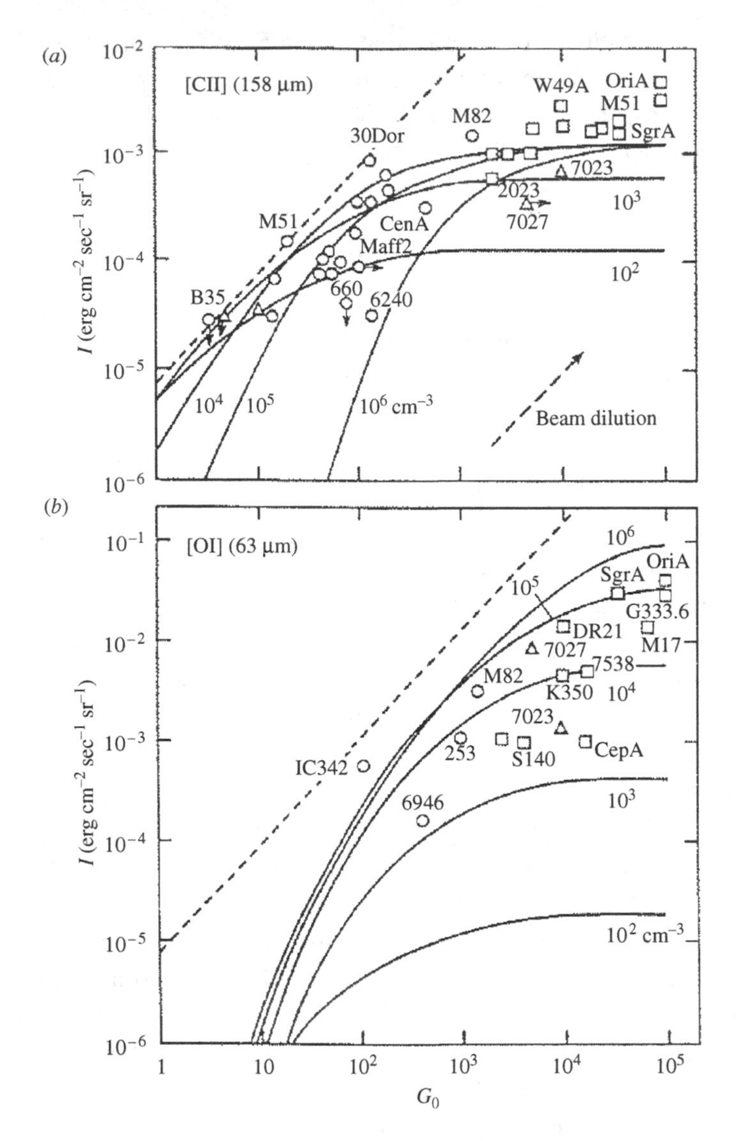

Figure 9.11 Comparison of observed and calculated line intensities of the [CII] 158 μm (*a*) and [OI] 63 μm (*b*) lines as a function of the incident FUV field, G_0. Symbols represent PDRs associated with HII regions (squares), reflection nebulae, dark clouds, and planetary nebulae (triangles), and galactic nuclei (circles). The models are labeled by their density. The dashed line indicates an efficiency of 3% in converting incident FUV energy into gas cooling. Figure reproduced with permission from D. Hollenbach, T. Takahashi, and A. G. G. M. Tielens, 1991, *Ap. J.*, **377**, p. 192.

Table 9.1 *Conditions in the Orion Bar*

Parameter	Simple-minded analysis				Diagnostic diagrams	
G_0	4 (4)[a]			2 (4)[b]	8 (3)[c]	
n (cm^{-3})	1.5 (5)[d]	2 (5)[e]	5 (4)[f]	1 (5)–1 (7)[g]	8 (4)[d]	1 (5)[h]
T (K)	230[i]	360[j]	450[k]			
M_H ($M_\odot$)	2[l]	0.6[k]				

[a] From far-IR.
[b] From [CII]/FIR and CO/FIR.
[c] From [CII]/[OI] and ϵ_{gas}.
[d] From ϵ_{gas}, assuming $G_0 = 4 \times 10^4$ and $T = 300$ K.
[e] From [CII]/[OI].
[f] From the observed spatial structure of the PDR.
[g] High-level CO.
[h] From [OI] versus G_0 ($= 4 \times 10^4$).
[i] From [OI].
[j] From [CII].
[k] From pure rotational H_2.
[l] From [CII].

the location of the H/H_2 transition region and thus the peak in H_2^* emission is displaced inwards from the ionization front by $\Delta A_v \simeq 2$ and the C^+/C/CO transition occurs another $\Delta A_v \simeq 2$ deeper into the cloud (cf. Fig. 9.7). Using standard dust parameters, the observed spatial scale translates then in a density of 5×10^4 cm^{-3}.

The observations are summarized in Table 9.2 and can be analyzed according to the simple-minded procedure outlined in the previous section. The results are summarized in Table 9.1. The density estimated from the heating efficiency and from the [CII]/[OI] ratio are in reasonable agreement. However, the bright high-level CO emission indicates the presence of denser gas as well. The H_2 rotational lines probably provide the best temperature estimate. The [OI] is likely subthermal while the [CII] is only marginally optically thick. The H_2 mass does not account for the atomic gas present in the PDR, which accounts for a factor 2 difference in mass. The diagnostic diagrams yield somewhat lower G_0s, mainly because they do not take the edge-on geometry of this PDR into account. The densities are in very good agreement with those from the simple-minded analysis discussed above.

These observations can also be compared to detailed PDR models to determine the physical conditions in the emitting gas. Figure 9.12 shows a comparison of the observed IR spectrum (1 μm–1000 μm) from a 1′ beam centered on the Trapezium stars of Orion with a theoretical PDR model. Derived physical conditions include the density ($\sim 10^5$ cm^{-3}), the incident FUV flux ($G_0 \sim 10^5$), and either a relatively

Table 9.2 *A model for the Orion Bar PDR*

Line	Observed intensity (erg cm^{-2} s^{-1} sr^{-1})	Model[a] (erg cm^{-2} s^{-1} sr^{-1})
[OI] 63 μm	4 (−2)	7 (−2)
[OI] 146 μm	2 (−3)	5 (−3)
[CII] 158 μm	6 (−3)	7 (−3)
[SiII] 34 μm	9 (−3)	2 (−2)
[CI] 609 μm	5 (−6)	6 (−6)
[CI] 370 μm	(3 (−5))[b]	4 (−5)
CO $J=1-0$	4 (−7)	2 (−7)
CO $J=7-6$	3 (−4)	2 (−7)
CO $J=14-13$	1 (−4)	3 (−7)

[a] Model results for $G_0 = 4\times10^4$ and $n = 5\times10^4$ cm^{-3}. The calculated temperature in the atomic gas is ~500 K.
[b] Measured towards a position near the base of the Orion Bar.

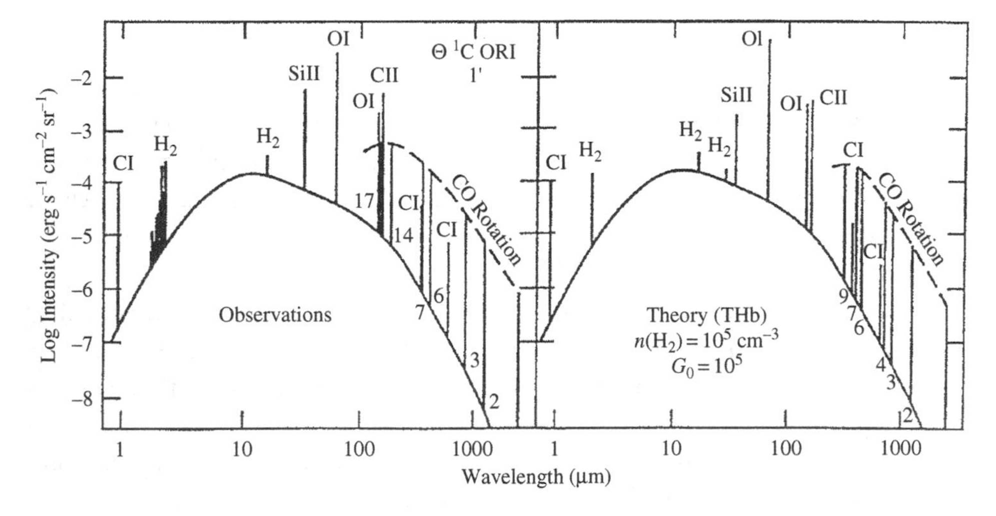

Figure 9.12 The emission spectrum calculated for conditions appropriate to Orion (cf. Fig. 9.7) is compared to the observed spectrum. Emission lines originating in the ionized gas are not shown. Figure reproduced with permission from D. Hollenbach and A. G. G. M. Tielens, 1999, *Rev. Mod. Phys.*, **71**, p. 173.

high gas-phase carbon abundance $\gtrsim$2–3$\times10^{-4}$ or a relatively low FUV extinction by the dust in this dense cloud. The observations of the Orion Bar PDR, which is located some 2′ to the south-east of the Trapezium stars, are compared to the results of a homogeneous model in Table 9.2. The agreement is quite reasonable and the model parameters derived this way are very comparable to those derived from the simple-minded analysis outlined above. The high-level CO lines form an exception to this general agreement. In view of their high critical density and excitation energy, their high intensity betrays the presence of small amounts of warm ($\simeq 10^3$ K), dense (10^6–10^7 cm^{-3}) molecular gas.

9.10 Physical conditions in PDRs

Table 9.3 summarizes relevant observations of PDRs associated with a variety of types of objects. Because of their high critical densities and high excitation energies, the bright emission in high-level CO transitions indicates the presence of high-density warm molecular gas in the PDR. In most PDRs, such clumps are also directly evident in the intensity distribution of many PDR tracers. In modeling, we are forced to use multicomponent models with dense clumps embedded in a more tenuous interclump medium. Of course, this also affects the radiative transfer; FUV photons can penetrate much deeper in a clumpy medium than in a homogeneous medium (with the same mass). As a result, the PDR will be more extended than expected. The atomic fine-structure lines – in particular, [CII] – as well as fluorescent molecular hydrogen are generally very extended. Typical

Table 9.3 *Physical conditions in PDRs*

Object type	NGC 2023 reflection nebula	Orion Bar HII regions	NGC 7027 planetary nebula	Sgr A galactic nucleus	M82 star burst nucleus
Observations					
[OI] 63 μm	4 (−3)	4 (−2)	1 (−1)	2 (−2)	1 (−2)
[OI] 146 μm	2 (−4)	2 (−3)	5 (−3)	7 (−4)	5 (−4)
[CII] 158 μm	7 (−4)	6 (−3)	1 (−2)	2 (−3)	2 (−3)
[SiII] 34 μm	2 (−4)	9 (−3)	—	2 (−2)	4 (−3)
CO $J = 7-6$	5 (−5)	2 (−4)	—	1.5 (−3)	5 (−5)
CO $J = 14-13$	—	3 (−4)	—	3 (−4)	—
F_{IR}	8 (−1)	5 (0)	4 (1)	5 (1)	6 (0)
PAHs	9 (−2)	1.5 (−1)	1.8 (0)	—	1.4 (−1)
G_0	1.5 (4)	4 (4)	6 (5)	1 (5)	1 (4)
Interclump conditions					
n (cm^{-3})	7.5 (2)	5 (4)	1 (5)	1 (5)	1 (4)
T (K)	250	500	1000	500	400
M_a/M_m	0.2	0.6	0.3	0.04	0.1
Clump conditions					
n (cm^{-3})	1 (5)	1 (7)	1 (7)	1 (7)	—
T (K)	750	(2000)	(2000)	(2000)	—
f_v	1 (−1)	5 (−3)	5 (−2)	6 (−2)	4 (−4)

Observed line intensities are in units of erg cm^{-2} s^{-1} sr^{-1}.

densities of the interclump gas are 10^3–10^5 cm^{-3}. Gas temperatures are in the range 200–1000 K. The atomic mass, M_a, in the PDR is a sizeable fraction of the molecular mass in the associated cloud core, M_m. The clumps are typically overdense by two to three orders of magnitude. The filling factors of these dense clumps are, however, small.

9.11 Hydrogen IR fluorescence spectrum

Far-red and near-IR H_2 fluorescence spectra have been observed in a variety of classical PDRs such as those associated with NGC 2023, S140, NGC 7023, the Orion Bar, Orion A and B, and ρ Oph. Figure 9.13 compares the observed IR spectrum 160″ north of the illuminating star in NGC 2023 with model calculations for the pure fluorescence case, illustrating the good fit that is possible. The fit mainly depends on the adopted value for G_0 and the detailed geometry of this emission ridge. It is not sensitive to the density for $n \lesssim 10^4$ cm^{-3}. H_2 spectra have also been obtained for the bright H_2 ridge, 78″ south of the star, which has a more thermal 1–0 S(1)/2–1 S(1) ratio, and the observed spectrum is consistent with emission by higher density gas ($\simeq 10^5$ cm^{-3}).

Figure 9.14 summarizes the column densities derived for the different H_2 levels from the ISO/SWS mid-IR spectrum and available ground-based data for the PDR in S140. The low J levels indicate emission from gas at 500 K with a column density of 1.6×10^{20} cm^{-2}. Higher pure rotational levels start to deviate,

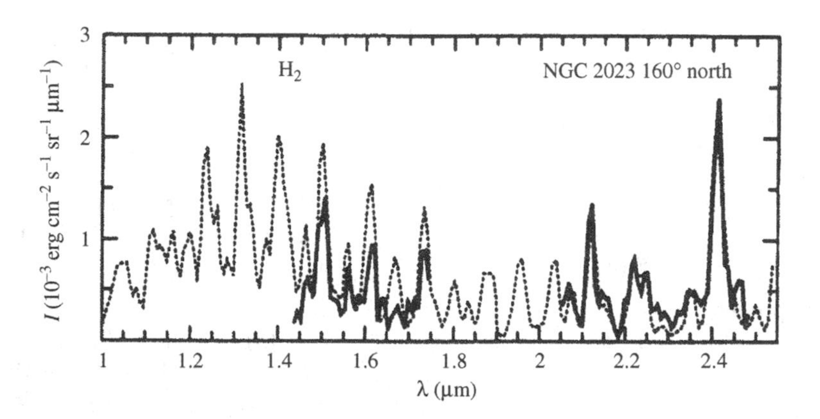

Figure 9.13 Comparison of the observed (solid) and calculated (dashed) H_2 emission spectra in the NGC 2023 reflection nebula. Figure reproduced with permission from D. Hollenbach and A. G. G. M. Tielens, 1999, *Rev. Mod. Phys.*, **71**, p. 173. The models are adapted from J. H. Black and E. F. van Dishoeck, 1987, *Ap. J.*, **322**, p. 412. The observations are taken from I. T. Gatley, *et al.*, 1987, *Ap. J.*, **318**, p. L73.

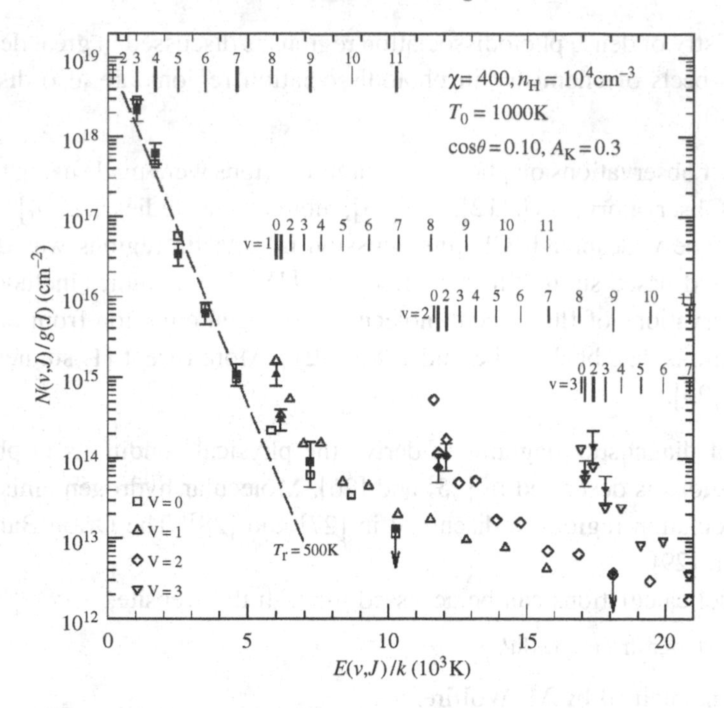

Figure 9.14 H_2 column densities, derived from observations of S140 (filled symbols) and calculations (open symbols) assuming optically thin emission, divided by the statistical weight of the level, are plotted as a function of the excitation energy. Adopted parameters are listed in the top right corner. The intensity, χ, of the FUV radiation field is in slightly different units from habings ($\chi = 1.7G_0$), θ is the inclination angle of the PDR normal to the line of sight, A_K is the extinction in the K band. A temperature profile, characterized by a peak temperature, T_0, was adopted. Figure reproduced with permission from R. Timmermann, *et al.*, 1996, *A. & A.*, **315**, p. L281.

because of the contribution by FUV pumping. Populations of the vibrational levels are displaced towards the right because of UV pumped fluorescence, but the lower J levels in these vibrationally excited states still remember the rotational temperature of the vibrational ground-state temperatures of 500 K.

9.12 Further reading

The seminal paper on photodissociation regions is [1] and much of the detail can be found in the appendixes to this paper. A recent review on the physics and chemistry of photodissociation regions is [2] and on photodissociation regions and the interstellar medium of galaxies [3].

Time-dependent effects of H_2 in photodissociation regions have been considered in [4], [5], and [6]. The effects on the emitted spectrum are discussed in [7].

The chemistry of dense photodissociation regions is discussed in great detail in [8]. Various aspects of chemistry in photodissociation regions are also discussed in [9] and [10].

The earliest observations on photodissociation regions were made using the Kuiper Airborne Observatory, [11], [12], and [13]; more recent studies are [14], [15], [16], and [17]. The widespread [CI] line emission in galactic regions was discovered using ground-based submillimeter telescopes, [18]. Later studies include [19] and [20]. Observations of fluorescent molecular hydrogen emission from photodissociation regions date back to the mid 1980s, [21]. More recent H_2 studies are [22], [23], and [24].

The use of diagnostic diagrams to derive the physical conditions in photodissociation regions is described in [25] and [26]. Molecular hydrogen emission from photodissociation regions is discussed in [27] and [28]. The Orion Bar model is taken from [29].
PDR model calculations can be accessed through the website,

http://dustem.astro.umd.edu/,

which is maintained by M. Wolfire.

References

[1] A. G. G. M. Tielens and D. Hollenbach, 1985, *Ap. J.*, **291**, p. 722
[2] D. Hollenbach and A. G. G. M. Tielens, 1997, *Ann. Rev. Astron. Astrophys.*, **35**, p. 179
[3] D. Hollenbach and A. G. G. M. Tielens, 1999, *Rev. Mod. Phys.*, **71**, p. 173
[4] F. Bertoldi and B. T. Draine, 1996, *Ap. J.*, **458**, p. 222
[5] O. Goldschmidt and A. Sternberg, 1995, *Ap. J.*, **439**, p. 256
[6] D. Hollenbach and A. Natta, 1995, *Ap. J.*, **455**, p. 143
[7] H. Störzer and D. Hollenbach, 1998, *Ap. J.*, **495**, p. 853
[8] A. Sternberg and A. Dalgarno, 1995, *Ap. J. S.*, **99**, p. 565
[9] D. Jansen, E. F. van Dishoeck, J. H. Black, M. Spaans, and C. Sosin, 1995, *A. & A.*, **302**, p. 223
[10] A. Fuente, A. Rodriguez-Franco, S. Garcia-Burillo, J. Martin-Pintado, and J. H. Black, 2003, *A. & A.*, **406**, p. 899
[11] G. Melnick, G. E. Gull, and M. Harwitt, 1979, *Ap. J.*, **227**, p. L29
[12] R. W. Russell, G. Melnick, G. E. Gull, and M. Harwitt, 1980, *Ap. J.*, **240**, p. L99; R. W. Russell, *et al.*, 1981, *Ap. J.*, **250**, p. L35
[13] J. W. V. Storey, D. M. Watson, and C. H. Townes, 1979, *Ap. J.*, **233**, p. 109
[14] R. T. Boreiko and A. L. Betz, 1996, *Ap. J.*, **464**, p. L83
[15] F. Hermann, S. C. Madden, T. Nikola, *et al.*, 1997, *Ap. J.*, **481**, p. 343
[16] D. T. Jaffe, S. Zhou, J. E. Howe, *et al.*, 1995, *Ap. J.*, **436**, p. 203
[17] K. Mizutani, *et al.*, 1994, *Ap. J. S.*, **91**, p. 613
[18] T. G. Phillips and P. J. Huggins, 1981, *Ap. J.*, **251**, p. 533
[19] J. Keene, G. A. Blake, T. G. Phillips, P. J. Huggins, and C. A. Beichman, 1985, *Ap. J.*, **299**, p. 967

[20] R. Plume, D. T. Jaffe, K. Tatematsu, N. J. Evans, and J. Keene, 1999, *Ap. J.*, **512**, p. 768
[21] I. Gatley, *et al.*, 1987, *Ap. J.*, **318**, p. L73
[22] M. G. Burton, M. Bulmer, A. Moorhouse, T. R. Geballe, and P. W. J. L. Brand, 1992, *M. N. R. A. S.*, **257**, p. 1
[23] E. Habart, F. Boulanger, L. Verstraete, *et al.*, 2003, *A. & A.*, **397**, p. 623
[24] R. Timmermann, *et al.*, 1996, *A. & A.*, **315**, p. L281
[25] M. G. Wolfire, A. G. G. M. Tielens, and D. Hollenbach, 1990, *Ap. J.*, **358**, p. 116
[26] M. J. Kaufman, M. G. Wolfire, D. Hollenbach, and M. L. Luhman, 1999, *Ap. J.*, **527**, p. 795
[27] J. H. Black and E. F. van Dishoeck, 1987, *Ap. J.*, **322**, p. 412
[28] B. T. Draine and F. Bertoldi, 1996, *Ap. J.*, **468**, p. 269
[29] A. G. G. M. Tielens, M. M. Meixner, P. van der Werf, *et al.*, 1993, *Science*, **262**, p. 86

10
Molecular clouds

10.1 Introduction

Molecular clouds differ in a number of important aspects from the atomic clouds discussed in Chapter 8. They tend to be denser and have a higher column density. As a result, the intensity of the dissociating FUV radiation field is lower and gas-phase chemistry in these clouds is primarily driven by cosmic-ray ionization. Accretion on grains is another process that becomes more important at these higher densities and ice mantles – absent in the diffuse ISM – become prevalent inside molecular clouds. In fact, the interaction between the gas phase and the solid state becomes one of the driving forces of molecular diversity in molecular clouds. Molecular clouds are also much cooler than diffuse clouds ($\simeq$10 K versus $\simeq$100 K) due to the enhanced cooling associated with the small energy spacing of molecular rotational levels and the higher densities of molecular clouds. A final difference with diffuse clouds is the importance of self-gravity in molecular clouds. While gravitational collapse is outside the scope of this book, new stars are only formed in molecular clouds and such embedded protostars can influence their environment through shocks driven by their powerful outflows. Protostars will also heat the surrounding dust and this energy is coupled to the gas through gas–grain collisions. This heating of the dust can evaporate previously accreted ice mantles when the temperature is raised above the sublimation temperature.

In this chapter, we will discuss the physical and chemical processes that dominate the characteristics of molecular clouds. We will start off discussing the ionization balance and the energy balance. In Section 10.4, we will discuss the gas-phase inventory of molecular clouds and the chemistry that plays a role in the formation of these species. The formation of interstellar ices is described in Section 10.5. Various aspects of gas–grain interaction are analyzed in Section 10.6. Molecular observations provide a powerful tool for the analysis of the physical and chemical conditions in molecular clouds and this is described in Section 10.7.

10.2 The degree of ionization

Ions play a central role in gas-phase chemistry of molecular clouds. The cosmic-ray ionization of molecular hydrogen is quickly passed on to other (protonated) molecular species (Fig. 10.1). The electrons, on the other hand, are "mopped up" by large PAH molecules. PAH anions can be neutralized by photodetachment and by recombination with metal ions and molecular ions. The photodetachment rate of a PAH with N_c carbon atoms is given by $k_{pd} \simeq 4 \times 10^{-10}\, N_c \exp[-0.25 A_v]\,\mathrm{s}^{-1}$, where the exponential factor describes the penetration of photons with energy $h\nu \sim 1\,\mathrm{eV}$ – the electron affinity of PAHs – and A_v is the visual extinction. The neutralization coefficient for molecular cations and PAH anions is approximately $k_n = 3 \times 10^{-7}\,\mathrm{cm}^3\,\mathrm{s}^{-1}$ (Section 6.5.5). Thus, PAH neutralization is dominated by the absorption of visible photons for $A_v \leq 15$ magnitudes. Since the typical extinction through a molecular cloud is only $\simeq$8 magnitudes, this includes most of the molecular gas in the Galaxy.

In dense cores with $A_v > 15$ magnitudes, PAH anions are the dominant negative charge carrier (cf. Fig. 10.2). The degree of ionization is then approximately given by

$$x \simeq \left(\frac{\zeta_{CR}}{k_n n}\right)^{1/2} \simeq \frac{10^{-5}}{\sqrt{n}}, \tag{10.1}$$

where we have adopted a primary cosmic-ray ionization rate, ζ_{CR}, of $3 \times 10^{-17}\,\mathrm{s}^{-1}$, somewhat lower than for diffuse clouds (cf. Section 8.8). Thus, the denser the gas, the lower the degree of ionization. This has direct consequences for processes such as ambipolar diffusion and hence the stability of cloud cores.

In the absence of PAHs, the ionization is transferred to metals, and radiative recombination of metal ions and electrons balance cosmic-ray ionization. Because this is a slow process, compared with recombination with PAH anions, the degree of ionization is considerably higher when PAHs are absent (Fig. 10.2). The importance of this charge transfer will depend on the metal abundance. Comparing the metal ion–PAH anion recombination rate ($\simeq 2\times10^{-7}\,\mathrm{cm}^3\,\mathrm{s}^{-1}$) with the radiative recombination rate ($\simeq 3\times10^{-11}\,\mathrm{cm}^3\,\mathrm{s}^{-1}$), we can conclude that metal neutralization will predominantly take place with PAH anions unless the PAH anion abundance is less than 10^{-4} of the electron abundance. Likewise, PAH anions will dominate the recombination of molecular ions if the PAH anion abundance is ten times that of the electrons. It is also easy to show

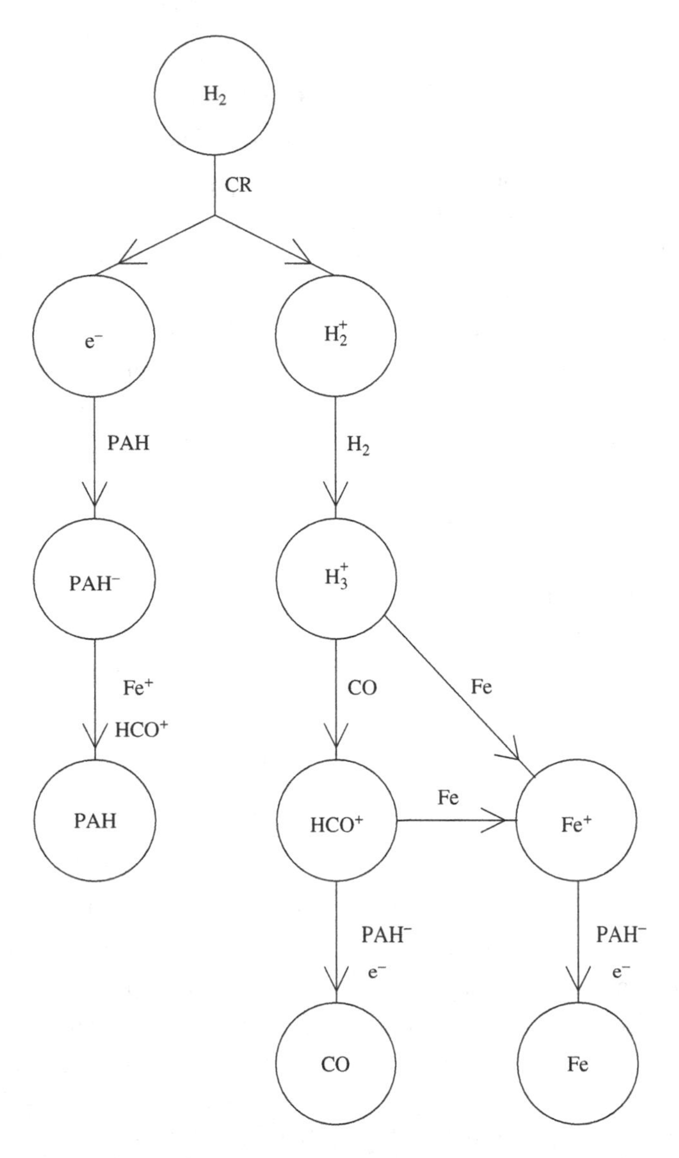

Figure 10.1 The flow of ionization in molecular clouds. Cosmic rays ionize molecular hydrogen. Reaction with H_2 rapidly forms H_3^+. The latter transfers its proton to other molecular species (HCO^+ is shown as the prime example of this reaction). H_3^+ and other molecular ions can charge transfer with trace metal atoms such as iron. The electron will be quickly soaked up by large molecules with high electron affinities such as PAHs. Eventually, the charge is lost through recombination between molecular or metallic cations with PAH anions.

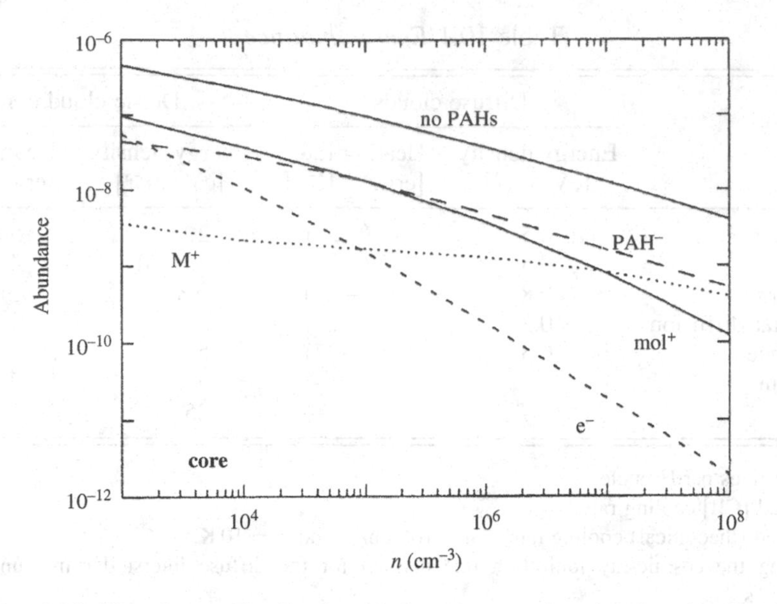

Figure 10.2 The charge distribution in dense cores of molecular clouds calculated including PAHs, metal ions, and molecular ions. The adopted PAH and metal abundances are 10^{-7} and 10^{-6}, respectively. The degree of ionization in the absence of PAHs is shown for comparison.

that charge transfer from molecular ions to metal ions only becomes important when the metal abundance is some $300X(\mathrm{PAH}^-)$. This only happens at a high density (cf. Fig. 10.2). At slightly higher densities than this, the positive and the negative charge will accumulate on grains. Of course, at such high densities, depletion and coagulation may have had an overriding influence on the charge balance in molecular clouds. Molecular abundances are very much tied to the degree of ionization. Thus, models without PAHs where the ionization balance is controlled by metals will have much higher molecular ion abundances than models with PAHs. To agree with the observations, models without PAHs have to adopt low-metal abundances (where the ionization balance is controlled by molecular ions). Such low-metal abundance–no PAH models for molecular clouds have become entrenched as one branch of astrochemistry studies.

10.3 Energy balance

Heating of molecular cloud cores has been discussed in Section 3.11.3. Various heating rates for the gas inside dense molecular cloud cores are summarized in Table 10.1. These heating rates are compared with those in diffuse clouds (cf. Table 1.2). Obviously, the interstellar radiation field is of no importance.

Table 10.1 *Energy balance*

Source	Diffuse clouds		Dense cloud cores	
	Energy density [eV cm^{-3}]	Heating rate[a] [erg s^{-1} H^{-1}]	Energy density [eV cm^{-3}]	Heating rate[a] [erg s^{-1} H^{-1}]
Thermal	0.5	−5 (−26)[b]	20	−5 (−28)[c]
UV	0.5	5 (−26)	4 (−6)	4 (−31)
Cosmic ray	0.8	3 (−27)[d]	0.8	5 (−28)[e]
Ambipolar diffusion	0.2	2 (−27)	5	1 (−28)
Turbulence	0.3	1 (−27)	5	3 (−28)
Gas–grain	—	—	3	3 (−29)
Gravity	—	—	25	4 (−28)[f]

[a] Heating rates per H nucleus.
[b] Observed [CII] cooling rate.
[c] Estimated (theoretical) cooling rate at $n = 10^4$ cm^{-3} and $T = 10$ K.
[d] Adopting the cosmic-ray ionization rate derived for the diffuse interstellar medium ($\zeta_{CR} = 2 \times 10^{-16}$ s^{-1}).
[e] Adopting the cosmic-ray ionization rate derived for dense clouds ($\zeta_{CR} = 3 \times 10^{-17}$ s^{-1}).
[f] Gravitational energy released during free-fall at $n = 10^4$ cm^{-3}.

By the same token, heating due to collisions of gas species with dust grains – heated by penetrating longer wavelength radiation – is not important for dark cores. Near protostars, gas–grain collisions are, however, efficient couplers of the dust and gas temperatures (see Section 3.11.3). Cosmic rays can still penetrate such cores and their heating rate is appreciable. Likewise, turbulence can be an important heating source for the gas. At the highest densities, ambipolar diffusion will be important. For collapsing clouds, gravitational heating dominates.

Cooling of the interstellar gas occurs through molecular rotational transitions of various molecules, and this has been discussed in Section 2.6.2. At a density of 10^4 cm^{-3}, the cooling rate is well represented by $4 \times 10^{-30}\, T^{2.4}$ erg s^{-1}(H$_2$)$^{-1}$ for low temperatures. Thus, a molecular cloud ($n \sim 10^3$ cm^{-3}) heated only by cosmic rays will be some 5 K. This is much less than the gas temperature derived from CO observations. Essentially, the CO line becomes optically thick near the surface of the cloud, which is still heated by incident FUV photons. Molecular cloud cores heated by cosmic rays and dissipation of turbulence will have temperatures of some 8 K. Even including the release of gravitational energy, this only leads to a temperature of 10 K. Indeed, a core will only heat up when a central object forms, and the core is optically thick to (cooling) radiation. The main energy coupling, then, is through gas–grain collisions. This will tend to keep the gas temperature locked to that of the dust. Finally, cooling time scales are fairly short in molecular cloud cores due to the high cooling efficiency of molecules and the high density. In particular, at 10^4 cm^{-3}, the cooling time scale is $\simeq 5 \times 10^4$ years at 10 K and decreases to only 7×10^3 years at 40 K.

10.4 Gas-phase chemistry

10.4.1 Gas-phase composition of molecular clouds

Over the last three decades, rotational spectroscopy in the millimeter wavelength region (cf. Fig. 10.3) has revealed that molecular clouds contain a large variety of molecules (Table 10.2). Given the overabundance of hydrogen in the ISM, one might expect the molecular composition to be dominated by saturated hydrides such as H_2O, NH_3, and CH_4. Simple hydrides and oxides are indeed present in molecular clouds (cf. Table 10.2). However, interstellar molecular clouds also contain a variety of unsaturated radicals and ions. These radicals are very reactive under terrestrial laboratory conditions and indeed some were discovered in space

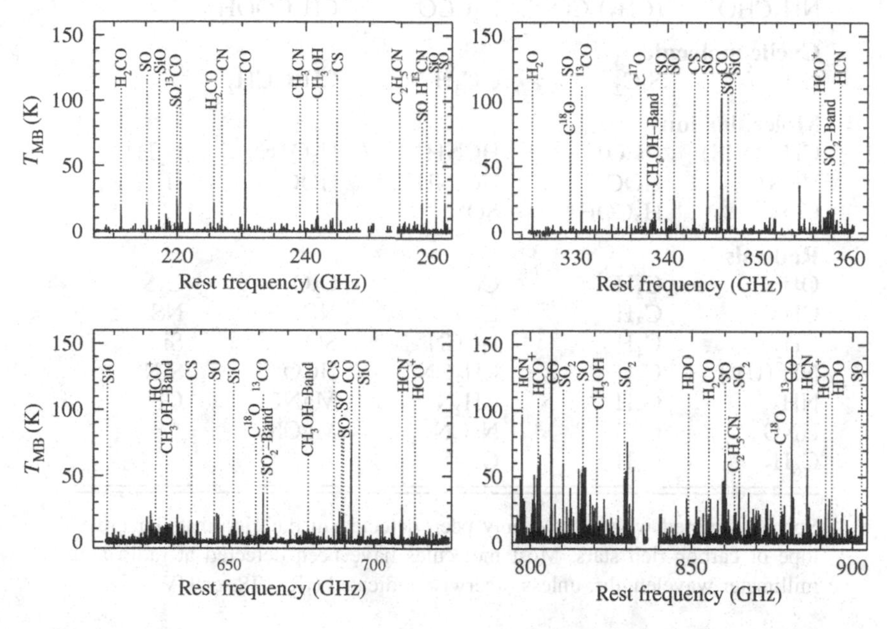

Figure 10.3 Line surveys of the Orion KL star-forming region in the four submillimeter atmospheric windows, centered at 240, 340, 650, and 850 GHz, reveal a multiple of rotational lines of a wide variety of species. Some of the lines detected have been labeled. Figure courtesy of C. Comito. Data taken from Sutton, *et al.* (1985, *Ap. J. S.*, **58**, p. 341; 230 GHz), G. Blake, *et al.* (1987, *Ap. J.*, **315**, p. 621; 230 GHz), P. Schilke, *et al.* (1997, *Ap. J. S.*, **108**, p. 301; 350 GHz), P. Schilke, *et al.* (2001, *Ap. J. S.*, **132**, p. 281; 650 GHz), and C. Comito (2003, Ph.D. thesis, M. P. I. f. R. Bonn; 850 GHz).

Table 10.2 *Identified interstellar and circumstellar molecules*

Simple hydrides, oxides, sulfides, halogens				
H_2 (IR,UV)	CO	NH_3	CS	NaCl*
HCl	SiO	SiH_4^* (IR)	SiS	AlCl*
H_2O	SO_2	C_2 (IR)	H_2S	KCl*
N_2O	OCS	CH_4 (IR)	PN	AlF*
HF				
Nitriles and acetylene derivatives				
C_3 (IR,UV)	HCN	CH_3CN	HNC	$C_2H_4^*$ (IR)
C_5^* (IR)	HC_3N	CH_3C_3N	HNCO	C_2H_2 (IR)
C_3O	HC_5N	CH_3C_5N ?	HNCS	
C_3S	HC_7N	CH_3C_2H	HNCCC	
C_4Si*	HC_9N	CH_3C_4H	CH_3NC	
	$HC_{11}N$	CH_3CH_2CN	HCCNC	
	HC_2CHO	CH_2CHCN		
Aldehydes, alcohols, ethers, ketones, amides				
H_2CO	CH_3OH	HCOOH	CH_2NH	CH_2CC
H_2CS	CH_3CH_2OH	$HCOOCH_3$	CH_3NH_2	CH_2CCC
CH_3CHO	CH_3SH	$(CH_3)_2O$	NH_2CN	
NH_2CHO	$(CH_3)_2CO$	H_2CCO	CH_3COOH	
Cyclic molecules				
C_3H_2	SiC_2	c-C_3H	CH_2OCH_2	
Molecular ions				
CH^+ (VIS)	HCO^+	$HCNH^+$	H_3O^+	N_2H^+
HCS^+	$HOCO^+$	HC_3NH^+	HOC^+	H_3^+ (IR)
CO^+	H_2COH^+	SO^+		
Radicals				
OH	C_2H	CN	C_2O	C_2S
CH	C_3H	C_3N	NO	NS
CH_2	C_4H	HCCN*	SO	SiC*
NH (UV)	C_5H	CH_2CN	HCO	SiN*
NH_2	C_6H	CH_2N	MgNC	CP*
HNO	C_7H	NaCN	MgCN	
C_6H_2	C_8H	C_5N*		

Species denoted with $*$ have only been detected in the circumstellar envelope of carbon-rich stars. Most molecules have been detected at radio and millimeter wavelengths, unless otherwise indicated (IR, VIS, or UV).

before they were even identified in the laboratory. The degree of unsaturation is particularly striking and, clearly, the molecular composition is far from local thermodynamic equilibrium. This attests to the importance of kinetics over thermodynamics in interstellar molecular clouds.

Perusing Table 10.2, it is clear that interstellar chemistry is rich and diverse. Except for molecular hydrogen formation, much of the chemistry occurs in the gas phase and is driven by penetrating FUV photons and cosmic rays. Molecular species do accrete onto dust grains where they can undergo further reactions. In this way an ice mantle is formed (cf. Section 10.5). The composition of hot cores – warm ($\sim$200 K), dense (10^6 cm^{-3}) regions around massive protostars – is quite different from that in dark cloud cores and is characterized by a preponderance of saturated species. Evidently, other chemical routes have opened up under these conditions. This is generally attributed to the chemistry of ice molecules evaporating into the warm gas through the heating action of the protostar (Section 10.6.3).

10.4.2 Chemistry of oxygen, carbon, and nitrogen

In evaluating the gas-phase chemistry of molecular clouds, it is important to keep the simple rules from Section 4.1.10 in mind: for ions, reactions with H_2 are most important unless they are inhibited by an activation barrier. Dissociative recombination then takes over. For neutrals, proton transfer from cationic species such as H_3^+, HCO^+, and H_3O^+ is important, while reactions with He^+ often provide a loss channel. If proton transfer is inhibited, neutral–neutral reactions with small radicals have to be considered. If all else fails, the general consensus is that the "scoundrel" rule applies and grain-surface chemistry may be invoked (Section 10.5). At the moment, that is only generally accepted for the formation of molecular hydrogen and for the formation of hydrides – such as H_2O, NH_3, CH_4, H_2CO, and CH_3OH – which are abundant in ices.

Near the surface of a molecular cloud, the chemistry is really that of diffuse clouds (Section 8.7.1). Here we will focus on the chemistry in the dark dense cores where penetrating UV photons play no role and the chemistry is driven by cosmic-ray ionization. We have already seen in Section 10.2 that cosmic-ray ionization of H_2 leads to the formation of the H_3^+ ion. This ion is the cornerstone of ion–molecule gas-phase chemistry. It will react with atomic O to form OH^+, which reacts with H_2 to form H_2O^+ and then H_3O^+. That is where reactions with H_2 will stop and dissociative electron recombination takes over, forming H_2O and OH in about a 1 to 2 ratio. These O-bearing species are lost through reaction with hydrocarbon radicals, which burn them to (eventually) CO. The radical OH may also react with atomic O to form O_2 but that reaction may not be as efficient as once thought.

The similar chain starting with C^+ stops at CH_3^+, because the next H transfer from H_2 is endothermic with a rate of 10^{-14} cm^3 s^{-1}. CH_3^+ will radiatively associate with H_2 to form CH_5^+ at a much slower rate (10^{-13} cm^3 s^{-1}) than typical ion–molecule reactions (10^{-9} cm^3 s^{-1}). Also, dissociative electron recombination of CH_5^+ produces CH_3, and CH_4 has to be formed through proton transfer (e.g., $CH_5^+ + CO \longrightarrow CH_4 + HCO^+$). Thus, unlike H_2O, the chemical routes towards CH_4 are inhibited in molecular clouds. In contrast, the carbon is readily transferred to CO. The small hydrocarbon ions, formed by the steps described above, react with O and OH to form CO^+ (which reacts with H_2) and HCO^+. The formyl ion dissociatively recombines with electrons to form CO. Almost all of the carbon is locked up this way in this stable molecule.

The chemistry of nitrogen is very different. Atomic N does not accept a proton from H_3^+ at low temperatures. Atomic N will react with OH to form NO, which itself will react again with N to form N_2. This is a very stable molecule and models predict that most of the nitrogen in molecular clouds is locked up in N_2. Reaction of N_2 with H_3^+ forms N_2H^+, which dissociatively recombines to N_2, closing this loop without breaking the N out of N_2. The reaction pathway to NH_3 in the gas phase is not well understood. The route analogous to that for oxygen-bearing hydrides – starting with proton transfer to N followed by rapid hydrogen abstraction from H_2 to form, eventually, NH_4^+, which dissociatively recombines to form NH_3 – has two bottlenecks: the initial step, forming NH^+, is endothermic, while the last hydrogen abstraction reaction, forming NH_4^+, has an activation barrier. Of course, formation of NH_3 will have to start with the break up of N_2 through reaction with cosmic-ray-produced He^+, forming N^+ and N. The latter is unreactive towards H_3^+, but NH_4^+ formation may also be initiated with the former. Reaction of N^+ with para H_2 has a small activation barrier of $\simeq$200 K, but N^+ may react more efficiently with ortho H_2, which lies about 170 K above the para ground state of H_2. The reaction may also be assisted by any excess translational energy that N^+ acquired in the breaking of the N_2 by He^+ or by electronic excitation energy of N^+, if the latter is efficiently formed in the excited fine-structure levels of the ground state. The radiative decay rate of these levels is very slow ($A \simeq 10^{-6}\ \mathrm{s}^{-1}$) and unimportant for dense gas. Once NH^+ is formed, subsequent reactions with H_2 quickly produce NH_3^+. The next hydrogen abstraction reaction has an activation barrier of some 1000 K. Fortunately, tunneling can assist this reaction and the measured rate coefficient is $10^{-12}\ \mathrm{cm^3\ s^{-1}}$ at 10 K. In the competition between dissociative electron recombination and hydrogen abstraction from H_2, the latter will win when the electron abundance is small ($x \ll 10^{-6}$; e.g., $n \gg 10^2\ \mathrm{cm}^{-3}$). So, in dense gas, NH_3 can be effectively formed if the first step can be taken. The formation rate of NH_3 then depends critically on the ortho-to-para ratio of H_2 in molecular clouds, which is not well known. Molecular hydrogen will be formed on grain surfaces in the high temperature ortho-to-para ratio (3), which will be favorable for NH_3 formation. However, reactions with cosmic-ray-produced H^+ will drive this ratio to the very low LTE ortho-to-para ratio at the gas temperature ($9\exp[-170/T]$). It is clear that N chemistry is a confluence of many ill-understood processes.

Finally, one should consider time-dependent effects. Gas-phase chemistry of molecular clouds requires the presence of molecular hydrogen, which is formed on grain surfaces on a time scale of

$$\tau_{\text{chem}}(\text{H}_2) \simeq \frac{7 \times 10^8}{n} \text{ years}, \quad (10.2)$$

or about 10^6 years for a typical molecular cloud. Once H_2 is abundant, chemistry can cycle the species around. The rate limiting step is then the ionization of H_2, which initiates the chemistry. Essentially, each ionization leads to molecule formation. Thus, the time scale to cycle the gas through molecules is then

$$\tau_{\text{chem}} = \frac{\mathcal{A}_{\text{O}}}{\xi_{\text{CR}}} \simeq 3 \times 10^5 \text{ years}, \quad (10.3)$$

with $\mathcal{A}_{\text{O}}$ the oxygen abundance. Cosmic rays also ionize helium and He^+ is generally a molecular destruction agent, which slows down these atom-to-molecule conversion processes somewhat. Also, once the gas has a large molecular fraction, the ionization may merely cycle CO to HCO^+ with H_3^+ and back again to CO with an electron and the conversion time scale will increase. In general, the fraction of the ionizations that are effective in driving the chemical conversion has to be taken into account in these time scale estimates.

10.4.3 Complex molecule formation

Molecular clouds contain a multitude of simple hydrocarbon species and Table 10.3 summarizes a selection of those towards two dark clouds, TMC-1 and L134N. TMC-1 provides an extreme example of abundant hydrocarbon chains while L134N has more typical abundances. The species listed in Table 10.3 can be broken up into various classes. Carbon monoxide and the simple ions and radicals (OH, HCO^+, N_2H^+) have very similar abundances in these two dark clouds. These two clouds are also alike in the oxygen-bearing species, H_2CO, CH_3OH, HCOOH, and CH_3CHO. Probably, these two classes of species have very similar production routes and rates in these two dark cloud cores. The first group are undoubtedly gas-phase products but the latter group might originate from grains. The abundances of the hydrocarbon chains, on the other hand, are very different in these two clouds, which becomes more pronounced when going to longer chains in the various series (C_nH, HC_nN). Whatever the formation route is, it is much more efficient in TMC-1 than in L134N and the physical origin for this is not agreed upon.

Inside dense molecular clouds, most of the carbon is locked up in CO and this impedes the formation of complex species. Cosmic rays provide a means by which carbon can be broken out of CO. Primary and secondary cosmic-ray

Table 10.3 *The molecular composition of dark cores*

Species	TMC-1	L134N
CO	8 (−5)	8 (−5)
OH	3 (−7)	8 (−8)
HCO^+	8 (−9)	8 (−9)
N_2H^+	5 (−10)	5 (−10)
NH_3	2 (−8)	2 (−7)
H_2CO	2 (−8)	2 (−8)
CH_3OH	2 (−9)	3 (−9)
HCOOH	2 (−10)	3 (−10)
CH_3CHO	6 (−10)	6 (−10)
CH_3CN	1 (−9)	1 (−9)
C_2H	5 (−8)	5 (−8)
C_3H	5 (−10)	3 (−10)
C_4H	2 (−8)	1 (−9)
C_5H	3 (−10)	—
C_6H	1 (−10)	—
HCN	2 (−8)	4 (−9)
HC_3N	6 (−9)	2 (−10)
HC_5N	3 (−9)	1 (−10)
HC_7N	1 (−9)	2 (−11)
HC_9N	3 (−10)	—
HNC	2 (−8)	6 (−9)
H_2O	<7 (−9)	

electrons can excite H_2, which cascades down through the Lyman–Werner bands. The FUV photons produced this way photodissociate CO (Section 4.1.1) and the resulting C is then available for the build-up of larger species. Equating the CO dissociation rate with the chemical build-up time scale (Eq. (10.3)), we find

$$\frac{\mathrm{C}}{\mathrm{CO}} = \frac{1-\omega}{p_{\mathrm{CO}}\mathcal{A}_{\mathrm{O}}} \simeq 6 \times 10^{-3}. \tag{10.4}$$

This C/CO ratio is some two orders of magnitude larger than without cosmic-ray produced photons. Cosmic rays also ionize helium and the He^+ ion can destroy molecules. In particular, He^+ breaks C^+ out of CO and makes it available for the formation of more complex hydrocarbon species. However, this is much less efficient than the cosmic-ray induced photons.

Complex molecule formation in dark cloud cores then proceeds through the insertion of C to form, for example, various carbon chains. This starts with the formation of small hydrocarbon radicals;

$$\mathrm{H_3^+ + C \longrightarrow CH^+ + H_2}. \tag{10.5}$$

The CH^+ radical can then react sequentially with H_2 to form the hydrocarbon ions CH_3^+ and CH_5^+ and, after dissociative electron recombination or proton transfer, to CH_3 and CH_4 (Section 10.4.2). More complex hydrocarbons can now be formed through an insertion reaction with C^+,

$$C^+ + CH_4 \longrightarrow C_2H_3^+ + H \qquad (10.6)$$

$$\longrightarrow C_2H_2^+ + H_2. \qquad (10.7)$$

The latter reacts (somewhat slowly) with H_2 (10^{-11} cm^3 s^{-1}) to form $C_2H_3^+$. Dissociative recombination then leads to the formation of acetylene, C_2H_2. Condensation reactions are also of importance in the build-up of molecular complexity,

$$C_2H_2^+ + C_2H_2 \longrightarrow C_4H_3^+ + H \qquad (10.8)$$

$$\longrightarrow C_4H_2^+ + H_2. \qquad (10.9)$$

Finally, neutral–neutral reactions are a key in the chemistry as well. Specifically,

$$C + C_2H_2 \longrightarrow C_3H + H \qquad (10.10)$$

and

$$C_2H + C_2H_2 \longrightarrow C_4H_2 + H \qquad (10.11)$$

are important builders of carbon chains. The reaction of the cyanogen radical, CN, with acetylene forms an efficient route towards cyanoacetylene,

$$CN + C_2H_2 \longrightarrow HC_3N + H, \qquad (10.12)$$

which has a rate of 10^{-10} cm^3 s^{-1} at 10 K. So, using these and similar reactions, a fairly complex reaction network for hydrocarbon formation can be built up, which leads to a preponderance of radicals and acetylene derivatives. Nevertheless, it all starts with breaking the C out of CO and it is clear that such complex species will be much less abundant than CO. Furthermore, the unusually high abundance of hydrocarbon chains in TMC-1 requires a very efficient and unique mechanism to break C out of CO in this source.

10.4.4 Deuterium fractionation

Originally, observational studies of interstellar deuterated molecules were inspired by the constraints provided by the deuterium abundance on cosmology and on stellar nucleosynthesis. However, observations quickly revealed high levels of deuteration (cf. Table 10.4) and it was realized that deuterated molecules instead provide an important probe of interstellar chemistry. These high deuteration levels reflect the slightly higher stability of deuterated species as a result of the zero-point

Table 10.4 *Observed deuteration in molecular clouds*

Species	TMC-1[a] 10 K	Orion[b] 50 K	Orion hot core/compact ridge[c] 200/150 K
DCO^+	0.015	0.002	0.002
N_2D^+	< 0.045		0.003/0.06
DCN	0.023	0.02	0.001/0.03
DNC	0.015	0.01	0.01
C_2D	0.01	0.045	0.05
NH_2D	< 0.02		
c–C_3HD	0.08		
C_4D	0.004		
DC_3N	0.015		
DC_5N	0.013		
HDCO	0.015	0.02	0.14
D_2CO			0.02
CH_3OD			0.03
CH_2DOH			0.04

[a] Dark cloud core in Taurus.
[b] The ridge in Orion.
[c] The hot core and compact ridge in the region of massive star formation in Orion.

energy difference; i.e., deuterium sits slightly deeper in the potential well than hydrogen.

The reactions of interest are of the form

$$XH^+ + HD \rightleftarrows XD^+ + H_2 + \Delta E. \tag{10.13}$$

The energy difference, $\Delta E/k$, is of the order of a few hundred degrees and these reactions therefore play a role only at low temperatures. Specifically of interest is

$$H_3^+ + HD \rightleftarrows H_2D^+ + H_2 + \Delta E = 178\,\mathrm{K}. \tag{10.14}$$

Ignoring other species for the moment, we would find in local thermodynamic equilibrium,

$$\frac{H_2D^+}{H_3^+} = \frac{HD}{H_2}\exp[\Delta E/kT], \tag{10.15}$$

and at a temperature of 10 K, this ratio would be much larger than 1 (the elemental D abundance is 1.5×10^{-5}). Of course, molecular clouds are not in LTE, and the deuterium fractionation in these species is limited by electron recombination and proton or deuteron transfer to other species. Indeed, in one third of the collisions, the enhanced fractionation of H_2D^+ is passed on to neutral molecules with deuterium affinities larger than H_2. In particular,

$$H_2D^+ + CO \rightarrow DCO^+ + H_2 \tag{10.16}$$

is important and leads to the formation of DCO^+. Figure 10.4 summarizes the various reactions involved.

We find, for the deuterium fractionation,

$$\frac{X(H_2D^+)}{X(H_3^+)} = \frac{k_f X(HD)}{k_b X(H_2) + k_e x + k_{CO} X(CO) + \cdots} \tag{10.17}$$

and

$$\begin{aligned} \frac{X(DCO^+)}{X(HCO^+)} &= \frac{X(H_2D^+)}{3X(H_3^+)} \\ &= \frac{2}{3} \frac{\mathcal{A}_D}{\exp[-\Delta E/kT] + 193x + 1.2X(CO) + \cdots}, \end{aligned} \tag{10.18}$$

where $\mathcal{A}_D$ is the elemental deuterium abundance and we have used the reaction rates $k_f = 1.4 \times 10^{-9}\ \text{cm}^3\ \text{s}^{-1}$, $k_e = 6 \times 10^{-8}\sqrt{300/T}\ \text{cm}^3\ \text{s}^{-1}$, and $k_{CO} = 1.7 \times 10^{-9}\ \text{cm}^3\ \text{s}^{-1}$. For typical electron abundances ($x = 10^{-8} - 10^{-6}$) and CO

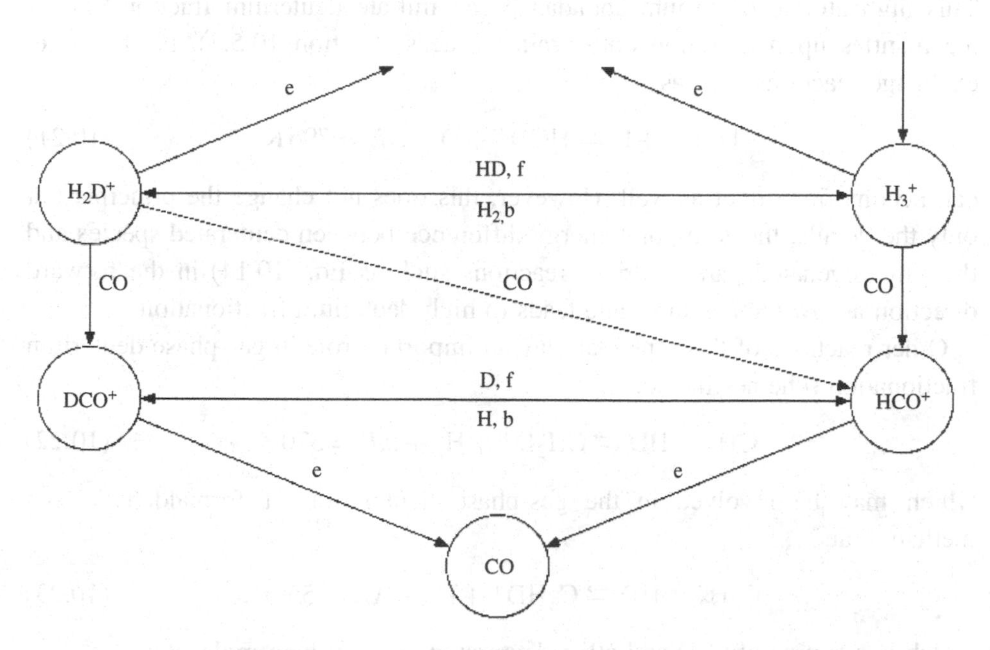

Figure 10.4 Simplified reaction scheme involved in the deuteration of HCO^+ through the H_2D^+ channel. Various other protonated and deuterated ions are similarly formed. Forward and backward reactions are indicated with f and b.

abundances ($10^{-5} - 10^{-4}$), the fractionation of deuterated molecules can much exceed ($X(\mathrm{DCO^+})/X(\mathrm{HCO^+}) \sim 3\times10^{-2}$) the elemental deuterium abundance.

Even higher deuterium fractionation can occur when the gas phase is heavily depleted. The absence of molecules such as CO shuts off the loss channel for H_2D^+ (Eq. (10.16)). Actually, H_3^+ and its isotopes then become the dominant cation and its recombination will dominate the charge balance as well (Section 10.2). We can then combine the deuterium balance with the ionization balance to find

$$\frac{X(\mathrm{H_2D^+})}{X(\mathrm{H_3^+})} = k_\mathrm{f} X(\mathrm{HD}) \left(\frac{n}{\zeta_\mathrm{CR} k_\mathrm{e}}\right)^{1/2} = 7\times10^{-1}\left(\frac{n}{10^4\,\mathrm{cm^{-3}}}\right)^{1/2}. \tag{10.19}$$

In fact, in this case, sequential reactions of the isotopes of H_3^+ with HD can drive the deuteration of this species all the way to D_3^+. But the principle will still be the same. In a heavily depleted gas, the deuterium fractionation of H_3^+ can reach very high values. Of course, there will be very few heavy species around to pass this enhanced fractionation on to and the effects of this deuteration – and high gas-phase depletion – may best be studied through observations of H_3^+ and its isotopes.

The high abundance of DCO^+ leads to a high abundance of atomic deuterium,

$$\frac{X(\mathrm{D})}{X(\mathrm{H})} = \frac{X(\mathrm{DCO^+})}{X(\mathrm{HCO^+})}. \tag{10.20}$$

This high atomic deuterium abundance can initiate deuterium fractionation of ice mantles upon accretion onto grain surfaces (Section 10.5.3). Furthermore, exchange reactions such as

$$\mathrm{DCO^+ + H \rightleftarrows HCO^+ + D} + \Delta E = 796\,\mathrm{K} \tag{10.21}$$

can become important as well. However, this does not change the principle but only the details; the zero-point energy difference between deuterated species and their hydrogenated parents drives reactions such as Eq. (10.14) in the forward direction at low temperatures and leads to high deuterium fractionation.

Other reactions of this type that play an important role in gas-phase deuterium fractionation schemes include

$$\mathrm{CH_3^+ + HD \rightleftarrows CH_2D^+ + H_2} + \Delta E = 370\,\mathrm{K}, \tag{10.22}$$

which may be involved in the gas-phase deuteration of formaldehyde and methanol, and

$$\mathrm{C_2H_2^+ + HD \rightleftarrows C_2HD^+ + H_2} + \Delta E = 550\mathrm{K}, \tag{10.23}$$

which is a source of C_2D and other deuterated carbon-chain molecules.

These gas phase deuteration schemes predict that deuterium fractionation is a sensitive function of temperature. In particular, at temperatures exceeding 25 K, the exponential term in Eq. (10.7) takes over, driving the fractionation back to the elemental deuterium abundance. It is therefore surprising that the warm gas around protostars shows high deuterium fractionation (cf. Table 10.4). This will be discussed further in Section 10.6.3.

10.4.5 Summary

Gas-phase models of interstellar clouds have been very successful. Among the key successes are explanations for the high abundance of CO, the presence of ions such as H_3^+ and HCO^+, the presence of radicals such as OH and CH, the presence of isomers such as HNC/HCN, and the high observed deuteration levels in species such as DCO^+. Species that have notoriously been difficult to explain include CH_3OH, H_2CO, and NH_3. Qualitatively, the agreement between gas-phase models and the observations is very good. Quantitatively, however, the agreement is wanting. The most extensive reaction networks contain some 4000 reactions among 400 species, and reach an order of magnitude agreement with observed abundances for some 80% of the species. Partly, this is because our chemical knowledge is imperfect: less than 20% of the reactions included have actually been studied in the laboratory. The remainder are educated guesses. It is not even known whether all the important chemical routes are included and whether perhaps some routes have been included that are inhibited at low temperature. Moreover, this "agreement" is somewhat misleading because often observations (particularly of the complex hydrocarbons) are compared to models that relate to an early time ($\sim 10^5$ years) in a time-dependent calculation. Gas-phase models really have difficulty explaining observations of complex hydrocarbons at late times because almost all the carbon tends to be locked up in the unreactive CO. At early times, this problem is circumvented because these models start with all the carbon initially available as C^+. Once some H_2 is formed, hydrocarbon radical chemistry really takes off, but in less than 10^6 years all of these hydrocarbons are converted into CO. While molecular clouds probably evolve from diffuser clouds where all the carbon is in C^+, the time scale for this evolution is around 10^7 years rather than 10^5 years and it is unlikely that in this form such time-dependent models hold much relevance for interstellar chemistry. However, they do illustrate the point and there are alternative solutions to this problem. One class of models has C/O elemental abundance ratios that are larger than 1. In that case, the formation of CO does not consume all of the C and hydrocarbon chemistry will be very efficient. Such a large C/O ratio is sometimes attributed to the formation of ice mantles, because, observationally, O seems to be locked more in interstellar ices than C. But in no case is this a major effect that would affect the C/O ratio so drastically. Some gas-phase models use the ices as repositories, where (early time) molecules are kept to be judiciously injected into the gas phase when needed; an approach that has been very effective in explaining hot cores, but ices are somewhat far-stretched to use as early-time storage houses. Other models invoke the effects of shocks or turbulence to drive the chemistry away from CO towards hydrocarbons. Be that as it may, at present, it seems fair to say that these various models are just a stand-in for an efficient way of breaking the carbon out of CO and it seems clear that, if such a key ingredient of the models is not well understood, a good agreement between models and observations cannot be expected. Finally, grain-surface chemistry, itself, is at best poorly understood (cf. Section 10.5) and the gas–grain interaction even less so. This too may have some influence on the comparison. Except for species such as CH_3OH, H_2CO, and NH_3, and except for special environments such as hot cores (Section 10.6.3), this seems of lesser importance, given the good agreement between the gas-phase models and the type of species observed to be present in the gas phase. In all, this summary should be read as a positive statement: qualitatively, we do understand gas-phase chemistry and the ion–molecule and neutral–neutral reaction networks outlined in the previous sections are very important. Understanding the quantitative aspects of astrochemistry may bring new insights in the physics of molecular clouds.

10.5 Grain-surface chemistry

The composition of the accreting gas sets the stage for grain-surface chemistry. Here, we will focus on dense cloud cores that are shielded from ambient FUV photons; except for H_2 formation, grain-surface chemistry and ice-mantle formation are limited in the more diffuse phases of the ISM because of the effects of photodesorption (Section 10.6.2). While hydrogen is mainly in the form of H_2, cosmic-ray ionization keeps a low level of atomic hydrogen around. Assuming each ionization eventually delivers two H atoms, which are lost through accretion onto grains, we have

$$n(\mathrm{H}) = \frac{2.3\zeta_{\mathrm{CR}}\, n(\mathrm{H}_2)}{k_{\mathrm{d}}} \simeq 2\,\mathrm{cm}^{-3}, \qquad (10.24)$$

where the accretion rate of H atoms has been set to $k_{\mathrm{d}} = 3 \times 10^{-17}\, n\, \mathrm{s}^{-1}$ (cf. Section 8.7.1.2). This is independent of the density. Oxygen will be predominantly in atomic form while carbon is in the form of CO and the abundance of atomic C is very small (cf. Section 10.4.2). Nitrogen is largely in the form of N_2. Thus, the composition of ice mantles is largely determined by reaction among H, O, and CO, with traces of C and N. Further, at low densities ($<10^4\,\mathrm{cm}^{-3}$), atomic H will be the dominant reaction partner and the abundance of hydrogenated species will be high. At high densities, on the other hand, hydrogenation reactions will play little role.

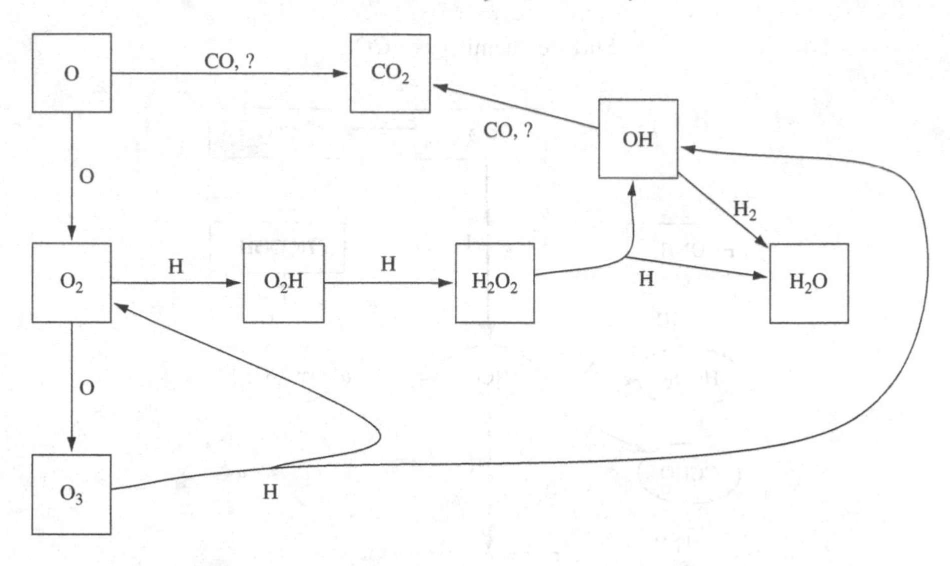

Figure 10.5 The chemistry of oxygen and the formation of H_2O on grain surfaces. Atomic oxygen reacts with O to form O_2 and O_3. Reaction of O_3 with H reforms O_2 and OH. O_2 can be lost through reactions with H, which lead through peroxides to H_2O and OH. The OH can react with H_2 to form H_2O or it might perhaps form CO_2 and H_2 with CO. CO_2 might also result from direct reaction of O with CO.

10.5.1 Grain-surface chemistry of oxygen

Figure 10.5 shows the chemistry involved in the formation of H_2O on grain surfaces. Generally, there are reaction partners for atomic O and H on a grain surface and the direct reaction of atomic O and H, leading eventually to H_2O, is ineffective (e.g., neither species will "wait" for accretion of the other). Rather, H_2O formation proceeds through reaction of atomic O with O_2 to O_3. Ozone is unstable to H attack, leading to OH (while reforming O_2). The OH then forms H_2O (actually by reacting with H_2, which is prevalent ($\theta(H_2) \sim 0.2$) on interstellar grain surfaces inside molecular clouds). Molecular oxygen can also be attacked by atomic H and this peroxide route converts it to H_2O.

10.5.2 Grain-surface chemistry of carbon

Atomic carbon can be hydrogenated on grain surfaces through sequential atomic H in addition to CH_4. Intermediate hydrocarbon radicals in this route can react with atomic O and N to form, for example, CH_3OH and HCN. However, only during the early diffuse phases of a cloud will C be abundant and ice formation is then inhibited by photodesorption (cf. Section 10.6.2). Most of the accreted carbon in molecular clouds is in the form of CO and the surface chemistry of this species

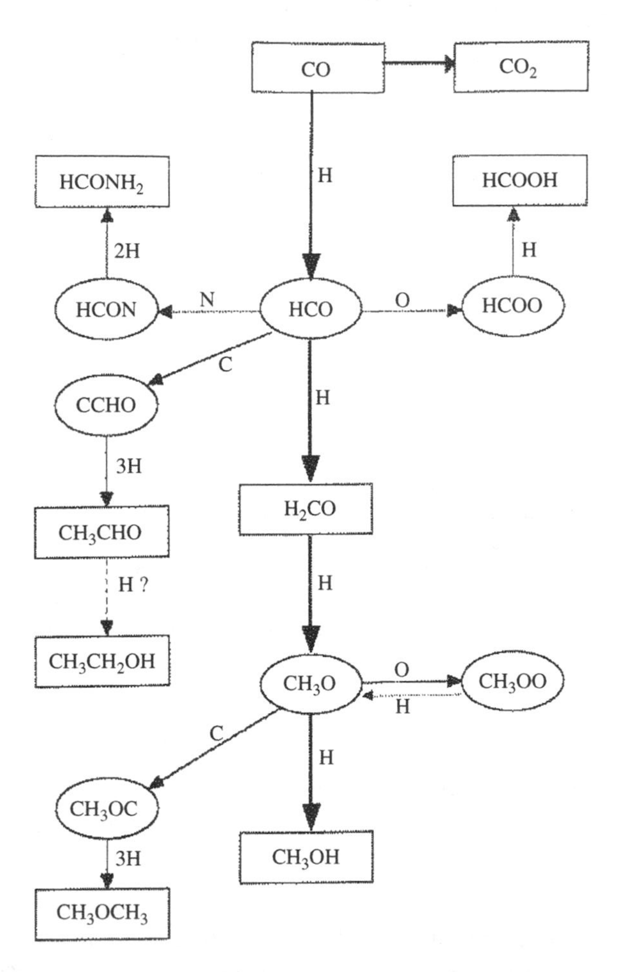

Figure 10.6 The reaction network involved in the hydrogenation and oxidation of CO on grain surfaces. O or OH may react with CO to form CO_2. Atomic H converts CO into H_2CO and CH_3OH. Traces of C and N can lead to ethanol or acetaldehyde, dimethyl ether, and formamide, while atomic oxygen may form formic acid. The peroxide radical CH_3OO is probably unstable to H attack.

is illustrated in Fig. 10.6. The reaction of CO with H has an activation barrier but will still proceed efficiently. The formyl radical is converted into formaldehyde by the next accreted H. Formaldehyde itself is also "unstable" to H attack leading to the methoxy radical, which will form methanol with the next accreting atomic H. The intermediate radicals, HCO and CH_3O, can react with atomic C, N, or O if they are the next accreting radical. In this way, formamide, ethanol, dimethyl ether, and formic acid can be formed. In this scheme, the efficiency of these routes depends on the accretion rate of these species relative to H. Atomic C and N are not expected to be very abundant in the gas phase. Carbon is largely locked up in CO while nitrogen is in the form of N_2. Photodissociation by cosmic-ray-generated FUV photons and ion–molecule reactions driven by cosmic-ray-produced He^+ are the dominant sources of these atoms with expected abundances of $\sim 1\%$. Hence, these complex molecules are not expected to be very abundant in ice mantles. Atomic oxygen is abundant and formic acid, HCOOH, can be important when the atomic O and H abundance in the accreted gas are comparable. This tends to occur at a density around 10^4 cm^{-3}. The peroxide route formed through the reaction of CH_3O with O may be very susceptible to attack by H.

Carbon monoxide can also form carbon dioxide through reactions with oxygen-bearing species. The precise mechanism for CO_2 formation on grain surfaces is controversial. Some laboratory experiments suggest that CO can react with atomic O. Others – sometimes done by the same laboratory – suggest that this reaction does not proceed in an ice matrix. Part of this discrepancy may reflect the presence of a small reaction barrier in the entrance channel for this reaction. As for reactions of atomic H, a reaction barrier may be of no concern – after all, the time available for reactions is $\sim$1 day. Thus, possibly, when CO is the dominant species on the grain, the O + CO reaction may well proceed anyway. However, when other species are (also) abundant, atomic O may be preferentially lost to other channels. Some of the laboratory studies on CO_2 formation may have suffered from this problem because of the presence of large abundances of O_2; the reaction $O + O_2 \longrightarrow O_3$ is known to proceed rapidly at low temperatures in an icy matrix. An alternative, attractive route towards CO_2 on grain surfaces is presented by the reaction $CO + OH \longrightarrow CO_2 + H$. OH as an intermediary in H_2O formation may be fortuitously close to a CO and CO_2 may result. In any case, in contrast to gas-phase chemistry, there may be various ways that CO can be efficiently converted into complex organic species.

10.5.3 Grain-surface chemistry of deuterium

The formation of deuterated species on grain surfaces largely follows from the reaction of accreted atomic D. Recall that in the gas phase, high deuterium fractionation levels can occur in species such as H_2D^+, which is then passed on to other ions such as DCO^+. Dissociative electron recombination then results in high atomic D to atomic H abundance ratios (Section 10.4.4). After accretion, the chemistry of atomic D on grain surfaces is very similar to that of atomic H. Atomic D will be able to scan the surface rapidly for coreactants. As for atomic H, reactions with activation barriers present little hindrance to reactions. Thus, any species that can be hydrogenated on a grain surface can also be deuterated and we expect that the deuterium fractionation will reflect, in some way, the ratio of atomic D to atomic H in the accreted gas.

For species that are formed through sequential addition of atomic H, this relationship between fractionation and D/H ratio in the gas will be straightforward. Consider the case where only reactions with D and H are important and the probability for accretion of (and reaction with) D is p_D. The probability for a reaction with H is then $(1 - p_D)$. The abundance ratio of NH_2D/NH_3 is, thus, $3p_D/(1 - p_D)$, where the factor 3 accounts for the three chances to deuterate this species. The expected fractionation of doubly deuterated ammonia (to NH_3) is, similarly, $3p_D^2/(1 - p_D)^2$. Thus, in a dark cloud with an observed gas phase DCO^+/HCO^+ ratio of 0.01 (Table 10.4), the gas-phase atomic D/H ratio will also be 0.01 (cf. Eq. (10.20), Section 10.4.4). High deuterium fractionations on a grain surface are thus also likely with, for example, an expected NH_2D/NH_3 ratio of 0.03 in the ice. When the gas phase is heavily depleted, the atomic D/H ratio can reach very high values $\sim$1 (cf. Section 10.4.4). Of course, when this occurs, further accretion will be limited and this enhanced gas-phase deuterium fractionation will be difficult to transfer to ice species.

The fractionation of daughter species of CO is interesting as well. Figure 10.7 shows the various routes towards deuterated formaldehyde and methanol on a grain surface. In this case, the expected fractionation of daughter molecules is somewhat more involved. In this scheme, the fractionation ratio, CH_3OD/CH_3OH, will still directly reflect the accretion rates of atomic D and H. But for the isotopes with the D on the carbon, reactions with activation barriers are involved and accreted D and H may not select their coreactants in entirely the same way. The barriers for the reaction of D with CO and H_2CO will be similar to those for the reaction with atomic H, but, of course, D is heavier than H and hence the tunneling rate is slower. While these reactions with D will still occur on a 1 day time scale, they have to compete with all other reactions possible for atomic D. Specifically, because the probability for reaction of D (and H) with these species is weighted by the tunneling rate (cf. Section 4.2.6), atomic D will favor reactions with low barriers more than atomic H will. The expected fractionation depends, then, on all the coreactants available to D and H on the grain surface and their respective reaction barriers. Thus, if D were to prefer the reaction with CO over that with H_2CO (compared to H), formaldehyde could be more highly fractionated than the atomic D/H ratio would indicate. This does not, however, carry over into the fractionation of methanol: if D were to prefer CO then H would prefer H_2CO and methanol formation requires these reactions sequentially. Of course, if other channels are open, this becomes even more involved because atomic H might preferentially "leak" away to other species.

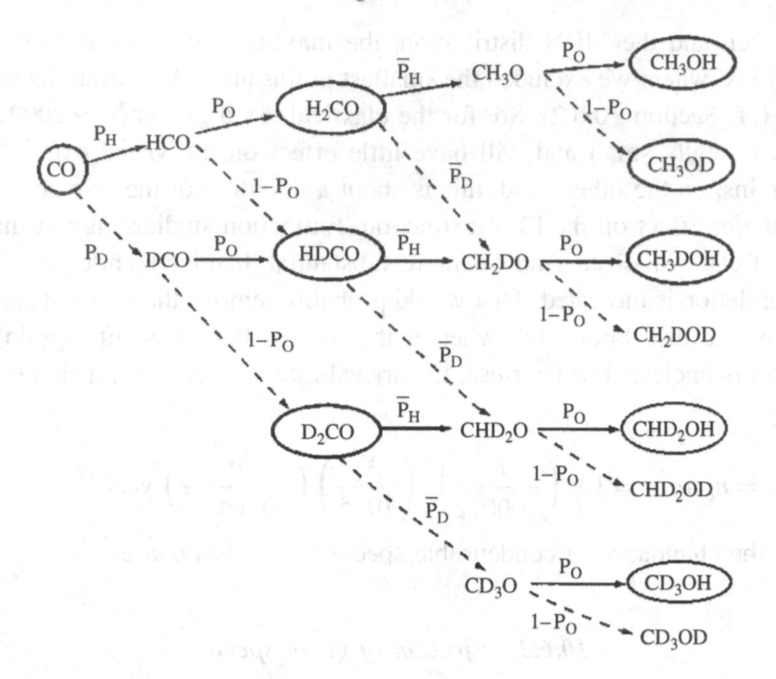

Figure 10.7 The reaction network involved in deuterium fractionation of CO on grain surfaces. Species that have been observed in the warm gas around the low-mass protostar, IRAS 16293, are enclosed in an ellipse. Reactions of radicals with H and D have a probability set by the relative accretion rates, which are assumed in this figure to add up to 1. The reactions of H and D with CO and H_2CO possess activation barriers and their reaction probabilities will depend on the accretion rates, surface concentrations of (all) reactants, and the barriers involved.

10.6 Gas–grain interaction

10.6.1 Accretion

For the increase in mass of a grain due to the formation of an ice mantle, we can write,

$$\frac{\mathrm{d}m_\mathrm{d}(a)}{\mathrm{d}t} = 4\pi a^2 \rho_\mathrm{s} \frac{\mathrm{d}a}{\mathrm{d}t} = \pi a^2 n_\mathrm{c} \overline{v}_\mathrm{c} \overline{m}_\mathrm{c}, \tag{10.25}$$

with ρ_s the specific density of the grain material, n_c the gas-phase density of condensables, $\overline{v}_\mathrm{c}$ their mean velocity, and $\overline{m}_\mathrm{c}$ their mean mass. Thus, the increase in grain size due to accretion of gas-phase species on a grain is independent of the size of the grain and all grains grow the same size mantle. Because the interstellar grain size distribution is so strongly weighted towards small grains, $n(a) \propto a^{-3.5}$ (Section 5.4), this implies that most of the ice volume is on the smallest grains. Conversely, the increase in grain size due to the formation of ice is only modest. Adopting 3×10^{-4} as the abundance of the condensables relative to hydrogen and the MRN distribution, the maximum increase in grain size is about 175 Å, where we excluded the smallest grains and PAHs from the accretion process (cf. Section 10.6.2). So, for the classical-sized grains ($a \sim 2000$ Å), this increase is really small and will have little effect on the visual extinction. For small grains, on the other hand, this is about a doubling of the size and this will have a major effect on the FUV extinction. Extinction studies suggest that grain growth of classical-sized grains is more substantial than ice mantles would allow and coagulation is indicated. That would probably remove the smallest sizes from the grain size distribution but whether this occurs before or after grain mantle formation is unclear. For later use, the arrival rate of species on a single grain is given by

$$k_{\mathrm{ar}} = n_{\mathrm{c}} \pi a^2 v = 10^2 \left(\frac{a}{1000\,\mathrm{\AA}}\right)^2 \left(\frac{X_{\mathrm{c}}}{10^{-4}}\right) \left(\frac{n}{10^4\,\mathrm{cm}^{-3}}\right) \mathrm{year}^{-1}, \qquad (10.26)$$

with X_{c} the abundance of condensable species in the gas phase.

10.6.2 Ejection of grain species

Sublimation

For a pure ice, molecules will start to evaporate when the dust temperature reaches the sublimation temperature. This sublimation will depend on the binding energy of the species in the ice and on the time scale involved. The sublimation rate coefficient is given by

$$k_{\mathrm{ev}} = \nu \exp[-E_{\mathrm{b}}/kT], \qquad (10.27)$$

where the frequency factor has been discussed in Section 4.2.3. Typically, sublimation temperatures are measured in a laboratory setting where the relevant time scale is minutes. In contrast, in space, the relevant time scale is set by the dynamical time scale of the region and may be as large as 10^5 years. As a result, the sublimation temperature of a pure substance will be much lower in space. Some relevant sublimation temperatures have been compiled in Table 10.5. However, ices in space seem invariably to be a mixture of multiple components. Such mixed ices may show a very different sublimation behavior. For an ice consisting of two components with comparable abundances but very different sublimation behavior, the two species will show independent evaporation. However, volatile species present as impurities in a less volatile matrix (e.g., a small amount of CO trapped in H_2O ice) can remain trapped until the matrix of the main component starts to loosen up, typically at 1/2 the sublimation temperature. For H_2O ice, this is typically some 50 K where phase transitions between different amorphous ice structures become important. Furthermore, in H_2O ice, small amounts of impurities (at the few percent level) can remain trapped until very high temperatures in

Table 10.5 *Sublimation temperatures of pure ices*

Species	Laboratory (K)	Space (K)
N_2	22	13
O_2	22	13
CO	25	16
CH_4	30	18
CO_2	83	50
NH_3	95	55
CH_3OH	140	80
H_2O	150	90

the form of a clathrate structure where the ice crystal forms "cages" around these foreign species. The encapsulated species will then not sublimate until the main clathrate–ice structure breaks up. Finally, for H_2O/CH_3OH ice mixtures, warming up will lead to segregation of the two components into separate regions of H_2O crystals and CH_3OH crystals before sublimation.

Photodesorption

For large grains, the most important desorption mechanism at the edge of molecular clouds is photodesorption. The decrease in the mass of a grain due to desorption is given by

$$\frac{dm_d}{dt} = 4\pi a^2 \rho_s \frac{da}{dt} = \pi a^2 Y_{pd} 4\pi \mathcal{N}_{ISRF} \overline{m}_s, \tag{10.28}$$

with ρ_s the density of the material ($\simeq 1\,g\,cm^{-3}$ for ice) and $\overline{m}_s$ the mean mass per sputtered species (2.9×10^{-23} g for H_2O). The photodesorption yield, Y_{pd}, is defined here as the desorption of a molecule per FUV photon absorbed by the grain. Equating this to the grain accretion rate and assuming a one-dimensional average interstellar FUV radiation field incident attenuated by an optical depth $\tau_{UV} = 1.8\,A_v$ and a density of "condensables" of 3×10^{-4}, we find that grains will be kept "clean" from accreting ices when

$$A_v < 4.1 + \ln\left[G_0\left(\frac{Y_{pd}}{10^{-2}}\right)\left(\frac{10^4\,cm^{-3}}{n}\right)\right]. \tag{10.29}$$

Photodesorption yields are notoriously difficult to measure and estimates in the astronomical literature range from 10^{-2}–10^{-5}. The best estimates give $Y_{pd} \sim 3 \times 10^{-3}$ for an H_2O ice film and $Y_{pd} \sim 10^{-2}$ for a rough H_2O ice surface and that has been adopted here. So, photodesorption should be able to keep a surface clean from physically adsorbed species in the diffuse ISM and actually throughout most of the molecular cloud. In fact, only in dense star-forming cores do we expect ice mantles to grow. When G_0 is higher, this ice-free outer volume of a molecular cloud becomes even larger. In a PDR associated with a reflection nebula or HII region ($G_0/n \simeq 1$), this limited extinction is 12 magnitudes, much larger than the depth of the PDR.

Equation (10.28) can also be used to determine the lifetime of an ice grain in the outer regions of a molecular cloud or in the ISM. We have

$$\frac{\mathrm{d}a}{\mathrm{d}t} = 2.2 \times 10^{-2} \left(\frac{Y}{10^{-2}} \right) \text{Å year}^{-1}. \tag{10.30}$$

With a maximum grain mantle thickness of 175 Å, this corresponds to a maximum lifetime of 10^4 years in the diffuse ISM. The photodesorption of an H_2O molecule on a clear silicate or carbonaceous surface may differ from that adopted here and a monolayer of H_2O may remain. The first shock encountered by a grain would, however, sputter the surface clean.

Sputtering in shocks

FUV photons do not penetrate dense cloud cores, but other processes may become important. Sputtering of grains has been discussed in Section 5.2.4. The threshold for sputtering of ice molecules from a grain is much lower than for more refractory grain materials due to the lower binding energies involved. The measured sputtering yield is very high at ion energies as low as 10 eV. Hence, a 100 km s^{-1} shock will completely sputter a 1000 Å grain and the calculated lifetime is only 10^6 years in the diffuse ISM (Chapter 13). Low velocity C-shocks driven by powerful stellar outflows may be an effective way of returning ice species to the gas phase inside dense clouds. The molecules injected into the gas phase will stay around for $3 \times 10^5\, f_\mathrm{O}$ years, with f_O the fraction of the O initially locked up in ices before it is broken down again (Section 10.4.2). So, shock processing may lead to a temporary phase of ice-mantle species and their daughter products in the gas phase (see Section 10.6.3).

Thermal spikes

As has been extensively discussed in Chapter 5, small species such as PAHs and PAH clusters will become very warm upon absorption of a single UV photon and this may lead to ejection of any species that are loosely bound to the grain. Similarly, the heat of reaction may be sufficient to cause ejection of the newly formed species. We can calculate the rate for this process using the unimolecular rates derived in Section 6.6 and compare those with the IR relaxation rate. PAHs smaller than a critical size will then become so hot that they will lose the newly formed species. The results for such a calculation are shown in Fig. 10.8.

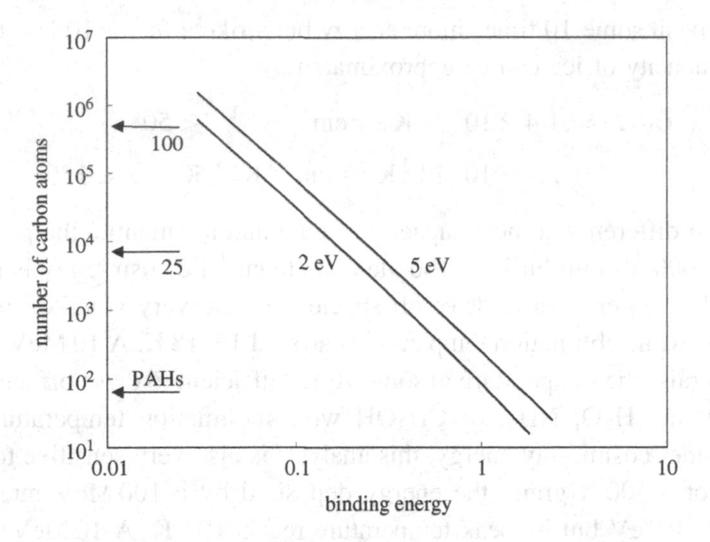

Figure 10.8 The critical "grain" size for mantle growth as a function of binding energy (in eV) of the adsorbed species. For sizes smaller than this critical size, an adsorbed species will evaporate when the grain adsorbs a visible photon (2 eV) or upon release of reaction heat (5 eV). Typical grain sizes (in Å) are indicated.

For a typical binding energy of 0.2 eV, molecules will be ejected from clusters with less than 10^4 C atoms ($\simeq$30 Å for a three-dimensional grain). Unreactive species could still directly adsorb from the gas phase but as long as there are a reasonable number of atoms or radicals around, this process would serve to keep small grains mantle-free.

Cosmic-ray-driven mantle explosions

The final process to be considered is cosmic-ray-driven evaporation. Upon collision, a heavy cosmic ray (Fe, C, N, O) with energy between 10 and 100 MeV per nucleon will deposit energy into a grain at a rate of $\simeq 4\times10^{11}\,(\mathrm{MeV}/E)\,\mathrm{eV\,cm^{-1}}$. This will heat up the grain to a high temperature (50–200 K), leading to evaporation of volatile ice-mantle species. The Fe cosmic-ray flux in this energy range is very uncertain but estimates place it at $\mathcal{N}_{\mathrm{CR}} = 10^{-4}$ particles $\mathrm{cm^{-2}\,s^{-1}\,sr^{-1}}$, around 100 MeV per nucleon (cf. Section 1.3.3). The time scale for a CR hit is then

$$\tau_{\mathrm{CR}} = (4\pi\,\mathcal{N}_{\mathrm{CR}}\,\pi a^2)^{-1} = 8\times10^4 \left(\frac{1000\,\text{Å}}{a}\right)^2 \text{ years.} \qquad (10.31)$$

So, over the lifetime of a molecular cloud, all grains larger than 50 Å will have been hit at least once by such a cosmic ray. At 100 MeV, iron cosmic rays will deposit about $\Delta E_{\mathrm{dep}} = 5\times10^4\,(a/1000\,\text{Å})$ eV in the grain. A 10 MeV Fe cosmic ray will deposit some 10 times more energy but strike a factor 10 less frequently. The heat capacity of ice can be approximated by

$$C_V(T) = 1.4 \times 10^3\, T^2\, \mathrm{K\,erg\,cm^{-3}\,K^{-1}}\; T < 50 \tag{10.32}$$

$$= 2.2 \times 10^4\, T^{1.3}\, \mathrm{K\,erg\,cm^{-3}\,K^{-1}}\; 50 < T < 150. \tag{10.33}$$

Ignoring the difference in heat capacity of core and ice mantle, the peak temperature of a 1000 Å grain hit by a 100 MeV nucleon^{-1} Fe cosmic ray is then about 35 K; barely high enough to drive off species that are very volatile, such as CO, N_2, and O_2, with sublimation temperatures around 15–18 K. A 10 MeV nucleon^{-1} Fe CR will raise the temperature to some 70 K, sufficient to drive off these volatile species but not H_2O, NH_3, or CH_3OH with sublimation temperatures around 100 K. Besides cosmic-ray energy, this analysis is also very sensitive to the grain size. So, for a 300 Å grain, the energy deposited by a 100 MeV nucleon^{-1} Fe CR is 1.5×10^4 eV but its peak temperature reaches 75 K. A 10 MeV nucleon^{-1} Fe CR will heat a 300 Å grain to 190 K, warm enough to eject even H_2O ice.

Typically, if $T \gg T_s$, most of the energy deposited will be used to evaporate volatile species and we have, then, for the total number of species injected into the gas, $N_{\rm ej,CR} = \Delta E_{\rm dep}/\Delta E_{\rm b}$. Thus, when the temperature exceeds some 135 K, H_2O will start to evaporate. In addition, when the temperature drops below the H_2O sublimation temperature, highly volatile species can still evaporate. Typical values are summarized in Table 10.6. Comparing these numbers of evaporated species with the total number of species in the ice mantle, we see that a large grain will lose less than about 10% of its accreted ice mantle this way. A small grain, on the other hand, may lose a substantial fraction of its mantle (15–60 %).

We have ignored here the effects of spot heating by the cosmic ray. Initially, the energy will be deposited in a cylinder with a radius of some 50 Å and it takes a finite time for this energy to diffuse. During that time, the spot will be very hot and even tightly bound species can evaporate near the surface. More importantly, we have focussed here on the direct heating of a grain by a cosmic ray. In addition, cosmic rays can also produce a weak FUV radiation field deep inside clouds through (electron) excitation of H_2 with a mean intensity of $\sim 4 \times 10^2$ photons cm^{-2} s^{-1} sr^{-1} (Section 4.1.1). These photons are not sufficiently numerous to prevent the formation of an ice mantle through photodesorption. However, these photons will photolyze ice species, producing radicals (e.g., OH, H), and about half of these radicals are trapped in the ice matrix, which prevents their recombination (e.g., the H atom flies off to neighboring sites, while the OH remains behind). In this way, CR energy is converted into FUV photon energy, which is used to create chemical energy, which is stored in ice mantles and can accumulate over time. If we take 10^5 years as a typical time scale for CR hits,

Table 10.6 *Cosmic-ray-driven desorption*[a]

		100 MeV			10 MeV		
Size	N_{tot}	$T_p(K)$	N_{ej}	k_{ej} (molecule year^{-1})	T_p (K)	N_{ej}	k_{ej} (molecule year^{-1})
1000	4 (7)	35	5.0 (5)^b	6^b	70	4 (6)^b	6^b
		135	2.0 (7)^b	2.5 (2)^b	150	8 (5)	1
300	1 (6)	75	1.5 (5)^b	0.2	190	1 (5)	0.02
		150	4.0 (4)	6 (−2)	225	2 (5)	0.03

^{a} For each size grain, the first line indicates H_2O evaporation driven by Fe cosmic-ray hit at the indicated energy. The second line includes the effect of stored chemical energy. Note that in addition to the evaporation of H_2O, a 1000 (300) Å grain can lose an additional 2×10^7 (5×10^5) CO molecules when T drops below the H_2O evaporation temperature. See text for details.
^{b} Evaporation of CO only.

a concentration of 10^{-2} of radicals is built up (somewhat less than the maximum concentration of 0.03) corresponding to 0.05 eV per molecule in stored chemical energy. After a cosmic-ray hit, these stored radicals become mobile and recombine, releasing their chemical energy. Again, the grain will cool through sublimation of some species. Taking $\Delta a = 100$ Å for the thickness of the ice mantle, this adds about 5×10^4 eV to the total energy in a 300 Å grain and this effect will be relatively small. For a 1000 Å grain, the stored chemical energy amounts to 2×10^6 eV; about 40 times more than the energy directly deposited by the cosmic ray. The evaporation driven by release of stored chemical energy is also summarized in Table 10.6.

Using Eq. (10.31), these evaporation values can be translated to give an "average" rate. These rates should be compared with the accretion rate (Eq. (10.26)), about 200 and 9 year^{-1} for a 1000 and 300 Å grain, respectively. It is obvious that without the release of stored chemical energy, the gas phase will rapidly accrete out onto grains until the density of condensables has decreased to $\sim 10^{-1}$ cm^{-3}. Including stored chemical energy in this balance, this density is about an order of magnitude larger. Note also that this implies that the abundance of molecules in the gas phase will scale with n^{-1}. Thus, cosmic-ray driven ejection will tie the effective depletion time scale to the dynamical time scale of the region. We also note that cosmic-ray driven ejection in the presence of stored chemical energy will lead to a segregation of the ice mantle composition with grain size. Specifically, small grains will lose accreted H_2O ice while large grains will not. So, due to outgassing, large grains will acquire grain mantles containing predominantly H_2O (if there is H around to hydrogenate the accreted O). The mantles on small grains, on the other hand, will be characterized by a larger ratio of highly volatile to volatile species (e.g., a larger CO/H_2O ratio).

10.6.3 Hot cores

Many luminous protostars are surrounded by a region of warm (>100 K) and dense (>10^6 cm^{-3}) material, so-called hot cores, with a composition resembling that of interstellar ices with high abundances of hydrogenated species (e.g., H_2O, NH_3, CH_3OH; cf. Section 10.7.4). This is generally taken to reflect the sublimation of ices – formed during the preceding cold cloud phase – because the newly formed luminous star has warmed the grains above 50–100 K. This is supported by the high observed levels of deuterium fractionation in this warm gas; high deuteration is a signpost for low temperature chemistry (cf. Sections 10.4.4 and 10.4.2) and hence indicates that the composition of these hot cores originates from cold storage and has not had sufficient time yet to accommodate to the new, warm environment. This chemical time scale is $\simeq 3\times 10^5$ years (cf. Section 10.4.2). With a typical size of 10^{16} cm and the low velocities of these regions, these chemical time scales are shorter than any dynamical time scale.

These hot cores around massive stars also contain a multitude of more complex species such as ethyl alcohol, dimethyl ether, methyl formate, methyl cyanide, and ethyl cyanide. The origin of these species is less obvious. Initially, like H_2O and CH_3OH, they were thought to represent grain-surface products – and putative chemical routes have been identified – but, because of their low abundance, confirmation through ice absorption studies is difficult to obtain. Presently, it is generally thought that these species represent active chemistry in the warm gas once high abundances of simple (H_2O, CH_3OH, NH_3) molecules in the ices have been injected into the gas. In particular, protonated methanol readily transfers methyl groups to other species. Dimethyl ether results from the following reaction sequence

$$CH_3OH \xrightarrow{H_3O^+} CH_3OH_2^+ \xrightarrow{CH_3OH} CH_3OCH_4^+ \xrightarrow{e} CH_3OCH_3, \qquad (10.34)$$

where H_3O^+ results from the reaction of H_3^+ with H_2O. The reaction of protonated methanol with formaldehyde, also thought to evaporate from grains, leads, after electron recombination, to methyl formate (CH_3COOH). The time scale for the build-up of molecular complexity is $\simeq 3 \times 10^4$ years. Models based upon these chemical schemes are in good agreement with the observations.

Low-mass stars also have associated hot-core-like regions characterized by warm, dense gas and high abundances of H_2O, CH_3OH, and H_2CO. Furthermore, the deuterium fractionations observed in these hot cores are even more extreme than for the hot cores around high mass protostars; singly, doubly, and triply deuterated methanol and ammonia as well as singly and doubly deuterated formaldehyde have been observed at abundances that approach that of the fully hydrogen bearing species. In this case, the dynamical time scale of a surrounding envelope would be only some 10^2 years and, possibly, this warm gas is associated with the warm surface layer of protoplanetary disk or another long-lived structure.

As for the complex molecules, the high deuteration also provides a chemical clock. The chemistry in the warm gas – proton transfer followed by dissociative electron recombination – will drive the deuteration of a species, XD, back to the gas phase value compatible with the temperature on a time scale

$$\tau_{\rm chem} = (X(H_3^+)\, n\, k_{\rm XD})^{-1} \simeq 3\times10^4\,\text{years}, \tag{10.35}$$

where we have adopted an H_3^+ abundance appropriate to the high density in such a region.

10.7 Observations

The multitude of rotational lines due to a myriad of molecular species – each with their own characteristic critical density and excitation energy – provides a powerful tool to study the physical properties of molecular clouds. Figure 10.9 provides an overview of often-used molecular diagnostics. Rotational lines of CO probe regions with densities of $\simeq 3\times10^3\,J^3\,\text{cm}^{-3}$ and temperatures of $\simeq 3\,J^2$ (cf. Chapter 2). Thus, the 1–0 and 2–1 transitions are well suited for the study of general molecular clouds. The mid-J CO transitions, such as $J = 7$–6, are sensitive to lukewarm gas near YSOs or in PDRs, while 14–13 and higher transitions generally trace warm gas associated with shocks. CO rotational–vibrational emission in the fundamental (4.6 μm) or overtone (2.3 μm) region of the spectrum originates from dense, hot gas near a star or in a planet-forming disk. Because of the larger dipole moment, rotational lines of trace species such as CS, HCN, and HCO^+ have higher critical densities than CO (cf. Chapter 2) and thus probe dense cores in molecular clouds. Molecular hydrogen has only electric quadrupole transitions and its low-J rotational transitions are readily excited in low density gas. However, the large energy spacing of H_2 does require high temperatures. Near-IR rotational–vibrational transitions of H_2 are indicative of warm ($\sim$1000 K), dense ($\sim 10^4\,\text{cm}^{-3}$) gas in shocks. The metastable and non-metastable inversion lines of NH_3 are excited in medium density gas (10^4–$10^6\,\text{cm}^{-3}$), while the far-IR rotational transitions of hydrides, such as H_2O, probe warm and denser (10^6–$10^9\,\text{cm}^{-3}$) gas.

10.7.1 Mass

Because of the large rotational spacing of the rotational levels of H_2 and the low Einstein A values of the rotational transitions, molecular hydrogen does not provide a good tracer of the mass of molecular clouds. Rather, observations of

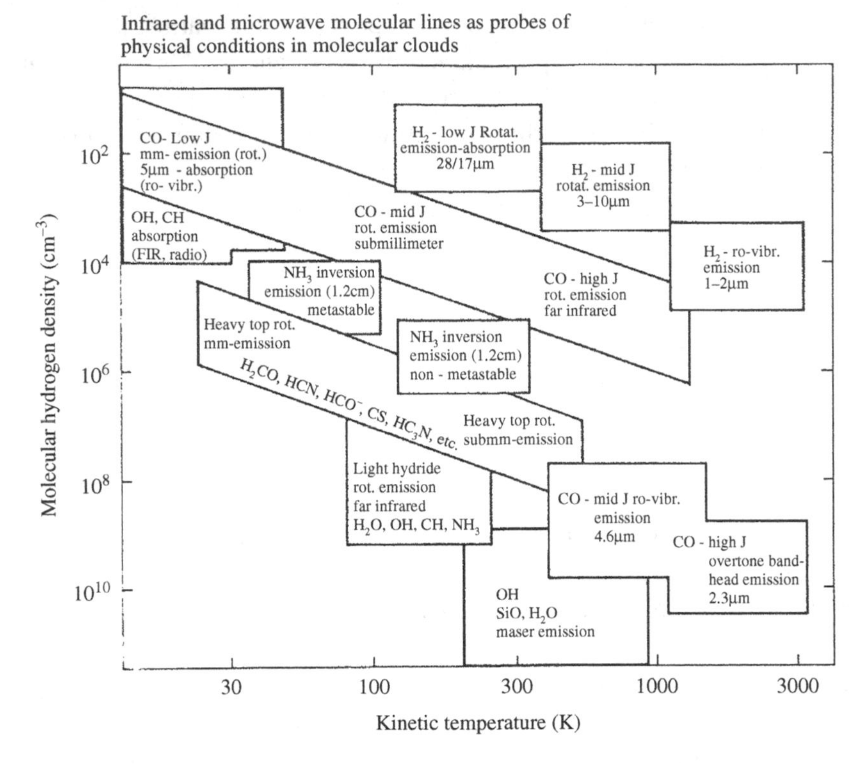

Figure 10.9 An overview of the molecular lines and the range of physical conditions in molecular clouds for which they are effective probes. Figure reproduced with permission from R. Genzel, 1991, in *The Physics of Star Formation and Early Stellar Evolution*, ed. C. J. Lada and N. D. Kylafis (Dordrecht: Kluwer), p. 155.

other molecules have to be used. There are two commonly used ways to determine the mass of molecular clouds: optically thin lines to determine the column of emitting gas and the optically thick ^{12}CO line in combination with the virial theorem.

CO isotopes and the mass of molecular clouds

Using optically thin lines, the total number of emitting molecules can be determined directly from the observations. Assuming an abundance, this can then be translated into a total mass within the beam. Generally, this is done using observations of one of the isotopes of CO. For an optically thin line, we have (cf. Chapter 2)

$$I_{\rm ul} = \frac{N_{\rm u}\, A_{\rm ul}\, h\nu_{\rm ul}}{4\pi}, \tag{10.36}$$

where N_u, A_{ul}, and E_{ul} are the column density in the upper level, the Einstein A coefficient of the transition, and the energy level separation of the two levels involved, respectively. Using the parameters for CO listed in Table 2.5, the CO J= 1 − 0 intensity is

$$I_{10} = 4.35 \times 10^{-24} N_1 \,\mathrm{erg\,cm^{-2}\,s^{-1}\,sr^{-1}}. \tag{10.37}$$

Commonly, intensities in the microwave region are presented in terms of integrated brightness temperatures,

$$T_B \Delta v = 10^{-5} \left(\frac{c^3}{2k\, \nu_{ul}^3} \right) \left(\frac{I_{ul}}{\mathrm{erg\,cm^{-2}\,s^{-1}\,sr^{-1}}} \right) \mathrm{K\,km\,s^{-1}}. \tag{10.38}$$

For the J= 1 − 0 transition, we have, then,

$$T_B \Delta v = 2.78 \times 10^{-15} N_1 \,\mathrm{K\,km\,s^{-1}}. \tag{10.39}$$

The column density in the first level can be converted to the total CO column density using the partition function,

$$N_1 = g_1 \left(\frac{kT_x}{hcB} \right) N_{\mathrm{CO}} \exp[-E_1/kT_x], \tag{10.40}$$

where g_1 is the statistical weight of the level $(2J+1)$, E_1 the energy of level $J = 1$, the factor in brackets is the inverse of the partition function (with B the rotational constant in units of $\mathrm{cm^{-1}}$), and it is tacitly assumed that the level populations can be described by an excitation temperature, T_x. Various assumptions can be made about the excitation of CO. Observed temperatures of the main isotope – which is optically thick – are typically 10 K. The rarer isotopes will refer to gas deeper in the molecular cloud where the gas will be cooler, perhaps only 5 K. Also, the higher CO levels have higher critical densities and hence may be subthermally excited. Here, we will assume $T_x = 10$ K and this yields

$$N_{\mathrm{CO}} = 7.5 \times 10^{14} \left(\frac{T_B \Delta v}{\mathrm{K\,km\,s^{-1}}} \right) \mathrm{cm^{-2}}. \tag{10.41}$$

The abundance of the CO isotope used in this analysis is required to turn this into an H_2 column density. These CO isotope abundances have been determined by comparing observed CO intensities with H nuclei column densities derived from extinction measurement through a number of well-studied dark clouds. These extinctions can be determined through either reddening studies of background stars or through star count techniques. In either case, the reddening is converted into H-nuclei column densities using standard dust-to-gas ratios. The fractional abundance of $^{12}C^{18}O/H_2$ is then measured to be 1.7×10^{-7}, which corresponds to an abundance of the main isotope of $^{12}C^{16}O/H_2 = 8.5 \times 10^{-5}$ and a $^{13}C^{16}O/H_2$ of 1.2×10^{-6}. This then yields, finally,

$$N(H_2) = 4.4 \times 10^{21} \left(\frac{T_B \, \Delta v}{\mathrm{K\,km\,s^{-1}}} \right) \mathrm{cm}^{-2}. \tag{10.42}$$

The same can be done for the $J = 2-1$ transition, which yields

$$N(H_2) = 3.3 \times 10^{21} \left(\frac{T_B \, \Delta v}{\mathrm{K\,km\,s^{-1}}} \right) \mathrm{cm}^{-2}. \tag{10.43}$$

There are a number of uncertainties in this approach. Principally, the excitation of CO and the abundance. Of course, through the normalization procedure, these uncertainties may largely cancel as long as the structures (n, T, abundance of CO isotopes) of the molecular clouds studied are comparable to those used in the normalization procedure. One last point should be made here. The derived abundance of ^{12}CO is quite low. Chemical studies invariably show that essentially all of the gas-phase carbon should be in the form of CO in molecular clouds. In the diffuse medium, the gas-phase carbon abundance is 1.4×10^{-4}, about a factor of 3 larger than the observed CO abundance. Some of the CO is condensed out in icy grain mantles in dense clouds. However, for the Taurus molecular cloud, even including solid CO_2, the solid ice fraction amounts to only 0.7 of the gas-phase CO and, hence, a considerable fraction of the carbon seems to be missing.

^{12}CO and the mass of molecular clouds

Various techniques have been used to determine the mass of molecular clouds observationally. Predominant among these is the correlation between extinction studies and CO line intensity discussed above. Others include γ-ray studies of molecular clouds and virial masses. Empirically, these studies show that there is a correlation between the integrated ^{12}CO $J = 1-0$ line intensity and the mass of a molecular cloud. This correlation is equivalent to

$$N(H_2) = 3.6 \times 10^{20} \left(\frac{T_B \, \Delta v}{\mathrm{K\,km\,s^{-1}}} \right) \mathrm{cm}^{-2}. \tag{10.44}$$

At first sight, such a correlation for an optically thick line is curious. The CO luminosity of a cloud is given by

$$L_{CO} = d^2 \int I_{CO} \, d\Omega = T_B \, \Delta v \, \pi R^2, \tag{10.45}$$

where R is the radius of the cloud. Now, for clouds in virial equilibrium, the line width is given by

$$\Delta v = \left(\frac{GM}{R} \right)^{1/2}, \tag{10.46}$$

with M the mass of a cloud. This leads to

$$L_{\rm CO} = \left(\frac{3\pi G}{4\rho}\right)^{1/2} T_{\rm B}\, M. \tag{10.47}$$

Thus, if clouds have similar densities and temperatures (more specifically, if $T_{\rm B}/\rho^{1/2}$ is constant), then the CO luminosity is a direct measure of the cloud mass. To phrase it differently, because the ^{12}CO line is optically thick, the observed flux measures the total radiating surface area. In virial equilibrium, the linewidth increases linearly with the size of the cloud and, hence, the integrated luminosity is a measure for the mass of the cloud.

10.7.2 Abundances

Observed molecular emission lines can be converted into abundances in the optically thin limit. Expressing the observed intensity again in terms of integrated brightness temperatures, $T_{\rm B}\,\Delta v$, we have, for the column density in the upper level of the transition (cf. Chapter 2.3.2),

$$N_{\rm u} = \frac{1.94\times 10^3}{A_{\rm ul}}\left(\frac{T_{\rm B}\,\Delta v}{\rm K\,km\,s^{-1}}\right)\left(\frac{\nu_{\rm ul}}{\rm GHz}\right)^2 \rm cm^{-2}. \tag{10.48}$$

Assuming that the excitation can be described by a single excitation temperature, $T_{\rm X}$, we have, for the total molecular column density;

$$N_{\rm mol} = \frac{1.94\times 10^3}{A_{\rm ul}}\left(\frac{\Phi(T_{\rm X})}{g_{\rm u}\exp[-E_{\rm u}/kT_{\rm X}]}\right)\left(\frac{T_{\rm B}\,\Delta v}{\rm K\,km\,s^{-1}}\right)\left(\frac{\nu_{\rm ul}}{\rm GHz}\right)^2 \rm cm^{-2}. \tag{10.49}$$

The partition function, Φ, has been discussed in Section 2.1.3. Characteristics of some molecular lines are presented in Table 2.4. The observed intensity of a single line can be converted into a molecular column density, adopting an excitation temperature. If many lines have been observed, the excitation temperature can be determined through a comparison of the line intensities. Often that is done through a rotational diagram. In that case, observed intensities are converted into column densities in the upper levels through Eq. (10.48). A plot of $\ln[N_{\rm u}/g_{\rm u}]$ versus $E_{\rm u}$ will then be a straight line and the slope provides the excitation temperature, while the total molecular column density is given by the intercept. We also encountered this diagram in the discussion of the excitation of molecular hydrogen in photodissociation regions (Chapter 9). There are several pitfalls with this technique. Optically thick lines will spoil the relation. Also, some of the lines may be subthermally excited and, in that case, there may not be a unique excitation temperature. Furthermore, the emission may not fill the beam. If the source is smaller than the beam for all lines, the derived column density is a beam-averaged column density, which can be converted into a "true" column density assuming a source size. However, if the lines have been observed at very different frequencies or with different telescopes, beam-filling effects may affect the various transitions very differently. Often there is no alternative, and the rotational diagram approach has found widespread application in molecular astrophysics. Once the molecular column densities have been derived, abundances are calculated by comparison to the molecular hydrogen column density either derived from CO isotope studies ($C^{18}O$ or $C^{17}O$) adopting an isotope abundance (cf. Section 10.7.1) or from dust measurements (generally emission at submillimeter wavelengths). There are various assumptions involved in this latter step. These do not influence the intercomparison of abundances within a cloud but do affect cloud-to-cloud comparisons. Examples of such abundance studies are presented in Tables 10.3 and 10.4.

10.7.3 Physical conditions

Molecular lines with their rotational ladders form ideal probes of the physical conditions in the emitting gas. The various rotational transitions originate from levels at different energies and have different critical densities. As a result, their relative intensities can be used to derive the density and the temperature of the gas. To extract this information from the observed lines, model calculations are required, which take the molecular cloud structure, the excitation, and the line radiative transfer properly into account. This is a formidable task and often simplifying assumptions are made to get a first impression of the implications of the observations, based on diagnostic diagrams. A set of diagnostic diagrams is illustrated in Fig. 10.10. In each of the panels, the ratio of two lines of a single species is shown in the form of contour plots. Line intensities have been calculated for a homogeneous cloud of a given density and kinetic temperature. The results have then been converted into a contour plot in the $n - T$ plane. Perusing these diagrams and recalling the critical densities and energies of the levels involved, we realize that these line ratios are sensitive probes of the density and temperature of the gas in the range between the critical densities and excitation energies of the levels involved. The power of molecular rotational spectroscopy is evident. By properly selecting the species and lines involved, gas under any condition can be traced. Of course, in general, molecular clouds will be highly inhomogeneous with spatial structure at all scales. Conversely, this implies that a line will trace the gas that best fits its critical density or energy level separation. Also, in using these diagrams, the optically thin approximation should be kept in mind. Often main isotopes of molecules are optically thick and then less abundant isotopes have to be used to derive the physical conditions in the gas.

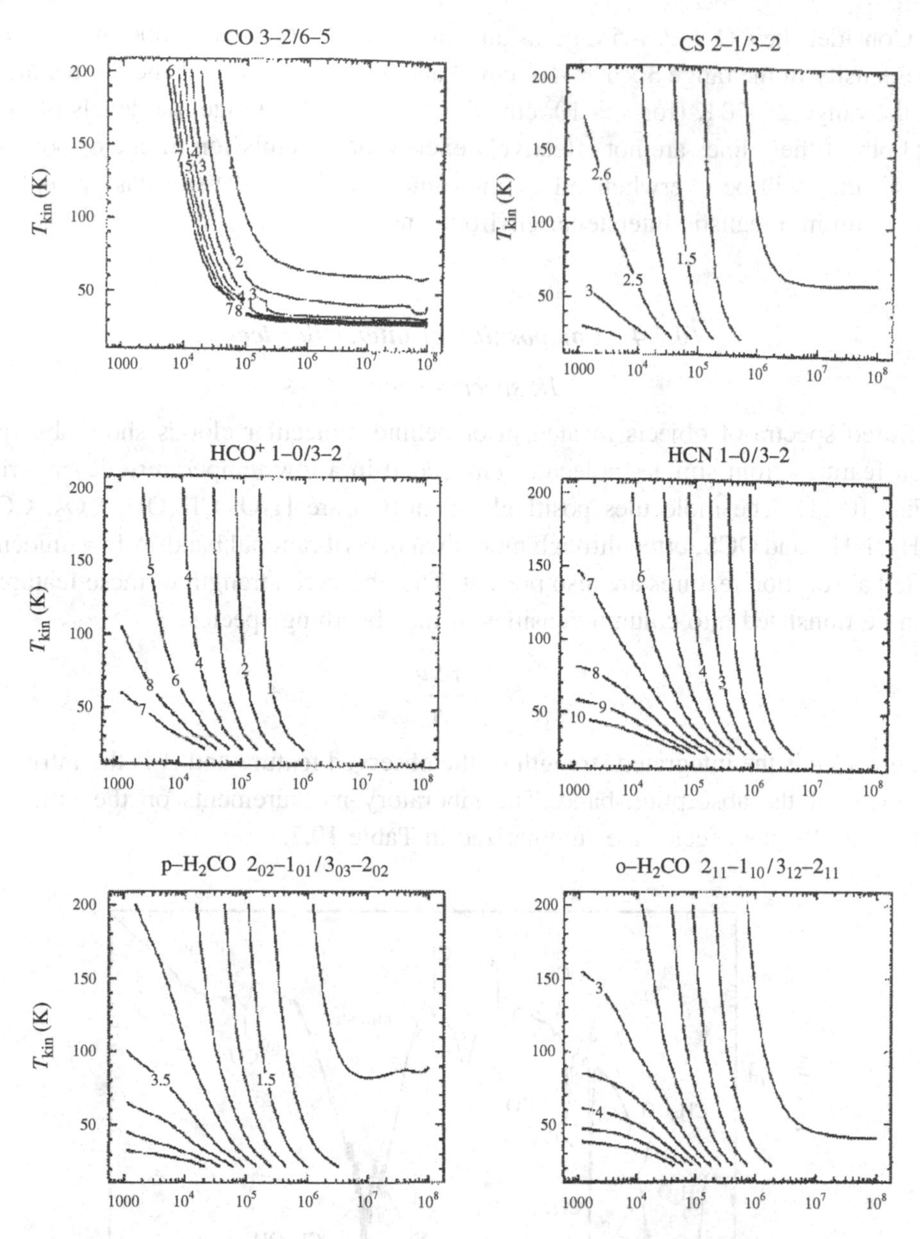

Figure 10.10 Contour plots of the expected ratios of integrated line intensities calculated in the optically thin limit using a large velocity gradient model. Contours are spaced linearly. The lines involved are indicated above each panel. Reproduced with permission from D. Jansen, 1996, Ph.D. thesis, Leiden.

Consider the CO 3–2/6–5 ratio as an example. This ratio is a good measure of the density in the range 3×10^{3}–10^{5} cm^{-3} and a good measure of the temperature in the range 20–70 K (for $n > 10^{5}$ cm^{-3}). Outside of this range, the levels of one or both of these lines are not effectively excited or the emission in one or both of these lines will be overwhelmed by more plentiful lower-density gas present in the beam in a realistic interstellar environment.

10.7.4 Composition of interstellar ices

IR spectroscopy

Infrared spectra of objects located in or behind molecular clouds show absorption features from simple molecules embedded in a low-temperature, icy matrix (Fig. 10.11). The molecules positively identified are H_2O, CH_3OH, CO_2, CO, CH_4, NH_3, and OCS, often through more than one vibrational band. A few unidentified absorption features are also present. The observed strength of these features can be translated into column densities of the absorbing species,

$$N = \frac{\tau\Delta\nu}{A}, \tag{10.50}$$

where $\tau\Delta\nu$ is the integrated strength of the observed feature and A is the intrinsic strength of the absorption band. The laboratory measurements on the intrinsic strength of some species are summarized in Table 10.7.

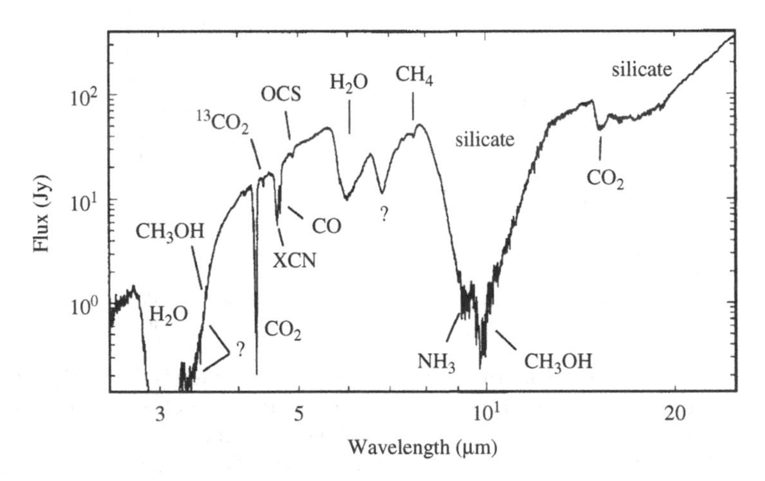

Figure 10.11 The mid-infrared spectrum of the protostar W33A observed with the Short Wavelength Spectrometer on board the Infrared Space Observatory shows a variety of absorption features. Except for the 10 μm and 18 μm silicate bands, these features are due to simple molecules in an ice mantle. Reproduced with permission from E. Gibb, *et al.*, 2000, *Ap. J.*, **536**, p. 347.

Table 10.7 *Integrated strength of ice bands*

Species	ν (cm^{-1})	A (cm molecule^{-1})
H_2O	3275	2.0 (−16)
H_2O	1670	1.0 (−17)
H_2O	750	2.8 (−17)
CO	2138	1.1 (−17)
CO_2	2340	7.6 (−17)[a]
CO_2	656	1.1 (−17)
CO_2	3708	1.4 (−18)
CO_2	3600	4.5 (−19)
CH_4	3010	1.0 (−17)
CH_4	1300	7.3 (−18)
NH_3	3375	1.3 (−17)
NH_3	1070	1.3 (−17)
CH_3OH	3250	1.3 (−16)
CH_3OH	2982	2.1 (−17)
CH_3OH	2828	5.4 (−18)
CH_3OH	1450	1.2 (−17)
CH_3OH	1026	1.9 (−17)
OCS	2042	1.5 (−16)
OCN^-	2160	1.3 (−16)[b]

[a] This band can be strongly affected by small particle scattering effects, which enhance its intrinsic strength (cf. Section 5.2.1).

[b] Uncertain. Depends on the assumed efficiency of the production of this species in UV irradiation experiments.

The observed species are in good agreement with those expected on the basis of the grain-surface chemistry discussed in Section 10.5. Table 10.8 summarizes abundances derived for these molecules relative to H_2O ice. This table also includes abundances observed in comets. Comets are thought to be rather pristine, icy reservoirs of interstellar cloud material. The good agreement in the kind of molecules present and their relative abundances supports this.

Dependence on A_V

The abundances of ices can also be measured by comparing ice-band optical depth with visual extinction measurements towards the same stars. For background field stars, the visual extinction can be measured through the colors when the spectral type is known as well as through measurements of the optical depth in the silicate feature. In either case, standard dust parameters (Sections 5.3.1 and 5.3.5) have to be adopted to convert this to an abundance. For absorption studies towards

Table 10.8 *Observed interstellar and cometary ice composition*[a]

Species	NGC 7538 IRS 9	W33A	Hale–Bopp comet
H_2O	100	100	100
CO (total)	10	1	23
CO (polar)	3	0.7	
CO (apolar)	7	0.3	
CO_2	16	3	6
CO_2 (polar)	9	1	
CO_2 (annealed)	7	2	
CH_3OH	9	10	2
H_2CO	3	2	1
HCOOH	2	0.5	0.1
NH_3	10	4	0.7
CH_4	1	0.4	0.6
OCS	0.1	0.05	0.4
OCN^-	0.8	1	

[a] Relative to $H_2O = 100$. Adopted H_2O column densities are 10^{19} and 4×10^{19} cm^{-2} for the massive protostars, NGC 7538 IRS 9 and W33A.

embedded objects, only the silicate feature method is generally available and even that is uncertain because radiative transfer effects in this band have to be taken into account. Furthermore, complicated geometries, such as circumstellar disks, may further hamper this analysis for embedded objects because, due to scattering effects, the sight-lines may not be exactly the same at all wavelengths. For a very limited number of embedded sources, ice optical depths have been directly compared to hydrogen column densities measured through the very weak near-IR rotational–vibrational H_2 lines in absorption.

Figure 10.12 illustrates this technique. Focussing on H_2O ice, there is a good correlation between the strength of the 3.08 μm ice band and A_v measured for background stars behind the Taurus molecular cloud. As expected, the scatter is much larger for embedded objects. For background stars, the observed correlation is well represented by

$$\tau(H_2O) = 9.3 \times 10^{-2}(A_v - 3.3) \qquad A_v > 3.3 \text{ magnitudes.} \qquad (10.51)$$

Thus, there is a constant abundance of H_2O ice in the Taurus molecular cloud ($X(H_2O) = 1.0 \times 10^{-4}$ with respect to H-nuclei and assuming standard gas-to-dust values [Section 5.3.5]). For solid CO, this relation is

$$\tau(CO) = 8 \times 10^{-2}(A_v - 6.1) \qquad A_v > 6.1 \text{ magnitudes,} \qquad (10.52)$$

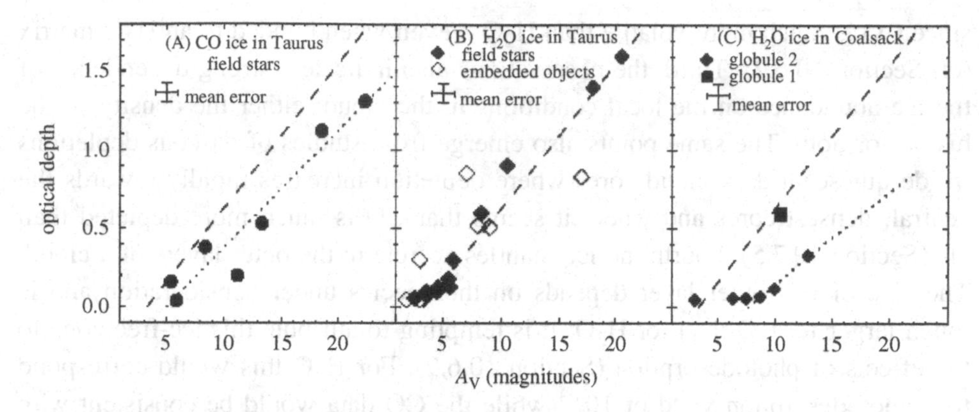

Figure 10.12 The relation between ice optical depth and total visual extinction. (A) Solid CO towards background field stars. (B) Solid H_2O towards field stars and embedded objects. (C) Solid H_2O towards background field stars behind two globules in the Coalsack molecular cloud. The dashed line in all three panels is the best fit relation for H_2O in Taurus. The dotted line in panel A presents the best fit for solid CO in Taurus. The dotted line in panel C presents the best fit for H_2O ice in the Coalsack globule 2. Data taken from D. C. B. Whittet, *et al.*, 1988, *M. N. R. A. S.*, **233**, p. 321 and 1989, *M. N. R. A. S.*, **241**, p. 707, J. Chiar, *et al.*, 1995, *Ap. J.*, **455**, p. 234 and R. G. Smith, *et al.*, 2002, *M. N. R. A. S.*, **330**, p. 837.

corresponding to a solid CO abundance deep in the cloud of $X(CO) = 2.1 \times 10^{-6}$ with respect to H-nuclei. For comparison, the gaseous CO measured towards these same lines of sight (but in a bigger beam) in the $J = 1-0$ submillimeter transition of $C^{18}O$ is

$$N_{gas}(CO) = 8.3 \times 10^{16}(A_v - 1.3) \qquad A_v > 1.3 \text{ magnitudes}, \qquad (10.53)$$

or a gaseous CO abundance of 4.4×10^{-5}, where a solar $^{16}O/^{18}O$ (= 490) abundance has been assumed.

A number of points emerge from this. First, after H_2, the most abundant molecule is H_2O ice, locking up some 30% of the available atomic O. Solid CO, on the other hand, is much less abundant than gaseous CO (solid CO/gaseous CO $\simeq 0.05$). Gaseous CO itself locks up only 30% of the available atomic C (the atomic C and O abundances in the diffuse ISM are 1.4×10^{-4} and 3.2×10^{-4}, respectively; Section 5.3.6). The inventories of these two elements are thus incomplete. There are indications that O may be mainly in the form of atomic oxygen in the gas phase in molecular clouds. The reservoir of the gaseous carbon that is not accounted for in CO is, however, unclear. Second, these $\tau - A_v$ relations are different for different species and clouds (Fig. 10.12). Clearly, depletion depends on the molecule involved. Because all species should stick at the temperatures of the dust in molecular clouds ($T_d < 10$ K), this probably reflects a difference in the return rate from the ice to the gas. This is reasonable since CO is much more volatile than H_2O – even when bonded to an H_2O matrix (cf. Section 10.6.2). Third, the observations also indicate a strong dependence of the ice abundance on the local conditions in the cloud; either the density or the history or both. The same points also emerge from studies of gaseous depletions inside quiescent dark cloud cores where depletion increases rapidly towards the central, densest cores and where it seems that CO is much more depleted than N_2 (Section 10.7.5). Fourth, no ice mantles accrete in the outer layers of a cloud. The size of this outer layer depends on the species under consideration and is much larger for CO than for H_2O. It is tempting to attribute this ice-free zone to the effects of photodesorption (Section 10.6.2). For H_2O this would correspond to a photodesorption yield of 10^{-3} while the CO data would be consistent with a yield of 2×10^{-2}. The required rate for H_2O is about an order of magnitude less than observed in the laboratory for rough H_2O surfaces, but what the yield would be for a single H_2O molecule adsorbed on silicate or graphitic surfaces is not known.

Spectral profiles and the ice composition

In addition to abundances, analysis of the observed profiles of interstellar ice bands also provides information on the matrix that the absorbed species are trapped in. This is very obvious for H_2O, which, when isolated in an inert environment, gives rise to sharp features around 2.7 μm but in an H_2O ice produces a very strong and broad absorption feature at 3.08 μm (Section 2.1.2). It is the latter that actually dominates the observed interstellar ice spectra (cf. Fig. 10.11) and H_2O is in an H_2O-rich matrix in space. The spectral effects are more subtle for non-hydrogen bonding species such as CO, but present IR instrumentation can extract this information as well. As an example, Fig. 10.13 shows the observed profile of the interstellar 2139 cm^{-1} band of solid CO. Analysis of a large sample of such profiles has shown that they can all be fitted by two main components: a narrow (3.5 cm^{-1}) band centered at about 2139.9 cm^{-1}, which generally dominates the observed profile, and a broad (10.6 cm^{-1}) underlying component at 2136.5 cm^{-1} (which dominates in only a few sources, but is present in many more). The narrow component is well fitted by the spectrum of pure solid CO and this is corroborated by the peak position and profile of the solid ^{13}CO ice band. Along most lines of sight 60–90% of the solid CO is in this component. The weaker and broader CO component is not fully understood. Probably it is related to traces of CO trapped in a hydrogen-bonding environment but CO in H_2O ice seems to be excluded. In any case, while H_2O is observed to dominate the bulk composition of the ices along all lines of sight, there is also a (minor) ice component that consists largely of CO (perhaps with other volatile species such as N_2) and has very little embedded H_2O. This dichotomy in the ice

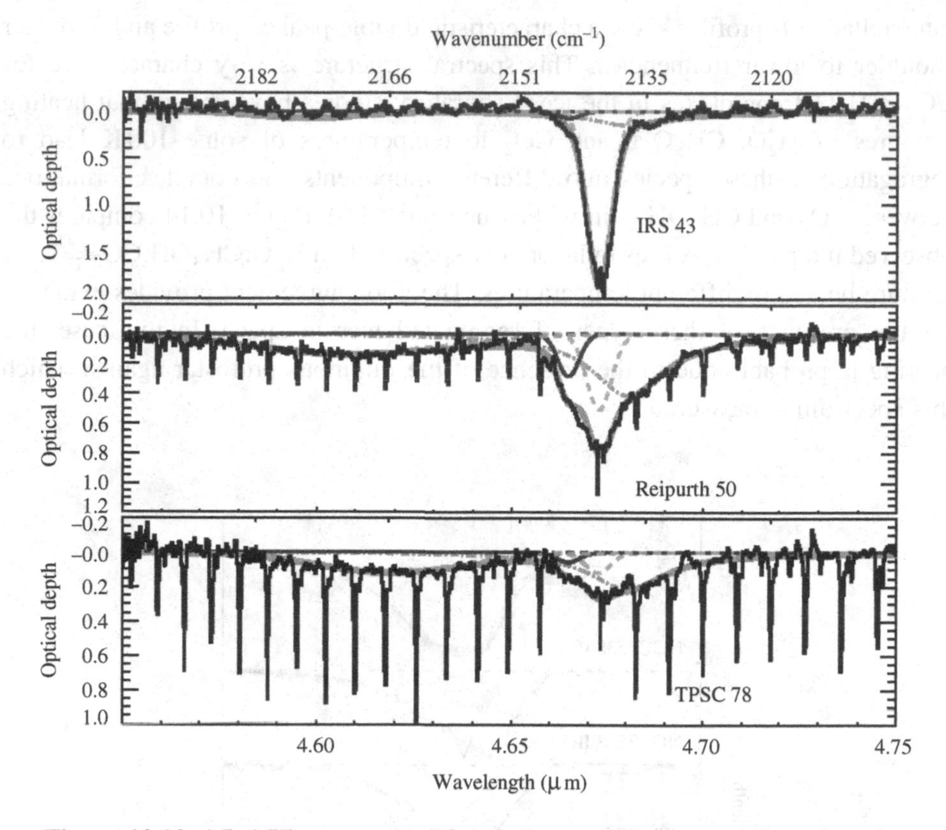

Figure 10.13 4.5–4.75 μm spectra of three low-mass protostars show the relatively broad solid CO absorption at about 4.67 μm and a forest of gas-phase CO rotational–vibrational transitions (cf. Section 2.1.2). In addition, this spectral region contains the broad absorption feature at 4.62 μm due to OCN^-. The profile of the interstellar solid CO band consists of three independent components: a narrow band at about 4.67 μm, which dominates the top spectrum. The profile in the bottom panel is dominated by a much broader band peaking at 4.68 μm. The middle panel is a combination of these two extreme profiles. In addition, the solid CO band is accompanied by a weak band at shorter wavelengths. Figure courtesy of K. M. Pontoppidan; adapted from K. M. Pontoppidan, *et al.*, 2003, *A. & A.*, **408**, p. 981.

composition may reflect the effects of volatility, perhaps driven by cosmic-ray heating events, or it may reflect a segregation of ices along the line of sight with denser regions producing ices with little H_2O (the atomic H abundance in the accreting gas depends strongly on the density). This segregation may also reflect the formation of different types of ices on grains of different sizes due to outgassing (cf. Section 10.6.2).

The solid CO_2 bending mode at 660 cm^{-1} provides another telling example of the power of IR spectroscopy to probe the characteristics of interstellar ice. Some interstellar CO_2 profiles show a characteristic double-peaked profile and a weaker shoulder to lower frequencies. This spectral structure is very characteristic for CO_2–CH_3OH complexes in the ice. Laboratory studies have shown that heating mixtures of H_2O, CH_3OH, and CO_2 to temperatures of some 100 K lead to segregation of these species into different components and complex formations between CO_2 and CH_3OH (still within the same film). Figure 10.14 compares the observed interstellar profiles to laboratory spectra of an H_2O:CH_3OH:CO_2 =1:1:1 mixture heated to different temperatures. The good agreement provides evidence for the presence of these types of segregated ices in space. In this case, the heating is probably due to the presence of the luminous protostar against which this spectrum is measured.

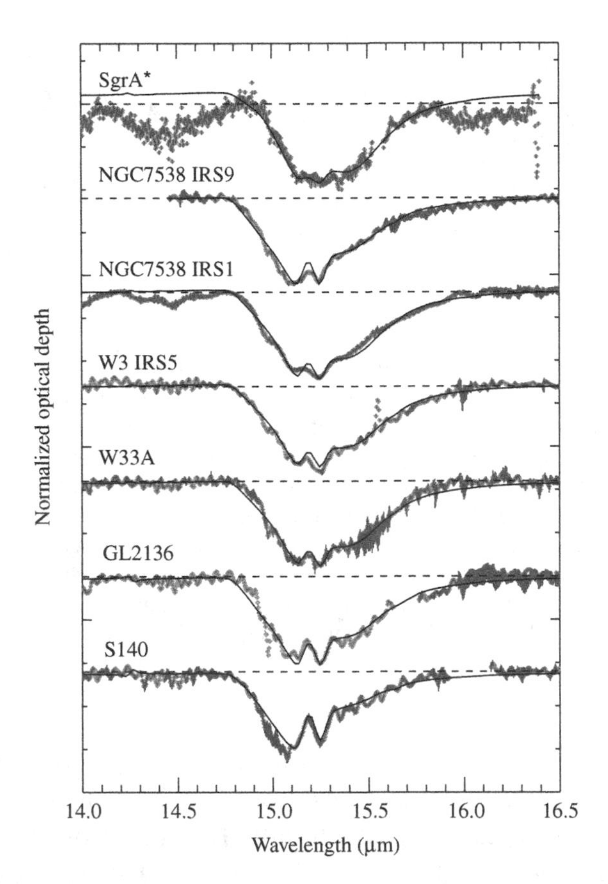

Figure 10.14 The profile of the interstellar solid CO_2 bending mode at 15.2 μm shows strong variations from source to source. The solid lines are laboratory spectra of a mixture of H_2O:CH_3OH:CO_2 = 1:1:1 at different temperatures. Temperature increases downwards from about 100 to 120 K. Figure adapted from P. Gerakines, *et al.*, 1999, *Ap. J.*, **522**, p. 357.

10.7.5 Depletions

In recent years, the depletion of species from the gas phase has been measured directly. In particular, column densities determined through optically thin isotopic lines of CO (e.g., $C^{18}O$ and $C^{17}O$) as well as other species can be compared with dust column densities derived from submillimeter continuum observations. These studies show that CO and CS are generally depleted by one to two orders of magnitude near the center of starless cores whenever the density is above $2 \times 10^4\,\text{cm}^{-3}$. The starless core of IC 5146 and the small Bok globule, B68, are a case in point, with measured depletions of a factor of 3 and of more than 100, respectively. The observations of deuterium depletion support the high inferred depletions of CO (cf. Section 10.4.4). The depletion time scale for CO on grains is $3 \times 10^5 (10^4/n\,[\text{cm}^{-3}])$ years, with n the density of hydrogen nuclei. This time scale should be compared with relevant dynamical time scales. The free–free time scale is very similar at this density, $10^6 (10^4/n\,[\text{cm}^{-3}])^{1/2}$ years. In contrast, ambipolar diffusion time scales tend to be longer by about an order of magnitude. Thus, if grain-to-gas ejection processes are inefficient, depletion measurements and ice formation provide a "clock" with which the dynamical evolution of molecular clouds can be timed and, in that case, free-fall seems to be the more relevant time scale. However, as discussed in Section 10.6.2, cosmic-ray processing of interstellar ices may tie the depletion time scale to the dynamical evolution of the region (e.g., $X_c \propto n^{-1}$, with X_c the abundance of condensables).

Apparently, not all molecules are affected in the same way. Many of these starless cores still show high abundances of the molecular ion N_2H^+ but not of HCO^+. It seems that N_2H^+ is unaffected by depletion until densities of at least 3–$10 \times 10^5\,\text{cm}^{-3}$. In B68, where CO is depleted by a factor of 10^2, the abundance contrast in N_2H^+ between low- and high-density regions is only a factor of 2. The chemistry of N_2H^+ is very simple, involving proton transfer from H_3^+ to N_2 and dissociative electron recombination of N_2H^+. The difference in depletion behavior of this species is, therefore, often ascribed to the volatility of N_2. However, that seems unlikely to be the full story. The binding energies and sublimation temperatures of N_2 and CO are very similar, 900 and 1000 K and about 16 and 17 K, respectively. Both of these temperatures are much higher than the temperatures of dust grains in cold dense clouds (6–8 K) and both species should deplete similarly and sublimate similarly. Perhaps this implies an active role of grains in returning species back to the gas phase. In particular, N_2 is unreactive on grain surfaces but CO might be completely converted to methanol and formaldehyde. If the ejection mechanism depends strongly on the volatility of the species (Section 10.6.2), N_2 would be much more likely to be returned to the gas phase than H_2CO or CH_3OH. But that would then imply that much of the CO is present on grains in the form of methanol and formaldehyde and observations do not seem to support this solution either. It is clear that our understanding of the interaction between gas and grains in dense clouds is incomplete.

10.8 Further reading

The key role of ionization in interstellar chemistry was first realized in [1] and [2]. Conversely, the use of molecules to estimate the degree of ionization in molecular clouds has found widespread use (cf. [3], [4], [5]). The influence of large molecules on the ionization balance of molecular clouds is discussed in [6].

The time scales involved in interstellar chemistry are discussed in [7].

The energy balance of molecular clouds has been discussed in [8] and [9].

The chemistry of molecular clouds has been discussed in [10] and [11]. The chemical implications of the observed high deuterium fractionation, and the central role of H_3^+ in this, date back to [12]. Studies expanding on this are [13] and [14]. The importance of the internal FUV field generated by cosmic rays was realized in [15] and expanded upon by [16]. The importance of chaotic behavior for the chemistry of molecular clouds has been discussed in [17].

The discussion on interstellar grain-surface chemistry and, in particular, the importance of hydrogenation and oxidation reactions of CO has been taken from [18], as updated in [19]. Grain-surface chemistry of deuterated species has been discussed in [20] and [21]. There are very few supporting experimental studies ([22]) and most of those are actually of the accretion limit rather than the diffusion limit (cf. Section 4.2.9), which is not easily translated to astrophysical conditions. A number of theoretical studies have appeared over the years, merging gas-phase and grain-surface chemistry ([23], [24]). Invariably, these calculations adopted a rate-equation approach and the results have to be carefully re-evaluated. Some recent studies have attempted a stochastic approach for both gas-phase and grain-surface chemistry ([25]) but only for limited networks. A different view on the origin of interstellar ice mantles is presented in [26].

A recent experimental study on photodesorption is [27]. Sputtering of ices in shocks has been considered in [28]. Ejection of newly formed molecules because of thermal spiking was investigated in [29]. Cosmic-ray-driven mantle explosions were studied in [30]. The importance of chemical energy in stored radicals dates back to the laboratory studies in [31] and [32]. Studies on the role of ices in the chemistry of hot cores are [33], [21], [34], and [35].

A good overview of the study of molecular cloud masses is [36]. A very illustrative discussion of the power of molecular lines for the determination of physical conditions in molecular clouds is [37]. The pitfalls associated with the use of the rotational diagram are detailed in [38]. Molecular line radiative transfer codes in one and two dimensions can be found at

http://www.mpifr-bonn.mpg.de/staff/fvandertak/ratran/frames.html,

which is maintained by F. van der Tak and M. Hoogerheijde.

Observations of the chemical composition of molecular clouds are reviewed in [39], [40], and [41].

Infrared spectroscopy of interstellar ices is reviewed in [42]. The large impact of SWS/ISO studies on interstellar ice studies can be gleaned from the compilation in [43]. The abundance studies in individual clouds were pioneered in [44].

Databases of laboratory spectra of interstellar ice analogs are available at

http://www.strw.leidenuniv.nl/~lab/databases/,

and

http://web.ct.astro.it/weblab/optico.html.

A compilation of ice optical properties is [45]. Important articles are [46] on H_2O, [47], [48], and [49] on CO, and [47] and [50] on CO_2. Energetic processing of ices has been studied in [51] and [23] (UV), [52] and [53] (UV and ions), and [54] (ions). The vaporization behavior of interstellar ice analogs was studied in [55] and [56].

Observations of high gas phase depletions in cold pre-stellar cores are very recent (cf. [57] and [58]).

References

[1] W.D. Watson, 1973, *Ap. J.*, **182**, p. L69
[2] E. Herbst and W. Klemperer, 1973, *Ap. J.*, **185**, p. 505
[3] M. Guelin, W.D. Langer, R.L. Snell, and H.A. Wootten, 1977, *Ap. J.*, **217**, p. L165
[4] P. Caselli, C.M. Walmsley, R. Terzieva, and E. Herbst, 1998, *Ap. J.*, **499**, p. 234
[5] A. Wootten, R. Snell, and A.E. Glassgold, 1979, *Ap. J.*, **234**, p. 876
[6] S. Lepp and A. Dalgarno, 1988, *Ap. J.*, **324**, p. 553
[7] A.E. Glassgold and W.D. Langer, 1973, *Ap. J.*, **186**, p. 859
[8] A.E. Glassgold and W.D. Langer, 1973, *Ap. J.*, **179**, P. L147
[9] P.F. Goldsmith and W.D. Langer, 1978, *Ap. J.*, **222**, p. 881
[10] E.A. Bergin, P.E. Goldsmith, R.L. Snell and W.D. Langer, 1997, *Ap. J.*, **482**, p. 285
[11] T.J. Millar, C.M. Leung, and E. Herbst, 1987, *A. & A.*, **183**, p. 109
[12] W.D. Watson, 1974, *Ap. J.*, **188**, p. 35
[13] T.J. Millar, A. Bennett, and E. Herbst, 1989, *Ap. J.*, **340**, p. 906
[14] H. Roberts, E. Herbst, and T.J. Millar, 2003, *Ap. J.*, **591**, p. L41

[15] S. S. Prasad and S. P. Tarafdar, 1983, *Ap. J.*, **267**, p. 603
[16] R. Gredel, S. Lepp, A. Dalgarno, and E. Herbst, 1989, *Ap. J.*, **347**, p. 289
[17] J. Le Bourlot, G. Pineau des Forets, E. Roueff, and P. Schilke, 1993, *Ap. J.*, **416**, p. L87
[18] A. G. G. M. Tielens and W. Hagen, 1982, *A. & A.*, **114**, p. 245
[19] A. G. G. M. Tielens and L. J. Allamandola, 1987, in *Interstellar Processes*, ed. D. Hollenbach and H. Thronson (Dordrecht: Reidel), p. 397
[20] A. G. G. M. Tielens, 1983, *A. & A.*, **119**, p. 177
[21] S. Charnley, A. G. G. M. Tielens, and S. Rodgers, 1997, *Ap. J.*, **482**, p. L203
[22] K. Hiraoka, T. Sata, S. Sato, *et al.*, 2002, *Ap. J.*, **577**, p. 265
[23] L. d'Hendecourt, L. J. Allamandola, and J. M. Greenberg, 1985, *A. & A.*, **152**, p. 130
[24] T. I. Hasegawa, E. Herbst, and C. M. Leung, 1992, *Ap. J. S.*, **82**, p. 167
[25] S. B. Charnley, 2001, *Ap. J.*, **562**, p. L99
[26] E. A. Bergin, D. A. Neufeld, and G. J. Melnick, 1999, *Ap. J.*, **510**, p. L145
[27] M. S. Wesley, R. A. Baragiola, R. E. Johnson, and G. A. Baratta, 1995, *Nature*, **373**, p. 405
[28] D. R. Flower and G. Pineau des Forets, 1994, *M. N. R. A. S.*, **268**, p. 724
[29] M. Allen and G. W Robinson, 1977, *Ap. J.*, **212**, p. 396
[30] A. Léger, M. Jura, and A. Omont, 1985, *A. & A.*, **144**, p. 147
[31] L. d'Hendecourt, L. J. Allamandola, F. Baas, and J. M. Greenberg, 1982, *A. & A.*, **109**, p. L12
[32] L. d'Hendecourt, L. J. Allamandola, R. Grim, and J. M. Greenberg, 1986, *A. & A.*, **158**, p. 119
[33] P. Caselli, T. I. Hasagawa, and E. Herbst, 1993, *Ap. J.*, **408**, p. 548
[34] S. Charnley, A. G. G. M. Tielens, and T. Millar, 1992, *Ap. J.*, **399**, p. L71
[35] S. D. Rodgers and S. B. Charnley, 2003, *Ap. J.*, **585**, p. 355
[36] N. Scoville and D. B. Sanders, 1987, in *Interstellar Processes*, ed. D. Hollenbach and H. Thronson (Dordrecht: Reidel), p. 21
[37] J. G. Mangum and A. Wootten, 1993, *Ap. J. S.*, **89**, p. 123
[38] P. F. Goldsmith and W. D. Langer, 1999, *Ap. J.*, **517**, p. 209
[39] E. A. Bergin, H. Ungerechts, P. F. Goldsmith, *et al.*, 1997, *Ap. J.*, **482**, p. 267
[40] W. M. Irvine, P. F. Goldsmith, and A. Hjalmarson, 1987, in *Interstellar Processes*, ed. D. Hollenbach and H. Thronson (Dordrecht: Reidel), p. 561
[41] W. D. Langer, E. F. van Dishoeck, E. A. Bergin, *et al.*, 2000, in *Protostars and Planets IV*, ed. V. Manning, A. P. Boss, and S. S. Russell (Tucson: University of Arizona Press), p. 29
[42] A. G. G. M. Tielens and D. C. B. Whittet, 1997, in *Astrochemistry*, ed. E. F. van Dishoeck (Dordrecht: Kluwer), p. 45
[43] E. L. Gibb, D. C. B. Whittet, A. C. A. Boogert, and A. G. G. M. Tielens, 2004, *Ap. J. S.*, **151**, p. 35
[44] D. C. B. Whittet, M. F. Bode, A. J. Longmore, *et al.*, 1988, *M. N. R. A. S.*, **233**, p. 321
[45] D. M. Hudgins, S. A. Sandford, L. J. Allamandola, and A. G. G. M. Tielens, 1988, *Ap. J. S.*, **86**, p. 713
[46] W. Hagen, A. G. G. M. Tielens, and J. M. Greenberg, 1983, *A. & A.*, **117**, p. 132
[47] G. A. Baratta and M. E. Palumbo, 1988, *J. Opt. Soc. America A*, **15**, p. 3076; M. E. Palumbo and G. A. Baratta, 2000, *A. & A.*, **361**, p. 298
[48] J. Elsila, L. J. Allamandola, and S. A. Sandford, 1997, *Ap. J.*, **479**, p. 818
[49] S. A. Sandford, L. J. Allamandola, A. G. G. M. Tielens, and L. J. Valero, 1988, *Ap. J.*, **329**, p. 498

[50] P. Ehrenfreund, A. C. A. Boogert, P. A. Gerakines, A. G. G.M. Tielens, and E. F. van Dishoeck, 1997, *A. & A.*, **328**, p. 649
[51] L. J. Allamandola, S. A. Sandford, and G. J. Valero, 1988, *Icarus*, **76**, p. 225
[52] P. A. Gerakines and W. A. Schutte, 1996, *A. & A.*, **312**, p. 289
[53] R. Grim, J. M. Greenberg, M. S. Groot, *et al.*, 1991, *A. & A. S.*, **78**, p. 161
[54] G. Strazulla and M. E. Palumbo, 2001, *Adv. Space Re.*, **27**, p. 237; G. A. Baratta, G. Leto, and M. E. Palumbo, 2002, *A. & A.*, **384**, p. 343
[55] M. P. Collins, J. W. Dever, H. J. Fraser, M. R. S. McCoustra, and D. A. Williams, 2003, *Ap. J.*, **583**, p. 1058
[56] S. A. Sandford and L. J. Allamandola, *Icarus*, **87**, p. 188; 1990, *Ap. J.*, **355**, p. 357
[57] E. A. Bergin, J. Alves, T. Huard, and C. J. Lada, 2002, *Ap. J.*, **570**, p. L101
[58] P. Caselli, C. M. Walmsley, M. Tafalla, L. Dore, and P. C. Myers, 1999, *Ap. J.*, **523**, p. L165

11

Interstellar shocks

11.1 Introduction

Shock waves are common phenomena in the interstellar medium. Shocks occur whenever material moves at velocities exceeding the sound velocity in the surrounding medium and the upstream material cannot dynamically respond to the upcoming material until it arrives. The shock will then compress, heat, and accelerate the medium. The heated material cools through the emission of line photons, further compressing the medium.

In the interstellar medium, two types of shocks are of interest. First, for fast shocks, the gas is so suddenly stopped and heated to a high temperature that insignificant radiative and non-radiative relaxation can take place; e.g., the shock front is much thinner than the postshock relaxation layer. As a result, we can cast the mass, momentum, and energy conservation equations in terms of simple "jump" conditions that relate the preshock (front) and postshock (front) density, temperature, and velocity to each other. These shocks are called J-shocks (J for jump). Second, for weak shocks in a magnetized medium with a low degree of ionization, the shock work is done by trace ions drifting through the predominantly neutral medium. In such shocks, the shock front is much thicker than the cooling length scale and the temperature is set by the balance between heating and cooling. In this case, we cannot use jump conditions but, rather, have to solve the conservation equations for the shock front structure. These shocks are called C-shocks (C for continuous). Note that, while such shocks are thicker than J-shocks, they are still very thin by interstellar standards. Depending on the ionization fraction, the gas density, the chemical composition, the magnetic field strength, and the shock velocity, the shock may be either a C-shock or a J-shock. In general, C-shocks occur in gas that has a low degree of ionization, a high density and molecular fraction, at least a moderate B field, and low shock velocities. For molecular clouds – with an intrinsic low degree of ionization ($x \lesssim 10^{-6}$) – shocks with velocities less than

$\simeq 40\,\mathrm{km\,s^{-1}}$ will be C-type, while high velocity ($v_s \gtrsim 40\,\mathrm{km\,s^{-1}}$) shocks will be J-type.

11.2 J-shocks

11.2.1 Jump conditions

The structure of a J-shock is shown schematically in Fig. 11.1. Like a piston, a high pressure region – due to an expanding supernova, HII region, or stellar wind – is driving a shock into the surrounding medium. Now consider this in the frame of reference that is co-moving with the shock front; i.e., the shock is stationary in this frame. The swept-up unshocked gas interacts through elastic collisions with the shocked gas and is decelerated that way. For neutral gas, this occurs on an elastic-collision length scale, the shock front. For low-density plasmas, the shock width will actually be further limited by plasma instabilities. In the (generally very thin) shock front, much of the (ordered) kinetic energy of motion of the gas is thus converted into (random) thermal energy through elastic collisions. In the postshock relaxation layer, the postshock gas will use this thermal energy initially for molecular dissociation and ionization processes (if possible). Downstream, the gas may radiatively relax, recombine, or form molecules again. Generally, these

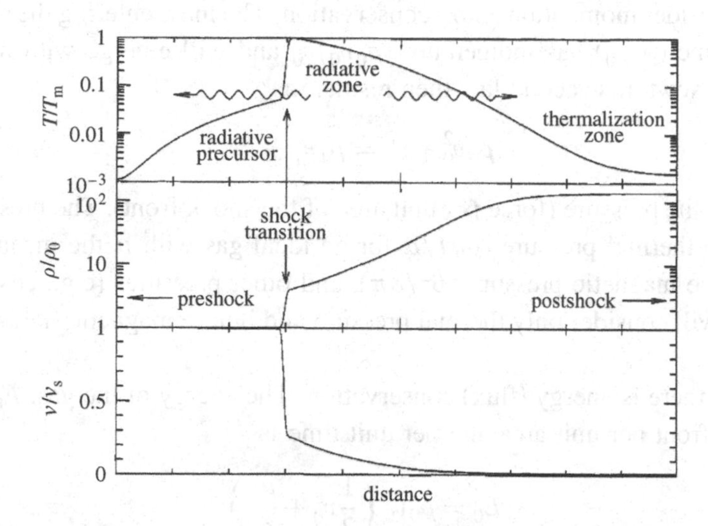

Figure 11.1 Schematically, the structure of a J-shock consists of a radiative precursor, the shock transition (shock front), and the postshock relaxation layer. The temperature, density, and velocity are normalized to the maximum temperature reached just behind the shock front, T_m, the preshock density, ρ_0, and the shock velocity, v_s. Figure courtesy of B. T. Draine and C. F. McKee; reprinted with permission from the *Ann. Rev. Astron. Astrophys.*, **31**, p. 373 ©1993 by *Ann. Rev.* (www.annualreviews.org).

competitive inelastic processes occur on a much longer time scale than the elastic conversion of kinetic energy into thermal energy. Shocks are supersonic with respect to the gas upstream and subsonic relative to the gas behind. However, while no sound waves can travel ahead of the shock, the shock may be preceded by a radiative precursor, affecting the preshock gas. In particular, high velocity shocks ($v_s > 50\,\text{km s}^{-1}$) will lead to ionization. The recombination photons originating in the relaxation layer may travel upstream and pre-ionize the preshock gas.

We will analyze the shock structure in the frame of the shock and assume steady flow (e.g., homogeneous upstream conditions and no slowing down of the shock while the gas flows through the postshock cooling zone). The thickness of the shock is generally very small compared to the hydrodynamic length scales, and the conservation equations (mass, momentum, and energy) take on the form of "jump" conditions – the Rankine–Hugoniot relations. Mass (flux) conservation implies that the mass inflow rate per unit area of the shock front must equal the mass outflow rate per unit area,

$$\rho_0 v_0 = \rho_1 v_1, \tag{11.1}$$

where ρ_j and v_j are the density and velocity and subscript $j = 0$ and $j = 1$ refer to upstream and downstream gas, respectively. These velocities are measured relative to the shock front.

Now consider momentum (flux) conservation. The mass entering the shock front per unit time $(\rho_0 v_0)$ has momentum $(\rho_0 v_0)v_0$ and will emerge with momentum $(\rho_1 v_1)v_1$. Newton's second law then results in

$$\rho_0 v_0^2 + P_0 = \rho_1 v_1^2 + P_1, \tag{11.2}$$

where P is the pressure (force per unit area of the shock front). The pressure is the sum of the thermal pressure ($\rho kT/\mu$ for an ideal gas with μ the mean mass per species), the magnetic pressure ($B^2/8\pi$), and other pressures (e.g., cosmic rays). Here, we will consider only thermal pressure and ignore magnetic fields and other pressures.

Finally, there is energy (flux) conservation. The energy of the gas, E_0, entering the shock front per unit area and per unit time is

$$E_0 = \rho_0 v_0 \left(\frac{1}{2} v_0^2 + \frac{u_0}{\rho_0} \right), \tag{11.3}$$

where the first term in brackets is the kinetic energy and the second term the internal energy density, both per unit mass. The internal energy per unit volume, u, is given by

$$u = \sum_i (u_i + n_i I_i), \tag{11.4}$$

where u_i is the energy of species i associated with internal degrees of freedom (rotational, vibrational, electronic) and I_i is the binding energy of species i (with number density n_i) relative to neutral atomic gas (e.g., for ionized hydrogen, $I_i = 13.6\,\mathrm{eV}$ and for molecular hydrogen, $I_i = -4.48\,\mathrm{eV}$). The expression for the energy of the gas downstream is similar. For the shock front, where cooling is unimportant, the energy difference has to equal the work done by the shock (per unit area); i.e.

$$E_1 - E_0 = P_0 v_0 - P_1 v_1. \tag{11.5}$$

Rearranging then yields for the energy conservation equation,

$$\left(\frac{1}{2}v_0^2 + \frac{u_0}{\rho_0} + \frac{P_0}{\rho_0}\right) = \left(\frac{1}{2}v_1^2 + \frac{u_1}{\rho_1} + \frac{P_1}{\rho_1}\right). \tag{11.6}$$

The term $(u+P)/\rho$ is the specific enthalpy of the gas. The mass and momentum flux conservation equations are valid immediately behind the shock front as well as more generally downstream. The energy flux conservation equation only holds as long as radiative cooling is unimportant. When radiative cooling (or heating) is important, then the energy exchange with the environment can be accounted for in the energy equation by including a term, F_x, on the right-hand side of the energy conservation equation (Eq. (11.6)).

Consider now an ideal gas with

$$P = \frac{\rho k T}{\mu}. \tag{11.7}$$

The sound speed is

$$C_\mathrm{s}^2 = \frac{\mathrm{d}P}{\mathrm{d}\rho} = \gamma\frac{P}{\rho}, \tag{11.8}$$

where the right-hand side holds for an adiabatic equation of state ($P \sim \rho^\gamma$ with $\gamma = C_\mathrm{p}/C_\mathrm{v}$ and the Cs are the specific heats at constant pressure and volume). For an ideal gas, the internal energy per unit mass is

$$u = \frac{P}{(\gamma-1)}. \tag{11.9}$$

A further point should be made here. In the shock front, only elastic processes play a role and hence γ has the value appropriate for a monatomic gas ($\gamma = 5/3$) even when the preshock gas is molecular. Downstream, when inelastic processes become important and internal degrees of freedom are excited, γ will adjust accordingly. In the energy conservation equation, the energy exchange term should include a term accounting for this. This energy exchange term should also allow for ionization and molecular dissociation,

$$F_x = F_{\rm rad} + \sum_i n_i v I_i + Pv\left(\frac{\gamma}{\gamma-1} - \frac{5}{2}\right), \tag{11.10}$$

where $F_{\rm rad}$ is the energy exchange (loss or gain) through radiation.

Define now the Mach number of the shock,

$$\mathcal{M} = \frac{v_0}{C_{\rm s,0}}, \tag{11.11}$$

expressed in the preshock values. For strong shocks, $\mathcal{M} \gg 1$. We can now use the mass conservation equation (Eq. (11.1)) to link the velocities (Eq. (11.2)) to the thermodynamic variables;

$$v_j^2 = \frac{1}{\rho_j^2}(P_1 - P_0)\left(\frac{1}{\rho_0} - \frac{1}{\rho_1}\right)^{-1}, \tag{11.12}$$

with $j = 0$ or 1. Substituting these expressions for the velocities in the energy equation (Eq. (11.6)), we can derive equations expressing the downstream thermodynamic variables into those upstream,

$$\frac{P_1}{P_0} = \frac{(\gamma-1)\rho_0 - (\gamma+1)\rho_1}{(\gamma-1)\rho_1 - (\gamma+1)\rho_0}, \tag{11.13}$$

or equivalently,

$$\frac{\rho_1}{\rho_0} = \frac{(\gamma+1)P_1 + (\gamma-1)P_0}{(\gamma-1)P_1 + (\gamma+1)P_0}. \tag{11.14}$$

Using this latter equation to eliminate ρ_1 from the expression for v_0^2 (Eq. (11.12)), we have, after some rearrangement,

$$\frac{P_1}{P_0} = \frac{2\gamma}{\gamma+1}\mathcal{M}^2 - \frac{\gamma-1}{\gamma+1} \tag{11.15}$$

and

$$\frac{\rho_0}{\rho_1} = \frac{\gamma-1}{\gamma+1} + \frac{2}{\gamma+1}\frac{1}{\mathcal{M}^2}. \tag{11.16}$$

We can now come to some interesting general conclusions. In the strong shock limit ($\mathcal{M} \gg 1$), the jump conditions become

$$\frac{\rho_1}{\rho_0} = \frac{\gamma+1}{\gamma-1} \longrightarrow 4, \tag{11.17}$$

$$P_1 = \frac{2}{\gamma+1}\rho_0 v_0^2 \longrightarrow \frac{3}{4}\rho_0 v_0^2, \tag{11.18}$$

and

$$v_1 = \frac{1}{4} v_0, \tag{11.19}$$

where the values have been evaluated for $\gamma = 5/3$, as appropriate for a monatomic gas, and Eq. (11.19) follows from mass conservations (Eq. (11.1)). Thus, while the pressure is unbounded for a strong shock, the density jump is a factor of 4 (immediately behind the thin shock front). Immediately behind the shock front, the kinetic energy of the inflowing material is divided as 1/16 kinetic energy, 9/16 internal energy and 6/16 work done.

The postshock temperature can be evaluated from the ideal gas law,

$$T_1 = \frac{2(\gamma-1)}{(\gamma+1)^2}\frac{\mu v_0^2}{k} = \frac{3}{16}\frac{\mu v_0^2}{k}, \tag{11.20}$$

where the right-hand side has been evaluated for $\gamma = 5/3$. This also follows directly from equating the kinetic energy per particle converted into heat $(1/2\,\mu\,(v_0 - v_1)^2)$ with the thermal energy per particle $(3/2\,kT_1)$. Like the pressure, the temperature is unbounded as $\mathcal{M} \to \infty$. For a 0.1 helium fraction, we find

$$T_1 \simeq 1.4 \times 10^5 \left(\frac{v_0}{100\,\mathrm{km\,s^{-1}}}\right)^2 \quad \text{in fully ionized gas,} \tag{11.21}$$

$$T_1 \simeq 2.9 \times 10^5 \left(\frac{v_0}{100\,\mathrm{km\,s^{-1}}}\right)^2 \quad \text{in neutral atomic gas,} \tag{11.22}$$

and

$$T_1 \simeq 5.3 \times 10^5 \left(\frac{v_0}{100\,\mathrm{km\,s^{-1}}}\right)^2 \quad \text{in molecular gas.} \tag{11.23}$$

This is the temperature immediately behind the shock front. Downstream, the temperature of the postshock gas can be affected by various heating and cooling processes, including radiative energy loss. The effects of this can be written as

$$\frac{\mathrm{d}F_{\mathrm{rad}}}{\mathrm{d}z} = n^2\,\Lambda - n\,\Gamma - 4\pi\kappa\,J, \tag{11.24}$$

where z is the distance behind the shock front and $n^2\Lambda$ and $n\Gamma$ are the gas cooling and heating rates (cf. Chapters 2 and 3). The third term represents energy gain from radiation emitted by other parts of the shock. We can now relate the temperature at any point downstream to the upstream conditions by integrating Eq. (11.24).

Gas cooling has been discussed extensively in Chapter 2. For low velocity shocks, the temperature will decrease in the postshock relaxation layer because of radiative losses. For temperatures above $\simeq 10^5$ K, the cooling time scale $(\tau_{\mathrm{cool}} \sim kT/n\Lambda(T))$ increases with increasing temperature. Hence, for high enough shock temperatures, the cooling time scale is longer than the dynamical time scale. Hence, high velocity shocks lack this relaxation layer and are isothermal. The criterion separating these two regimes can be expressed in terms of the cooling column density,

$$N_{\rm cool} = n_0\, v_0\, \tau_{\rm cool}, \tag{11.25}$$

with n_0 the preshock number density and $\tau_{\rm cool}$ the cooling time behind the shock. At high temperatures, the gas is cooled by permitted or semiforbidden transitions and collisional de-excitation is unimportant ($n \ll n_{\rm cr}$, with $n_{\rm cr}$ the critical density; cf. Section 2.3.1). In that case, the cooling rate, $n^2\Lambda$, scales with n^2, $\tau_{\rm cool}$ scales with $1/n$, and the cooling column is independent of n. Specifically, in the range $60 < v_0 < 1000\ {\rm km\,s^{-1}}$, the cooling-column density at a postshock temperature of 10^4 K is, approximately,

$$N_{\rm cool} \simeq 2\times 10^{17}\left(\frac{v_0}{100\ {\rm km\,s^{-1}}}\right)^{4.2}\ {\rm cm^{-2}}. \tag{11.26}$$

The postshock column density has to be greater than this cooling column density for a radiative shock to develop.

If cooling is important, the temperature downstream from the shock will approach the preshock temperature. If the gas is only thermally supported, the density contrast between preshock (ρ_0) and gas downstream from the shock front (ρ_2) will be given by (e.g., Eq. (11.15) or (11.16) with $\gamma = 1$),

$$\frac{\rho_2}{\rho_0} = \mathcal{M}^2. \tag{11.27}$$

Strong isothermal shocks (in an unmagnetized medium, see below) are thus characterized by large density contrasts. For an isothermal shock, the momentum equation can be manipulated to yield

$$v_0\, v_2 = C^2. \tag{11.28}$$

Generally, the density contrast is limited by magnetic pressure. If a magnetic field is present, flux freezing conserves vB and B/ρ, where B is the magnetic field component perpendicular to the flow. The compression as the gas cools leads to a concomitant increase in the magnetic field and this will cushion the shock. For strong magnetic fields, the maximum postshock density, $\rho_{\rm max}$, can be found by equating the magnetic pressure to the dynamic pressure,

$$\frac{B_1^2}{8\pi} = \left(\frac{\rho_{\rm max}}{\rho_0}\right)^2 \frac{B_0^2}{8\pi} = \rho_0\, v_0^2, \tag{11.29}$$

which can be written as

$$n_{\rm max} \simeq 80\, n_0^{3/2}\left(\frac{v_0}{100\ {\rm km\,s^{-1}}}\right)\left(\frac{1\ \mu{\rm G}}{B_0}\right)\ {\rm cm^{-3}}, \tag{11.30}$$

where n_{max} is the corresponding maximum number density allowed by the compressed magnetic field. The interstellar magnetic field scales approximately as $B_0 \simeq 1\, n_0^{1/2}\ \mu$G (for $n_0 \gtrsim 10\,\text{cm}^{-3}$) and the maximum compression is $n_{max}/n_0 \simeq 80(v_0/100\,\text{km s}^{-1})$. The temperature corresponding to this maximum density is

$$T_{max} = \frac{n_0\, \mu\, v_0^2}{n_{max} k} \simeq 10^4 \left(\frac{v_0}{100\,\text{km s}^{-1}}\right) \text{K}, \tag{11.31}$$

where the values are for a fully ionized gas. Thus, for temperatures above this maximum temperature, the postshock gas cooling is approximately isobaric, while below this density the cooling is approximately isochoric.

Finally, in fast J-shocks, the hot ($T \sim 10^5$ K) postshock gas can produce Lyman continuum photons that travel upstream to create radiative precursor ionization. These photons will also travel downstream and will create a $T \sim 10^4$ K ionized gas region; e.g., a plateau in the postshock temperature distribution.

11.2.2 Chemistry

Fast shocks lead to dissociation of molecules in the postshock gas. At high densities ($n \gtrsim 10^4\,\text{cm}^{-3}$), molecular hydrogen can be dissociated by a 25 km s^{-1} J-shock.[1] However, for low densities, the vibrational levels of molecular hydrogen are not in equilibrium and, for example, at $n \simeq 10^3\,\text{cm}^{-3}$, molecular hydrogen is mainly in $v = 0$. Because of the smaller dissociation cross sections from $v = 0$ compared with higher v levels, shock speeds of some 50 km s^{-1} are then required to dissociate H_2. The atomic hydrogen produced by the dissociation will quickly chemically attack other molecules and such fast shocks will be completely dissociative. Molecules can begin to reform in the cooled-down, warmish ($T < 10^4$ K) postshock gas.

The elevated temperature allows reactions to proceed among neutral species with considerable activation barriers (0.1–1 eV) and this leads to a considerably different composition from that produced by the ion–molecule chemistry characteristic of cold, dark clouds. Carbon and oxygen chemistry is of particular importance. Figure 11.2 summarizes the important chemical processes involved. Typically, the reaction rate coefficients involved are $10^{-10}\,\text{cm}^3\,\text{s}^{-1}$ (Chapter 4) for temperatures exceeding 1000 K, which is similar to collisional excitation rates. Hence, as the gas cools down from $\sim 10^4$ K – requiring many collisional excitations particularly when $n \gg n_{cr}$ (cf. Chapter 2) – reactive collisions will quickly form molecules (if H_2 has formed, see below). The chemical composition of the

[1] But note that such low velocity shocks into molecular gas are generally C-type because of the low degree of ionization.

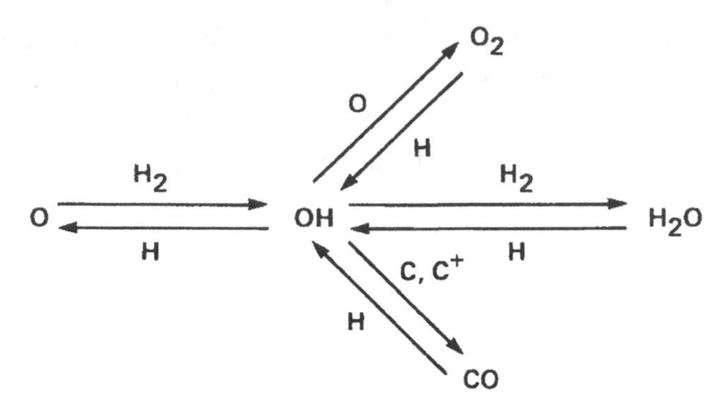

Figure 11.2 The important reactions involved in the chemistry of carbon and oxygen in shocks. Coreactants are indicated along the arrows. Figure reprinted with permission from D. Hollenbach and C. F. McKee, 1989, *Ap. J.*, **342**, p. 306.

warm postshock gas is then largely controlled by the ratio of atomic to molecular hydrogen. When molecular hydrogen is abundant, atomic oxygen will be converted into H_2O and O_2 while carbon is locked up in CO. In contrast, a high abundance of atomic H will keep the elements in atomic form as well. Now molecular hydrogen formation on grain surfaces is the rate-limiting step with a time scale of $\tau_{H_2} \sim 10^9/n$ years (cf. Section 8.7.1). The postshock column density corresponding to the molecular H_2 formation is

$$N_{H_2} = n_0\, v_0\, \tau_{H_2} \simeq 3 \times 10^{23} \left(\frac{n_0}{n}\right) \left(\frac{v_0}{100\,\mathrm{km\,s^{-1}}}\right) \simeq 4 \times 10^{22}\,\mathrm{cm^{-2}}, \tag{11.32}$$

where the last approximation assumes a magnetically cushioned shock. This is typically much longer than the cooling column. However, the H_2 formation process itself liberates 4.48 eV of chemical energy. How much of this energy will be carried away by the newly formed H_2 as vibrational energy and how much is transferred to the grain is unclear, but the vibrational energy may be $\sim$0.5 eV. If that is the case, then, for high enough densities ($n > 10^4\,\mathrm{cm^{-3}}$), heating by collisional de-excitation is important and molecular hydrogen formation can effectively keep the gas warm ($T \sim 500$ K). In summary, if the H_2 chemical energy is available for gas heating, a fast shock driven into dense gas will drive the chemistry to the right in Fig. 11.2. In low density gas, or if the H_2 chemical energy is not available for gas heating, the shocked gas will cool before producing much H_2, and therefore the warm chemistry is not significantly activated.

Besides OH and H_2O, shocks also produce other notable chemical tracers. In particular, the reaction

$$\mathrm{C^+ + H_2 \longrightarrow CH^+ + H} \tag{11.33}$$

is endothermic by about 0.4 eV and is inhibited at low temperatures. However, shocks driven into partially molecular gas may readily produce detectable amounts of CH^+. The OH produced by the reaction sequence outlined in Fig. 11.2 can react with atomic silicon present in the gas. This then leads to the formation of SiO. Efficient conversion of atomic sulfur in SH and H_2S will occur in molecular gas with the reaction sequence

$$\mathrm{S} \underset{9620\,\mathrm{K}}{\xrightarrow{\mathrm{H_2}}} \mathrm{SH} \underset{8050\,\mathrm{K}}{\xrightarrow{\mathrm{H_2}}} \mathrm{H_2S}, \tag{11.34}$$

where the values indicate the activation barriers in K. This sequence is inhibited in low temperature gas but leads to appreciable amounts of hydrogen sulfide in shocks. However, these species are rapidly destroyed in lukewarm (~300 K) gas with abundant atomic H (e.g., the first step in the reverse of Eq. (11.34) has a barrier of 860 K. The second step has no barrier.). Atomic sulfur reacts readily with abundant OH in shocks to produce SO and SO_2. The radical SH can also react quickly with atomic O to form SO.

Once the shocked gas cools down, the chemistry will evolve into the ion–molecule chemistry characteristic of dark clouds. The chemical signatures of the passage of the shock will, however, persist for some time. Water is destroyed through the proton transfer reaction with H_3^+ followed by dissociative electron recombination. The destruction time scale is then governed by the cosmic-ray ionization rate of H_2. The destruction time scale is approximately (cf. Chapter 10)

$$\tau_{\mathrm{H_2O}} \simeq \frac{X(\mathrm{H_2O})}{\zeta_{\mathrm{cr}}} \simeq 3 \times 10^5 \text{ years}, \tag{11.35}$$

with $X(H_2O)$ the water abundance and ζ_{cr} the cosmic-ray ionization rate. Photodissociation by FUV photons produced by cosmic-ray ionization of H_2 leads to a time scale that is about a factor of 2 longer than this (cf. Section 4.1.1). If the density is high enough, loss can also occur through accretion on grain surfaces, leading to ice mantle formation. The time scale is then (cf. Section 4.2.2)

$$\tau_{\mathrm{dep}} = (n\,\sigma_{\mathrm{d}}\,v)^{-1} \simeq \frac{3 \times 10^9}{n} \text{ years}. \tag{11.36}$$

This time scale equals that of the gas-phase ion–molecule chemistry when the (postshock) density is approximately 10^4 cm^{-3}. Since shocks are very compressive, this process is actually very important in molecular clouds.

11.2.3 The shock spectrum and shock diagnostics

We will first focus on radiative J-shocks. Gas cooling has been extensively discussed in Chapter 2. Because of the high electron velocities and cross sections for electronic excitation, the excitation of species behind a strong (ionizing) shock is generally dominated by inelastic electron collisions. For the excitation of neutral atomic fine-structure lines, proton collisions are also important. In neutral postshock gas, hydrogen atoms dominate the collisional excitation process (Sections 2.5 and 2.6.1). For temperatures in the range 10^4–5×10^5 K, collisional excitation of permitted lines of H and of various ionization stages of He, C, N, and O will dominate the cooling. The more abundant species tend to dominate until they are fully ionized. The complex interplay of ionization and radiative cooling processes will be affected by possible radiative precursors and this, in turn, will affect the spectrum. For low temperatures ($T < 5000$ K), cooling is dominated by fine-structure lines of neutral atoms or of ions with low ionization potentials (ionized by the ambient interstellar radiation field). Because of their lower critical densities (cf. Section 2.3.1), collisional de-excitation can now be important. At low densities ($n < 10^4$ cm^{-3}, the [CII] 158 μm line dominates while at higher densities, [OI] 63 μm takes over. For such cooler gas, molecule formation can become important as well and their signatures will appear in the shock spectra.

The total intensity of the cooling lines in a radiative shock is $I_{\rm tot} = \rho_0 v_{\rm s}^3/4\pi$ (assuming isotropic emission). For high velocity shocks ($v > 60$ km s^{-1}), much of the shock energy will ultimately emerge as HI Lyα photons, which are resonantly scattered and eventually absorbed by dust over a column of some 10^{20} cm^{-2} (cf. Section 7.2.4). Note that the preshock gas is velocity shifted by $3/4 v_{\rm s}$ to $v_{\rm s}$ relative to the Lyα-emitting postshock gas, which complicates the radiative transfer. The Lyα photons can also be absorbed in the postshock relaxation layer and hence Lyα diffusion can influence the shock structure. The hydrogen and helium optical lines are produced by recombination and, hence, their spectrum will be similar to that of photo-ionized gas. In the optical range, shock-ionized regions can be distinguished from photo-ionized gas (HII regions) through their strong lines from low excitation species such as [OI] (6300 Å and 63 μm), [OII] (3727 and 3729 Å), [NI] (5198 and 5200 Å), [NII] (6548 and 6584 Å), and [SII] (4069, 4076 Å, 6717 and 6731 Å) relative to Hβ. A high [FeII] 1.644 μm to Brγ ratio is a similarly good indicator of shocks, because Fe is rapidly ionized to Fe^{2+} or Fe^{3+} in HII regions. Essentially, in shocks, the recombination zone exceeds the fully ionized zone by a factor of some 100 in the column. Temperature-sensitive line ratios such as [OIII] 4363/5007 are also good indicators of shock activity since shocks can readily attain much higher temperatures while photo-ionization invariably produces 10^4 K gas (cf. Chapter 7). Shocks also produce a wider range of ionization stages of species than photo-ionized gas. Figure 11.3 illustrates the use of line ratios to discriminate between shocked and photo-ionized regions. In the UV, shocks produce strong resonance lines of CII 2326 Å, CIII 1909 Å, CIV 1549 Å, NIII 1750 Å, NV 1240 Å, OIII 1663 Å, and OIV 1402 Å. The shock

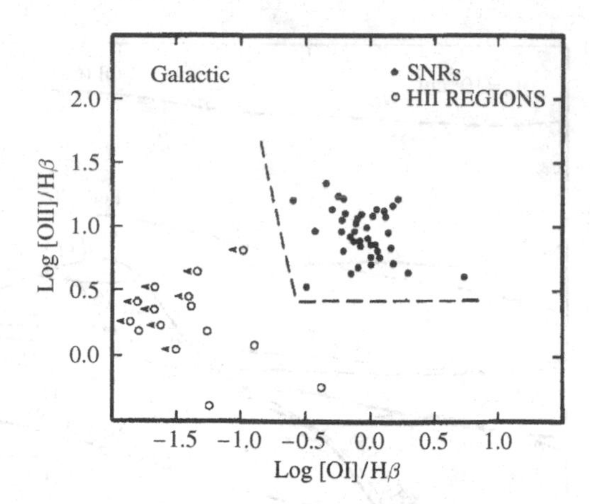

Figure 11.3 The [OII]/Hβ and [OI]/Hβ ratios provide an easy discriminant between shocked and photo-ionized gas. Figure reprinted with permission from R. A. Fesen, W. P. Blair, and R. P. Kirshner, 1985, *Ap. J.*, **292**, p. 29.

velocity can be diagnosed using Hα, [OI] 6300 Å, [OII] 3727 and 5007 Å, OIII 1662Å, and OIV 1400 Å.

In the infrared, shocked gas can be distinguished from photoheated gas (PDRs are now more confusing than HII regions) through the presence of a strong [SI] 25 μm line from shocked gas (in PDRs, sulfur will be photo-ionized in the warm surface layers capable of exciting the [SI] 25 μm line) and the absence of PAH emission characteristic of gas illuminated by FUV photons (cf. Chapter 9). The [OI] 63 μm to [CII] 158 μm ratio is also a good discriminant for shockheated versus photoheated gas, particularly if the H_2 chemical energy is available for heating an extended column density of gas to some 500 K, in which carbon is mainly in the form of CO while oxygen is largely atomic.

For fast shocks ($v_s > 200\,\mathrm{km\,s^{-1}}$) moving into low-density ($n_0 \sim 1\,\mathrm{cm^{-3}}$) gas, the cooling time exceeds the dynamical time scale and such non-radiative shocks produce relatively faint optical and UV emission. After passage through the shockfront, a neutral atom is rapidly collisionally ionized by the surrounding hot gas. At the same time, a neutral atom can also be collisionally excited and, if the temperature is high, the number of excitations per ionization is constant. For neutral hydrogen, for example, every ionization produces about 0.25 Hα photons. The rapid ionization limits the size of such collisionally excited emission zones ($\sim 10^{14}$ cm). The size of the emitting zones in trace elements may be larger but these lines will be fainter than the H lines by the abundance ratio. Both the small size and the dominance of the H lines in the visible spectrum are characteristics for non-radiative shocks. At other wavelengths, X-ray emission is,

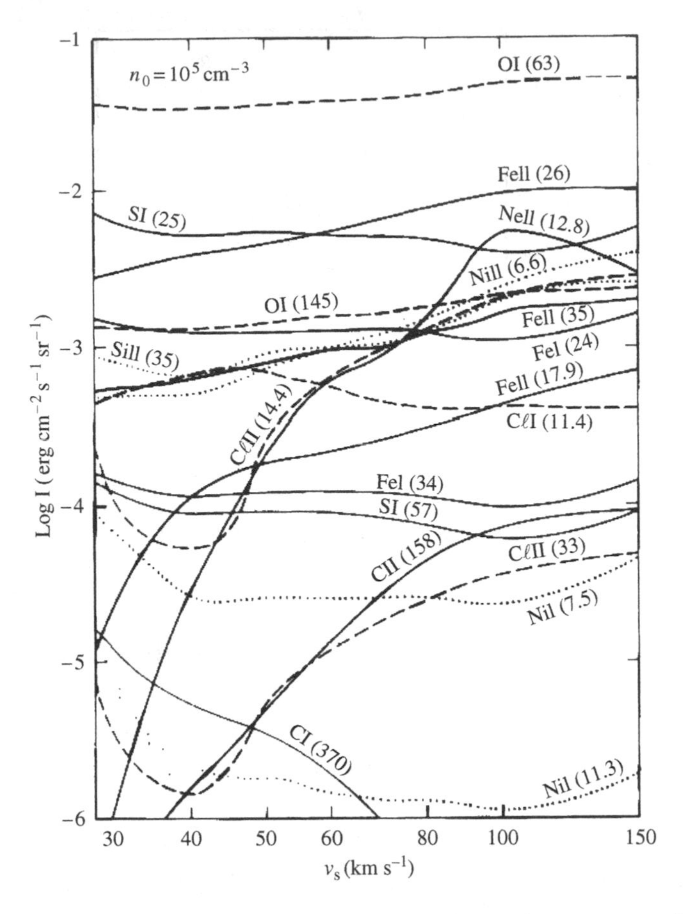

Figure 11.4 Calculated intensities of important infrared fine-structure lines as a function of shock velocity for a preshock density of $10^5\,cm^{-3}$. Figure reprinted with permission from D. Hollenbach and C. F. Mckee, 1989, *Ap. J.*, **342**, p. 306.

of course, a dead ringer for hot gas ($\sim 10^6$ K) and, hence, collisionally (shock) ionized gas.

11.3 C-shocks

In the discussion of J-shocks, the role of magnetic fields was only minor. For strong shocks, interstellar magnetic fields cushion a shock and limit its compression to a factor of $\sim 10^2$. However, the presence of a magnetic field in a plasma allows a different set of shocks. Here we will be mainly concerned with a gas with a small ionization fraction, x. In the diffuse ISM, the density may be $10^2\,cm^{-3}$ with an ionization fraction of 5×10^{-4}. In molecular clouds, the density is higher, $10^4\,cm^{-3}$, and the ionization fraction is lower, 10^{-7}. In these cases, low velocity shocks will be C-type.

11.3.1 Magnetohydrodynamics

Interstellar plasma will allow two sets of Alfvén waves. The first are pure transverse Alfvén waves, which transport energy along the magnetic field, but there is no particle transport involved. The second set of waves are partly longitudinal and partly transverse. These magnetosonic waves do involve variations in the density and magnetic field. These waves travel at the Alfvén velocity,

$$v_{\rm A} = \left(\frac{B^2}{4\pi\rho}\right)^{1/2} = 1.3\times 10^5 \left(\frac{B}{1\,\mu{\rm G}}\right)\left(\frac{{\rm cm}^{-3}}{n}\right)^{1/2} \ {\rm cm\,s^{-1}}, \qquad (11.37)$$

with n the molecular hydrogen density. Possibly because of flux freezing, in molecular clouds the magnetic field is observed to scale with the square root of the density,

$$B \simeq \left(\frac{{\rm cm}^{-3}}{n}\right)^{1/2} \ \mu G. \qquad (11.38)$$

Hence, the Alfvén speed in molecular clouds is typically 1.3 km s^{-1}. In a partially ionized plasma, the neutrals and ions are coupled through ion–neutral collisions with a typical Langevin rate, $k_{\rm L} = 10^{-9}$ cm^3 s^{-1} independent of the temperature. When the frequency of an Alfvén wave is much less than the frequency of collisions of neutrals with ions, $k_{\rm L}\, n_i$ (with n_i the ion density), the wave will propagate at an Alfvén speed given by the total density rather than the ion density because of this coupling. The neutrals will not quite follow the motion of the ions but lag behind by a distance given by the ion speed divided by the collision frequency. This is the process of ambipolar diffusion. When the frequency of the Alfvén wave is much higher than the collision frequency of ions with a neutral species, the neutrals will make little attempt to follow the motion of the ions. But the collisions of the neutrals with the ions will damp the plasma wave at a rate, $1/2\, k_{\rm L}\, n$. It is these latter (short wavelength) waves that are important for our shock analysis. The propagation speed of these fast modes is, for practical purposes (e.g., low degree of ionization), the Alfvén velocity of the ion fluid,

$$v_{\rm A,i} = (B^2/4\pi\rho_{\rm i})^{1/2}, \qquad (11.39)$$

with $\rho_{\rm i}$ the density of ions and, in a dense molecular cloud, this is approximately 1.3×10^8 cm s^{-1}.

Recapitulating, we have a medium consisting of two fluids, neutrals and ions. Each of these fluids can sustain short wavelength waves. For the neutrals, these are just sound waves that propagate with the sound speed,

$$C_{\rm s} = \left(\frac{5kT}{3m}\right)^{1/2} = 2.7\times 10^4 \left(\frac{T}{10\,{\rm K}}\right)^{1/2} \ {\rm cm\,s^{-1}}. \qquad (11.40)$$

Shocks will occur in the neutral fluid whenever the wave speed exceeds the Alfvén speed of the neutral fluid, $v_A = 1.3\,\mathrm{km\,s^{-1}}$. In the ionized fluid, short wavelength disturbances propagate at essentially the Alfvén speed of the ions. Thus, compressional waves, which travel slower than this ion Alfvén velocity but faster than the neutral Alfvén speed, can propagate signals upstream through the ionized fluid, "warning" the neutrals that a shock is coming. Such waves are rapidly damped though and, hence, "travel" with the disturbance. This warning signal compresses and accelerates the ions in the upstream gas in a smooth and continuous fashion. The drifting ions will accelerate and heat the neutrals through ion–neutral collisions on a time scale that is long compared to the cooling time scale. As a result, the shock front and relaxation layer are "merged" into one layer and, in C-shocks, the density and temperature will vary continuously. Now, when the ion fraction increases, the C-shock thickness decreases and the heating rate increases, and so eventually the shock will develop into the J-type. An increased preshock magnetic field will limit the compression of the ions and, hence, lead to larger columns of cooler gas. As a very general point, the postshock temperature has to be limited to less than some 3000 K to keep molecular hydrogen from dissociating. Once the molecules are dissociated, cooling is limited; e.g., a warm atomic gas will have to cool through Lyα excitation and, at temperatures where this becomes efficient, collisional ionization also becomes important. Thus, because of the increase in the cooling length scale relative to the heating length scale and the decrease of the ion–neutral mean free path because of the increased ionization, the shock front and the relaxation layer will segregate and the shock will transit to a J-type shock, characterized by sudden "jumps" in the physical conditions.

11.3.2 The equations of motions

Compared with a J-shock, the equations describing the structure of a C-shock are more involved. First, because cooling occurs on a time scale that is short compared to the heating time scale, the jump conditions cannot be applied, but rather, we must solve for the shock structure. Second, there are separate sets of equations for each fluid: the neutral, the ions, and possibly the grains. Third, these sets of equations are connected through exchange terms. For the continuity equations, we have to account for ionization and recombination processes. The fluids also exchange momentum through collisions. Energy is transferred through ion–neutral collisions both because of the friction involved and because of the temperature difference that will exist between the fluids. Moreover, thermal energy can be lost through excitation of internal degrees of freedom followed by radiative decay as well as through chemical reactions. Fourth, the momentum and energy equations for the ions contain terms for the magnetic pressure and energy (which, except for magnetic cushioning, we ignored in the J-shock discussion). The variation of the magnetic field follows from Maxwell's equation, $B = B_0 v_0 / v_i$, where B_0 is the preshock magnetic field and v_0 and v_i the shock and ion velocity.

Writing the continuity, momentum, and energy equations in a somewhat different form, we have

$$\frac{d}{dr}(\rho_n v_n) = S_n \tag{11.41}$$

$$\frac{d}{dr}(\rho_n v_n^2 + P_n) = F_n \tag{11.42}$$

$$\frac{d}{dz}\left(\rho_n v_n \left(\frac{1}{2} v_n^2 + \frac{u}{\rho} + \frac{5P}{2\rho}\right)\right) = G_n + F_n v_n - \frac{1}{2} S_n v_n^2, \tag{11.43}$$

where z is the distance along the propagation direction, S_n is the net rate per unit volume at which ion mass is converted into neutral mass, F_n is the rate per unit mass at which momentum is added to the neutral fluid through the interaction with the ion fluid, and G_n is the rate per unit volume at which internal energy is added to the neutral fluid. A similar set of equations pertains for the ions and electrons, except that the pressure in the momentum equation also contains a magnetic term $(B^2/8\pi)$. We can schematically represent the ionization and recombination processes as neutralization reactions between electrons and ions at a rate $k_{i,n}$, $S_n = \mu_n n_i n_e k_{i,n}$, with μ_n the mean mass per neutral particle created. Other processes can be similarly included. For non-polar molecules, the collisional interaction between ions and neutrals is well described by the Langevin rate, k_L (Section 4.1.3, Eq. (4.12)). We can then write

$$F_n = n_n n_i k_L (v_i - v_n). \tag{11.44}$$

The energy addition term due to ion–neutral collisions is

$$G_n = n_n n_i k_L \left(\mu_i (v_i - v_n)^2 + 3k(T_i - T_n)\right), \tag{11.45}$$

where the first term in the brackets on the right-hand side of this equation describes the frictional heating due to ion–neutral slip and the second term represents thermal energy exchange. This G term should also include terms for radiative energy exchange as well as other heating processes (cf. Eq. (11.24)).

11.3.3 C-shock structure

The calculated structure for a representative C-shock is illustrated in Fig. 11.5. The ions are rapidly accelerated to the shock velocity, while the neutral velocity varies more gradually. So, the relative velocity reaches values as high as ~35 km s^{-1} for

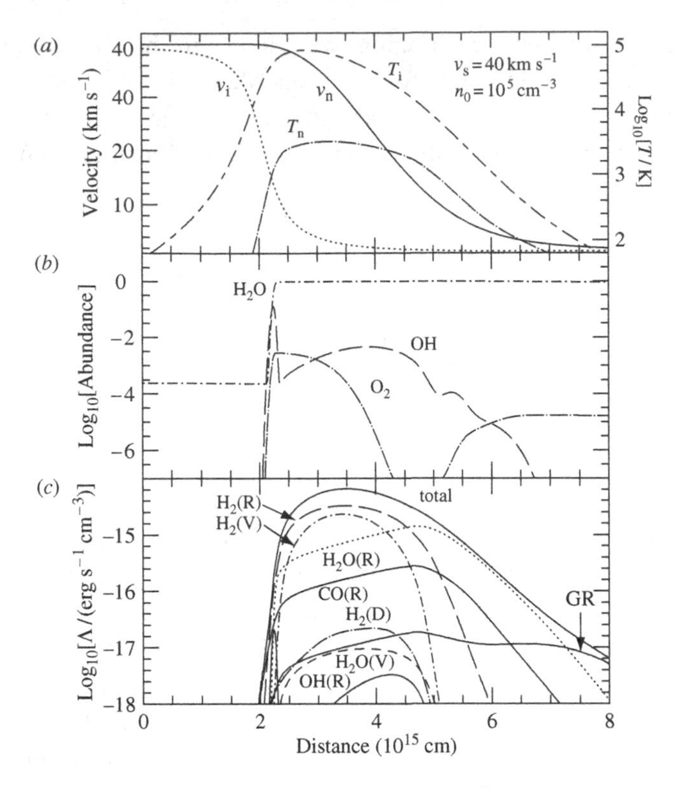

Figure 11.5 The calculated structure for a $40\,\mathrm{km\,s^{-1}}$ C-shock propagating into a gas of density $n_0 = 10^5\,\mathrm{cm^{-3}}$ and a magnetic field of $(n_0)^{1/2}\,\mu G$. (*a*) The ion (v_i, T_i) and neutral (v_n, T_n) velocity and temperature. The densities follow from the continuity equation ($\rho \sim 1/v$). (*b*) Abundances of the major O-bearing species normalized to the abundance of O not in CO. (*c*) Contribution to the total cooling (total) of the neutral gas by various processes: rotational (R) and vibrational (V) cooling by H_2, H_2O, CO, and OH, the dissociation of H_2 (D), and cooling through collisions with grains (GR). Figure kindly provided by M. J. Kaufman. Adapted from M. J. Kaufman and D. A. Neufeld, 1996, *Ap. J.*, **456**, p. 611.

this particular case. Both far upstream and far downstream, the ions and neutrals are flowing with the same velocity. The characteristic length scale over which the neutral velocity changes can be evaluated from the momentum equation, ignoring the thermal and inertial pressure, and boils down to

$$L \simeq \frac{v_d}{n_i\, k_L}, \tag{11.46}$$

where we have assumed that the mass of an ion well exceeds the mass of a neutral and v_d is the ion–neutral drift velocity. Now, the ions will have already reached their maximum compression, $n_i/n_{i,0} \simeq v_s/v_A$ (with v_A the Alfvén velocity in the preshock gas), when the C-shock kicks in for the neutrals and, hence, if we set $v_d \simeq v_s/2$, we have

$$L \simeq \frac{v_A}{2\, n_{i,o}\, k_L}. \tag{11.47}$$

If we assume that the preshock magnetic field scales with the square root of the density (cf. Eq. (11.38)), this results in

$$L \simeq \frac{1.1 \times 10^5}{n_{i,o}\, k_L}. \tag{11.48}$$

For the shock shown in Fig. 11.5, large PAHs are the dominant negative charge carriers. With $k_L \simeq 10^{-7}\ \mathrm{cm^3\ s^{-1}}$ (cf. Chapter 6) and $x = 10^{-8}$, we have $L \simeq 10^{15}$ cm. The thickness of C-shocks varies inversely with the square root of the density through the dependence of the degree of ionization on the density (cf. Chapter 10).

Over most of the shock, the frictional heating between ions and neutrals is low enough for radiative cooling to keep the temperature relatively constant at a value of only some 3000 K, much less than the 85 000 K expected for a $40\ \mathrm{km\ s^{-1}}$ J-shock (cf. Eq. (11.23)). It is clear that for this C-shock, the shock front and the radiative relaxation layer have merged (cf. Fig. 11.5). The average heating rate is then

$$n\Gamma \simeq \frac{\rho_0\, v_s^3}{L}, \tag{11.49}$$

which also follows from evaluating the ion–neutral friction heating (cf. Eq. (11.45)), ignoring the thermal energy exchange term. At these velocities and densities, the cooling is dominated by molecular hydrogen. Rotational cooling by H_2 is well represented by local thermodynamic equilibrium for such dense, warm gas ($n_{crit} \simeq 10^2\ \mathrm{cm^{-3}}$ for a temperature of 10^3 K),

$$n^2\, \Lambda(T) = 2.5 \times 10^{-33}\, n\, T^{3.82}\ \mathrm{erg\ cm^{-3}\ s^{-1}}. \tag{11.50}$$

Ignoring other cooling agents, the energy balance then yields

$$T = 1.2 \times 10^3 \left(\frac{v_s}{10\ \mathrm{km\ s^{-1}}}\right)^{0.79} \left(\frac{10^{15}\ \mathrm{cm}}{L}\right)^{0.26} \simeq 2700\ \mathrm{K}, \tag{11.51}$$

where we have evaluated this for a postshock density of twice the preshock density. Thus, the postshock temperature is not very sensitive to the preshock density, $T \sim n^{0.13}$ ($L \sim n_i^{-1} \sim n^{1/2}$ since $x \sim n^{1/2}$; cf. Section 10.2), but does depend on the velocity to the $\sim$0.8 power. Because ion-inelastic cross sections are much smaller than ion-elastic cross sections, the ion fluid will heat up until collisional energy transfer to the neutrals balances the compressional heating,

$$k(T_i - T_n) = \frac{1}{3} m_n v_d^2. \tag{11.52}$$

For a drift velocity of $20\,\mathrm{km\,s^{-1}}$, the temperature difference between the two fluids amounts to some 3×10^4 K. Because molecular hydrogen is not dissociated, the chemistry in C-shocks is dominated by reactions of H_2 with other species. In particular, as for J-shocks, the warm molecular gas in C-shocks is perfect to drive atomic oxygen into H_2O (Fig. 11.5; cf. Section 11.2.2). Because of its high critical density ($\sim 10^7$–$10^9\,\mathrm{cm^{-3}}$ depending on T and optical depth; cf. Chapter 2), water becomes the dominant coolant at densities $\gtrsim 10^6\,\mathrm{cm^{-3}}$.

11.3.4 The spectrum of C-shocks

The dominant emission lines of C-shocks are shown in Figs. 11.6 and 11.7. The atomic fine-structure lines are particularly important at low shock velocities in dense gas. The relative intensities of the rotational lines of molecular hydrogen (e.g., 0–0 S(1) and S(9) at 17.03 and 4.694 μm) are good velocity indicators for moderately fast shocks. C-shocks propagating into dense molecular gas will

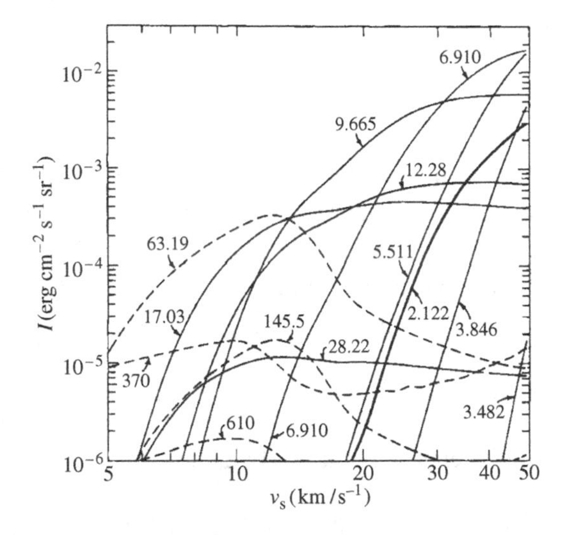

Figure 11.6 Line intensities calculated for C-shocks propagating into a molecular cloud with a density of $10^6\,\mathrm{cm^{-3}}$, a magnetic field of 10^3 μG, and an ionization fraction of 10^{-8}. The atomic fine-structure lines of [OI] (63 and 146 μm) and [CI] (370 and 610 μm) are indicated by dashed lines. The solid lines are the pure rotational lines of H_2 and the rotational–vibrational transition, 1–0 S(1) (2.12 μm). Figure reprinted with permission from B. T. Draine, W. G. Roberge, and A. Dalgarno, 1983, *Ap. J.*, **264**, p. 485.

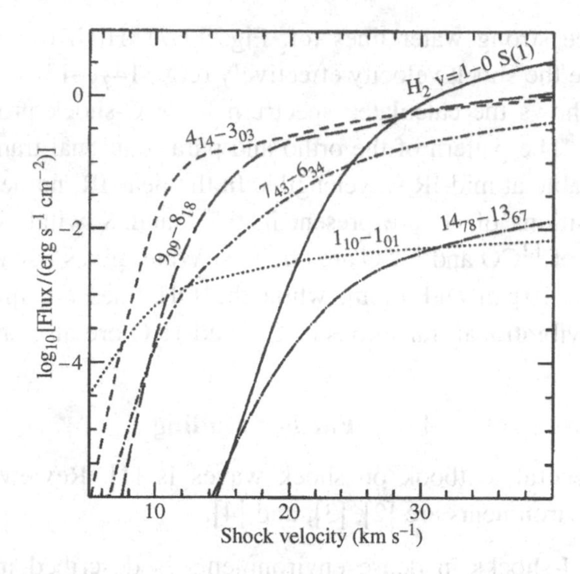

Figure 11.7 Water line intensities calculated for C-shocks propagating into a molecular cloud with a density of 10^6 cm^{-3}, a magnetic field of 10^3 μG, and an ionization fraction of 10^{-8}. Figure reprinted with permission from M. J. Kaufman and D. A. Neufeld, 1996, *Ap. J.*, **456**, p. 611.

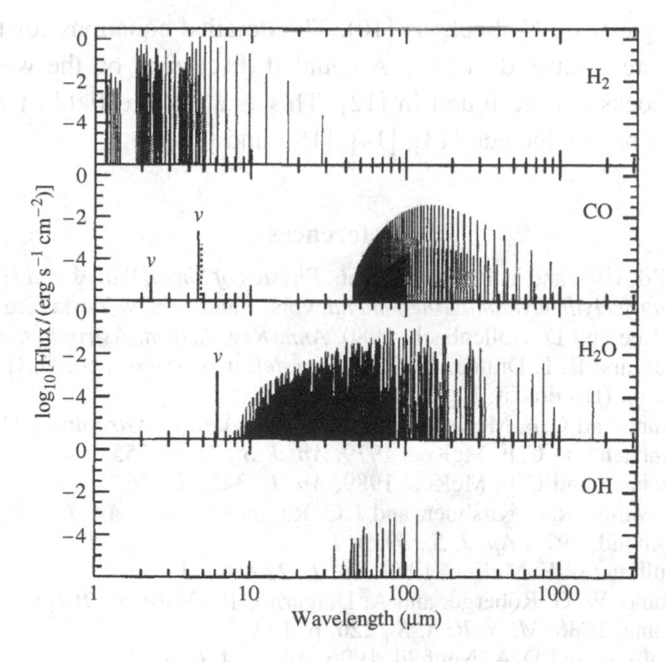

Figure 11.8 Calculated molecular spectra for a 40 km s^{-1} C-shock propagating into a molecular cloud with a density of 10^5 cm^{-3}, a magnetic field of 447 μG, and an ionization fraction of 3×10^{-8}. Figure reprinted with permission from M. J. Kaufman and D. A. Neufeld, 1996, *Ap. J.*, **456**, p. 611.

typically produce strong water lines (cf. Fig. 11.7). High rotational levels are required to probe the shock velocity effectively (e.g., 14_{78}–13_{67} versus 4_{14}–3_{03}).

Figure 11.8 shows the calculated spectrum for a C-shock propagating into a molecular cloud. The pattern of the ortho and para rotational transitions of H_2 is readily recognizable at mid-IR wavelengths. In the near-IR, numerous rotational–vibrational transitions of H_2 are present in the J and K windows. For CO, the separate ladders of ^{12}CO and ^{13}CO are obvious. Water gives rise to a dense forest of lines between 10 μm and 1 mm, while the OH lines are more sparse. The emission in the vibrational transitions of CO and H_2O are also shown.

11.4 Further reading

An excellent general textbook on shock waves is [1]. Reviews on shocks in astrophysical environments are [2], [3], and [4].

The physics of J-shocks in dense environments is described in [5]. Molecule formation and predicted line intensities can be found in [6], which includes detailed appendixes with relevant atomic and molecular data. Fast shocks in more tenuous environments are discussed in [7], [8], and [9].

The seminal paper on C-shocks is [10]. The detailed equations for the structure of C-shocks are discussed in [11]. A detailed discussion on the water emission from such shocks can be found in [12]. This is an active field of research and other relevant papers include [13], [14], [15], and [16].

References

[1] Y. B. Zel'dovitch and Y. P. Raizer, 1966, *Physics of Shock Waves and High Temperature Hydrodynamic Phenomena*, vols. 1 and 2 (New York: Academic Press)
[2] C. F. McKee and D. Hollenbach, 1980, *Ann. Rev. Astron. Astrophys.*, **18**, p. 219
[3] J. M. Shull and B. T. Draine, 1987, in *Interstellar processes*, ed. D. Hollenbach and H. Thronson (Dordrecht: Reidel), p. 283
[4] B. T. Draine and C. F. McKee, 1993, *Ann. Rev. Astron. Astrophys.*, **31**, p. 373
[5] D. Hollenbach and C. F. McKee, 1979, *Ap. J. S.*, **41**, p. 555
[6] D. Hollenbach and C. F. McKee, 1989, *Ap. J.*, **342**, p. 306
[7] R. A. Chevalier, R. P. Kirshner, and J. C. Raymond, 1980, *Ap. J.*, **235**, p. 186
[8] J. C. Raymond, 1979, *Ap. J. S.*, **39**, p. 1
[9] J. M. Shull and C. F. McKee, 1979, *Ap. J.*, **227**, p. 131
[10] B. T. Draine, W. G. Roberge, and A. Dalgarno, 1983, *Ap. J.*, **264**, p. 485
[11] B. T. Draine, 1986, *M. N. R. A. S.*, **220**, p. 133
[12] M. J. Kaufman and D. A. Neufeld, 1996, *Ap. J.*, **456**, p. 611
[13] D. R. Flower, J. Le Bourlot, G. Pineau des Forets, and S. Cabrit, 2003, *M. N. R. A. S.*, **341**, p. 70
[14] D. R. Flower and G. Pineau des Forets, 1998, *M. N. R. A. S.*, **297**, p. 1182
[15] D. R. Flower and G. Pineau des Forets, 1999, *M. N. R. A. S.*, **308**, p. 271
[16] D. A. Neufeld and A. Dalgarno, 1989, *Ap. J.*, **340**, p. 869; 1989, *Ap. J.*, **344**, p. 251

12

Dynamics of the interstellar medium

12.1 Introduction

The dynamics of the interstellar medium can be described by the continuity, momentum, and energy equations. For a plane-parallel slab, these are given by

$$\frac{1}{\rho}\frac{\partial \rho}{\partial t} + \frac{v}{\rho}\frac{\partial \rho}{\partial r} + \frac{\partial v}{\partial r} = 0 \tag{12.1}$$

$$\frac{\partial v}{\partial t} + v\frac{\partial v}{\partial r} + \frac{1}{\rho}\frac{\partial P}{\partial r} = 0 \tag{12.2}$$

$$\frac{\partial}{\partial t}\left(u + \frac{1}{2}\rho v^2\right) + \frac{\partial}{\partial r}\left(\rho v\left(u + \frac{1}{2}v^2\right) + Pv\right) = \rho Q. \tag{12.3}$$

These five equations (the momentum equation is vectorial) link the density, ρ, pressure, P, internal energy, u, and (three components of the) velocity, v, as a function of time, t, and position, r. A fourth equation has to be added; the equation of state, linking the internal energy to the pressure and density. Also, the energy addition or loss per unit mass per unit time, Q, has to be specified as a function of the density and temperature. These equations have to be integrated with the appropriate initial and boundary conditions.

It is important to realize that the fluid dynamic equations do not contain an inherent length scale. There are, of course, scale lengths on a microscopic level, due to interaction forces between the atoms or ions of the fluid. For example, a shock front will have a certain thickness, dictated by the viscosity of the fluid, in which the kinetic energy of the motion is converted into thermal energy. Likewise, there are cooling scale lengths associated with the downstream relaxation processes. However, on length scales that well exceed these microscopic length scales, the fluid equations are scale invariant. As long as the boundary conditions do not introduce a separate length scale, the solution of the equations describing the dynamics of the medium must be scale invariant as well. Such motions are called self-similar. For self-similar motions, the fluid dynamics equation can be expressed in ordinary differential equations in reduced time and position coordinates, which are often readily solved. Obviously, for self-similar motions, the distribution of pressure, density, and temperature remain constant and only the scale changes. This provides a very powerful tool for the analysis of the dynamics of the interstellar medium. In particular, it is then advantageous to adopt power-law time dependences for the physical parameters. Insertion of these into the equations of continuity, momentum, and energy provides, then, an easy way to determine these dependencies. We will encounter several examples of this in the sections of this chapter and sample a number of slightly different approaches to solve these.

In this chapter, we will examine the dynamical effects of massive stars on their environment. We will discuss, successively, the expansion of HII regions, supernova remnants, and stellar winds. All of these processes impart kinetic energy onto the interstellar medium, and their relative importance will be summarized at the end.

12.2 The expansion of HII regions

The ionization of the interstellar medium by stellar photons has been discussed in detail in Chapter 7. Here we will focus on the dynamical effects of the ionized gas on its surroundings. First, we will examine the general characteristics of ionization fronts (Section 12.2.1). We will then apply this to the formation and evolution of an HII region in a homogeneous medium. In particular, when a massive star is suddenly "turned on" in a homogeneous environment, an ionization front will expand so rapidly into the surrounding medium that the ionized gas has no time to react dynamically (Section 12.2.2). Because the photon flux reaching the ionization front will drop as a result of spherical dilution, the ionization front will eventually change character. At that point, the high pressure of the ionized gas will drive the expansion of the HII region into the neutral medium. The slowly expanding ionization front is then preceded by a shock front compressing the neutral medium (Section 12.2.3). Now, cursory inspection of optical images of HII regions shows large and systematic deviations from this simple theoretical picture. In reality, HII regions are characterized by large density variations – bright blisters at the surface of molecular clouds as well as the presence of "elephant trunks" and globules inside the ionized gas. Indeed, the stellar formation process itself is expected to lead to large density variations and this influences the ensuing dynamical evolution of the region. In subsequent sections, we will examine some of these effects, starting with the ionization of neutral globules in the ionized gas (Section 12.2.4), the ionization of accretion flows and stellar winds (Section 12.2.5), and the formation of a blister HII region in a cloud with a strong density gradient (Section 12.2.6). We will also briefly discuss ultracompact HII regions (Section 12.2.7). Finally, O stars generally have strong stellar winds, which can sweep up their environment and thereby influence the evolution of the ionized gas volume. This is discussed in Section 12.5.3.

12.2.1 The dynamics of ionization fronts

Because ionization fronts are very thin compared with the size of the HII region (cf. Chapter 7), we can approximate them as plane-parallel. As for shock fronts, the upstream and downstream conditions of the gas are connected through jump conditions. Choosing a coordinate system that is co-moving with the ionization front (see Fig. 12.1), the continuity equation is

$$\rho_0 v_0 = \rho_1 v_1, \tag{12.4}$$

and the momentum equation,

$$\rho_0 v_0^2 + P_0 = \rho_1 v_1^2 + P_1, \tag{12.5}$$

where the 0 and 1 indexes indicate upstream and downstream values. In the energy equation, stellar photons set the temperature since the radiative heating and cooling time scales are many orders of magnitude faster than the adiabatic expansion and

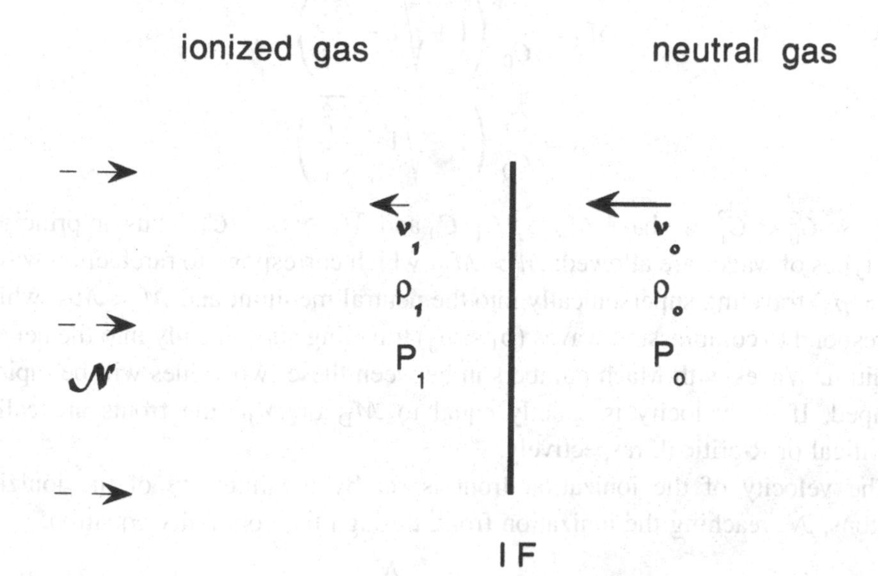

Figure 12.1 A schematic diagram of the structure of an ionization front (IF). A flux of ionizing photons, $\mathcal{N}$, shines upon neutral gas with density and pressure ρ_0 and P_0. The ionized gas density and pressure are ρ_1 and P_1. The velocities are in the frame of the ionization front.

compression time scales. Hence, we will assume that the gas in the ionization front is instantaneously heated to the same temperature as in the HII region ($\simeq$8000 K, largely independent of the characteristics of the region or the star; cf. Section 7.3). In contrast, in the surrounding neutral PDR, the temperature may range from 100 to 1000 K, depending on the ionization parameter, $G_0 T^{1/2}/n_e$, with G_0 the intensity of the incident FUV field and n_e the electron density (cf. Section 9.3).

The continuity and momentum equations can be combined to yield

$$\frac{\rho_1}{\rho_0} = \frac{1}{2}\frac{C_0^2}{C_1^2}\left[(\mathcal{M}^2+1) \pm \left((\mathcal{M}^2+1)^2 - 4\mathcal{M}^2\frac{C_1^2}{C_0^2}\right)^{1/2}\right], \tag{12.6}$$

with C_i the sound speed in medium i (typically 10 km s^{-1} in the ionized gas and 1–3 km s^{-1} in the neutral medium), and $\mathcal{M} = v_0/C_0$ is the Mach number of the ionization front in the neutral gas. These relations are somewhat different from those for adiabatic shocks because the pressures and densities are now linked through the independently set temperatures. The condition that the square root portion has to be positive yields

$$\mathcal{M}^2 - 2\mathcal{M}\frac{C_1}{C_0} + 1 > 0. \tag{12.7}$$

Thus, there are two critical Mach numbers, $\mathcal{M}_R$ and $\mathcal{M}_D$, given by

$$\begin{aligned} \mathcal{M}_R &= \frac{C_1}{C_0}\left(1 + \sqrt{1 - \frac{C_0^2}{C_1^2}}\right) \\ \mathcal{M}_D &= \frac{C_1}{C_0}\left(1 - \sqrt{1 - \frac{C_0^2}{C_1^2}}\right). \end{aligned} \tag{12.8}$$

Because $C_0^2 \ll C_1^2$, we have $\mathcal{M}_R \simeq 2C_1/C_0$ and $\mathcal{M}_D \simeq C_0/2C_1$. Thus, in principle, two types of waves are allowed: $\mathcal{M} > \mathcal{M}_R$, which correspond to rarefaction waves ($\rho_1 > \rho_0$) traveling supersonically into the neutral medium; and $\mathcal{M} < \mathcal{M}_D$, which correspond to compression waves ($\rho_1 < \rho_0$) traveling subsonically into the neutral medium. Waves with Mach numbers in between these two values will be rapidly damped. If the velocity is exactly equal to $\mathcal{M}_D$ or $\mathcal{M}_R$, the fronts are called D-critical or R-critical, respectively.

The velocity of the ionization front is set by the intensity of the ionizing photons, $\mathcal{N}$, reaching the ionization front, through the continuity equation,

$$v_0 = \mu\, m_H \frac{\mathcal{N}}{\rho_0}, \tag{12.9}$$

with μ the mean atomic weight per ion (= 1.40 when He stays neutral and = 1.27 when He is ionized). The density jump across the ionization front is then given by the smaller of the two possible solutions in Eq. (12.6); e.g., the weak R-front with the minus sign and the weak D-front with the plus sign in this equation.

12.2.2 The initial phase of rapid ionization

Consider a uniform density medium and a central star that is suddenly "turned on" to a constant ionizing photon luminosity, $\mathbb{N}_{\rm Lyc}$. Because, initially, recombination is slow, the size of the ionized region, $R_{\rm IF}$, is set by

$$\mathbb{N}_{\rm Lyc} t = \frac{4\pi}{3} n R_{\rm IF}^3(t). \tag{12.10}$$

The velocity of the ionization front is thus

$$\frac{\mathrm{d}R_{\rm IF}}{\mathrm{d}t} = \frac{1}{3}\left(\frac{3\mathbb{N}_{\rm Lyc}}{4\pi n}\right)^{1/3} t^{-2/3}. \tag{12.11}$$

So, as an example, for an HII region with a density of 10^3 cm^{-3} ionized by an O4 star, the velocity of the ionization front is $\sim 4\times 10^3$ km s^{-1} at $t = 100$ years. The velocity of the ionization front cannot exceed the velocity of light but we are ignoring propagation effects. On a time scale set by recombination, $\tau_{\rm rec} = (\beta_{\rm B} n)^{-1}$, with $\beta_{\rm B}$ the recombination coefficient to levels with $n \geq 2$, the ionization front will slow down because more and more photons will be used to keep the gas ionized inside the ionized region than to propagate the ionization front. This time scale is about 100 years at a density of 10^3 cm^{-3}.

For times approaching this recombination time scale, we have to include recombinations in our analysis as well. The change in the stellar ionizing photon flux at position r due to ionization of a shell dr is given by

$$\mathrm{d}(4\pi r^2 \mathcal{N}_\star(r)) = -4\pi r^2 \beta_{\rm B} x^2 n^2 \mathrm{d}r, \tag{12.12}$$

with x the degree of ionization, and we have made the on-the-spot approximation. For a constant density and (almost) complete ionization ($x \simeq 1$), this results in

$$4\pi r^2 \mathcal{N}_\star(r) = \mathbb{N}_{\rm Lyc} - \frac{4\pi}{3} r^3 \beta_{\rm B} n^2 = \frac{4\pi}{3}\beta_{\rm B} n^2 (\mathcal{R}_{\rm s,o}^3 - r^3), \tag{12.13}$$

where we have assumed that the density jump across the ionization front is very small. We have also introduced $\mathcal{R}_{\rm s,o}$ as the radius of the Strömgren sphere for the density of the medium and the total stellar photon luminosity, $\mathbb{N}_{\rm Lyc}$ (cf. Section 7.2). Note that the radius of the ionized region will only approach this Strömgren radius near the end of this phase (see below). We find, then, for the velocity of the ionization front (Eq.(12.9)),

$$4\pi R_{\rm IF}^2 \frac{\mathrm{d}R_{\rm IF}}{\mathrm{d}t} = \frac{4\pi\beta_{\rm B}\, n}{3}(\mathcal{R}_{\rm s,o}^3 - R_{\rm IF}^3). \tag{12.14}$$

The evolution of the ionized volume is now given by

$$R_{\mathrm{IF}}^3(t) = \mathcal{R}_{\mathrm{s,o}}^3(1 - \exp[-t/\tau_{\mathrm{rec}}]). \tag{12.15}$$

In the limit of $t \ll \tau_{\mathrm{rec}}$, this reduces to $R_{\mathrm{IF}} = \mathcal{R}_{\mathrm{s,o}}(t/\tau_{\mathrm{rec}})^{1/3}$ and $v_{\mathrm{IF}} = 1/3 R_{\mathrm{IF}}/t$. Resubstituting the expression for the Strömgren radius, we recover Eq. (12.11). Thus, the essence is that the ionized gas volume will rapidly expand until a time equal to the recombination time scale when the ionization front approaches the value of the Strömgren radius corresponding to the density and stellar ionizing photon luminosity (cf. Section 7.2.1).

Consider, again, an HII region with a density of 10^3 cm^{-3} ionized by an O4 star; the Strömgren radius is 1.2 pc, the recombination time scale is $\simeq 10^2$ years and the velocity of the ionization front is $\sim 10^4$ km s^{-1} (Fig. 12.2). It is now easy to check that the density jump across the ionization front is indeed small; assuming $v_0 \gg C_1$ and C_0 and expanding the square-root portion of Eq. (12.6) to the second order in C_1/v_0 yields $\rho_1/\rho_0 = 1 + C_1^2/v_{\mathrm{IF}}^2$. The velocity difference across the ionization front is also small, $v_{\mathrm{IF}} - v_1 \simeq C_1^2/v_{\mathrm{IF}}^2 \ll 1$. Obviously, initially, the gas has no time to react dynamically to the ionization. Only when the size of the HII regions approaches the Strömgren size will the ionization front slow down ($t \sim 4 \times \tau_{\mathrm{rec}}$). At that point, a shock front will develop ahead of the ionization front and the evolution of the HII region will transit to that described in the next section.

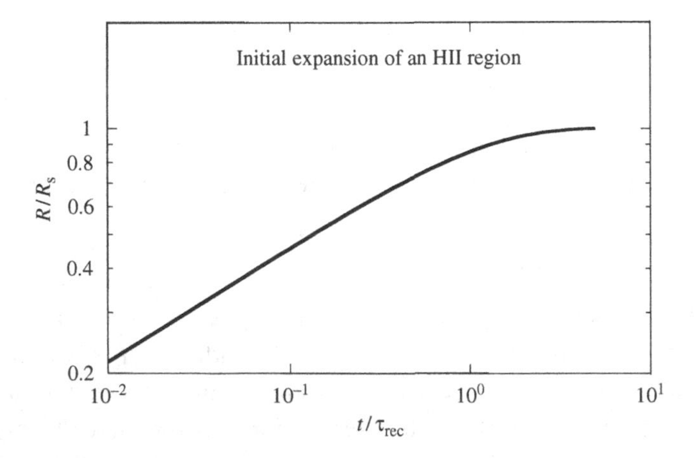

Figure 12.2 The initial expansion of the ionization front after the sudden "turn on" of a massive star in a uniform density cloud. The size is plotted (relative to the Strömgren radius) as a function of the time in units of the recombination time scale, $\tau_{\mathrm{rec}} = (\beta_{\mathrm{B}} n)^{-1}$.

12.2.3 The pressure-driven expansion of an HII region

After the initial phase of rapid expansion of the ionized gas volume, the HII region and surrounding neutral gas (PDR) are separated by a very thin (relative to the size of the HII region) ionization front, in which the degree of ionization changes rapidly from essentially completely ionized to essentially completely neutral (cf. Section 7.2). Because of the increased temperature and density, the ionized gas will have a pressure very much exceeding that of the neutral medium. As a result, a shock will be driven into the surrounding medium, which will sweep up the surrounding gas, increasing its density and hence its pressure. Because of the expansion, the average density in the HII region will drop and, as a result, the number of stellar photons required to keep the gas ionized decreases. The excess ionizing photons will drive the expansion of the ionization front, increasing the mass of ionized gas (cf. Section 7.2.1). We will ignore here the complex dynamical phase separating this pressure-driven stage from the rapid initial ionization stage and consider here the pressure-driven evolution of an HII region in a homogeneous environment.

Because of the high density, cooling behind the shock front will be fast and the swept-up gas will form a thin, dense, neutral shell – bounded on the outside by the shock front and on the inside by the ionization front – moving as a unit (recall that, because of continuity, the velocity of the shocked, cooled gas scales inversely with the density jump across the shock front; Section 11.2.1). During this phase, most of the ionizing photons will be absorbed by the (largely ionized) gas inside the HII region and the number of ionizing photons arriving at the ionization front is small. Unlike a shock front, the number density of the gas entering the ionization front is prescribed by the number of ionizing photons arriving at the ionization front (Eq. (12.9)). The number of ionizing photons arriving at the ionization front depends on the number of recombinations in the ionized gas between the star and the ionization front and thus the density of the ionized gas. The density of gas entering the ionization front is set by the surrounding medium as modified by the shock. Because of the small number of photons arriving at the ionization front and the high density in the swept-up neutral shell, the velocity of the ionization front will be small relative to this shell.

We will consider the expansion of this shell in the shock frame. The shock has brought the pressure in the shell up to the pressure of the ionized gas. If we ignore the preshock pressure of the surrounding medium and assume that the velocity of the gas behind the ionization front is small (but see below), then we can write for the momentum equation of the combined shock front and ionization front (cf. Eq. (12.2))

$$\rho_0 v_s^2 = 2n_i kT = \rho_{II} C_{II}^2, \tag{12.16}$$

with $\rho_{\rm II}$ the density of the ionized gas, $C_{\rm II} = \sqrt{P_{\rm II}/\rho_{\rm II}}$ the sound speed in the ionized gas, and $v_{\rm s}$ the velocity of the neutral gas in the frame of the shock front. We will assume that the HII region is in pressure equilibrium, which is not entirely correct. With this assumption, the expansion is uniform and at all times the density of the ionized gas, $\rho_{\rm II}$, is independent of radius; but this density decreases, of course, with time. Because ionization and recombination occur on a very rapid time scale relative to the expansion time scale, the density and size of the HII region are related through the Strömgren equation (cf. Section 7.2.1),

$$\mathcal{R}_{\rm s}(t) = \left(\frac{3\,\mathbb{N}_{\rm Lyc}}{4\pi n^2(t)\beta_{\rm B}}\right)^{1/3}, \tag{12.17}$$

with $n(t)$ the number density of hydrogen in the ionized gas. We can also write this as

$$\mathcal{R}_{\rm s}(t) = \left(\frac{\rho_0}{\rho_{\rm II}(t)}\right)^{2/3} \mathcal{R}_{\rm s,o}, \tag{12.18}$$

with $\mathcal{R}_{\rm s,o}$ the initial size of the HII region when it had the density of the surrounding medium. In the thin shell approximation, $R_{\rm IF} = R_{\rm SF} = \mathcal{R}_{\rm s}$ and $\mathrm{d}R_{\rm IF}/\mathrm{d}t = \mathrm{d}R_{\rm SF}/\mathrm{d}t = \mathrm{d}\mathcal{R}_{\rm s}/\mathrm{d}t$, where these velocities are measured in the frame of the star (and the stationary neutral medium). Realizing that $v_{\rm s} = -\mathrm{d}\mathcal{R}_{\rm s}/\mathrm{d}t$, we can combine Eqs. (12.16) and (12.18) to yield

$$\frac{\mathrm{d}\mathcal{R}_{\rm s}}{\mathrm{d}t} = C_{\rm II}\left(\frac{\mathcal{R}_{\rm s,o}}{\mathcal{R}_{\rm s}}\right)^{3/4}, \tag{12.19}$$

which can be integrated to

$$\frac{\mathcal{R}_{\rm s}}{\mathcal{R}_{\rm s,o}} = \left(1 + \frac{7}{4}\frac{t}{t_0}\right)^{4/7}, \tag{12.20}$$

where $t_0 = \mathcal{R}_{\rm s,o}/C_{\rm II}$.

As an example, consider an O4 star formed in a region with a density of $10^5\ \mathrm{cm^{-3}}$; the initial radius of the HII region is $\mathcal{R}_{\rm s,o} = 0.056$ pc (cf. Section 7.2.1). With a sound speed of $10\ \mathrm{km\,s^{-1}}$, we have $t_0 \simeq 5500$ years. After 10^6 years, the HII region has expanded to 1.5 pc and the density in the ionized gas has dropped to approximately $700\ \mathrm{cm^{-3}}$. At that point, the shock speed is approximately $0.8\ \mathrm{km\,s^{-1}}$. For an electron temperature of 8000 K, the pressure of the ionized gas is $\simeq 1.4 \times 10^7\ \mathrm{K\,cm^{-3}}$. For comparison, assuming a temperature in the surrounding medium of 30 K, the pressure of the neutral gas is approximately $3 \times 10^6\ \mathrm{K\ cm^{-3}}$, still somewhat less than the HII region pressure. However, the sound speed in the neutral medium is also about $0.8\ \mathrm{km\,s^{-1}}$ and the expansion becomes subsonic, mediated by sound waves. Of course, by that time, the compression of the shock

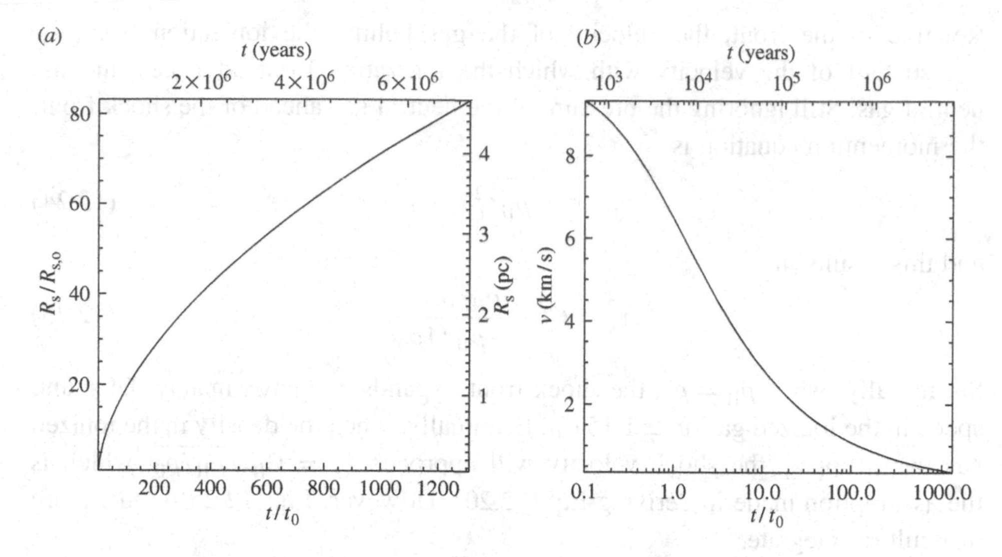

Figure 12.3 The pressure-driven expansion phase of an HII region. (*a*) Size as a function of time (on a linear scale). The top and right-hand axes refer to an HII region formed in a cloud with an initial density of 10^5 cm^{-3}. (*b*) Expansion velocity as a function of time (on a logarithmic scale).

and the thin shell approximation break down. So, by the end of the massive star's lifetime, pressure equilibrium with the surrounding medium is approximately reached.

In this analysis, we have made the approximation that the velocity of the gas behind the ionization front is small and this allowed us to simplify the momentum equation. Initially, however, this assumption is not correct. In the frame of the star, the velocity of the ionization front, $v_{\rm if}$, is (cf. Eq. (12.18))

$$v_{\rm if}(t) = -\frac{2\mathcal{R}_{\rm s}(t)}{3\rho_{\rm II}(t)}\frac{{\rm d}\rho_{\rm II}(t)}{{\rm d}t}. \tag{12.21}$$

With the assumption of uniform density in the ionized gas, the continuity equation (in spherical geometry) for the HII region is

$$\frac{1}{r^2}\frac{\partial\, r^2 v}{\partial r} = \frac{1}{\rho_{\rm II}}\frac{{\rm d}\rho_{\rm II}}{{\rm d}t}. \tag{12.22}$$

This leads to uniform expansion (Hubble's law) with v proportional to r at any time. The velocity of the gas, $V_{\rm i}$, directly behind the ionization front (relative to the star) is then

$$V_{\rm i} = -\frac{\mathcal{R}_{\rm s}}{3\rho_{\rm II}}\frac{{\rm d}\rho_{\rm II}}{{\rm d}t}. \tag{12.23}$$

Relative to the front, the velocity of the gas behind the ionization front, $v_{\rm i}$, is also half of the velocity with which the ionization front advances into the neutral gas. Still ignoring the pressure of the neutral gas ahead of the shock front, the momentum equation is

$$\rho_0 v_{\rm s}^2 = \rho_{\rm II} v_{\rm i}^2 + P_{\rm II}, \tag{12.24}$$

and this results in

$$v_{\rm s}^2 = C_{\rm II}^2 \frac{\rho_{\rm II}/\rho_0}{1 - \rho_{\rm II}/4\rho_0}. \tag{12.25}$$

So, initially, when $\rho_{\rm II} = \rho_0$, the shock front expands at approximately the sound speed in the ionized gas, $v_{\rm s} \simeq 1.15\, C_{\rm II}$. Eventually, when the density in the ionized gas has dropped, the shock velocity will approach $v_{\rm s} = C_{\rm II}\sqrt{\rho_{\rm II}/\rho_0}$, which is the assumption made in deriving Eq. (12.20). However, Eq. (12.25) is just more difficult to integrate.

The ionization front will lag a little bit behind the shock front. Applying the continuity equation across the shock front (in the frame of the shock) and across the ionization front (in the frame of the ionization front) and assuming that the density of the swept-up shell is uniform, the ratio of the gas velocity behind the shock to the gas velocity entering the ionization front is

$$\frac{v_{\rm s,1}}{v_{\rm i,o}} = \frac{\rho_0\, v_{\rm s}}{\rho_{\rm II} v_{\rm i}} = \frac{2\rho_0\, v_{\rm s}}{\rho_{\rm II}\, V_{\rm i}}. \tag{12.26}$$

In the frame of the star (and the stationary neutral medium), the velocity difference between the shock front and ionization front is

$$\frac{V_{\rm s} - V_{\rm i}}{V_{\rm s}} = \frac{v_{\rm s,1}}{V_{\rm s}}\left(1 - \frac{\rho_{\rm II} v_{\rm i}}{2\rho_0 v_{\rm s}}\right) = \frac{1}{\mathcal{M}}\left(1 - \frac{\rho_{\rm II}\, v_{\rm i}}{2\rho_0 v_{\rm s}}\right), \tag{12.27}$$

where we have used the density contrast for isothermal shocks (Eq. (11.27)) in the last step. Thus, as long as the shock is strong (e.g., $P_{\rm II} \gg P_0$), the two velocities will be nearly equal and the swept-up shell very thin. Eventually, upon approaching pressure equilibrium with the surrounding medium, the ionization front will start to lag considerably compared to the shock and the density in the shell will approach that of the surrounding medium.

Finally, we are interested in the flow of the neutral atomic and molecular gas through the swept-up shell of gas into the ionization front. In the frame of the ionization front, we can write for the momentum equation

$$\rho_{\rm pdr} v_{\rm pdr}^2 + \rho_{\rm pdr} C_{\rm pdr}^2 = \rho_{\rm II} v_{\rm i}^2 + \rho_{\rm II} C_{\rm II}^2, \tag{12.28}$$

with $\rho_{\rm pdr}$ and $v_{\rm pdr}$ the density and velocity of the gas in the swept-up shell relative to the ionization front. For simplicity, we will set $v_{\rm i}$ equal to the sound speed in the ionized gas and, using the continuity equation across the ionization front ($\rho_{II} v_i = \rho_{pdr} v_{pdr}$), we find from Eq. (12.28),

$$v_{pdr} = \frac{C_{pdr}^2}{2C_{II}}. \tag{12.29}$$

We have used this expression in Section 9.5.1 to estimate the effects of advection on the chemistry in PDRs.

Despite the many approximations, this analytical model gives a reasonable representation of numerical calculations for expansion in an initially homogeneous cloud. The effects of inhomogeneities can, however, be large and we will address these in the next sections.

12.2.4 *Ionization of globules*

Many HII regions contain small, dense globules of neutral gas. Such dense blobs may be associated with newly formed, low-mass stars such as the proplyds in the Orion HII region. Or they may be dense remnants of the surrounding cloud, which were left standing when the rest of the cloud was ionized "away" by the massive stars, as the well-known "elephant trunks" in the Eagle Nebula. In either case, an ionization front will advance slowly into the neutral material, while the freshly ionized gas can stream away freely. The ionized gas will typically reach velocities a few times the sound speed in the ionized gas. In contrast, the velocity of the ionization front is small. Hence, globules are characterized by bright rims: the ionization fronts, which are eating their way into the globule. Most of the stellar photons arriving at the globule's surface will be absorbed by neutral atoms resulting from recombinations in the expanding, mainly ionized, gas flow from the surface of the globule and only a few will make it to the ionization front.

Consider a spherical globule of mass M_g, and radius R_g, at a distance, d, from the ionizing star. The density, mass, and size are thus related through

$$R_g = \left(\frac{3M_g}{4\pi n_g m_H}\right)^{1/3}. \tag{12.30}$$

So, a globule with a mass of 10^{-2} M$_\odot$ and a radius of 10^{-2} pc has a density of 10^5 cm^{-3}, an average H column of 4×10^{21} cm^{-2}, and spans an angular size of $\simeq 4''$ at the distance of Orion. We will assume that the globule is not immediately completely ionized; e.g.,

$$\mathcal{N}_\star < 2R_g \beta_B n_g^2, \tag{12.31}$$

with the incident stellar photon flux, $\mathcal{N}_\star = \mathrm{N}_{Lyc}/4\pi d^2$. In this equation, we have ignored absorption in the HII region, made the on-the-spot approximation, and ignored absorption by dust in the globule. We will focus here on Orion powered by $\Theta^1 C$; an O6.5 star for which we will adopt a total ionizing photon luminosity of 2×10^{49} photons s^{-1} (cf. Table 7.1). Hence, for a given mass and size, the globule has to be farther from the star than $d_{\rm min}$, given by

$$d_{\rm min} = 3 \times 10^{-2} \left(\frac{\mathbb{N}_{\rm Lyc}}{2 \times 10^{49} {\rm s}^{-1}} \right)^{1/2} \left(\frac{10^{-2}\, M_\odot}{M_{\rm g}} \right) \left(\frac{R_{\rm g}}{10^{-2}\,{\rm pc}} \right)^{5/2} {\rm pc}, \qquad (12.32)$$

or some 20″ at the distance of Orion.

The ionization front will drive a shock front into the globule. The globule will be compressed, reach pressure equilibrium with the HII region, and settle into a cometary configuration on a dynamical time scale. The shape of the surface of the globule will depend on the fraction of incident photons absorbed in this evaporating surface layer (by recombined atoms). Conversely, the density distribution in the evaporating layer will depend on the shape of the ionization front. While this problem can be solved, we will focus on the evaporation lifetime of the globule as a whole, approximating it with the properties of the tip where we can adopt an approximate spherical geometry. We will adopt a coordinate system centered on the globule with a radius for the ionization front of $R_{\rm if}$. Because of the compression, this radius will be less than the initial radius, while the neutral gas density will be higher. Ignoring the internal structure of the globule, this radius and the neutral gas density are related through the mass of the globule,

$$R_{\rm if} = \left(\frac{3 M_{\rm g}}{4\pi n_{\rm i} m_{\rm H}} \right)^{1/3}. \qquad (12.33)$$

The flow through the ionization front is set by the number of ionizing photons, $\mathcal{N}$, arriving at the ionization front (Eq. (12.9)), for which we will write $\mathcal{N} = n(R_{\rm if}) C_{\rm II}$, with $n(R_{\rm if})$ the density at the ionization front and $C_{\rm II}$ the sound speed in the gas. Thus, we approximate that the velocity of the ionized gas relative to the ionization front is of order $C_{\rm II}$. The number of ionizing photons arriving at the ionization front is equal to the incident photon flux where the outflowing ionized gas merges with the HII region (e.g., the number of ionizing photons that would be at this position in the nebula in the absence of the globule), $\mathcal{N}_\star$, minus the number of photons required to keep this ionized flow from the globule ionized; i.e.,

$$\mathcal{N}_\star = \int_{R_{\rm if}}^{\infty} \beta_{\rm B} (n(R))^2 {\rm d}R + n(R_{\rm if}) C_{\rm II}. \qquad (12.34)$$

Assuming spherical divergence from the surface of the globule, this yields

$$\mathcal{N}_\star = \frac{1}{3} \beta_{\rm B} (n(R_{\rm if}))^2 R_{\rm if} + n(R_{\rm if}) C_{\rm II}. \qquad (12.35)$$

Ignoring the last term (e.g., (almost) all the incident EUV photons are required to keep the flow ionized), this can be simplified to

$$n(R_{\rm if}) = \left(\frac{3\mathcal{N}_\star}{\beta_{\rm B} R_{\rm if}}\right)^{1/2}. \tag{12.36}$$

In the momentum equation for the ionization front, we will ignore the flow term associated with the motion of the ionization front into the compressed neutral globule and replace the terms on the ionized side by $\rho_{\rm II} C_{\rm II}^2$, which yields $n_{\rm I} C_{\rm I}^2 = n(R_{\rm if}) C_{\rm II}^2$. We can then express the parameters of the ionized flow in terms of the mass and temperature of the globule,

$$n(R) = \left(\frac{3\mathcal{N}_\star}{\beta_{\rm B}}\right)^{3/5} \left(\frac{4\pi m_{\rm H} C_{\rm II}^2}{3M_{\rm g}\, C_{\rm I}^2}\right)^{1/5} \tag{12.37}$$

and

$$R_{\rm if} = \left(\frac{3M_{\rm g} C_{\rm I}^2}{4\pi C_{\rm II}^2 m_{\rm H}}\right)^{2/5} \left(\frac{\beta_{\rm B}}{3\mathcal{N}_\star}\right)^{1/5}. \tag{12.38}$$

Consider, as a numerical example, a globule at a distance of 0.1 pc from an O6.5 star; e.g., $\mathcal{N}_\star \simeq 1.7\times 10^{13}$ photons cm^{-2} s^{-1}. For a dense PDR illuminated by a strong FUV field, the temperature ratio between ionized and neutral gas is a factor $\simeq$10. This then yields

$$n(R) \simeq 1.2\times 10^5 \left(\frac{10^{-2}\, M_\odot}{M_{\rm g}}\right)^{1/5} \ {\rm cm}^{-3}, \tag{12.39}$$

while the neutral gas density is a factor of 10 higher. The radius of the globule is

$$R_{\rm if} \simeq 4.4\times 10^{-3} \left(\frac{M_{\rm g}}{10^{-2}\, M_\odot}\right)^{2/5} \ {\rm pc}, \tag{12.40}$$

and the neutral column through the globule is typically 2×10^{22} cm^{-2}. The ionizing flux that reaches the ionization front is

$$\mathcal{N} \simeq 1.2\times 10^{11} \left(\frac{10^{-2}\, M_\odot}{M_{\rm g}}\right)^{1/5} \ {\rm photons\,cm^{-2}\,s^{-1}}. \tag{12.41}$$

For a background density of ionized gas of 3×10^3 cm^{-3}, the ionized gas flow from the globule has to expand by a factor of 6 to a radius of $\simeq 3\times 10^{-2}$ pc to merge.

We will approximate the mass loss from the front surface of the globule by

$$\frac{dM_g}{dt} = -\pi R_{if}^2 \mathcal{N} m_H, \tag{12.42}$$

which ignores the detailed shape of the ionization front. This can be written as

$$\frac{dM_g}{dt} = -\pi R_{if}^{3/2} \left(\frac{3\mathcal{N}_\star}{\beta_B}\right)^{1/2} C_{II} m_H, \tag{12.43}$$

which leads to

$$\frac{dM_g}{dt} \simeq -1.8 \times 10^{-6} \left(\frac{M_g}{10^{-2} M_\odot}\right)^{3/5} M_\odot\, year^{-1}. \tag{12.44}$$

Thus, globules survive only a short time in an HII region ($\tau \simeq 6 \times 10^3 \left(M_g/10^{-2} M_\odot\right)^{2/5}$ years), unless they are much more massive – e.g., more centrally condensed (or contain a more condensed object) – or further away from the ionizing star.

The proplyds in the Orion HII region represent a special case of such globules. These proplyds are planetary-disk-like structures surrounding newly formed, low-mass stars, which are being ionized by the Trapezium stars. In this case, the effect of gravity due to the low-mass star at the center of the disk has to be taken into

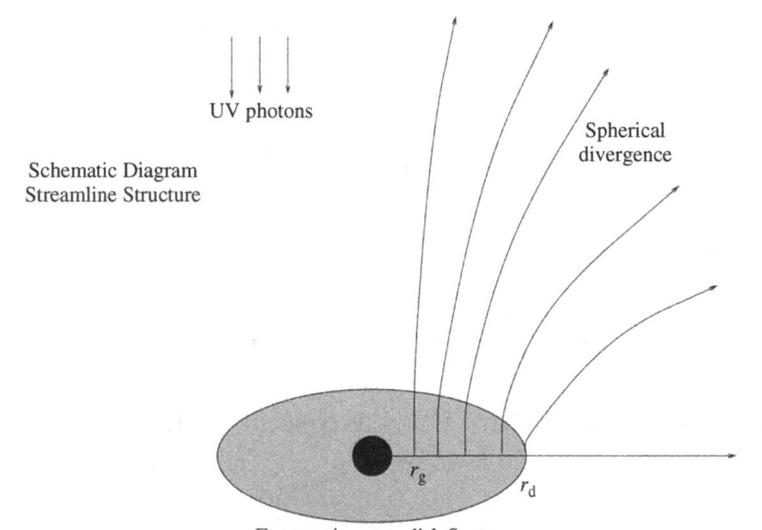

Figure 12.4 A schematic diagram of the structure of the flow from the surface of a proplyd within an HII region. While near the surface, the flow will be cylindrical; eventually, the flow will become spherically symmetrical. Figure reproduced with permission from D. Johnstone, D. Hollenbach, and J. Bally, 1998, *Ap. J.*, **499**, p. 758.

account. Evaporation can then only occur for regions outside of a critical disk radius,

$$r_{\rm d,cr} = \frac{GM_\star}{C_{\rm II}^2}, \tag{12.45}$$

where the thermal energy of the ionized gas exceeds the gravitational energy. This critical radius is about 10^{14} cm (6 AU) for a 1 $M_\odot$ star. The ionizing flux will set up an evaporative ionized flow with a mass loss rate of

$$\frac{dM_g}{dt} \simeq 6\times10^{-9}\left(\frac{\mathbb{N}_{\rm Lyc}}{2\times10^{49}\,\rm s^{-1}}\right)^{1/2}\left(\frac{10^{17}\,\rm cm}{d}\right)\left(\frac{r_d}{10\,\rm AU}\right)^{3/2} M_\odot\, year^{-1}, \tag{12.46}$$

with r_d the size of the disk. More extensive studies, taking the acceleration through the sonic point into account, result in a smaller critical disk radius, $r_{\rm d,cr} = 1/5GM_\star/C_{\rm II}^2$, and a concomitantly larger mass loss rate.

Here we have assumed that the ionization front is located near the disk surface. Actually, if the flux of FUV photons at the disk surface is large enough, the heating by these lower energy photons (produced by the photo-electric effect on PAHs and small dust grains) may set up a neutral evaporative flow for radii larger than about $1/5\times10^{15}$ cm (cf. Eq. (12.45) for the neutral sound speed at $\sim10^3$ K), originating in the PDR layers of the disk. The gas, flowing at about the speed of sound of the neutral gas ($\simeq 3\,\rm km\,s^{-1}$), is then ionized at about several disk radii above the disk and, at that point, further accelerated to the sound speed in the ionized gas. In this case, the gas flowing away from the disk surface will attenuate the FUV flux, which is required to heat the gas to the escape velocity. Setting the attenuation column to $N_{\rm PDR} = 10^{21}\,\rm cm^{-2}$, typical for PDRs, and assuming spherical divergence, we have for the density at the base of the neutral evaporative flow, $n(R_b) = N_{\rm PDR}/R_b$. If ionization is important, then the neutral gas will flow through an ionization point well above the disk and is subsequently accelerated to the sound speed in the ionized gas and the discussion follows the lines set out above. For purely neutral flow, we would have a mass loss rate that is now $\dot{M}_{\rm PDR} = 2\pi N_{\rm PDR}(GM_\star R_{\rm out})^{1/2}$, with $R_{\rm out}$ the outer radius of the disk. For an outer radius of 10^{15} cm, the neutral mass loss rate would be $6\times10^{-8}\,M_\odot$ year^{-1}. Figure 12.5 illustrates the calculated structure of the neutral–ionized flow from the surface of a protoplanetary disk. The carbon ionization front is near the surface of the disk, while the hydrogen ionization front stands off at several radii. The gas is accelerated between the two ionization fronts by the strong pressure gradient because of spherical expansion. Further acceleration takes place when the gas crosses the hydrogen ionization front. Note also the cometary geometry set up by the flow, pointing away from the star.

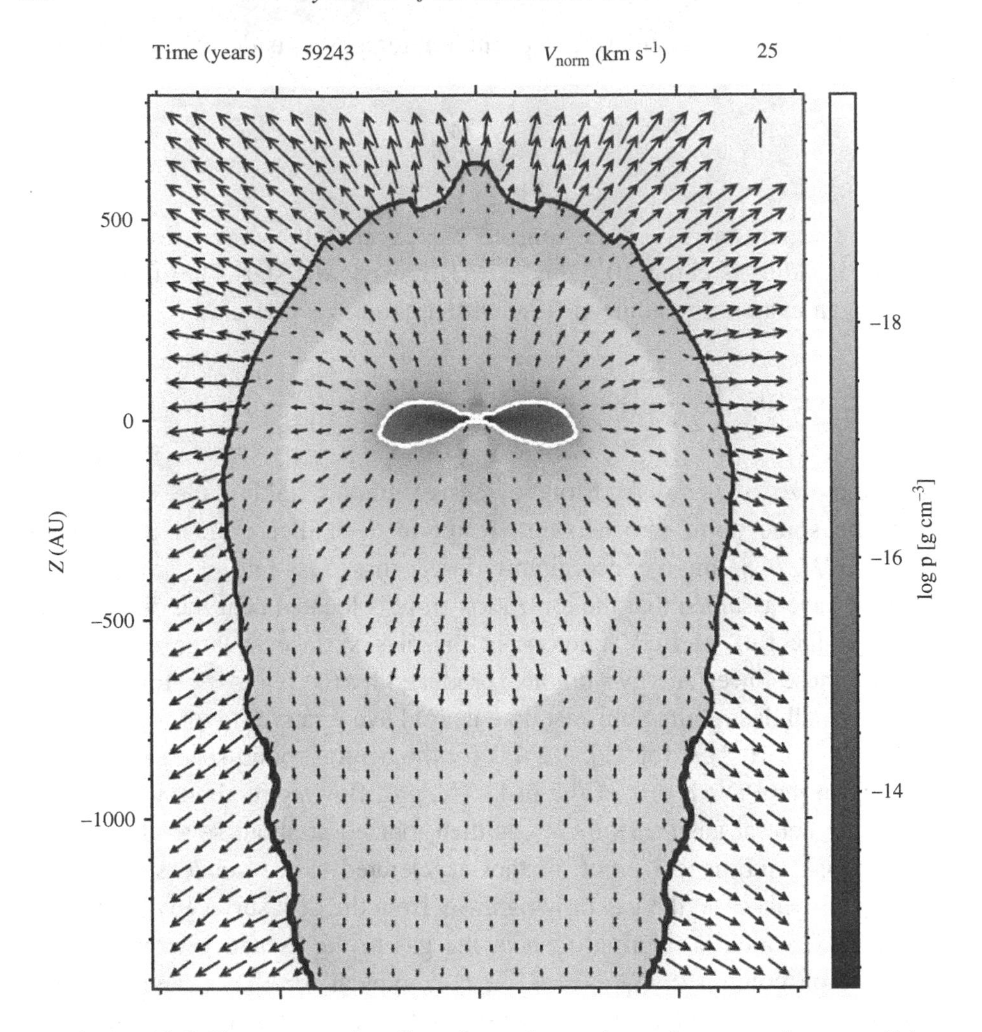

Figure 12.5 The evaporative flow from the surface of a protoplanetary disk owing to external illumination by an EUV–FUV radiation field of a star at a distance of 0.1 pc along the symmetry axis at a time of $\sim 6 \times 10^4$ years. The size scale is in AU. The gray scale represents the density while the arrows show the velocity field. The irregular dark line marks the location of the H-ionization front, separating the ionized gas on the outside from the photodissociation material on the inside. The white contour is the location of the C-ionization front. Figure reproduced with permission from S. Richling and H. W. Yorke, 2000, *Ap. J.*, **539**, p. 258.

The models for a PDR-initiated ionized flow are in good agreement with the observed characteristics of the proplyds in Orion. Typical derived mass loss rates are $\simeq 10^{-7}$ $M_\odot$ year^{-1}. With an estimated mass in the protoplanetary disk of $\sim 10^{-2}$ $M_\odot$, the lifetime is $\sim 10^5$ years. If there is no neutral flow (e.g., the PDR temperature is not high enough to cause the initial neutral gas evaporation), the ionization front is located near the disk surface and the disk lifetime against evaporation is $\sim 2 \times 10^6$ years.

12.2.5 The ionization of winds and accretion flows

The formation of massive stars is expected to lead to an accretion flow in which the density increases inwards with $r^{-3/2}$. Likewise, O-type stars can also have strong, constant velocity winds, which set up an r^{-2} density law. Indeed, the environment of newly formed massive stars may be characterized by infall in some directions and outflow in others. The presence of a density distribution will influence the evolution of the ionized gas volume. Consider a region with a density distribution given by

$$n(r) = n_0 \left(\frac{r}{r_0} \right)^{-\alpha}. \tag{12.47}$$

We can evaluate the size of the HII region in the on-the-spot approximation,

$$\mathbb{N}_{\rm Lyc} = 4\pi\beta_{\rm B} \int_{r_0}^{r_{\rm II}} r^2\, n^2 {\rm d}r, \tag{12.48}$$

with $\mathbb{N}_{\rm Lyc}$ the Lyman photon luminosity flowing through the surface at r_0 (corrected for the advection of gas). For an accreting envelope ($\alpha = 3/2$), this yields

$$\frac{r_{\rm II}}{r_0} = \exp\left[\left(\frac{\dot{M}_{\rm cr}}{\dot{M}} \right)^2 \right], \tag{12.49}$$

where $\dot{M}$ is the accretion rate at r_0,

$$\dot{M} = 4\pi r_0^2 n_0 m_{\rm H} v, \tag{12.50}$$

and the critical mass accretion rate is given by

$$\dot{M}_{\rm cr} = \left(\frac{4\pi r_0 \mathbb{N}_{\rm Lyc}}{\beta_{\rm B}} \right)^{1/2} m_{\rm H} v = \left(\frac{8\pi G M \mathbb{N}_{\rm Lyc}}{\beta_{\rm B}} \right)^{1/2} m_{\rm H}. \tag{12.51}$$

Here, we have inserted the free-fall velocity to arrive at the right-hand side. Thus, when the mass accretion rate is much larger than this critical mass accretion rate, the ionized volume is trapped in the infalling envelope. A similar situation occurs for a wind. With $\alpha = 2$, the size of the ionized volume is

$$\frac{r_{\rm II}}{r_0} = \left(1 - \left(\frac{\dot{M}_{\rm cr}}{\dot{M}} \right)^2 \right)^{-1}. \tag{12.52}$$

Here, again, if the mass loss rate exceeds the critical mass loss rate, the ionized volume is trapped in the wind. Thus, in both situations – mass accretion or mass loss – only when the mass accretion or loss rate is less than the critical rate can the ionizing photons escape the immediate stellar environment and ionize the surrounding medium.

Inserting numerical values for an O5 star ($M = 40\,M_\odot$, $\mathbb{N}_{\rm Lyc} = 3 \times 10^{49}$ photons s^{-1}), we arrive at a critical mass accretion or loss rate of $\simeq 10^{-4}$ $M_\odot$ year^{-1}. The ionizing luminosity is a strong function of the stellar mass. Thus, initially, during the formation process, the ionized volume will be trapped near the stellar photosphere. Once the photon luminosity has grown large enough, or the accretion rate has dropped, or a stellar wind has "cleaned" the environment, an ionization front will rapidly expand, as described in Section 12.2.2. Eventually, the pressure of the HII region takes over driving the expansion of the ionized gas volume (Section 12.2.3).

12.2.6 Blister HII regions

Here, we will consider the evolution of an HII region when a massive star is formed near the boundary of a molecular cloud. Initially, the ionization front will proceed as described before with a phase in which an R-type ionization front will rapidly ionize the immediate surroundings of the star (Section 12.2.2), followed by a phase in which the over-pressure of the ionized gas drives the expansion (Section 12.2.3). When the ionization front reaches the (near) boundary of the cloud, in view of the density contrast, an R-type ionization front will rapidly move into the surrounding intercloud medium. The pressure differential between cloud and intercloud medium will drive a strong isothermal shock into the intercloud medium, while a rarefaction wave will move into the ionized cloud gas (Fig. 12.6). As a result, the ionized cloud material will start to flow into the intercloud medium; the onset of the so-called "champagne" flow. This rarefaction wave will set up an exponential density drop in the ionizing gas. Initially, the ionization front moving into the cloud at the other end will not notice these changes in the structure of the HII region. However, once the rarefaction wave has reached the material between the star and the cloud, the ionized gas density will drop. As a result, the number of ionizing photons reaching the ionization front will increase and this front will speed up. Eventually, the whole HII region will have been evacuated and the cloud gas flowing through the ionization front will immediately expand into the surrounding intercloud medium.

We can analyze the conditions in analogy to a centered rarefaction wave in an isothermal shock tube (Fig. 12.6). Consider a one-dimensional tube with a high density region with ρ_0 for $r > 0$, which is separated by a contact discontinuity

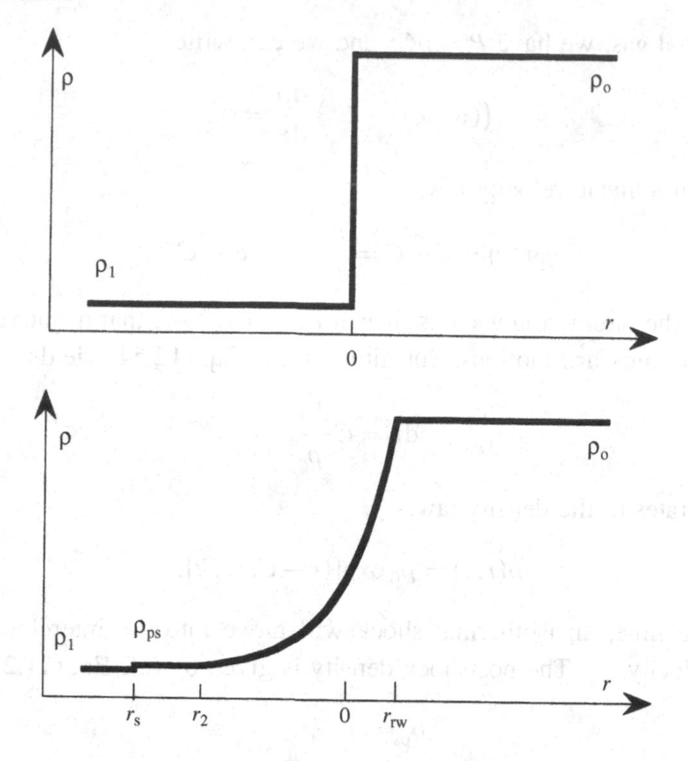

Figure 12.6 A one-dimensional representation of the density profile in a blister HII region upon breakout (top) and at a later time (bottom). The locations of the shock front (r_s) and rarefaction wave (r_{rw}) are indicated. r_2 is the location of the contact discontinuity.

from a low density region ρ_1 ($<\rho_0$) for $r<0$. Because heating is dominated by ionization of hydrogen, we will assume that the gas is isothermal. The gas is at rest at $t=0$. For $t>0$, a centered rarefaction wave will move at constant velocity into the dense cloud gas. This flow is self-similar (there is no preferred length or time scale in this ideal situation). Because the boundary and initial conditions do not contain inherent length or time scales, the flow can only depend on the combination r/t. Introducing the independent variable $\xi = r/t$ and realizing that $t\partial/\partial r = \mathrm{d}/\mathrm{d}\xi$ and $t\partial/\partial t = -\xi\mathrm{d}/\mathrm{d}\xi$, the equation of continuity and momentum for the flow (Eq. (12.1)) can be written as

$$(v-\xi)\frac{\mathrm{d}\rho}{\mathrm{d}\xi} = -\rho\frac{\mathrm{d}v}{\mathrm{d}\xi} \tag{12.53}$$

and

$$(v-\xi)\rho\frac{\mathrm{d}v}{\mathrm{d}\xi} = -\frac{\mathrm{d}P}{\mathrm{d}\xi}. \tag{12.54}$$

For isothermal gas, we have $P = \rho C^2$ and we can write

$$\left((v-\xi)^2 - C^2\right)\frac{\mathrm{d}\rho}{\mathrm{d}\xi} = 0. \tag{12.55}$$

We find then a linear velocity law,

$$v(r) = \xi - C = \frac{r}{t} - C \quad \xi < C. \tag{12.56}$$

The head of the rarefaction wave is then at $r_{\rm rw} = Ct$. Note that negative velocities correspond to leftward motions. Substitution into Eq. (12.54) yields

$$\mathrm{d}v = C\frac{\mathrm{d}\rho}{\rho}, \tag{12.57}$$

which integrates to the density law,

$$\rho(r,t) = \rho_0 \exp[(r - Ct)/Ct]. \tag{12.58}$$

At the same time, an isothermal shock will move into the intercloud gas with constant velocity, $v_{\rm s}$. The postshock density is given by (cf. Eq. (11.27))

$$\rho_{\rm ps} = \rho_1 \mathcal{M}^2, \tag{12.59}$$

with $\mathcal{M} = V_{\rm s}/C$. The postshock velocity, $v_{\rm ps}$, is readily determined from the equation of continuity evaluated in the frame of the shock and transformed to the frame of rest. Expressed relative to the sound speed, this gives

$$\mathcal{M}_{\rm ps} = -\mathcal{M} + \frac{1}{\mathcal{M}}. \tag{12.60}$$

The density and velocity of the rarefaction wave and the shock wave will join smoothly at a contact discontinuity located at r_2, where

$$\begin{aligned} \rho_1 \mathcal{M}^2 &= \rho_0 \exp[(r_2 - Ct)/Ct] \\ \frac{r_2}{Ct} - 1 &= -\mathcal{M} + \frac{1}{\mathcal{M}} \end{aligned} \tag{12.61}$$

(cf. Eqs. (12.58) and (12.59) and Eqs. (12.56) and (12.60)). These equations can be combined to relate the Mach number of the shock to the initial density contrast,

$$\frac{\rho_0}{\rho_1} = \mathcal{M}^2 \exp\left[\mathcal{M} - \frac{1}{\mathcal{M}}\right]. \tag{12.62}$$

This relation is shown in Fig. 12.7. If we assume that before ionization the cloud and intercloud had a density contrast of 10^3, then the Mach number of the shock is $\simeq 4$.

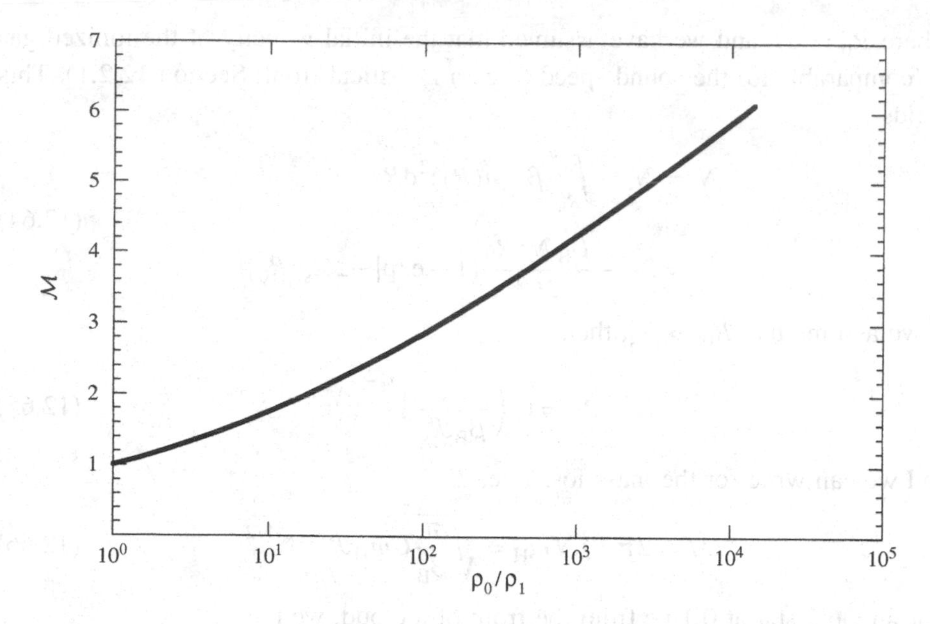

Figure 12.7 The Mach number of the shock expanding into the intercloud medium as a function of the initial cloud–intercloud density contrast.

Thus, the pressure gradient will accelerate the ionized gas to velocities of about four times the sound speed in the ionized gas ($\simeq 30\,\mathrm{km\,s^{-1}}$). The result is a blister-type HII region: ionization-bounded on the side of the cloud and density-bounded in the direction of the "champagne" flow. The ultimate size reached by the HII region will be determined by the actual density profile. In principle, it could become as large as ~100 pc into the intercloud direction while it may still be less than a pc in the cloud direction. Thus, these blisters will provide effective "channels" for the propagation of ionizing photons into the intercloud medium. Both the size and the velocity are much larger in a blister geometry than in the classical pressure-driven expansion. These simple considerations are in good agreement with numerical studies.

Once the rarefaction wave reaches the ionization front propagating into the cloud, the blisters will also very effectively erode the molecular cloud and impart kinetic energy to the gas. This can be evaluated in a manner very similar to the erosion of a globule. Let us define again the stellar flux that would reach the ionization front in the absence of absorptions as $\mathcal{N}_\star = \mathrm{N}_{\mathrm{Lyc}}/4\pi R_s^2$. The actual flux reaching the ionization front is again determined by solving the ionization balance in the intervening gas. We will adopt an exponential density gradient between the star and ionization front (Eq. (12.58)),

$$n(R) = \frac{\mathcal{N}}{C}\exp[(R - \mathcal{R}_s)/R_0], \qquad (12.63)$$

where $R_0 = Ct$, and we have assumed that the initial velocity of the ionized gas is (comparable to) the sound speed (e.g., a D-critical front; Section 12.2.1). This yields

$$\mathcal{N} = \mathcal{N}_\star - \int_{R_{\rm if}}^{\infty} \beta_{\rm B}(n(R))^2 {\rm d}R$$
$$= \mathcal{N}_\star - \frac{\beta_{\rm B}\mathcal{N}^2 R_0}{2C^2}(1 - \exp[-2\mathcal{R}_{\rm s}/R_0]). \tag{12.64}$$

If we assume that $R_0 \gg \mathcal{R}_{\rm s}$, then

$$\mathcal{N} = C\left(\frac{\mathcal{N}_\star}{\beta_{\rm B}\mathcal{R}_{\rm s}}\right)^{1/2}, \tag{12.65}$$

and we can write for the mass loss rate,

$$\dot{M} = 2\pi\mathcal{R}_{\rm s}^2\mathcal{N} m_{\rm H} = \sqrt{\frac{\pi}{\beta_{\rm B}}} C m_{\rm H} \mathcal{R}_{\rm s}^{1/2} \mathrm{N}_{\rm Lyc}^{1/2}. \tag{12.66}$$

For an O6.5 star at 0.1 pc from the front of a cloud, we have

$$\dot{M} \simeq 2\times 10^{-4}\left(\frac{\mathcal{R}_{\rm s}}{0.1\ {\rm pc}}\right)^{1/2}\left(\frac{\mathrm{N}_{\rm Lyc}}{2\times 10^{49}\ {\rm s}^{-1}}\right)^{1/2} M_\odot\, year^{-1}. \tag{12.67}$$

Of course, as the blister evolves, the mass loss rate for the cloud will increase and can reach values of $10^{-3}\ M_\odot$ year^{-1} over a substantial portion of the main sequence lifetime of the star ($\sim 3\times 10^6$ years). So, the total molecular cloud mass affected could be some $3\times 10^3\ M_\odot$. The total kinetic energy input into the ionized gas is, then, $E_{\rm k} \simeq 3\times 10^{49}$ erg.

12.2.7 Ultracompact HII regions

Observations of HII regions have been extensively discussed in Chapter 7. In that chapter, HII regions were divided into different classes based on their density and size. Here, we focus on one class, the ultracompact HII regions with densities and sizes of $10^4\ {\rm cm}^{-3}$ and 0.05 pc, respectively. These HII regions are still deeply embedded in their parental cloud and much of the stellar luminosity is therefore converted into IR radiation. The IRAS mission and follow-up radio studies have provided a census of these objects throughout the Milky Way. Radio observations reveal a range of morphologies (Fig. 12.8), ranging from cometary to shell-like sources and also including irregular, spherical or unresolved, and core-halo sources. The large number of ultracompact HII regions discovered by IRAS indicates lifetimes of 10^5–10^6 years, which is incompatible with the pressure-driven expansion age of HII regions (Section 12.2.3). Such an HII region is only expected to last some 5000 years (cf. Eq. (12.20)), which is less than 2×10^{-3} of

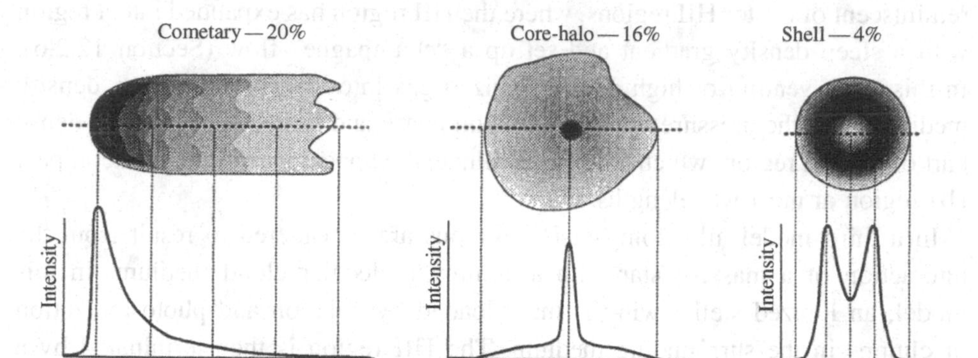

Figure 12.8 Schematic representation of the various morphologies of ultracompact HII regions. The percentages give the relative number of sources in each class. Figure reproduced with permission from D. O. S. Wood and E. Churchwell, 1989, *Ap. J. S.*, **69**, p. 831.

the lifetime of the star. Various solutions have been proposed for this confinement problem.

Many of these ultracompact HII regions have a cometary appearance with a dense rim in one direction and a more diffuse tail in the other. Such a structure may result from a star moving supersonically into the molecular cloud. If the star has a strong stellar wind, this will create a bow shock on the leading edge and a tail in the other direction. The "size" of the HII region is then essentially the stand-off distance for the shock, where the ram pressure of the stellar wind, $\rho_w v_w^2$, and the interstellar medium, $\rho_0 v_\star^2$, are equal (with ρ_w the wind density at that distance and $v_\star$ the stellar velocity). For a spherically expanding wind this can be written as

$$R_{bs} = \sqrt{\frac{\dot{M}_w v_w}{4\pi\rho_0 v_\star^2}}. \tag{12.68}$$

For typical parameters ($\dot{M}_w = 10^{-6}\,M_\odot\,\text{year}^{-1}$, $v_w = 2000\,\text{km s}^{-1}$, $n_0 = 10^4\,\text{cm}^{-3}$, and $v_\star = 10\,\text{km s}^{-1}$; Section 12.2.6), the stand-off distance is $\simeq 0.04\,\text{pc}$, comparable to the size of the ultracompact HII regions. Of course, for an ultracompact HII region to last for 10^5 years, the dense ($10^4\,\text{cm}^{-3}$) core has to extend 1 pc in size.

Alternatively, the cometary morphology of ultracompact HII regions is also reminiscent of blister HII regions, where the HII region has expanded into a region with a steep density gradient and set up a "champagne" flow (Section 12.2.6). In this case, venting of high density ionized gas into a surrounding low density medium takes the pressure off the expansion phase and limits the size of the dense part of the HII region, which tends to dominate the appearance of the ultracompact HII region at most wavelengths.

In a third model, ultracompact HII regions are considered to result from the interaction of a massive star with a clumpy molecular cloud medium. In this model, an ionized stellar wind is mass loaded by ablation and photo-ionization of clumps in the surrounding medium. The HII region is then terminated by a recombination front. Adopting an appropriate mass loading rate, the ionized region can then be "long-lived." With a mass loss rate per clump of $10^{-6}\,M_\odot$ year^{-1} (cf. Section 12.2.4), about 150 of such clumps are needed to keep the size of the HII region to $\sim 10^{17}$ cm. The different observed morphologies of ultracompact HII regions can be reproduced by varying the distribution of clumps and the density distribution in the environment. With a lifetime of ultracompact HII regions of 10^5 years, the mass of such globules has to be of order 1 $M_\odot$.

Finally, evaporative flows from massive neutral disks surrounding the ionizing star provide a natural size scale and lifetime for ultracompact HII regions. Disk sizes of $>10^{15}$ cm and masses of 2–10 $M_\odot$ are required to explain the typical characteristics of ultracompact HII regions. The observed morphologies of the bipolar HII regions, S 106 and NGC 7538 IRS 1, provide very convincing support for such a model. However, for other sources, the ionizing star is clearly displaced from the densest structures and this does not seem to be a good general model.

In summary, many models seem to be able to explain the observed characteristics of ultracompact HII regions. To some extent, this reflects the difficulty associated with observing these types of objects at the high spatial and spectral resolution required to differentiate between these models. However, it should also be recognized that this class of objects is not very homogeneous and this, by itself, suggests more than one possible origin.

12.3 Supernova explosions

At the end of its life, a high mass ($M > 8M_\odot$) star ejects a large fraction of its mass (1–5 $M_\odot$) in an explosive event. Velocities of the ejecta are some 10^4 km s^{-1} and the typical kinetic energy associated with this explosion is some 10^{51} erg. The ejecta will expand into the surrounding medium. The interaction of the ejecta with the surrounding medium can be divided into three stages. In the first stage, the mass of the ejecta greatly exceeds that of the swept-up interstellar medium and the interstellar medium has little effect. In the second stage, when the mass of swept-up interstellar material exceeds that of the ejecta, the expansion will start to slow down but the expansion is still rapid and the shocked gas has no time to cool. Expansion is then adiabatic. In the third phase, the expansion has slowed down so much that radiative energy losses become important and the shock is essentially isothermal. During these three phases, the pressure of the ISM is negligible compared to the pressure of the supernova remnant but, eventually, the supernova remnant will have slowed down so much that its pressure equals that of the surrounding medium. It will then merge with the ISM. Each of these three phases is described by different relations between the characteristics of the flow and those of the surrounding medium. In this section, we will analyze the flow characteristics for a supernova ejecting M_s with kinetic energy E_{sn} (10^{51} erg) into a homogeneous intercloud medium with density ρ_0 (1.1×10^{-24} g cm^{-3}; e.g., an atomic hydrogen number density of $n_0 = 0.5\,\text{cm}^{-3}$) and temperature T_0 (8×10^3 K). In the next section, we will examine the physical processes that become important when the interstellar medium is inhomogeneous and their effect on the dynamics of supernova remnants. The role of supernovae in regulating the global structure of the interstellar medium has already been discussed in Section 8.5.

12.3.1 The initial phase

During the first phase, the ejecta completely overpower the surrounding ISM and the velocity of the ejecta does not change. This phase ends when the mass of the swept-up material equals that of the ejecta,

$$R_s(\text{phase I}) \simeq \left(\frac{3M_s}{4\pi\rho_0}\right)^{1/3}. \tag{12.69}$$

The thickness of the shell of swept-up material is

$$\frac{\Delta R_s}{R_s} = \frac{\rho_0}{3\rho_1} = \frac{1}{12}, \tag{12.70}$$

where we have adopted the density jump for a strong shock (Eq. (11.17)). The shell of swept-up material is thus quite thin. For an ejecta mass of 1 $M_\odot$, R_s(phase I) is 2.4 pc. With an ejecta velocity of

$$v_{sn}(\text{phase I}) = 10^4 (E_{sn}/10^{51})(1\,M_\odot/M)^{1/2}\,\text{km s}^{-1}, \tag{12.71}$$

this phase ends some 240 years after the explosion. At this point, a reverse shock will have traveled through the ejecta, conveying the presence of an interstellar medium. The kinetic energy of the ejecta has now been thermalized and has created a bubble of hot gas.

12.3.2 The adiabatic expansion phase

In this second phase – the so-called Sedov–Taylor phase – the bubble of hot gas will expand because of its high pressure. As long as the shock velocity is above $\sim 250\,\mathrm{km\,s^{-1}}$ – corresponding to a temperature of 10^6 K – gas cooling is inefficient because the most abundant atoms are fully ionized (see Chapter 2). During this phase, the accumulating mass of swept-up material will slow down the expansion but the total energy is conserved. To gain insight in the characteristics of this phase, we will therefore start by examining the energy conservation law. The kinetic energy of the swept-up material is

$$E_\mathrm{k} = \frac{1}{2}\int_0^{R_\mathrm{s}} 4\pi R^2 \rho v^2 \mathrm{d}R = \frac{2\pi}{3} R_\mathrm{s}^3 \rho_0 v_\mathrm{s}^2\, \phi_\mathrm{k}, \tag{12.72}$$

where ϕ_k is defined as the average kinetic energy in units of Mv_s^2,

$$\phi_\mathrm{k} = \frac{1}{Mv_\mathrm{s}^2}\int_0^{R_\mathrm{s}} 4\pi R^2\, \rho\, v^2\, \mathrm{d}R, \tag{12.73}$$

with M and v_s the mass of the supernova remnant and the shock velocity and where we have ignored the mass of the ejecta relative to that of the swept-up interstellar medium. The internal energy of the swept-up gas is

$$E_\mathrm{i} = \int_0^{R_\mathrm{s}} 4\pi R^2 \frac{P}{\gamma - 1}\mathrm{d}R = \frac{4\pi}{3} R_\mathrm{s}^3 \rho_0 v_\mathrm{s}^2 \frac{2}{\gamma^2 - 1}\phi_\mathrm{i}, \tag{12.74}$$

where ϕ_i is the average pressure of the gas in the hot bubble in units of the postshock pressure, P_1,

$$\phi_\mathrm{i} = \frac{3}{4\pi R_\mathrm{s}^3 P_1}\int_0^{R_\mathrm{s}} 4\pi R^2 P \mathrm{d}R, \tag{12.75}$$

and we have used $P_1 = 2\rho_0 v_\mathrm{s}^2/(\gamma + 1)$. The ratio of the kinetic to the internal energy of the blast wave is then

$$\frac{E_\mathrm{k}}{E_\mathrm{i}} = \frac{(\gamma^2 - 1)\phi_\mathrm{k}}{4\phi_\mathrm{i}}. \tag{12.76}$$

Both the kinetic and the internal energy of the hot gas in the bubble have now been cast in terms of the kinetic energy of a mass, M $(= 4\pi/3 R_\mathrm{s}^3 \rho_0)$, of hot gas expanding at the shock velocity, v_s. From energy conservation ($E_\mathrm{sn} = E_\mathrm{k} + E_\mathrm{i}$), we then find that

$$R_\mathrm{s}^3 v_\mathrm{s}^2 = \left(\frac{2}{5}\right)^2 \xi_0 \frac{E_\mathrm{sn}}{\rho_0}, \tag{12.77}$$

with ξ_0 given by

$$\xi_0 = \frac{75}{8\pi}\left(\phi_\mathrm{k} + \frac{4}{(\gamma^2 - 1)}\phi_\mathrm{i}\right)^{-1}. \tag{12.78}$$

If we assume that ϕ_i and ϕ_k are constants during this phase, we find that the expansion velocity will scale with $R_s^{-3/2}$.

Realizing that v_s is just dR_s/dt, this equation can be integrated to give

$$R_s(t) = \left(\frac{\xi_0 E_{sn}}{\rho_0}\right)^{1/5} t^{2/5}, \tag{12.79}$$

and the shock velocity,

$$v_s(t) = \frac{2}{5}\left(\frac{\xi_0 E_{sn}}{\rho_0}\right)^{1/5} t^{-3/5}. \tag{12.80}$$

The temperature corresponding to the average pressure of the hot bubble is then

$$T_1 = \left(\frac{\phi_i}{\gamma+1}\right)\frac{8\mu}{25k}\left(\frac{\xi_0 E_{sn}}{\rho_0}\right)^{2/5} t^{-6/5}. \tag{12.81}$$

Thus, the expansion of the supernova remnant is set by the initial conditions, the total energy of the explosion and the density of the ambient medium, and by the constants ϕ_k and ϕ_i or, equivalently, the mean pressure and mean kinetic energy of the gas in the hot bubble.

A large body of analytical and numerical theoretical studies as well as experiments have shown that these simple considerations lead to the correct dependence of the velocity and radius on time. Essentially, after the thermalization of the ejection energy in the first phase, the expansion will have lost all memory of the ejection event and the flow will be self-similar (cf. Section 12.1). The evolution of the flow is just governed by the total energy and the density of the ambient medium. Since no new length scales are introduced, the dependence of all physical quantities (pressure, temperature, velocity) on radius is invariant under the expansion. That immediately implies that the mean pressure and mean kinetic energy of the hot gas are constant relative to $M v_s^2$ (i.e., ϕ_k and ϕ_i are constant during this phase).

To fully specify the solution, the values of these parameters, or, equivalently, ξ_0, have to be determined. These depend on the detailed velocity and pressure distribution of the shocked gas in the bubble. A variety of approximations, of increasing sophistication, can be made to determine these parameters. Reviewing these is beyond the scope of this book. Suffice it to say that detailed numerical and analytical studies have revealed that for a blast wave expanding in a homogeneous medium with a constant ratio of specific heats of 5/3, the value of ϕ_k is 0.417 and of ϕ_i is 0.470. This results in a value of 2.026 for ξ_0. The ratio of kinetic to thermal energy is then 0.394. The size, shock velocity, and postshock temperature of the supernova remnant are then given by

$$R_s(t) \simeq 14.5 \left(\frac{E_{sn}}{10^{51}\,\mathrm{erg}}\right)^{1/5} \left(\frac{0.5\,\mathrm{cm}^{-3}}{n_0}\right)^{1/5} \left(\frac{t}{10^4\,\mathrm{years}}\right)^{2/5} \mathrm{pc},$$

$$v_s(t) \simeq 570 \left(\frac{E_{sn}}{10^{51}\,\mathrm{erg}}\right)^{1/5} \left(\frac{0.5\,\mathrm{cm}^{-3}}{n_0}\right)^{1/5} \left(\frac{10^4\,\mathrm{years}}{t}\right)^{3/5} \mathrm{km\,s^{-1}}, \quad (12.82)$$

$$T_1 \simeq 8\times 10^6 \left(\frac{E_{sn}}{10^{51}\,\mathrm{erg}}\right)^{2/5} \left(\frac{0.5\,\mathrm{cm}^{-3}}{n_0}\right)^{2/5} \left(\frac{10^4\,\mathrm{years}}{t}\right)^{6/5} \mathrm{K},$$

where n_0 is the number density of hydrogen. Assuming that the non-radiative phase ends when the shock velocity drops to $250\,\mathrm{km\,s^{-1}}$ ($T \simeq 10^6\,\mathrm{K}$), we find that this occurs at $t \simeq 3.9\times 10^4$ years when the supernova remnant is $R_s \simeq 25\,\mathrm{pc}$ in size. The total swept-up mass is then approximately $1100\,\mathrm{M}_\odot$ at an ambient density of $0.5\,\mathrm{cm}^{-3}$. Note that at this point, the temperature, T_1, corresponding to the average pressure of the gas in the hot bubble is still slightly higher than 10^6 K.

12.3.3 The radiative expansion phase

When the gas temperature drops below some 10^6 K because of adiabatic expansion, radiative cooling becomes important and the shock becomes, for practical purposes, isothermal.

The supernova remnant still has an over-pressure relative to the interstellar medium and this drives further expansion. However, energy is no longer conserved, because the newly shocked gas quickly radiates its internal energy away. Because of this rapid cooling, the density contrast across the shock is very large (cf. Eq. (11.27)) – even when limited by a magnetic field, this is still almost a factor of 100. The recently shocked material is then almost stationary relative to the shock and a dense, thin shell of shocked gas will develop.

If we assume that all the mass in the supernova remnant is located in a dense shell at R_s, moving into the ISM at velocity v_s, the dynamics of the supernova remnant during this so-called pressure-driven snowplow phase is governed by the momentum equation,

$$\frac{d(Mv_s)}{dR_s} = \frac{4\pi R_s^2}{v_s}\left(\overline{P} - P_0\right), \quad (12.83)$$

with $\overline{P}$ the mean pressure of the hot gas and P_0 the external pressure. Ignoring radiative energy losses of the interior hot gas, the interior will still lose energy because it pushes this shell into the ISM. Now, if we can ignore the external pressure, this is again a self-similar expansion. Realizing that for the pressure of the hot gas we have $\overline{P} \sim V^{-\gamma}$, with V the volume of the supernova remnant and γ the ratio of the specific heats, for $\gamma = 5/3$, this results in $\overline{P} \sim R^{-5}$. The mass of swept-up gas is $4\pi/3R_s^3\rho_0$. A self-similarity analysis with $R = R_0 t^\eta$ and $v_s = \eta R_0 t^{\eta-1}$ results in $\eta = 2/7$ and $\overline{P} = 1/6\rho_0 v_s^2$. We can normalize these results to the mass, M_0, and velocity, v_0, at the end of phase II. For the velocity, we will adopt 250 km s^{-1}, while the mass scaling is given by (Section 12.3.2)

$$M_0 = 1400 \left(\frac{E_{\rm sn}}{10^{51}\,{\rm erg}}\right)^{11/45} n_0^{4/15}\, M_\odot. \tag{12.84}$$

The size is then given by

$$R_s(t) = 21.6 \left(\frac{E_{\rm sn}}{10^{51}\,{\rm erg}}\right)^{11/45} n_0^{-11/45} \left(\frac{t}{t_0}\right)^{2/7} {\rm pc}, \tag{12.85}$$

corresponding to a mass,

$$M(t) = M_0 \left(\frac{t}{t_0}\right)^{6/7} M_\odot, \tag{12.86}$$

while the velocity is

$$v_s(t) = 250 \left(\frac{t}{t_0}\right)^{-5/7} {\rm km\,s^{-1}}. \tag{12.87}$$

The normalization time, t_0, is given by

$$t_0 = 2.4 \times 10^4 \left(\frac{E_{\rm sn}}{10^{51}}\right)^{11/45} n_0^{-11/45} {\rm years}. \tag{12.88}$$

Eventually, when the pressure of the hot gas driving the expansion has dropped to the ambient pressure, the shell will coast at constant radial momentum. At that point, the size of the supernova remnant will evolve with $t^{1/4}$. However, numerical simulations show that at the transition of phase II to phase III, the remnant is over-pressured relative to these analytical calculations. As a result, during the pressure-driven snowplow phase, the time dependence of the size corresponds to an effective η of $\simeq 0.3$, which is slightly larger than 2/7. This memory of the previous phase also prevents the development of the momentum-conserving phase of the evolution. Furthermore, during this radiative phase, the temperature of the hot gas in the bubble will drop rapidly because of adiabatic expansion and at some point radiative cooling of this gas becomes important. The resulting loss of pressure inside the supernova remnant will now reverse the flow and compress the supernova remnant. We will ignore these complications here and let the radiative expansion phase end when the shock wave velocity approaches the ambient sound speed, $C_0 = 12\,\mathrm{km\,s^{-1}}$ (e.g., $\tau/t_0 = 71$). This corresponds to a time,

$$\tau = 1.7 \times 10^6 E_{\rm sn}^{11/45} n_0^{-11/45} \,\text{years}, \tag{12.89}$$

when the size of the supernova remnant is $\simeq$74 pc, or

$$V_{\rm snr}(\tau) = 1.7 \times 10^6 E_{\rm sn}^{11/15} n_0^{-11/15} \,\mathrm{pc}^3, \tag{12.90}$$

with $V_{\rm snr}(\tau)$ the final supernova remnant volume.

During this phase, the energy of the supernova is radiated away. The presumption in this analysis is that this energy loss occurs directly behind the shock front. This energy loss is then given by

$$\frac{\mathrm{d}E}{\mathrm{d}t} = -4\pi R_{\rm s}^2 \left(\frac{1}{2}\rho_0 v_{\rm s}^3\right). \tag{12.91}$$

The energy is thus also self-similar,

$$\frac{E(t)}{E(t_0)} = \left(\frac{t}{t_0}\right)^{5\eta-2}. \tag{12.92}$$

This could also be written as $E \sim R^{-k_{\rm E}}$, with $k_{\rm E}$ equal to 2. The efficiency with which the supernova energy is converted into kinetic energy of the gas in the ISM is

$$\epsilon_{\rm snr} = \frac{M_{\rm snr} v_{\rm s}^2}{2E_{\rm sn}}, \tag{12.93}$$

with $M_{\rm snr}$ the final mass of the supernova remnant. Inserting the values from above,

$$\epsilon_{\rm snr} = 7.7 \times 10^{-2} \left(\frac{10^{51}}{E_{\rm sn}}\right)^{4/15} n_0^{4/15}. \tag{12.94}$$

The internal energy of the supernova remnant is

$$\frac{E(t)}{E_{\rm sn}} = 0.12 E_{\rm sn}^{-4/15} n_0^{4/15}. \tag{12.95}$$

A number of approximations have been made here, which are not fully justified. Energy loss through radiative cooling of the hot gas as well as the external pressure of the medium will introduce deviations from self-similarity. In a more general sense, the transition from one phase in the expansion of a supernova remnant to the next one will be smooth rather than abrupt. These transitions will be characterized by readjustment of the internal structure of the remnant and a concomitant change in the detailed behavior of the expansion law. Nevertheless, these general trends are well preserved in detailed numerical calculations.

12.4 Supernovae and the interstellar medium

In the previous section, we have focussed on the expansion of a supernova remnant in a homogeneous medium. However, the ISM is highly inhomogeneous with clouds embedded in warm and hot intercloud media (cf. Chapter 8). Given a constant pressure in the supernova remnant and that the postshock pressure scales with $\rho_0 v_s^2$, the supernova will rapidly expand in the lowest density phase, while dense clouds are subsequently slowly crunched by much lower velocity shocks. Also energy can be conducted from the hot gas filling the SNR to embedded clouds. This will lead to evaporation of these clouds. The exchange of mass, energy, and momentum between the different phases governs, then, the dynamical evolution of the supernova. As a result, in a highly inhomogeneous environment, the supernova evolution will reflect the structure of the surrounding interstellar medium.

12.4.1 Cloud crushing

While the supernova remnant expands adiabatically at high velocity into the low density medium, the shock driven into the cloud will be at much lower velocity and the shocked cloud gas may be able to radiate away its thermal energy. Specifically, the cloud shock velocity, v_c, is related to the intercloud shock velocity, v_s, by

$$v_c = \left(\frac{\rho_0}{\rho_c}\right)^{1/2} v_s, \tag{12.96}$$

with ρ_c the cloud density. For a typical cloud/warm intercloud density contrast of 10^2, this amounts to a factor of ten and whenever the intercloud shock drops below 2000 km s^{-1}, crunching of clouds will drain away the energy of the supernova remnant. Cloud crushing will only be important if the shock crossing time through the cloud is less than the age of the supernova remnant; e.g., when the size of the supernova remnant exceeds a critical radius, R_{cc}, given by equating these two time scales,

$$R_{cc} = \left(\frac{\rho_c}{\rho_0}\right)^{1/2} \frac{r_c}{\eta} = \frac{3}{4\eta}\left(\frac{\rho_c}{\rho_0}\right)^{1/2} \lambda_c f_0, \tag{12.97}$$

where η is the exponent describing the time dependence of the SNR (e.g., $R \sim t^\eta$ and $v_s \sim \eta t^{\eta-1}$) and we have written the cloud size in terms of the mean free path between clouds, λ_c, and the cloud filling factor, f_0. Thus, when $R_s \gtrsim R_{cc}$, clouds are crushed and the energy loss is given by

$$\frac{dE}{dt} = 4\pi R_s^2 \, v_s \overline{P} \, f_0 = 3 f_0 (\gamma - 1) \frac{E}{R_s}. \tag{12.98}$$

The energy scales, then, with $R^{-k_{\rm E}}$, with $k_{\rm E} = 3f_0(\gamma - 1)$, which is typically much less than 1. The radius of the SNR scales approximately with $t^{1/3}$. For the ISM, the cloud-filling factor is $\simeq 0.01$, the cloud to intercloud density contrast is $\simeq 10^2$, and the mean free path between clouds is $\simeq 200$ pc. We find, then, that $R_{\rm cc} \simeq 45$ pc and $k_{\rm E} \simeq 0.02$. So energy loss through cloud crushing is not very important with these adopted parameters.

12.4.2 Cloud evaporation

The dynamics of the supernova remnant in an inhomogeneous medium are very much influenced by the evaporation of embedded clouds. Thermal conduction from the hot gas filling the supernova remnant into a cloud will be balanced by an evaporation mass flux from these cold clouds. Heat conduction is a very sensitive function of the temperature. During the earliest, hottest phases of a supernova remnant's evolution, the heat flux is very high and this mass loading will dominate the mass balance of the remnant. The heat is conducted by electrons and the classical expression for the heat conduction has to include the effects of the electric fields required to maintain the electric current at zero. This results in

$$K(T) = 1.8 \left(\frac{2}{\pi}\right)^{3/2} \frac{k^{7/2}}{m_{\rm e} e^4 \ln \Lambda} T^{5/2} = K_0 T^{5/2}. \qquad (12.99)$$

The logarithmic factor is a slowly varying function of $n_{\rm e}$ and T, which we will set equal to 30, resulting in $K_0 \simeq 6.1 \times 10^{-7}$ erg s^{-1} deg$^{-7/2}$ cm^{-1}. The energy conservation equation reads

$$\nabla \left(\rho v \left(\frac{u}{\rho} + \frac{1}{2} v^2\right)\right) + K(T) \nabla T = 0, \qquad (12.100)$$

with u the internal energy (cf. Eq. (11.9)). We will assume that the Mach number of the flow is small and we will consider "stationary" mass loss from the cloud surface. Essentially, we have now decoupled the equations of motion and energy. In spherical geometry, with the above approximations, we can rewrite the energy equation as

$$\frac{5}{2} C^2 \dot{M} \left(1 + \frac{1}{5} \mathcal{M}^2\right) = 4\pi r^2 K(T) \frac{{\rm d}T}{{\rm d}r}, \qquad (12.101)$$

where $\dot{M} = 4\pi r^2 \rho v$. Assuming $\mathcal{M}$ is small, this can be integrated to yield

$$\frac{5}{2} \dot{M}_{\rm c} C^2 = 4\pi r^2 K(T) \frac{{\rm d}T}{{\rm d}r}. \qquad (12.102)$$

This equation can be integrated with the boundary conditions that $T \simeq 0$ at the cloud boundary $r = R_c$ and $T = T_h$ when $r \to \infty$, with T_h the temperature of the hot gas filling the supernova remnant. This yields, for the mass loss rate,

$$\dot{M}_c = \frac{16\pi}{25}\frac{\mu}{k} K_0 T_h^{5/2} R_c. \tag{12.103}$$

Thus, the mass injection rate into the supernova remnant increases rapidly with the temperature and with the size of the clouds. We have

$$\dot{M}_c \simeq 2.3 \times 10^{-6} \left(\frac{R_c}{1\,\mathrm{pc}}\right) \left(\frac{T}{2 \times 10^6\,\mathrm{K}}\right)^{5/2} \mathrm{M}_\odot\ \mathrm{year}^{-1}, \tag{12.104}$$

and the lifetime of a cloud with a size of 1 pc and a density of $50\,\mathrm{cm}^{-3}$ is some 2×10^6 years, much longer than the evaporation phase of a supernova remnant (see below).

This expression for the classical heat conductivity assumes that the mean free path for electron energy exchange is small compared with the temperature gradient scale length. The mean free path for electrons is given by

$$\lambda_e = \frac{3}{4}\left(\frac{3}{\pi}\right)^{1/2} \frac{k^2 T_e^2}{n_e e^4 \ln \Lambda}, \tag{12.105}$$

and this condition becomes, approximately,

$$\frac{T^2}{nR_c} < 10^{-4}, \tag{12.106}$$

where we have set the temperature gradient scale length equal to the cloud radius, R_c. If this inequality is not fulfilled, the heat flux will saturate at $3/2 n_e k T_e v_e$, where v_e is a characteristic electron velocity, which will be of the order of the electron thermal velocity. We will not explore that limit here further.

Because the mass of the supernova remnant increases rapidly owing to evaporation while the total energy stays constant, the temperature will drop much faster than for the classical Sedov–Taylor expansion discussed in Section 12.3.2. Assuming that the total mass of the remnant is dominated by evaporation from clouds, we have

$$\frac{\mathrm{d}M}{\mathrm{d}t} = \frac{4\pi}{3} N_{cl} \dot{M}_c R_s^3, \tag{12.107}$$

with N_{cl} the density of clouds per unit volume. The energy is still conserved. The solution of the mass, momentum, and energy conservation equations for the remnant are again self-similar. Adopting $R \sim t^\eta$, we find that the mass is proportional to $M_{ev}(t) \sim t^{8\eta-5}$, and, hence, for the energy we have $E_{snr} \sim t^{10\eta-6}$.

Thus, with evaporation $\eta = 3/5$, which should be compared to the Sedov–Taylor result of 2/5, the radius of the remnant can be found with some tedious algebra,

$$R_s = 1.21 \left(\left(\frac{k}{\mu}\right)^{7/2} \frac{1}{K_0} \left(\frac{\gamma+1}{2\phi_i}\right)^{5/2} \left(\phi_k + \frac{4\phi_i}{\gamma^2-1}\right)^{-1} \left(\frac{E_{sn}}{N_{cl} R_c}\right)\right)^{1/10} t^{3/5}, \tag{12.108}$$

where ϕ_i and ϕ_k have been defined in Section 12.3.2. In this case, $\phi_i = 0.167$ and $\phi_k = 0.45$ and we have

$$R_s \simeq 18.3 \left(\frac{E_{sn}}{10^{51}\,\text{erg}}\right)^{1/10} \left(\frac{R_c}{1\,\text{pc}}\right)^{1/5} \left(\frac{0.01}{f_0}\right)^{1/10} \left(\frac{t}{10^4\,\text{years}}\right)^{3/5} \text{pc}, \tag{12.109}$$

with f_0 the cloud filling factor. The temperature of the hot gas is given by

$$T \simeq 2.0 \times 10^6 \left(\frac{E_{sn}}{10^{51}\,\text{erg}}\right)^{1/5} \left(\frac{R_c}{1\,\text{pc}}\right)^{2/5} \left(\frac{0.01}{f_0}\right)^{1/5} \left(\frac{10^4\,\text{years}}{t}\right)^{4/5} \text{K}. \tag{12.110}$$

The mass injection rate is

$$\frac{dM}{dt} \simeq 7.7 \times 10^{-3} \left(\frac{E_{sn}}{10^{51}\,\text{erg}}\right)^{4/5} \left(\frac{1\,\text{pc}}{R_c}\right)^{2/5} \left(\frac{f_0}{0.01}\right)^{1/5} \left(\frac{10^4\,\text{years}}{t}\right)^{1/5} M_\odot\,\text{year}^{-1}. \tag{12.111}$$

The evaporated mass is then

$$M_{ev} = 95 \left(\frac{E_{sn}}{10^{51}\,\text{erg}}\right)^{4/5} \left(\frac{1\,\text{pc}}{R_c}\right)^{2/5} \left(\frac{f_0}{0.01}\right)^{1/5} \left(\frac{t}{10^4\,\text{years}}\right)^{4/5} M_\odot. \tag{12.112}$$

The average hydrogen density of the hot gas in the supernova remnant is

$$<\rho> = 2.0 \times 10^{-25} \left(\frac{E_{sn}}{10^{51}\,\text{erg}}\right)^{1/2} \left(\frac{1\,\text{pc}}{R_c}\right) \left(\frac{f_0}{0.01}\right)^{1/2} \left(\frac{10^4\,\text{years}}{t}\right) \text{g}\,\text{cm}^{-3}, \tag{12.113}$$

which is, of course, a function of time.

These expressions will govern the expansion of the supernova remnant until the temperature has dropped so much that the swept-up interstellar mass starts to dominate the mass balance. For expansion into a medium with average density ρ_0, the swept-up mass is governed by

$$\frac{dM_{sw}}{dt} = 4\pi R_s^2 v_s \rho_0. \tag{12.114}$$

Thus, the accretion rate of interstellar matter will dominate the mass evaporation rate when

$$t > 1.5 \times 10^3 \left(\frac{E_{sn}}{10^{51}\,\text{erg}}\right)^{1/2} \left(\frac{1\,\text{pc}}{R_c}\right)^{1/5} \left(\frac{f_0}{0.01}\right)^{1/10} \left(\frac{0.5\,\text{cm}^{-3}}{n_0}\right) \text{years}. \tag{12.115}$$

At that point, the solution will transit to the Sedov–Taylor solution described before (Section 12.3.2). Thus, the evaporative phase in the expansion of a supernova remnant is only important if the supernova remnant expands into a very tenuous medium (e.g., into the HIM).

12.4.3 Summary

The expansion behavior of a supernova thus depends very strongly on the structure of the interstellar medium. Consider first a two-phase interstellar medium with cold clouds embedded in a warm intercloud medium. We will adopt here a cloud density of $50\,\mathrm{cm}^{-3}$, a cloud size of 1 pc, a filling factor of $f_0 = 0.01$, and an intercloud density of $n_0 = 0.25\,\mathrm{cm}^{-3}$. The average density is, therefore, $<n>= 0.75\,\mathrm{cm}^{-3}$. In such an interstellar medium, evaporation of the clouds into the hot gas of the supernova remnant is only important in the first few thousand years and the expansion is actually well described by the Sedov–Taylor solution for expansion in a medium with the intercloud density. The supernova will thus expand more rapidly than for an interstellar medium at the average density. While the intercloud medium is strongly shocked, the clouds will be much less affected by the supernova.

For a three-phase interstellar medium, we will adopt a hot intercloud medium with a density of $3 \times 10^{-3}\,\mathrm{cm}^{-3}$ and a filling factor of 0.5. We will fill the rest of the medium mainly with a warm intercloud phase of density $0.5\,\mathrm{cm}^{-3}$ and filling factor 0.49. We will adopt a typical size scale for the intercloud regions of 10 pc. For the cloud phase, we will adopt the same parameters as before. With these choices, the average density is thus again $<n>= 0.75\,\mathrm{cm}^{-3}$. In this case, the supernova will rapidly expand into the very tenuous hot medium. Evaporation from the warm intercloud medium into the hot medium will now be very important and will dominate the mass loading of the hot gas in the SNR during the whole evolution. Note that the supernova remnant now cools because of both adiabatic expansion and the mass loading. So, compared with a supernova expanding into a homogeneous hot intercloud medium, the remnant will now cool faster. Of course, compared with a supernova expanding into a warm intercloud medium, cooling will be slower. These results are summarized in Fig. 12.9.

The behavior of supernovae in these two different classes of models for the ISM can be summarized as follows. Ignoring cloud crushing and evaporation, during the Sedov–Taylor phase, the evolution of the supernova remnant in the two-phase medium described above is given by

$$R(t) = \left(\xi_0 \frac{E_{\rm sn}}{\rho_{\rm wnm}}\right)^{1/5} t^{2/5}, \qquad (12.116)$$

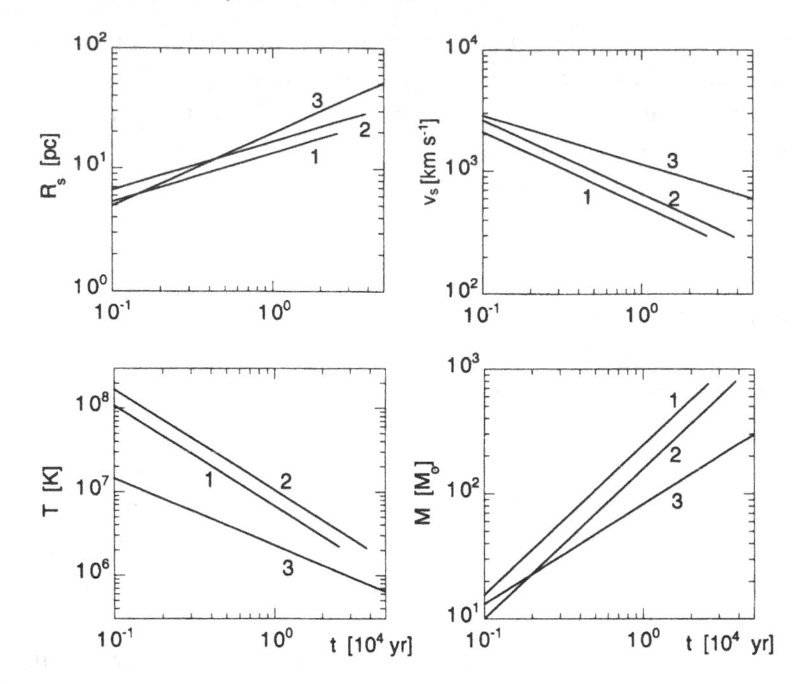

Figure 12.9 The characteristics of a supernova in a homogeneous medium, (1), a two-phase (2), and a three-phase (3) interstellar medium during the Sedov–Taylor expansion phase. In the two-phase medium, M is the mass of the shocked warm medium. In the three-phase medium, the evaporated mass is shown. Note that, because of the low density, the Sedov–Taylor phase lasts longer in the three-phase medium.

with $\rho_{\rm wnm}$ the density in the warm intercloud medium and $\xi_0 = 2.026$. For the three-phase interstellar medium, adopting the classical evaporation rate, we can write the evolution formally in the same way with $\xi_0 = 1.71$ but the density is now the average density of the SNR ($<\rho>$; Eq. (12.113)), which is actually a function of time. For the two-phase medium, the mass of warm intercloud material processed by a shock exceeding a velocity $v_{\rm s}$ is

$$\begin{aligned} M\,(v > v_{\rm s}) &= \frac{16\pi}{75}\frac{\xi_0 E_{\rm sn}}{v_{\rm s}^2} \\ &\simeq 1.1 \times 10^3 \left(\frac{E_{\rm sn}}{10^{51}\,{\rm erg}}\right)\left(\frac{250\,{\rm km\,s^{-1}}}{v_{\rm s}}\right)^2 M_\odot. \end{aligned} \tag{12.117}$$

This equation, of course, just expresses energy conservation. In a two-phase medium, the radiative cooling phase is important and the supernova remnant slows down much faster. Numerical calculations show that the mass shocked to velocities exceeding v_s is then given by

$$M(v > v_s) \simeq 2.5 \times 10^3 \left(\frac{E_{sn}}{10^{51}\,\text{erg}}\right)^{0.95} \left(\frac{\text{cm}^{-3}}{n_0}\right)^{0.1} \left(\frac{100\,\text{km s}^{-1}}{v_s}\right)^{9/7} M_\odot. \tag{12.118}$$

For a three-phase medium, we arrive at a similar expression,

$$M(v > v_s) = \frac{16\pi}{75} \frac{\xi_0 E_{sn}}{v_s} \frac{f_0 \rho_{wnm}}{<\rho>}. \tag{12.119}$$

Now, we are really interested in the processing of the WNM, which has the combination of a relatively high density and filling factor, and realizing that the pressure of the hot supernova remnant gases drives the shocks into the warm intercloud medium, $\rho_{wnm} v_{wnm}^2 = <\rho> v_s^2$, we can write this as

$$\begin{aligned} M(v > v_{wnm}) &= \frac{16\pi}{75} f_0 \frac{\xi_0 E_{sn}}{v_{wnm}^2} \\ &\simeq 1.1 \times 10^3 f_0 \left(\frac{E_{sn}}{10^{51}\,\text{erg}}\right) \left(\frac{250\,\text{km s}^{-1}}{v_{wnm}}\right)^2 M_\odot. \end{aligned} \tag{12.120}$$

So, the mass shocked to a given velocity is somewhat smaller in a multiphase ISM because the filling factor of the WNM is lower than in a two-phase ISM.

12.5 Interstellar winds

Massive stars inject mass and momentum into the interstellar medium through strong winds with velocities, v_w, from ~1500 to 3000 km s^{-1}. For main sequence stars, mass loss rates, $\dot{M}$, are in the range of 10^{-7}–$4 \times 10^{-6}\,M_\odot$ year^{-1}, while supergiants have somewhat more massive winds (3×10^{-6}–$10^{-5}\,M_\odot$ year^{-1}). The values $v_w = 2000$ km s^{-1} and $\dot{M} = 10^{-6}\,M_\odot$ year^{-1} correspond to a mechanical luminosity of the wind of

$$\begin{aligned} L_w &= \frac{1}{2} \dot{M} v_w^2 \\ &\simeq 1.3 \times 10^{36} \left(\frac{\dot{M}}{10^{-6}\,M_\odot\,\text{year}^{-1}}\right) \left(\frac{v_w}{2 \times 10^3\,\text{km s}^{-1}}\right)^2 \text{erg s}^{-1}, \end{aligned} \tag{12.121}$$

or 330 $L_\odot$, which is less than 1% of the stellar radiative luminosity. We will adopt 10^{36} erg s^{-1} as a typical value. When the dynamical pressure of this wind, $\rho_w v_w^2$, drops to the thermal pressure of the surrounding gas, a reverse shock will slow the wind down. In its turn, this shocked gas will drive a strong shock into the ambient interstellar medium, sweeping it up. The two shocked fluids are separated by a contact discontinuity. The resulting structure is called a bubble, because much of the interior is filled with a tenuous, hot gas.

This structure is illustrated in Fig. 12.10. The stellar wind expands freely to radius R_1, where it is shocked by the wind shock. The contact discontinuity and the shock driven into the ambient medium are located at radii of R_c and R_2, respectively. We will assume that the star is initially surrounded by a homogeneous medium with density ρ_0. With a wind velocity of some 10^3 km s^{-1}, the shocked wind material will be very hot, coupled with its low density, this gas cannot cool, and its expansion is adiabatic. In contrast, the swept-up interstellar medium gas is denser and, over most of the evolution of the bubble, is not shocked to such high temperatures. This material will therefore cool very effectively and will collapse into a thin, dense shell. The (constant) injection of energy through the wind makes this system slightly different from that of a supernova explosion. However, assuming a constant mechanical luminosity and a negligible mass injection (compared to the swept-up mass), the expansion is also self-similar and its behavior can be derived similar to that of supernova blast waves.

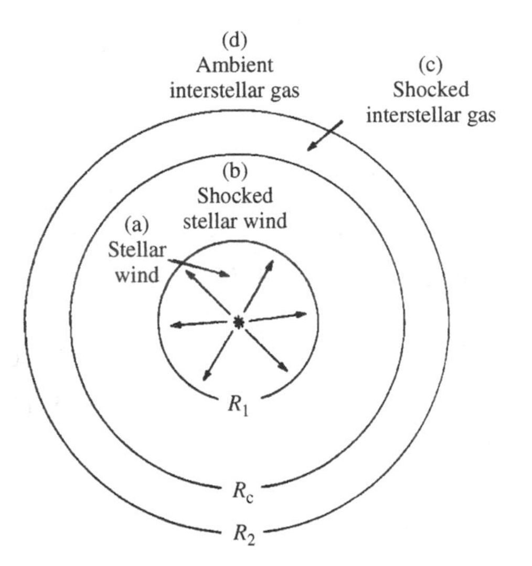

Figure 12.10 A schematic diagram of the various regions created by the stellar wind from a massive star. R_1 is the location of the wind shock, R_2 is the location of the shock in the ambient medium, and R_c is the location of the contact discontinuity. Note that the shell of shocked interstellar gas is cool and thin, while the shocked stellar wind is hot and tenuous. The sizes of the different regions are thus not to scale. Figure reproduced with permission from R. Weaver, R. McCray, J. Castor, P. Shapiro, and R. Moore, 1977, *Ap. J.*, **218**, p. 377.

12.5.1 The wind–ISM interaction

The high pressure of the hot (wind-shocked) gas will drive the expansion of the thin shell into the surrounding medium. We will assume that the shocked wind region is isobaric and that the internal energy is much larger than the kinetic energy, both of which are reasonable assumptions given the high temperature. During this phase, R_1 is much less than R_2. The internal energy of the shocked wind gas is related to this pressure through

$$E_{\rm i} = \frac{3}{2}\frac{4\pi}{3}R_2^3 P, \tag{12.122}$$

where we have assumed a γ of 5/3. The mass of the shell is

$$M_{\rm shell} = \frac{4\pi}{3}R_2^3 \rho_0. \tag{12.123}$$

Because this shock is isothermal, the velocity of the gas downstream is small in the frame of the shock front and – in contrast to a non-radiative blast wave from a supernova – the shell will essentially move as a whole. The momentum equation for the shell reads, then,

$$\frac{{\rm d}[M_{\rm shell} v_{\rm s}^2]}{{\rm d}t} = 4\pi R_{\rm s}^2 P. \tag{12.124}$$

The energy balance for the hot region is now

$$\frac{{\rm d}E}{{\rm d}t} = L_{\rm w} - 4\pi R_{\rm s}^2 P v_{\rm s}. \tag{12.125}$$

These equations are very similar to those of a supernova blast wave and can be solved in the same way. In contrast with the supernova blast wave, the role of the energy is now played by the wind luminosity. The other parameter controlling the structure of the region is again the density of the ambient medium. Because the flow is self-similar, we have $R_{\rm s} \sim t^\eta$, with η a constant, and $v_{\rm s} \sim t^{\eta-1}$. Also, the internal energy of the hot gas will increase linearly with time for a constant wind luminosity. Then, P has to scale with $\sim t^{1-3\eta}$ (Eq. (12.122)). Consequently, the momentum equation (Eq. (12.124)) yields $\eta = 3/5$. The constants relating the radius of the shell, the pressure of the medium, and the energy of the bubble to the time can be derived by proper substitution into the equations above and some algebraic manipulation. This yields

$$\begin{aligned} R_2(t) &= \left(\frac{125}{154\pi}\right)^{1/5} \left(\frac{L_w}{\rho_0}\right)^{1/5} t^{3/5}, \\ P(t) &= \frac{7}{(3850\pi)^{2/5}} L_w^{2/5} \rho_0^{3/5} t^{-4/5}, \text{ and} \\ E(T) &= \frac{5}{11} L_w t. \end{aligned} \tag{12.126}$$

For typical parameters, we find

$$\begin{aligned} R_2(t) &\simeq 32 \left(\frac{L_w}{10^{36}\,\mathrm{erg\,s^{-1}}}\right)^{1/5} \left(\frac{0.5\,\mathrm{cm^{-3}}}{n_0}\right)^{1/5} \left(\frac{t}{10^6\,\mathrm{years}}\right)^{3/5} \mathrm{pc}, \\ P(t) &\simeq 2.3 \times 10^{-12} \left(\frac{L_w}{10^{36}\,\mathrm{erg\,s^{-1}}}\right)^{2/5} \left(\frac{n_0}{0.25\,\mathrm{cm^{-3}}}\right)^{3/5} \\ &\quad \times \left(\frac{10^6\,\mathrm{years}}{t}\right)^{4/5} \mathrm{dyne\,cm^{-2}}, \text{ and} \\ E(T) &\simeq 1.4 \times 10^{49} \left(\frac{L_w}{10^{36}\,\mathrm{erg\,s^{-1}}}\right) \left(\frac{t}{10^6\,\mathrm{years}}\right) \mathrm{erg}. \end{aligned} \tag{12.127}$$

So, over an appreciable fraction of the lifetime of a massive star, the wind pressure will exceed that of the surrounding medium (ignoring cosmic-ray pressure, $P_0 \simeq 2.3 \times 10^{-12}$ dyne cm^{-2} or 1.6×10^4 K cm^{-3}; cf. Table 1.2). That will drive the expansion to sizes of 32 pc and swept-up masses of some 2000 M$_\odot$. The velocity of the shell is then some 20 km s^{-1}. The total kinetic energy imparted onto the interstellar medium is about 30% of the integrated mechanical wind luminosity. The remainder is stored in internal energy of the hot bubble gas.

12.5.2 The structure of the hot bubble

As discussed in Section 12.4.2 in the context of supernova remnants, heat conduction is important when (inter)cloud material is in contact with hot bubble gas. In this case, the heat flux into the dense shell will be balanced by a slow evaporation of this cold gas into the hot interior. This mass flux dominates the mass – and thereby the temperature – of the bubble. The mass loss from the cold gas is, however, unimportant for the mass balance of the shell and, hence, the dynamics and structure of this shell (discussed above) are not affected. Because the shell is very thin, we will assume plane-parallel steady flow. We will ignore cooling in the transition region, and adopt a coordinate system centered on R_s (e.g., $z = R_s - R$).

Equating the energy gained by the evaporating cold gas with the heat conducted from the hot bubble, we have

$$\rho v \frac{\mathrm{d}}{\mathrm{d}z}\left[\frac{u}{\rho} + \frac{P}{\rho}\right] = \frac{\mathrm{d}}{\mathrm{d}z}\left[K(T)\frac{\mathrm{d}T}{\mathrm{d}z}\right], \tag{12.128}$$

where ρv is the mass flux ($\dot{M}/4\pi R_{\mathrm{s}}^2$) and the heat conductivity is given by Eq. (12.99). This flow is again self-similar. The solution for the temperature is given by

$$T(z) = T_{\mathrm{b}}\left(\frac{z}{R_{\mathrm{s}}}\right)^{\eta}, \tag{12.129}$$

with $T = 0$ in the shell and $T = T_{\mathrm{b}}$ in the center of the bubble. The energy equation (Eq. (12.128)) then yields $\eta = 2/5$ and, for $\gamma = 5/3$,

$$\dot{M} = \frac{16\pi}{25}\frac{\mu}{k} K_0 T_{\mathrm{b}}^{5/2} R_{\mathrm{s}}. \tag{12.130}$$

The bubble is isobaric and the density structure is given by

$$\rho(z) = \rho_{\mathrm{b}} \frac{T_{\mathrm{b}}}{T(z)}. \tag{12.131}$$

We can rewrite the energy–pressure relation (Eq. (12.122)) as

$$E(t) \simeq \frac{3}{2} 0.42 \frac{M}{\mu} k T_{\mathrm{b}}, \tag{12.132}$$

where the factor 0.42 accounts for the density distribution in the bubble. We now have two equations linking the mass and the central temperature of the gas in the bubble. Realizing that the solution is self-similar, we substitute $M_{\mathrm{b}} = Bt^{\beta}$ and $T_{\mathrm{b}} = Dt^{\delta}$, which yields $\beta = -6/35$ and $\delta = 29/35$. The central temperature and mass of the bubble are then given:

$$\begin{aligned} T_{\mathrm{b}} &= 37.6 L_{\mathrm{w}}^{8/35} \rho_0^{2/35} t^{-6/35} \\ &\simeq 1.25 \times 10^6 \left(\frac{L_{\mathrm{w}}}{10^{36}\,\mathrm{erg\,s^{-1}}}\right)^{8/35} \left(\frac{n_0}{0.5\,\mathrm{cm^{-3}}}\right)^{2/35} \left(\frac{10^6\,\mathrm{years}}{t}\right)^{6/35} \mathrm{K}. \end{aligned} \tag{12.133}$$

The corresponding central atomic hydrogen density is then

$$\begin{aligned} n_{\mathrm{b}} &= 1.4 \times 10^{13} L_{\mathrm{w}}^{6/35} \rho_0^{19/35} t^{-22/35} \\ &\simeq 6.7 \times 10^{-3} \left(\frac{L_{\mathrm{w}}}{10^{36}\,\mathrm{erg\,s^{-1}}}\right)^{6/35} \left(\frac{n_0}{0.5\,\mathrm{cm^{-3}}}\right)^{19/35} \left(\frac{10^6\,\mathrm{years}}{t}\right)^{22/35} \mathrm{cm^{-3}}, \end{aligned} \tag{12.134}$$

where we have assumed a fully ionized gas with 10% helium.

12.5.3 The structure of the dense shell

The wind has, therefore, profound influence on the density structure of the surrounding interstellar medium. This will, in particular, affect the structure of the photo-ionized gas: the HII region. In Chapter 7, we have studied the structure of homogeneous HII regions. In a homogeneous medium, the size of such an HII region is given by the Strömgren radius, which we will label here $\mathcal{R}_0$,

$$\mathcal{R}_0 = \left(\frac{3\mathbb{N}_{\rm Lyc}}{4\pi n_0^2 \beta_{\rm B}} \right)^{1/3} . \tag{12.135}$$

The shocked and unshocked wind material is very tenuous and highly (collisionally) ionized and no EUV photons are required to maintain its ionization. The shell is, of course, dense and possesses a high column density. The HII region may then well be trapped in this shell. The column density of the shell is

$$N_{\rm shell} = \frac{nR_{\rm s}}{3}, \tag{12.136}$$

which is of the order of $4 \times 10^{19}\,{\rm cm}^{-2}$. This is much larger than the cooling column density (cf. Eq. (11.26)) and, hence, the temperature will quickly drop behind the shock front to some 8000 K if the gas is photo-ionized and some 100 K for atomic–molecular gas. The density of the shell is then given by the isothermal shock conditions (cf. Eq. (11.27)). The density in the shell will decrease with $t^{-4/5}$. The number of recombinations in the shell scales with $R_2^3 n \sim t^{41/35}$ – where we have used the fact that the total number of atoms in the shell scales with $R_2^3 n_0$ – and hence rapidly increases with time. Thus, while initially the star will be able to ionize a large region, eventually, the HII region will become trapped in the shell. Once the HII region is trapped in the shell, its size is governed by

$$\mathcal{R}_{\rm HII} = \left(\frac{n_0}{n} \right)^{1/3} \mathcal{R}_0 = \left(\frac{C_{\rm s}}{v_2} \right)^{2/3} \mathcal{R}_0, \tag{12.137}$$

where, now, we have ignored the ambient pressure term. The ionized shell will give a limb-brightening appearance to the HII region (cf. Section 12.2.7). The shell material outside $\mathcal{R}_{\rm HII}$ will recombine and form a dense layer of cold ($\simeq 10^2$ K) HI gas.

12.6 The kinetic energy budget of the ISM

Over its lifetime, a massive star may radiate about 2×10^{53} erg. Most of that, however, serves to ionize (or photodissociate) and heat the surrounding gas. Eventually, almost all of this energy is radiated away. In terms of imparting kinetic energy on the ISM, when an HII region is in the embedded phase, the

Table 12.1 *Kinetic energy input into the ISM*

phase	kinetic energy (erg)
YSO wind	2×10^{47}
Embedded HII regions	2×10^{46}
Blister HII regions	3×10^{49}
Stellar winds	10^{49}
Supernova explosions	6×10^{49}

pressure-driven expansion is very ineffective. If there is no blister phase, some $2000\,M_\odot$ of swept-up neutral gas is accelerated to $\sim 1\,\mathrm{km\,s^{-1}}$. For comparison, the strong stellar winds associated with embedded protostars may input a factor of ten times more mechanical energy in the surrounding molecular cloud. However, during a blister phase, an appreciable mass of surrounding gas ($\sim 3 \times 10^3\,M_\odot$) is (ionized and) accelerated to some $30\,\mathrm{km\,s^{-1}}$. A stellar wind from a massive star can also sweep up an appreciable mass of interstellar gas in a shell and accelerate it to $\sim 20\,\mathrm{km\,s^{-1}}$. Finally, as a supernova, massive stars again convert a small fraction of their energy into mechanical energy of the ISM. Overall, massive stars inject mechanical energy into the ISM over their full lifetime and blister HII regions and stellar winds are of comparable importance to supernova explosions in driving kinetic motions in the ISM.

12.7 Further reading

The concept of self-similarity is well explained in the general textbook on dynamics, [1]. While most studies apply these principles by inserting simple power laws (e.g., $R \propto t^\eta$) into the appropriate equations of motion (e.g., conservation of mass, momentum, and energy), [2] approaches the problem through the virial theorem. This paper discusses a range of related problems in a similar vein and is very instructive and illuminating. This paper is well recommended to the serious student in this area.

The principles of the dynamics of ionization fronts are discussed in [3]. The dynamics of the initial ionization and the pressure-driven phase are, for example, discussed in [4]. The analytical approach to the dynamics of blisters is discussed in [5]. Pioneering numerical studies were performed in [6]. An excellent review is provided in [7]. The ionization of globules and proplyds has been studied in [8], [9], [7], and [10]. A review can be found in [11]. A recent study of the ionization of accretion flows is provided in [12].

While the approach is somewhat different, the best paper for further study of the dynamics of supernova remnants is [2]. The physics of the interaction of supernova remnants with clouds is discussed in [13]. Of course, the best theoretical synthesis can be found in [14].

The seminal paper on the interaction of stellar winds with the interstellar medium is [15]. A relevant textbook is [16].

References

[1] Y. B. Zel'dovitch and Y. P. Raizer, 1966, *Physics of Shock Waves and High Temperature Hydrodynamic Phenomena*, vols. 1 and 2 (New York: Academic Press)
[2] J. Ostriker and C. F. McKee, 1988, *Rev. Mod. Phys.*, *60*, p. 1
[3] F. Kahn, 1975, in *Atomic and Molecular Physics and the Interstellar Matter* (Amsterdam: North Holland Publ. Co.), p. 533
[4] D. Osterbrock, 1989, *Astrophysics of Gaseous Nebulae and Active Galactic Nuclei* (Mill Valley: University Science Books)
[5] P. Bedijn and G. Tenorio-Tagle, 1981, *A. & A.*, **98**, p. 85
[6] G. Tenorio-Tagle, 1979, *A. & A.*, **71**, p. 59; P. Bodenheimer and R. Moore, 1977, *Ap. J.*, **233**, p. 85
[7] H. W. Yorke, 1986, *Ann. Rev. Astron. Astrophys.*, **24**, p. 49
[8] F. Bertoldi and C. F. Mckee, 1990, *Ap. J.*, **354**, p. 529
[9] D. Johnstone, D. Hollenbach, and J. Bally, 1998, *Ap. J.*, **499**, p. 758
[10] S. Richling and H. W. Yorke, 2000, *Ap. J.*, **539**, p. 258
[11] D. Hollenbach, H. W. Yorke, and D. Johnstone, 2000, in *Protostars and Planets IV*, ed. V. Manning, A. P. Boss, and S. S. Russell (Tucson: University of Arizona Press), p. 401
[12] E. Keto, 2003, *Ap. J.*, **599**, p. 1196
[13] L. Cowie and C. F. McKee, 1977, *Ap. J.*, **211**, p. 135
[14] C. F. McKee and J. Ostriker, 1977, *Ap. J.*, **218**, p. 148
[15] R. Weaver, R. McCray, J. Castor, P. Shapiro, and R. Moore, 1977, *Ap. J.*, **218**, p. 377
[16] H. J. G. L. M. Lamers and J. S. Cassinelli, 1999, *Introduction to Stellar Winds* (Cambridge: Cambridge University Press)

13

The lifecycle of interstellar dust

13.1 Introduction

The lifecycle of interstellar dust and the processes that play a role are summarized in Fig. 13.1. Dust is formed at high densities and temperatures in the ejecta from stars. This leads to the formation of high temperature condensates such as silicates, graphite, and carbides. Stardust grains with an isotopic composition betraying their birthsites have been isolated from meteorites (Chapter 5). In the interstellar medium, dust cycles many times between the intercloud and cloud phases. In the warm neutral and ionized intercloud media, dust is processed by strong shocks driven by supernova explosions (Section 12.3). The hot gases in the shock can sputter atoms from the grains. Also, high velocity collisions among grains can lead to vaporization, melting, phase transformation, and shattering of the projectile and target. These processes have been discussed in Section 5.2. In the denser media – diffuse and dense clouds – gas-phase species can accrete onto grains forming a mantle. In diffuse clouds, the accreted species may be predominantly bound by chemisorbed forces – partly because physisorbed species will be rapidly photodesorbed by the high flux of FUV photons (Section 10.6.1). In molecular clouds, the accretion process leads to the formation of an icy mantle consisting of simple molecules such as H_2O, CO, CO_2, and CH_3OH (see Section 10.5). These ices may be processed by UV photons and high-energy cosmic rays into larger, more complex species, which could be more tightly bound to the cores. While there is compelling experimental evidence for these processes, their importance in an interstellar setting is more controversial and, presently, there is little direct observational support. Coagulation may also play a role in increasing the grain size inside diffuse and dense clouds. A grain will cycle many times between the various phases of the interstellar medium on a typical time scale of $\simeq 3 \times 10^7$ years. If the grain is not destroyed by an interstellar shock, eventually, during one of these cycles, a grain may find itself in a dense cloud core when this core becomes gravitationally unstable against collapse. The grain may then wind up in the

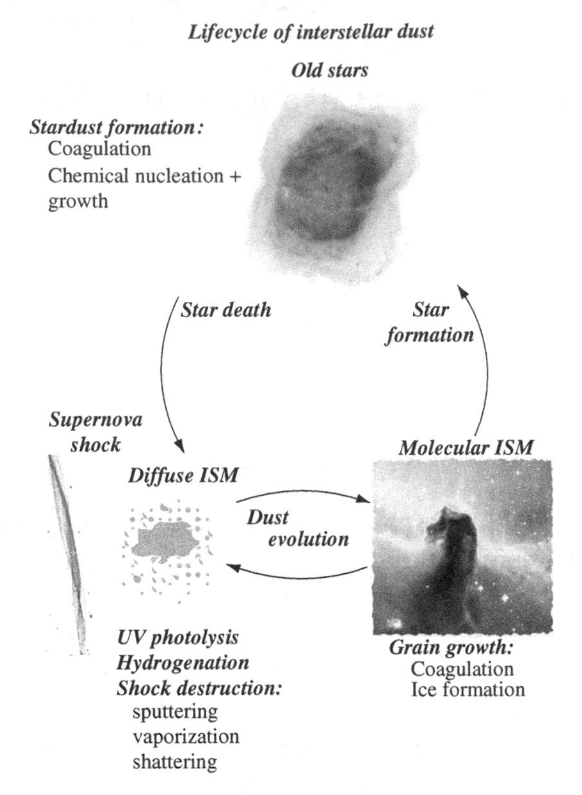

Figure 13.1 A schematic diagram of the evolution of interstellar dust, indicating some of the relevant processes. Dust is formed in stellar ejecta and mixes into the interstellar medium. In the interstellar medium, it cycles rapidly between the different phases. In the intercloud phases, strong shocks destroy dust and change its size distribution. In the denser phases, accretion of gas-phase species will form a mantle, which can be processed by FUV photons and energetic particles. Coagulation will also lead to an increase in the grain size. Eventually, in one of its sojourns into the dense media, dust grains will partake in the star and planet formation process.

star or in the surrounding protoplanetary disk. If disk evaporation by a nearby luminous star is important (see Section 12.2.4), dust grains in the outer disk may "escape." The strong protostellar winds of young stellar objects originate from the inner protoplanetary disk and it is probable that all interstellar dust grains are completely vaporized and recondensed as the Solar System condensates. At that point, any stardust heritage is completely lost. Some of these fresh condensates may become entrained in this wind and injected into the interstellar medium. The complete cycle from injection by a star until formation of a new star takes some 2×10^9 years.

In this chapter, we will discuss the evolution of interstellar dust in the diffuse interstellar medium, focussing on shock processing and the implications for the dust size distribution, the dust lifetime, and the depletion of the elements.

13.2 Shock destruction

The dust destruction processes discussed in Section 5.2.4 have to be convolved with the detailed structure of interstellar shocks to determine the destruction rate of grains in shocks of a given velocity. For low velocity shocks, destruction is dominated by inertial sputtering, while at high shock velocities, thermal sputtering becomes important. It is advantageous to consider these two regimes separately.

13.2.1 Low-velocity steady-state shocks

Figure 13.2 shows the density, temperature, and electron fraction for a typical shock ($v_s = 100$ km s^{-1}, $n = 0.25$ cm^{-3}, $B = 3$ μG) as a function of the column density, N_H, behind the shock front (see Chapter 11). The time, t, is related to N_H through $\log(t(\text{years})) = \log(N_H) - 13.9$. We recognize many of the aspects discussed in Section 11.2.1. At the shock front, the gas motion is converted into heat. Typically, behind the shock the gas will be flowing at 1/4 of the shock velocity. As the gas cools behind the shockfront, it compresses. The gas is ionized around $N_H = 10^{14}$ cm^{-2}. Recombination occurs around $N_H = 10^{18}$ cm^{-2}. Further compression at that time is limited because of magnetic cushioning.

This figure also shows the velocities of various-sized graphite grains. The results for silicate grains are very similar. Because of their inertia, grains will keep moving with respect to the gas at approximately 3/4 of the shock speed. Because grains are also charged, they will gyrate around the magnetic field lines. The Larmor radius, R_L, is approximately given by

$$\begin{aligned} R_L &= \frac{m_d v_d c}{Z_d e B} \\ &\simeq \frac{1.3\times10^{16}}{Z_d e^2/akT}\left(\frac{a}{1000\,\text{Å}}\right)^2\left(\frac{v}{100\,\text{km s}^{-1}}\right)\left(\frac{10^5\,\text{K}}{T}\right)\,\text{cm}, \end{aligned} \tag{13.1}$$

where the grain potential parameter, $Z_d e^2/akT$, is typically of the order of unity. Thus, for reasonable sizes (<0.3 μm), grains are position-coupled to the hot gas behind the shock. The magnetic moment, μ_d, of the grains,

$$\mu_d = \frac{m_d v_d^2}{2B}, \tag{13.2}$$

is conserved and, thus, as the gas compresses behind the shock because of cooling and the magnetic field strength increases because of flux freezing ($B \sim \rho$), the

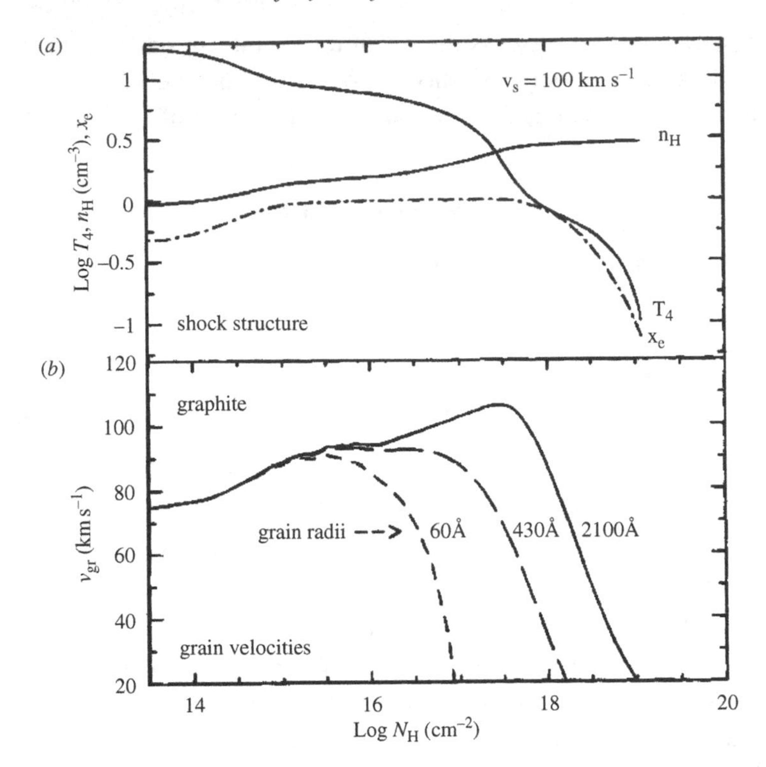

Figure 13.2 The structure of a 100 km s^{-1} shock. (*a*) The postshock density, n_H, temperature, T_4, in units of 10^4 K, and electron fraction, x_e, as a function of column density behind the shock front. (*b*) Velocities of different size graphite grains as a function of the column density behind the shock front. Figure reproduced with permission from A. P. Jones, A. G. G. M. Tielens, and D. Hollenbach, 1996, *Ap. J.*, **469**, p. 740.

grains will spin up. This process, called betatron acceleration, is counteracted by gas drag. As a result, grains will initially speed up before being dragged to a halt (Fig. 13.2). The effects of betatron acceleration depend on the grain size as well as the shock speed. In the standard shock, all grains are betatron accelerated during the initial compression around 3×10^{14} cm^{-2}. The larger grains experience a second burst around 10^{17} cm^{-2}, while the smaller grains have been stopped at that point already. Betatron acceleration is unimportant for low shock velocities ($\lesssim$50 km s^{-1}) and becomes progressively more important when the shock speed increases. Of course, betatron acceleration is unimportant for very fast, adiabatic shocks.

Because of betatron acceleration, sputtering of grains is more important than our earlier estimate (Section 5.2.4) would indicate. Moreover, because betatron acceleration acts differently on differently sized grains, the destruction rate is actually size-dependent. Thus, for the standard shock of Fig. 13.2, the fractional destruction of large grains by inertial sputtering exceeds that of small grains. However, shattering affects the grain size distribution the most, while thermal sputtering and vaporization by grain–grain collisions are relatively unimportant. That shattering dominates so much over vaporization reflects the much lower threshold pressure. As a result, the grain volume affected by shattering in any given collision is much larger than the volume vaporized ($V \sim P^{-1}$). For a shock with $v_s < 200$ km s^{-1}, shattering rapidly removes the large grains from the grain size distribution and puts the fragments into the small size bins ($\lesssim$300 Å; Fig. 13.3).

The fraction of the total grain mass destroyed increases with increasing shock velocity (cf. Fig. 13.3). Inertial sputtering dominates at velocities below $\simeq$200 km s^{-1}. Because of the high postshock temperatures, thermal sputtering takes over at high velocities. Vaporization is never important. At all velocities, shattering affects the grain size distribution the most. Typically, about 80% of the mass is shattered and a large fraction of that mass is turned into very small grains ($\lesssim$50 Å).

Processing of interstellar grains also depends on the material properties. In particular, the specific density of the grain affects the grain motion behind the shock. Also, sputtering, vaporization, and shattering thresholds are set by material bond parameters. For equal-sized grains, diamonds are more readily sputtered than graphite grains because of their larger mass. This is also the main cause for the larger destruction of silicate, SiC, and Fe grains. Ice grains, on the other hand, have very low destruction thresholds and are readily destroyed at all velocities. Because of their much lower shear strength, graphite grains are more readily shattered than SiC and diamond grains. Iron grains are also readily shattered but that mainly reflects their high specific weight (i.e., long stopping length). Ice grains are completely sputtered before shattering collisions can take effect.

13.2.2 Adiabatic shocks

For very fast shocks ($v_s > 250$ km s^{-1}), the postshock gas becomes very hot ($T > 10^6$ K) and cooling is ineffective (cf. Section 11.2.1). Moreover, the supernova remnant expands rapidly during this phase and, hence, cooling is mainly through adiabatic expansion (cf. Section 12.3.2). Evaluation of the effects of thermal sputtering in adiabatic shocks requires a model for the expansion of supernova remnants (SNR) in the ISM, which determines how long shocked gas stays hot. The expansion of an SNR in a two-phase as well as in a three-phase medium is self-similar with $R \sim t^{\eta}$ (Section 12.4.3). Consider an intercloud medium hit by a

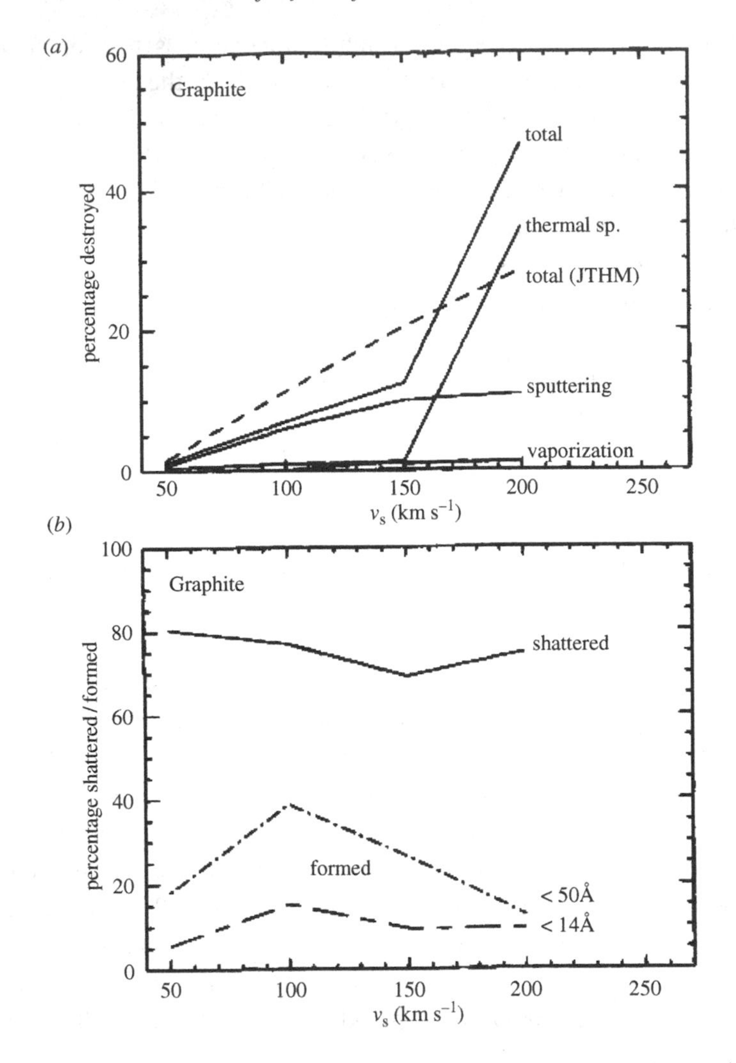

Figure 13.3 Graphite grain processing as a function of shock velocity. (*a*) The percentage of the total mass destroyed (i.e., returned to the gas phase) as a function of the shock velocity. Separate curves for inertial sputtering, thermal sputtering, and vaporization in grain–grain collisions are also shown. (*b*) The percentage of the total graphite grain mass shattered as a function of shock velocity (solid line). The dashed and dash-dotted lines give the percentage of the mass entered into bins smaller than 14 and 50 Å, respectively. See text for details. Figure reproduced with permission from A. P. Jones, A. G. G. M. Tielens, and D. Hollenbach, 1996, *Ap. J.*, **469**, p. 740.

shock at time t_0 of the expansion of the SNR. While the SNR expands, the grains in the shocked gas are thermally sputtered, and their mass decreases as

$$\frac{\mathrm{d}m_\mathrm{d}}{\mathrm{d}t} = 4\pi a^2 \rho_\mathrm{s} \frac{\mathrm{d}a}{\mathrm{d}t} = -\pi a^2 \, n \, \overline{vY} m_\mathrm{a}, \tag{13.3}$$

where ρ_s is the density of the grain material, m_a is the average mass of a sputtered atom, and the sputtering yield, Y, has been averaged over the Maxwellian velocity distribution, $f(v)$,

$$\overline{vY} = \int_0^\infty vY(v)\, f(v)\, dv. \tag{13.4}$$

The grain size then changes at a rate

$$\frac{1}{n}\frac{da}{dt} = -f(T), \tag{13.5}$$

where $f(T)$ is a function (essentially the averaged sputtering yield) that depends only on the temperature. The thickness of the removed grain layer is independent of grain size,

$$\Delta a = \int_{t_0}^{t_1} nf(T)dt. \tag{13.6}$$

Using the self-similarity solutions for a supernova remnant expanding in a two-phase or three-phase medium (cf. Section 12.4), we have for the size of a grain mantle removed from a grain shocked at time t_0,

$$\Delta a(t_0) = \frac{20 n_0 t_0}{6\eta T_s^\alpha} \int_0^{T_s} T^{\alpha-1} f(T)dT, \tag{13.7}$$

where n_0 is the preshock density of the warm medium, T_s is the initial shock temperature (i.e., at t_0), η is the self-similarity exponent, and $\alpha = (9\eta - 5)/6\eta$. For a two-phase medium (clouds with a small filling factor embedded in the intercloud medium), we have $\eta = 2/5$, $t_0 \simeq 2.3 \times 10^5 (100\,\mathrm{km\,s^{-1}}/v_s)^{5/3}$ years (Section 12.4.3; recall that $v_s = \eta R_s/t$), and

$$\Delta a(T_s) = 4.6 \times 10^5 \left(\frac{100\,\mathrm{km\,s^{-1}}}{v_s}\right)^{5/3} T_s^{7/12} \int_0^{T_s} T^{-19/12} f(T)dT\ \mathrm{cm}. \tag{13.8}$$

For a three-phase medium, the supernova remnant expands in the hot intercloud medium with velocity, $v_s = \eta\, R_s/t$. The hot gas in the supernova remnant drives shocks into the warm intercloud medium with a shock velocity given by $v_{\mathrm{wnm}} = (<\rho>/\rho_{\mathrm{wnm}})^{1/2}\, v_s$. In this case, adopting the three-phase parameters discussed in Section 12.4.3, we have $\eta = 3/5$, $t_0 \simeq 4.8 \times 10^4 (100\,\mathrm{km\,s^{-1}}/v_s)^{10/9}$ years, and

$$\Delta a = 1.1 \times 10^5 \left(\frac{100\,\mathrm{km\,s^{-1}}}{v_s}\right)^{10/9} T_s^{-1/9} \int_0^{T_s} T^{-8/9} f(T)dT\ \mathrm{cm}. \tag{13.9}$$

The results of these integrations are shown in Fig. 13.4. At high velocities, silicates lose a thicker surface layer than graphite grains, reflecting the difference in yields at high energies. At low velocities, graphite suffers more of a loss because of its lower threshold energy. The results also show that, for a given speed, adiabatic shocks in a two-phase medium are more destructive than in a three-phase medium, because SNRs expand faster in a three-phase medium than in a two-phase one.

13.3 Dust lifetimes

The destruction rate, $k_{\rm des}$, of interstellar dust by supernova shock waves is given by

$$k_{\rm des} M_{\rm ISM} = \frac{1}{\tau_{\rm SN}} \int \epsilon(v_{\rm s}) {\rm d}M_{\rm s}(v_{\rm s}), \tag{13.10}$$

where $M_{\rm ISM}$ is the total mass of the ISM (4.5×10^9 $M_{\odot}$), $\tau_{\rm SN}$ is the effective interval between supernova explosions, $\epsilon(v_{\rm s})$ is the fraction of dust destroyed by a shock of velocity $v_{\rm s}$, and $M_{\rm s}(v_{\rm s})$ is the mass shocked to a velocity of at least $v_{\rm s}$. The mass processed by an SNR depends on the structure of the ISM (cf. Section 12.4.3).

The supernova rate in our Galaxy is 2×10^{-2} year^{-1}. However, not every supernova interacts effectively with the ISM. About half of the supernovae are type Ia, which have low mass progenitors with a large galactic scale height. The correction factor for this is about 1/3. Half of the type II supernovae occur in associations. The first star to explode in an association will be very effective in destroying dust while sweeping up matter into a shell. However, subsequent supernovae will interact predominantly with the gas inside the shell and the shell itself and, therefore, do not contribute much to the dust destruction. The expansion of the supershell is too slow to destroy much dust. Given that a typical OB association yields 10–20 supernovae, the destruction per supernova in an OB association is only $\simeq$0.1 of that for field supernovae. The other half of type II supernovae are isolated – because of field stars and runaways. Here again a correction factor ($\simeq$0.6) for the increased scale height as compared with the warm ISM is needed. Hence, in summary, we find for the effective supernova time scale, $\tau_{\rm SN}$,

$$\begin{aligned} \frac{1}{\tau_{\rm SN}} &= \left(0.4 \times \frac{1}{2} + (0.5 \times 0.1 + 0.5 \times 0.6) \times \frac{1}{2}\right) \times 2 \times 10^{-2} \\ &\simeq 8 \times 10^{-3}\,{\rm year}^{-1}. \end{aligned} \tag{13.11}$$

Using the results for dust destruction as a function of shock velocity, the interstellar mass shocked to different velocities, and the effective supernova rate, the rate for dust destruction can be evaluated from Eq. (13.10). Adopting an MRN grain size distribution (see Section 5.4), graphite dust is destroyed at the equivalent of about 8 $M_{\odot}$ of interstellar gas per year. For comparison with the gas and dust budgets in Tables 1.4 and 13.6, this corresponds to 1.3×10^4 $M_\odot$ $Myear^{-1}$ kpc^{-2} for the rate at which the (gas) mass of the ISM is processed (i.e., cleaned of dust). For silicates, the values are slightly larger, 12 $M_\odot$ $year^{-1}$ and 2×10^4 $M_\odot$ $Myear^{-1}$ kpc^{-2}, respectively. For clarity, with a dust-to-gas ratio of 1% by mass, the rate at which dust mass is destroyed is a factor of 100 smaller. This results in calculated grain lifetimes against destruction (return to the gas phase) of 6×10^8 and 4×10^8 years for graphite and silicate grains, respectively.

Besides grain material, the dust destruction rate also depends on the grain size. The above-quoted values are averaged over the MRN grain size distribution and refer to the grain mass that is mainly contained in large grains ($\gtrsim$1000 Å). All velocity shocks contribute about equally to the destruction of large grains. In contrast, the destruction of small grains is dominated by thermal sputtering in adiabatic shocks. In a two-phase ISM, graphite or silicate grains of 200 Å or less are completely sputtered away in any shock with $v_s > 300$ km s^{-1} (cf. Fig. 13.4). So, all the small grains in the equivalent of 750 $M_\odot$ of warm interstellar gas are destroyed by a single SNR, equivalent to 6 $M_\odot$ $year^{-1}$ or, averaged over the Galaxy, 10^4 $M_\odot$ $Myear^{-1}$ kpc^{-2}. In a three-phase medium, a shock velocity in excess of 700 km s^{-1} is required for the destruction of a 200 Å graphite grain and this can be ignored. A silicate grain smaller than 200 Å will be destroyed by a shock faster than 450 km s^{-1}. In this case, about 140 $M_\odot$ is processed by a single SNR (1 $M_\odot$ $year^{-1}$ and 1.6×10^3 $M_\odot$ $Myear^{-1}$ kpc^{-2}). Thus, 200 Å grains are much better protected in a three-phase medium than in a two-phase medium because an SNR expends much of its energy expanding in the hot phase rather than the warm phase and because the expansion is faster. The lifetimes of

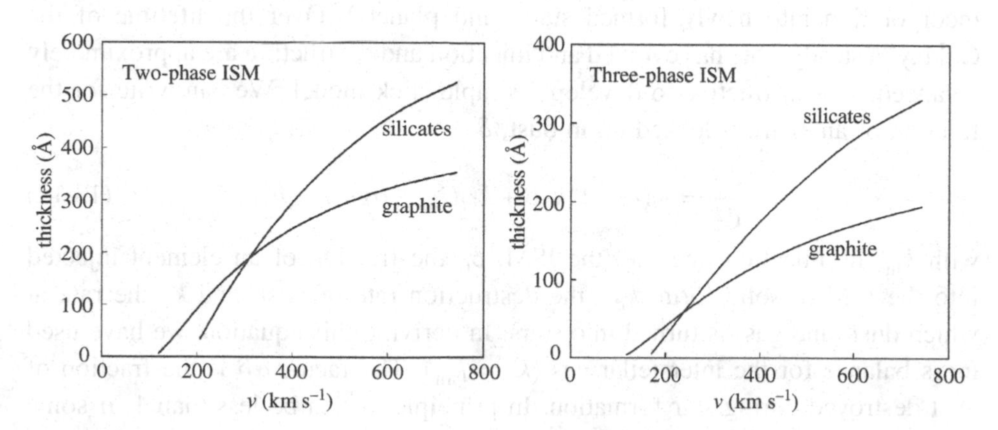

Figure 13.4 The thickness of the thermally sputtered layer as a function of the (adiabatic) shock velocity for an SNR expanding into a two-phase and three-phase ISM, respectively.

small grains are then 7.5×10^8 and 4.5×10^9 years in a two-phase and three-phase ISM, respectively. These are lifetimes against complete destruction of a 200 Å grain. Of course, removal of only 100 Å (which requires $v_s > 300$ km s^{-1} in a three-phase medium) would still return 7/8 of the grain mass to the gas and such a shock occurs about twice as often.

13.4 The grain size distribution

As the results discussed above demonstrate, shattering dominates grain processing by interstellar shocks. The grain size distributions after processing by different shocks are compared in Fig. 13.5. Obviously, most of the large grains are shattered into very small grains. The effect of shattering increases with increasing shock velocity because of the increased relative grain velocities. At high velocities ($\gtrsim 200$ km s^{-1}), betatron acceleration becomes less important, and the effects of shattering decrease somewhat. The size distribution of the fragments produced by individual grain–grain collisions is predicted to be somewhat less steep (−3.3; Section 5.2.4) than the MRN grain size distribution exponent (−3.5). Nevertheless, the size distribution produced by the shocks is slightly steeper than MRN. Essentially, the velocity of the (target) grains increases with size due to betatron acceleration. Hence, the largest grains, which contain most of the mass, are, relatively speaking, shattered into smaller fragments than somewhat smaller grains.

13.5 Dust abundances and depletions

Dust is injected into the ISM by stars and destroyed by stardust as well as by incorporation into newly formed stars (and planets). Over the lifetime of the Galaxy, a steady state has evolved and injection and destruction are approximately balanced. It is instructive to develop a simple stick model. We can write for the fraction of an element locked up in dust, δ,

$$\frac{\mathrm{d}\delta}{\mathrm{d}t} = -(\alpha - 1)k_* \delta + k_{\mathrm{in}}(\delta_0 - \delta) - k_{\mathrm{des}}\delta, \tag{13.12}$$

with k_{in} the injection rate into the ISM, δ_0 the fraction of an element injected into the ISM in solid form, k_{des} the destruction rate of dust, and k_* the rate at which dust (and gas) is turned into stars. In deriving this equation, we have used mass balance for the interstellar gas ($k_* = k_{\mathrm{in}}$). The factor $\alpha\delta$ is the fraction of dust destroyed during star formation. In principle, α can be less than 1, if some dust formed in the circumstellar disk is entrained in the protostellar outflow and ejected into the ISM, and greater than 1, if the shocks driven by the outflow destroy more dust than the outflow injects. Neither of these processes is well

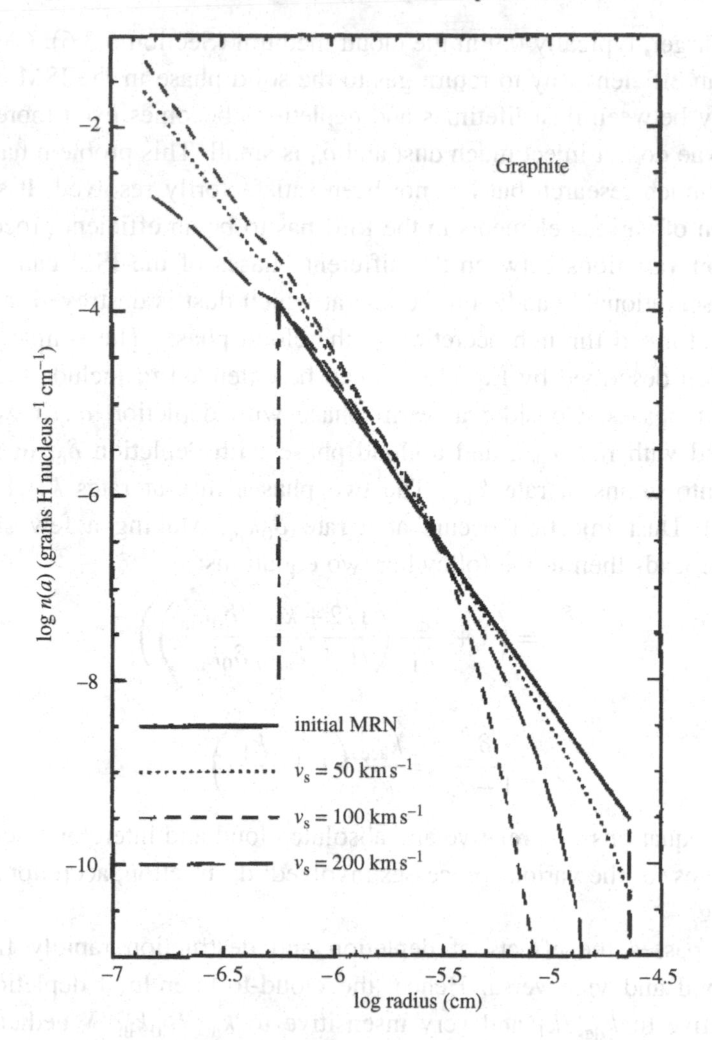

Figure 13.5 The grain size distribution for different shock velocities. The initial MRN grain size distribution ($n(a)$ $a^{-3.5}$, for $50 < a < 3000$ Å) is shown for comparison. Figure reproduced with permission from A. P. Jones, A. G. G. M. Tielens, and D. Hollenbach, 1996, *Ap. J.*, **469**, p. 740.

known and, for simplicity, we will adopt $\alpha = 1$. In that case, the steady-state fraction of an element locked up in dust is given by

$$\delta = \delta_0 \left(1 + \frac{k_{\text{des}}}{k_{\text{in}}}\right)^{-1}. \tag{13.13}$$

With a dust destruction time scale of $\simeq 5 \times 10^8$ years and an injection time scale of $\simeq 3 \times 10^9$ years, the fraction of an element locked up in stardust is about 0.14 of the fraction initially injected as dust. Observed depletions of Si, Fe, and Mg are much larger, typically 0.9 in the cloud medium (Section 5.3.6). Clearly there has to be an efficient way to return gas to the solid phase in the ISM itself. This discrepancy between dust lifetimes and depletions becomes even more poignant if supernovae do not inject much dust and δ_0 is small. This problem has been the subject of much research but has not been satisfactorily resolved. It seems that re-accretion of various elements in the ISM has to be an efficient process.

Depletion variations between the different phases of the ISM can be used to get an "observational" handle on the rate at which dust is destroyed in the warm ISM and reformed through accretion in the cloud phase. The simple model for the depletion described by Eq. (13.12) can be extended to include the presence of different phases. Consider a warm phase with depletion δ_i, in which dust is destroyed with rate k_des, and a cloud phase with depletion δ_c, in which gas accretes onto grains at rate k_acc. The two phases mix at rates k_1 (i $\rightarrow$ c) and k_2 (c $\rightarrow$ i). Dust injection occurs at a rate $\delta_0 k_\mathrm{in}$. Making a few simplifying assumption leads then to the following two equations:

$$\frac{\delta_\mathrm{c}}{\delta_\mathrm{i}} = \left(1 + \frac{k_\mathrm{des}}{k_1}\left(\frac{1/2 + k_\mathrm{acc}/\delta_0 k_\mathrm{in}}{1 + k_\mathrm{acc}/\delta_0 k_\mathrm{in}}\right)\right) \tag{13.14}$$

and

$$\frac{\delta_\mathrm{c}}{1-\delta_\mathrm{c}} = \frac{k_\mathrm{acc}}{k_2}\left(1 + \frac{k_1}{k_\mathrm{des}}\right). \tag{13.15}$$

These two equations link relative and absolute cloud and intercloud depletions to ratios of rates for the various processes involved: destruction, accretion, injection, and mixing.

Mixing passes the effects of depletion and destruction rapidly from cloud to intercloud and vice versa. Hence, the cloud-to-intercloud depletion ratio is very sensitive to k_des/k_1 and very insensitive to $k_\mathrm{acc}/\delta_0 k_\mathrm{in}$. Whether accretion or injection dominates dust formation is not terribly important for the relative depletion ratios. The observed $\delta_\mathrm{i}/\delta_\mathrm{c}$ of $\simeq$0.65 and $\simeq$0.95 for Si and Fe correspond to k_des/k_1 of 0.5 and 0.06, respectively. Obviously, substantial differences in depletion between the intercloud and cloud phases of the ISM directly imply that the destruction rate is comparable to the mixing time scale between the phases. These mixing time scales are much shorter than the stardust injection rate. In the plane, the cloud phase contains about ten times more mass than the intercloud phase, so that in steady state $k_1/k_2 = 10$. Cloud lifetimes are $\simeq 3\times10^7$ years, and this implies a lifetime against destruction averaged over the ISM phases of $\simeq 6\times10^7$ and $\simeq 5\times10^8$ years for silicon-bearing and iron-bearing interstellar dust, respectively. Thus, Si and Fe are in two distinct interstellar dust components with quite different destruction lifetimes. Comparison with the theoretical calculations discussed above imply binding energies of 1–2 and 5 eV for these two grain components, respectively. These differences in dust components are perhaps not too surprising since nucleosynthesis of iron has an important contribution from type Ia supernovae – presumably in the form of iron grains – while silicon is mainly produced by type II supernovae – probably in the form of magnesium silicates.

While relative depletion differences between the phases are largely controlled by the ratio of the destruction rate to the mixing rate, absolute depletion levels are set by the balance between accretion in the cloud phase and destruction in the intercloud phase. It is clear that accretion in the interstellar medium must dominate the dust depletion over injection by stars by a factor of $\simeq 5$ (cf. Fig. 13.6). Compared with the destruction time scales, the accretion time scales are more similar for these elements. Probably, accretion is limited by the collision rate with grains and not by the specific chemical interaction with the surface of an element. These observed cloud depletions refer mainly to ζ Oph, which is known to be unusually dense ($\simeq 300$ cm^{-3}), and this is reflected in the high derived accretion rates ($\simeq 2 \times 10^{-7}$ year^{-1}).

Of course, this is a very simplified model. In particular, accretion in the cloud phase will form mantles that can protect the underlying cores against the effects of sputtering in the intercloud medium. Nevertheless, these conclusions on the

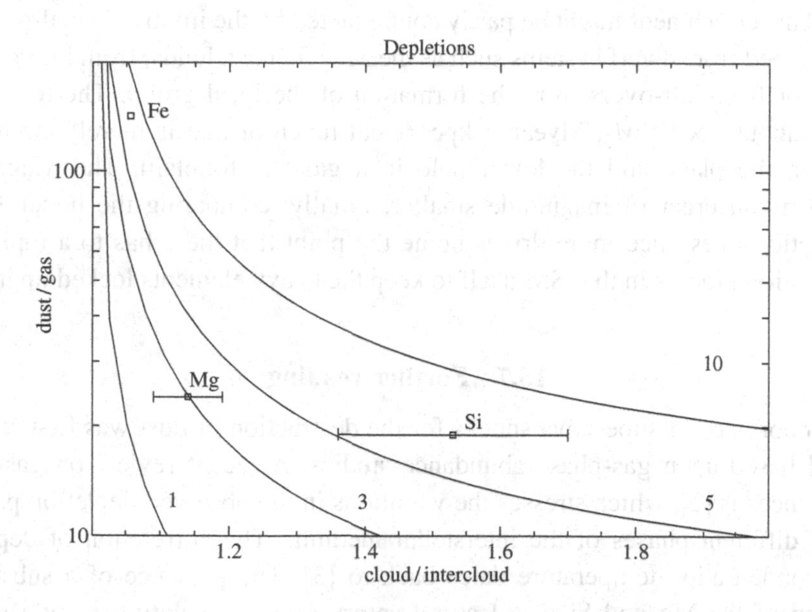

Figure 13.6 The relation between the dust-to-gas ratio and the cloud-to-intercloud depletion ratio for different ratios of the accretion rate over the cloud-to-intercloud mixing rate (k_{acc}/k_2). The points are the measured depletions for Fe, Si, and Mg. Depletion data taken from B. D. Savage and K. R. Sembach, 1996, *Ann. Rev. Astron. Astrophys.*, **34**, p. 279.

relative importance of destruction, accretion, and mixing and the differences between the different elements and their main "dust carriers" are expected to be very general.

13.6 Mass balance of interstellar dust

The gas and dust budget of the interstellar medium has been summarized in Table 1.4. Inspection of this table shows that, while the gas mass return rate is dominated by red giants, many different objects contribute substantially to the dust budget of the Galaxy. With a total interstellar gas mass of $4.5 \times 10^9 M_\odot$, the gas injection time scale is $\simeq 3 \times 10^9$ years. The total gas injection rate approximately balances the loss of gas to form stars and the ISM is in an approximate steady state. The dust injection rates are very uncertain. Thus, while supernovae do produce dust, the efficiency of dust formation in these objects is unknown. Likewise, protostars have powerful winds, ejecting some 30% of the accreted mass back into the molecular cloud, but the dust content of their jets and winds is unknown. The total dust injection rate corresponds to a dust-to-gas ratio in the ejecta of 1.5%. This is somewhat more than current estimates of the interstellar dust-to-gas ratio (1%), largely because of the synthesis of heavy elements in type II supernovae. This enrichment parallels the general elemental enrichment of the ISM. This enrichment might be partly counteracted by the infall of metal-deficient gas stripped from dwarf systems such as the Magellanic Clouds, from intergalactic space, or from left-overs from the formation of the local group. The total infall rate is about $7 \times 10^3 M_\odot$ Myear^{-1} kpc^{-2}, but much of that is merely circulation between the plane and the lower halo in a galactic fountain. The Magellanic Stream is an order of magnitude smaller. Finally, comparing the injection and destruction rates once more drives home the point that there has to a rapid dust reformation process in the ISM itself to keep the heavy elements locked up in dust.

13.7 Further reading

The importance of supernova shocks for the destruction of dust was first realized by [1] based upon gas-phase abundance studies. A recent review on gas-phase abundances is [2], which stresses the variations in the observed depletion patterns of the different phases of the interstellar medium. The correlation of depletion with condensation temperature dates back to [3]. The presence of a substantial fraction of the Mg and Si in a less refractory (but not volatile) mantle is well established observationally, [4] and [2].

Theoretical studies of dust destruction by supernova shocks have a long history. An early study is [5]. An early estimate of the effects on the grain size distribution

Table 13.1 *Interstellar dust budget*[a]

Source	Injection rate	Destruction rate
Silicates	17.2[b]	200
Carbonaceous dust	5.7[c]	130
Magellanic stream	0.3[d]	—
Star formation	10.0[e]	20

[a] Mass injection rates in units of $M_\odot$ Myear^{-1} kpc^{-2}. For comparison, the gas injection rate by stars is 1700 in these units. See also Table 1.4.

[b] Assuming supernovae inject all their Mg, Si, and Fe in the form of silicates. Otherwise, this value decreases to 3.

[c] Assuming supernovae inject all their C in the form of carbonaceous dust. Otherwise, this value decreases to 2.3.

[d] A dust-to-gas ratio of 0.002 is assumed (0.2 of the interstellar value).

[e] Assuming that 1/3 of the accreted mass is processed through a stellar wind and a 1% dust-to-gas mass ratio.

can be found in [6]. More recent studies are [7] and [8]. The interaction of young supernova remnants with interstellar dust is discussed in [9]. The infrared emission from dust in such strong shocks is [10]. Dust destruction in dense molecular clouds is discussed in [11]. A first study on the lifecycle was done in [12]. A more recent study is [13]. A somewhat different approach was taken in [14], a Monte Carlo simulation of the lifecycle of interstellar dust.

References

[1] P. M. Routly and L. Spitzer, Jr, 1952, *Ap.J.*, **115**, p. 227

[2] B. D. Savage and K. R. Sembach, 1996, *Ann. Rev. Astron. Astrophys.*, **34**, p. 279

[3] G. B. Field, 1974, *Ap. J.*, **187**, p. 453

[4] C. L. Joseph, 1988, *Ap. J.*, **335**, p. 157

[5] M. Barlow, 1978, *M. N. R. A. S.*, **183**, pp. 367 & 397

[6] C. G. Seab and J. M. Shull, 1983, *Ap. J.*, **275**, p. 652

[7] A. P. Jones, A. G. G. M. Tielens, D. Hollenbach, and C. F. McKee, 1994, *Ap. J.*, **433**, p. 797

[8] A. P. Jones, A. G. G. M. Tielens, and D. Hollenbach, 1996, *Ap. J.*, **469**, p. 740

[9] K. Borkowski and E. Dwek, 1995, *Ap. J.*, **454**, p. 254

[10] O. Vancura, J. C. Raymond, E. Dwek, *et al.*, 1994, *Ap. J.*, **431**, p. 188

[11] D. R. Flower and G. Pineau des Forets, 1995, *M. N. R. A. S.*, **275**, p. 1049

[12] E. Dwek and J. M. Scalo, 1980, *Ap. J.*, **239**, p. 193

[13] E. Dwek, 1998, *Ap. J.*, **501**, p. 643

[14] K. Liffman and D. D. Clayton, 1989, *Ap. J.*, **340**, p. 853

14
List of symbols

Nomenclature in astronomy is often very confusing if not intentionally obfuscating. Every subfield arbitrarily introduces its own symbols and a book encompassing many different subfields has to dance a fine line between consistency with common usage in the literature and avoiding internal confusion. I have tried to avoid using the same symbols for more than one quantity as much as possible but did not quite succeed. This glossary provides a list of symbols that are used throughout the book, often with a reference to the equation where the symbol is first introduced or best defined. In those cases where a symbol is used in a different way than listed here – and this is not obvious from the context – I have tried to redefine the symbol consistently in the text.

A_{ij} Einstein A coefficient (Eq. 2.17)
A_j The abundance of species j (Eq. 2.37)
$A_{\rm v}$ The dust extinction at visual wavelength, in magnitudes
A_λ The dust extinction at wavelength λ, in magnitudes (Eqs. (5.2) and (5.84))
a The grain size

$B(\nu, T)$ The Planck function at frequency ν and temperature T
B_{ij} Einstein B coefficient (Eq. (2.18))
$B_{\rm e}$ (also $B,\ A,\ C$) Rotational constant (Eqs. (2.4), (2.8), (2.9), (2.11))
B_0 The preshock magnetic field strength
B_1 The postshock magnetic field strength
b The Doppler line-width parameter

$C_{\rm V}$ The heat capacity of the grain material
$C_{\rm s}$ The speed of sound (Eq. (11.8))
$C_{\rm I}$ The speed of sound in neutral gas
$C_{\rm II}$ The speed of sound in ionized gas

c The speed of light

d The distance

E The energy
E_i The energy of level i
EM The emission measure (Eq. (7.62))
E_{B-V} The color excess between blue (B) and visual (V) wavelengths

$f(Z=0)$ The neutral fraction of PAHs (Eq. (6.58))

G The gravitational constant
G_0 The intensity of the radiation field in units of the average interstellar radiation field (the Habing field; Eq. (9.1))
$G(T)$ The fraction of the time a cooling species has a temperature T (Eq. (6.30))
g_i The statistical weight of level i

H The scale height of the interstellar medium (Eq. (8.24))
h Planck's constant

I The emergent intensity (Eqs. (2.46), (2.49–2.51))
$I(\nu, r)$ The intensity at frequency ν and position r (Eq. (7.28))
$I_{\rm neb}(\nu, r)$ The intensity of the diffuse ionizing radiation field at frequency ν and position r (Eq. (7.31))
$I_\star(\nu, r)$ The intensity of the stellar ionizing radiation field at frequency ν and position r (Eq. (7.29))
IP The ionization potential of a species (for PAHs: Eq. (3.11))

$J(\nu)$ The mean intensity at frequency ν
$J_{\rm ion}$ The collision rate of a charged particle (electron, ion) with a dust grain or PAH (Eq. (5.51))
$J_{\rm pe}$ The photo-electric ionization rate (Eq. (5.59))
$J_{\rm ul}$ The mean line intensity
$\tilde{J}$ The reduced collision rate of a charged particle with a grain or PAH (Eqs. (5.52), (5.56), (5.57))
$j(\nu, r)$ The emissivity at frequency ν and position r (Eq. (7.32))
$j_{\rm ff}$ The free–free emissivity

$K(T)$ The heat conductivity (Eq. (12.99))
k The chemical rate coefficient
$k_{\rm ac}$ The accretion rate coefficient of gas onto grains (Eq. (4.27))

k_{ar} The arrival rate coefficient of gas onto a specific grain (Eq. (10.26))
k_b The backward rate coefficient (Eq. (4.9))
k_d The rate coefficient for H_2 formation on grain surfaces
k_{des} The destruction rate of grains (Eq. (13.9))
k_f The forward rate coefficient (Eq. (4.9))
k_{ion} The ionization rate (Eq. (8.1))
k_L The Langevin reaction rate coefficient
k_{pd} The photodissociation rate (Eq. (4.4))
k_{pump} The FUV pumping rate of H_2
k_{rec} The electron recombination rate coefficient (Eq. (8.3))
k_{sn} The supernova rate per unit volume and unit time in the Galaxy
k_{UV} The UV absorption rate of PAHs (Eq. (6.4))
k_{ra} The radiative association rate coefficient

$\mathcal{L}$ The net cooling rate (Eq. (8.14))
$L_\star$ The stellar luminosity
L_w The mechanical luminosity of the wind from a star (Eq. (12.21))

M The mass of a cloud
$\mathcal{M}$ The Mach number of the shock or ionization front (Eqs. (11.11) and (12.6))
M_{HII} The mass of ionized gas (Eq. (7.18))
$M(v > v_s)$ The mass of WNM material processed by supernova shocks faster than v_s (Eqs. (12.117) and (12.119))
$\dot{M}$ Also, dM_g/dt The mass loss rate of a neutral globule or cloud because of ionization (Eqs. (12.43) and (12.66))
$\dot{M}_c$ The mass injection rate by a cloud into a supernova remnant through heat conduction (Eq. (12.103))
$\dot{M}_w$ The stellar wind mass loss rate
m_d The mass of a dust grain
m_e The mass of an electron

$\mathbb{N}_{Lyc}$ The total number of hydrogen-ionizing photons emitted by the star
$\mathbb{N}_{Lyc}(He)$ The total number of helium-ionizing photons emitted by the star
$\mathcal{N}_{ISRF}$ The mean photon intensity of the interstellar radiation field (Eq. (1.1))
$\mathcal{N}_\star$ The mean stellar photon intensity (Eq. (7.7))
$\mathcal{N}$ The mean photon intensity (Eq. (3.4))
$\mathcal{N}_{XR}$ The mean X-ray photon intensity (Eq. (3.32))
$N(i)$ The column density of species i
N_c The number of carbon atoms in a PAH

N_{cl} The number of evaporating clouds per unit volume in a supernova remnant
N_{cool} The cooling column density (Eq. (11.25))
N_H The column density of neutral hydrogen
n The gas number density
$n(i)$ The gas number density of species i
n_d The dust number density
n_e The electron number density
$n_{u,l}$ The number density of the upper, lower level
n_H The number density of neutral hydrogen
n_p The number density of protons
n_{cr} The critical number density (Eqs. (2.31), (2.32), (2.42))

P The pressure
P_t The turbulent pressure
P_0 The preshock gas pressure
P_1 The postshock gas pressure
p The polarization of light
p_m The photodissociation efficiency due to the cosmic-ray-induced photon field inside dense clouds (Eq. (4.7))

$Q_{abs}(a)$ The absorption efficiency of a dust grain of size a
$Q_{ext}(a)$ The extinction efficiency of a dust grain of size a
$Q_p(T_d)$ The dust absorption efficiency averaged over the Planck function at temperature T_d
$Q_{sca}(a)$ The scattering efficiency of a dust grain of size a

$\mathcal{R}_s$ The Strömgren radius of an HII region (Eq. (7.16))
R_s The radius of the shock front in a supernova remnant
$\mathcal{R}_{He}$ The equivalent Strömgren radius of the helium ionized zone (Eq. (7.36))
R_{IF} The radius of the ionization front (distance from the star)
R_{if} The distance from the center of a globule to the ionization front at its surface (Eq. (12.33))
R_V The ratio of total to selective extinction (Eq. (5.88))
r The distance to the star

S The source function (Eq. (2.21))
$S(T, T_d)$ The sticking coefficient of a species at gas temperature T and dust temperature T_d
s The number of degrees of freedom (Eq. (6.10))

T The gas temperature
T_B The brightness temperature (Eq. (7.70))
T_C The average energy per photon ionization of carbon (Eq. (3.6))
T_d The dust temperature
T_e The electron temperature
T_{eff} The stellar effective temperature
T_H The average energy per photon ionization of hydrogen (Eq. (3.6))
T_R The radiation field temperature
T_X The excitation temperature (Eq. (2.34))
t The time

U The ionization parameter (Eqs. (7.33) and (7.34))
u The internal energy of the gas (Eq. (11.4))

V_{snr} The final volume of a supernova remnant when it merges with the ISM (Eq. (12.90))
v_A The Alfvén wave velocity (Eq. (11.37))
v_s The shock velocity
v_w The velocity of the wind from a star
v_0 The velocity of the gas ahead of the shock or ionization front in the frame of the shock or ionization front
v_1 The velocity of the gas behind the shock or ionization front in the frame of the shock or ionization front

W The work function of the grain material (Eq. (3.9))
W_λ The equivalent width of a line (Eq. (8.92))

x The degree of ionization (Eq. (7.9))

Y The photo-ionization yield of the grain material (Eqs. (3.10) and (5.60))
Y_{pd} The photodesorption yield (Eq. (10.27))

Z The charge of an atom, PAH, or grain

α_C The photo-ionization cross section of carbon (Eq. (3.7))
α_{grain} The photo-ionization cross section of grains (Eq. (3.7))
α_H The photo-ionization cross section of hydrogen (Eq. (7.5))
α_{He} The photo-ionization cross section of helium (Eq. (7.35))
$\alpha(\nu, X_j^i)$ The photo-ionization cross section of species X_j in ionization stage i (Eq. (7.37))

$\overline{\alpha}_H$ The average photo-ionization cross section of hydrogen (Eq. (7.11))
α_i The photo-ionization cross section of species i (Eq. (3.3))

$\beta(\tau)$ The escape probability at optical depth τ (Eqs. (2.20), (2.40), (2.44), (2.45))
β_A The radiative recombination rate of H to all levels (Eq. (7.1))
β_B The radiative recombination rate of H to all levels with $n \geq 2$ (Eq. (7.10))
$\beta(X_j^i)$ The radiative recombination rate of species X_j in ionization stage i (Eq. (7.10))
$\beta_{ss}(N(H_2))$ The self-shielding factor for an H_2 column density $N(H_2)$ (Eq. (8.41))

Γ The gas heating rate coefficient
Γ_{ad} The gas heating rate due to ambipolar diffusion (Eq. (3.40))
Γ_{CI} The gas heating rate due to carbon ionization (Eq. (3.8))
Γ_{chem} The gas heating rate due to H_2 formation (Eq. (3.26))
Γ_{CR} The gas heating rate due to cosmic-ray ionizations (Eq. (3.30))
Γ_{g-d} The gas heating rate due to gas–grain collisions (Eq. (3.27))
$\Gamma_{gravity}$ The gas heating rate due to gravitational compression (Eq. (3.42))
Γ_{HI} The gas heating rate due to hydrogen ionization (Eq. (3.8))
Γ_{pd} The gas heating rate due to H_2 photodissociation (Eq. (3.18))
Γ_{H_2} The gas heating rate due to collisional de-excitation of FUV-pumped H_2 (Eq. (3.25))
Γ_{pe} The gas photo-electric heating rate due to PAHs or grains (Eqs. (3.15), (3.17))
$\Gamma_{turbulence}$ The gas heating rate due to decay of turbulence (Eq. (3.37))
γ The photo-ionization parameter $(G_0 T^{1/2}/n_e)$
γ_{ij} The collisional rate coefficient (Eqs. (2.28), (2.33), (2.66))
γ_0 The photo-ionization rate over the recombination rate (Eq. (6.59))

δ_d The dust depletion factor ($\delta_d = 1$ corresponds to standard dust abundances)
δ The fraction of an element locked up in dust (commonly referred to as the depletion factor; Eq. (13.11))

ϵ The photo-electric heating rate efficiency (Eq. (3.16))
ϵ_{snr} The efficiency with which supernova energy is converted into kinetic energy of interstellar gas (Eq. (12.93))

ζ_{CR} The primary cosmic-ray ionization rate
ζ_{XR} The primary X-ray ionization rate
θ The fractional surface coverage of grain surface sites

$\kappa_{\rm ff}$ The free–free absorption coefficient (Eq. (7.67))

Λ The gas cooling rate coefficient (Eq. (2.20))
$\Lambda_{\rm fb}$ The gas cooling rate coefficient due to H free–bound transitions (Eq. (2.63))
$\Lambda_{\rm ff}$ The gas cooling rate coefficient due to H free–free transitions (Eq. (2.64))
$\lambda_{\rm max}$ The wavelength of maximum polarization
$\lambda_{\rm ul}$ The line wavelength

μ The transition moment for rotational transitions (Eq. (2.12))
$\mu_{\rm e}$ The permanent electric dipole moment (Eq. (2.12))

$\nu_{\rm H}$ The photo-ionization threshold of hydrogen
$\nu_{\rm ul}$ The line frequency
ν_z The vibrational frequency of a species adsorbed on a grain surface perpendicular to the surface (Eq. (4.30))
ν_0 The vibrational frequency of a species adsorbed on a grain surface parallel to the surface (Eq. (4.32))

ρ The gas density
$\rho(E)$ The microcanonical density of states
$\rho_{\rm II}$ The ionized gas density
$\rho_{\rm s}$ The specific density of the grain material
ρ_0 The gas density ahead of the shock or ionization front
ρ_1 The gas density behind the shock or ionization front
$\rho_\star$ The density of stars in the Solar neighborhood

$\sigma_{\rm d}$ The geometric cross section of a dust grain
$\sigma_{\rm PAH}$ The geometric cross section of a PAH
$\sigma_{\rm UV}$ (PAH) The UV absorption cross section of a PAH

$\tau(r, \nu)$ The optical depth at position r and frequency ν (Eq. (7.8))
$\tau_{\rm ar}$ The arrival time scale for gas-phase species on a grain (Eq. (4.29))
$\tau_{\rm cool}$ The gas cooling time scale behind a shock front
$\tau_{\rm d}$ The dust optical depth
$\tau_{\rm ev}$ The evaporation time scale of species from a grain (Eq. (4.30))
$\tau_{\rm ff}$ The free-fall time scale (Eq. (3.43))
τ_{ij} The line-averaged optical depth (Eqs. (2.20), (2.43))
$\tau_{\rm m}$ The thermal hopping time scale (Eq. (4.35))
$\tau_{\rm rec}$ The radiative recombination time scale (Eq. (8.20))

$\tau_{\rm snr}$ The time scale for a supernova remnant to merge with the interstellar medium (Eq. (8.29))
$\tau_{\rm t}$ The tunneling time scale between sites for an adsorbed species (Eq. (4.34))
$\tau_{\rm th}$ The thermal hopping time scale (Eq. (8.18))

Φ (T) The partition function at temperature T
ϕ The grain potential (Eq. (5.83))
$\phi_{\rm c}$ The coulomb potential

$\xi_{\rm CR}$ The total cosmic-ray ionization rate (primary plus secondaries; Eq. (3.29))
$\xi_{\rm XR}$ The total X-ray ionization rate (primary plus secondaries; Eq. (3.29))

ψ The dissociation parameter, which regulates the dehydrogenation of PAHs (Eq. (6.84))

$\Omega(\mathrm{u,l})$ The collision strength (Eq. (2.66))
ω The albedo of the dust

Index of compounds

Index of polycyclic aromatic hydrocarbons

Index of Molecules

Index of objects

Index